D0984245

Paint Handbook

OTHER McGRAW-HILL HANDBOOKS OF INTEREST

Paint Handbook

Edited by

GUY E. WEISMANTEL

Chemical Engineering

McGRAW-HILL BOOK COMPANY

New York St. Louis San Francisco Auckland Bogotá
Hamburg Johannesburg London Madrid
Mexico Montreal New Delhi Panama
Paris São Paulo Singapore
Sydney Tokyo Toronto

Library of Congress Cataloging in Publication Data
Main entry under title:

Paint handbook.

    Includes index.
    1. Painting, Industrial—Handbooks, manuals, etc.
2. Paint—Handbooks, manuals, etc. I. Weismantel,
Guy E.
TT324.P34   667   80-12093
ISBN 0-07-069061-8

  567891011121314151617181920 VBVB 89876

The editors for this book were Harold B. Crawford and Beatrice E.
Eckes, the designer was Naomi Auerbach, and the production
supervisor was Paul A. Malchow. It was set in Baskerville by
Bi-Comp, Incorporated.

Contents

Contributors

Benson G. Brand (deceased) Formerly with Battelle Columbus Laboratories, Columbus, Ohio (Chapter 9, "Exterior Coatings for Wood")

Gary Boyd Charlesworth Colorite, Inc., Salt Lake City, Utah (Chapter 18, "Application Techniques")

Richard J. Dick Battelle Columbus Laboratories, Columbus, Ohio (Chapter 14, "Marine Paints")

Benjamin Farber Consultant, Lake Worth, Florida (Chapter 16, "Clear Coatings")

Neil B. Garlock Consultant, Arlington, Virginia (Chapter 19, "Troubleshooting and Inspection")

Howard W. Goetz Consultant, South Bend, Indiana (Chapter 10, "Interior Architectural Coatings")

Harry M. Herr Paint Engineering, Larkspur, California (Chapter 11, "Exterior Masonry Surfaces")

Kenneth K. Kaiser Gibson-Homans Company, Cleveland, Ohio (Chapter 12, "Roof Coatings")

Seymour I. Kawaller TSI, Inc., Simsbury, Connecticut (Chapter 13, "Fireproof Coatings")

William Lawrence Consultant, Brick Town, New Jersey (Chapter 17, "Specialized Functional Coatings")

Sidney B. Levinson D/L Laboratories, New York, New York (Chapter 15, "Specification Products")

W. M. McMahon Reliance-Universal, Houston, Texas (Chapter 7, "Coatings for Steel")

Donald R. Peshek PPG Industries, Springdale, Pennsylvania (Chapter 8, "Coatings for Metals Other Than Steel")

Jon Rodgers Reliance-Universal, Houston, Texas (Chapter 7, "Coatings for Steel")

Albert H. Roebuck Fluor Engineers and Constructors, Irvine, California (Chapter 20, "Economics")

Elias Singer Troy Chemical Corporation, Newark, New Jersey (Chapter 3, "Raw Materials")

Kenneth B. Tator KTA-Tator Associates, Coraopolis, Pennsylvania (Chapter 6, "Surface Preparation: Part II")

Paul E. Weaver Consulting Engineer, Houston, Texas (Chapter 4, "Selecting the Paint System")

Guy E. Weismantel *Chemical Engineering*, Houston, Texas (Chapter 5, "Surface Preparation: Part I"; Chapter 7, "Coatings for Steel"; Chapter 9, "Exterior Coatings for Wood"; Chapter 18, "Application Techniques"; Chapter 20, "Economics")

Trevellyan V. Whittington Aseptic-Thermo Indicator, Medical-Surgical Division, Parke, Davis & Co., North Hollywood, California (Chapter 1, "Paint Fundamentals"; Chapter 2, "Paint Testing")

Eric S. Wormser Gibson-Homans Company, Cleveland, Ohio (Chapter 12, "Roof Coatings")

Preface

The *Paint Handbook* is a compilation of practical information related to specifying paint and coatings. As an integral part of this topic, the book covers surface preparation, testing, and troubleshooting. The main emphasis is on architectural and industrial coatings that are field- or shop-applied. The book is meant to serve as a reference work for architects, corrosion engineers, specification writers, painting contractors, and operating companies' staff engineers who are responsible for the proper choice, application, and life expectancy of paints and coatings.

The editor and contributors have concentrated on paint specifications and not on paint formulation. Emphasis is placed on which finishes are best for a particular surface under varying environmental conditions. The book covers the ways in which these finishes are applied, the kinds of materials and equipment employed in the painting process, and the problems that may be encountered and the means of avoiding them.

This is not a textbook; it does not cover theory. It is intended for experienced and inexperienced practitioners who need a ready reference work for basic specification data. It is meant to serve the specialist in paint specification work and the layman who, of necessity, deals with coatings and must understand basic principles without becoming bogged down in minute details of fundamental paint chemistry.

The editor has tried to bring together a finely balanced collection of material written by experts in their specialized fields. Every attempt has been made to discuss paints and coatings generically, without favoring specific products. The *Handbook* is not meant to be a substitute for good judgment or for practical, proven field experience for coatings that have been used in the past, but it should act to supplement such pragmatic information. If some material is not included, it is only that practical limitations of space precluded the use of everything available.

The heart of this *Handbook* is the Index. The editor has made this section extensive so that the user can find a particular piece of information to answer a specific question with minimum effort. The Index is comprehensive and practical and contains numerous cross-references.

Obviously this *Handbook* is based not only on the practical experience of its editor and contributors but on data provided by numerous paint manufacturers, consultants, architects, engineering firms, painting contractors, and technical associations. The editor gratefully acknowledges their input as well as information from the personal files of the contributors. The latter have exercised extreme patience and understanding in helping the editor prepare and keep up to date this volume of technical information. A similar acknowledgment is appropriate for the publisher, whose staff has built flexibility into the editing and production schedule, facilitating last-minute additions to assure that the *Handbook* is as timely as possible.

GUY E. WEISMANTEL

Paint Handbook

Chapter **1**

Paint Fundamentals

TREVELLYAN V. WHITTINGTON

Aseptic-Thermo Indicator, Medical-Surgical Division, Parke, Davis & Co.

1.1   Introduction   The purpose of this chapter is to present a simplified and condensed survey of paint technology from the user's point of view. Focus is on the selection and use of paint. The chapter deals with the ingredients of paint so that the reader will understand why paint behaves as it does and why it must be used in certain definite ways. The material is presented as an introduction and a handy reference to the general principles of paint technology that are included in the following chapters.

This chapter is built on three concepts:

First concept. Paint should be defined and discussed in terms of its functions. These functions are decorative, protective, and specialized.

Second concept. Paint should be discussed in terms of its essential properties and its necessary specific properties. The essential properties are those which all paints must possess. The necessary specific properties are those properties which are needed by paints designed for certain specific end uses but which may not be needed by paints designed for other specific end uses.

Third concept. Paint should be treated as an engineering material. This means that it should be specified in accordance with the performance desired in its end use and that the conditions and techniques of its use should be specified and controlled.

1.2   Paint Properties   Paint differs from other engineering materials because its successful use requires a proper mixture of science and art. *Nonquantifiable* properties are important in the performance of paint. *Quantifiable* properties are those that can be measured accurately with good precision and can then be used accurately in specifications and measurements. In the field of rheology (flow of fluids under stress), for example, the process of quantification is only now gaining much momentum.

Paint also has inseparable functions. This means that decorative, protective, and specialized functions all are present simultaneously. Once the coating system is dried or cured, the separate layers cannot be cleanly separated to examine them. As a result, the effects on the different layers of the various factors that have influenced their formation and performance cannot be studied individually.

1.3   Definition and Classification of Paint   Paints are commonly called "surface coatings." "Paint" can be defined as a coating applied to a surface, or substrate, to decorate it, protect it, or perform some other specialized function.

1.3.1   Definition   We define a paint as a decorative, protective, or otherwise functional coating applied to a substrate. This substrate may be another coat of paint.

1.3.2 Classification   Paints can be classified by many methods. The method chosen depends on what the classifier is trying to accomplish.

The first purpose of classification is to group those paints which have the property being discussed and have it to the degree considered necessary for inclusion. In this way they are set apart from paints not having this property or not having it to the required degree.

The second purpose of classification is to group those paints which are used in the same way or for the same purpose or for the same type of application. They are thus set apart from other paints not used in the same way or for the same type of application.

As examples of paint classification, gloss paints have a shine like a mirror, while flat paints lack this shine. Industrial finishes are applied to manufactured objects (e.g., automobiles, appliances, and furniture) before they are sold to the user. Trade sales paints (e.g., house paints, wall paints, and kitchen enamels) are applied to completed articles by the owner or the owner's employees or by a painter hired by the owner.

1.4 Essential and Specific Properties   Certain *essential properties* (adhesion, ease of application, film integrity, and consistent quality) occur in all paints to assure their decorative, protective, and useful functions as an engineering material. *Specific properties* relate to end use.

1.4.1 Adhesion   Coatings must stick to the substrate to bring other properties into play. Even a coating that is designed to be stripped off later must adhere to the substrate until it is ready to be stripped off.

1.4.2 Ease of application   Paints must be easy to apply by the method indicated by the manufacturer or by some other approved method of application preferred by the user. For paints to be practical engineering materials, there must be a minimum of lost time or other extra expense during application. Paints must go onto the substrate in the specified film thickness, dry in the specified time to the desired appearance, and possess the necessary specific properties.

1.4.3 Film integrity   The cured or dried film of paint, when properly applied, must have all the film properties claimed for it by the paint manufacturer. This property is film integrity, or wholeness. There must also be no weak spots in the film caused by imperfect film drying or curing or "holidays" where the film thickness is less than that specified.

1.4.4 Consistent quality   To be usable as engineering materials, paints must be consistent in quality from can to can, batch to batch, shipment to shipment. Color, viscosity, application properties, durability, etc., must all be within specifications.

1.4.5 Specific properties   Specific properties should be taken into account when specifying the performance of paint for the particular end use. For example:

- Kitchen enamels must resist kitchen grease, heat, and repeated cleaning.
- Stucco paints must resist water, alkali, and sunlight and permit the passage of water vapor.
- Swimming-pool paints must have specific resistance to pool chemicals, especially acids and chlorine, and to water and sunlight.
- Exterior commercial-aircraft finishes must resist ultraviolet degradation, erosion by air and loss of adhesion at high speeds, rapid changes of temperature, chemical attack by the hydraulic fluids of the aircraft, and film rupture from the flexing of the film by the denting of the surface.
- All these paints must be made in the specified color, gloss, etc. They must also have the necessary ease of application and other essential properties.

1.5 Elements of a Good Paint Job   There are five elements of a good paint job:

- Correct surface preparation
- Choice of the proper paint system
- Good application with the right techniques and tools
- Correct drying or curing cycle
- Protection against water

1.5.1 Correct surface preparation   The first essential property of paint is adhesion. Good adhesion demands that the surface be properly prepared to receive and to hold paint. Although correct surface preparation varies somewhat with type of paint being applied, five features of surface preparation are necessary for all classes of paints:

Surface integrity   The surface must be knit together tightly enough to resist the shrinkage of the curing paint film. If it is loose or crumbly, it cannot resist the forces of paint shrinkage, and the whole paint system will lose adhesion and peel off.

Surface cleanness   The surface must be cleaned of any foreign material that will prevent the paint from flowing out on the surface and wetting it. If the paint does not flow out, it will not have film continuity. If it does not wet the surface, it will lack adhesion.

Surface smoothness   The surface to be coated with paint must be made smooth enough by filling, sanding, etc., to produce the required smoothness of the finished paint system. The paint system (combined coats of paint applied) will faithfully reproduce the contours of the substrate being coated unless a rough-textured paint has been deliberately chosen to hide surface imperfections. This reproduction of substrate contours by the paint system is termed "photographing."

On the other hand, minute surface roughness improves adhesion.

Sometimes good adhesion is difficult or impossible to obtain without this minute roughness, which need not be pronounced enough to affect the surface smoothness of the completed paint system. Mechanical anchorage due to the "tooth," or "anchor pattern," of the substrate is especially needed by paint systems that are subject to vibration, impact, and bending, particularly when the paint is exposed to exterior weathering. A good example of a paint system with such an exposure is that on the ordinary automobile.

Surface porosity   The suction, or porosity, of the substrate must be so reduced that the coating is not sucked into the substrate. If it is sucked in, the film continuity is destroyed and ease of application is impaired.

Substrate protection   Some substrates, especially plastics, are attacked and irreparably damaged by the solvents in coatings that are otherwise desirable for these substrates. In such cases, the composition of first coat, whatever it is called, must be such that its solvents will not attack the substrate. The first coat, in turn, must not be attacked by the solvents in the following coats.

1.5.2   Choice of the proper paint system   A paint system is a combination of coats of paint that will do the following:

- Perform the desired function or functions of paint: decoration, protection, specialized functions.
- Have all the essential properties of paint and the necessary specific properties for the specified end use.

We speak of paint systems because a single coat of paint normally will not do all that is required. Even when a single coat of paint is sufficient for a repaint job, we are still dealing with a paint system. The preceding coats of old paint being covered by the single coat of new paint are part of the total paint system and must be so considered. Let us think for a moment about some of the problems that can be encountered by not considering the old paint and the new paint as a combined total paint system.

- The new coating may attack the old coating, as occurs when lacquer is applied over oil paint. In this case a protective sealer should be used first.
- The old paint may not resist the shrinkage of the new paint, causing alligatoring. In this case an undercoat should be used first.
- The old paint may be too porous to seal the surface for the new paint, causing blotching, poor film continuity, and other film weaknesses. In this case a sealer should be used first to seal off the porosity.
- The total paint system may be too thick, causing cracking from thermal expansion and contraction. In this case the old paint must be removed or sanded down to a thinner film.

The following discussion deals with additional operations performed by paint systems. All these operations would never be performed by a single coat of paint.

Bind the substrate together   This provides the needed surface integrity. Applying a surface conditioner to a chalky stucco surface before applying an emulsion paint performs this operation.

Stain the substrate   This gives color without film buildup. A penetrating stain performs this operation.

Fill the surface   This is done by building up low spots in the surface, as by filling and sanding. This operation prevents photographing of the surface irregularities up through the paint system.

Seal the substrate   This is done by using a sealer or primer to reduce and even out the porosity of the substrate.

Protect and/or decorate without opacity or color   This is done by applying a clear coating, such as a clear varnish or an oil finish. This operation is usually desirable to preserve the existing color of a surface, such as an attractive wood grain.

Change the surface to improve adhesion   This is normally done by applying a primer. In extreme cases, as with some urethane coatings, two primers are needed so that each coat will stick to the preceding coat and also provide good adhesion for the following coat.

Secure the right color   This operation is usually performed by the topcoat alone. However, in some automobile finishes, in antiquing finishes, and in a few other finishes, the color of two or more coats is involved.

Obtain the right level of gloss   The presence of gloss, as in an enamel, or the absence of it, as in a flat paint, is due to the topcoat. Old paint or an improperly cured or formulated undercoat can affect the gloss of the topcoat.

Provide the necessary specific properties   This provision is usually a product of the total paint system. Necessary specific properties include resistance to all the destructive forces to which the paint system will be exposed in the end application.

Examples of paint systems   The following examples are chosen to illustrate the need for the paint-system approach. Only by thinking of a paint system from the beginning can we *always* be sure of selecting compatible component coats and also of considering all the necessary individual operations that the paint system must perform.

example 1: A sizable area of a metal alloy that presents an adhesion problem is to be subjected to corrosive fumes. A wash primer will adhere to the alloy, and a urethane coating will resist the corrosive fumes. However, the urethane coating adheres poorly to the wash primer. An intermediate primer which adheres well to the wash primer and to which the urethane coating also adheres well is the answer. The result is a three-coat system.

example 2: A stucco building has been sandblasted to remove a heavily chalked paint system which in places is eroding down to the stucco. The sandblasting shows that the stucco has begun to lose integrity. A test patch shows that a surface conditioner binds the stucco together well enough to give a paintable surface. However, the sandblasting has removed a small amount of the stucco, making the smooth stucco finish begin to look like a sand finish in some areas. Extensive patching of the stucco would be expensive and difficult. A quicker and less expensive solution is first to brush or roll a soft paste of stucco and the emulsion stucco paint over the roughened areas to smooth the surface and then to wipe the surface lightly to eliminate brush or roller marks. This should be allowed to dry overnight, and the surface conditioner then be applied over the entire surface to give it surface integrity. Finally, one or more coats of the emulsion stucco paint at normal thinning should be applied over the entire surface.

example 3: The kitchen of a large hospital needs repainting to renew its appearance. Thorough washing of a greasy area shows that the paint system underneath still has good film integrity, but it is discolored and the gloss of the original enamel has become spotty. The kitchen is to be closed down for the minimum time possible, and the odor of painting is to be minimized. What paint system will meet these requirements? First, a test patch shows that washing with a strong trisodium phosphate (TSP) solution removes the grease. It also dulls the surface enough, by minute etching, to receive and hold paint. An emulsion primer followed by a water-thinned enamel or an odorless alkyd enamel will give the required low odor, rapid drying, grease and heat resistance, and minimum downtime required. The test patch plays an important role in this system specification because emulsion primers often do not adhere to hard, glossy undercoats or previously painted glossy surfaces.

1.5.3 Good application with the right techniques and tools   Good paint application is that which enables the paint to do the job for which it was formulated and specified. Certain guidelines must be followed to satisfy these requirements.

Uniform wet (and dry) film thickness   Failure to get uniform film thickness can cause a lack of film continuity, leading to serious film defects. The most serious of these defects are:

- Loss of required resistance properties because of film porosity from too dry an application
- Loss of resistance properties from poorly cured spots caused by too thick an application or from thin spots caused by uneven application
- Nonuniform gloss or color, which spoils the decorative function

Correct number and sequence of applications   This information is necessary for each paint in the system. The manufacturer's directions should make this point clear. In any case, the applicator should clearly understand this point before actual application begins.

Other guidelines   Some of the most important guidelines to keep in mind are:

- Each paint must be applied in the proper number of wet applications to give the specified dry film thickness without trapping solvent.

- The proper drying or curing time between successive applications of the same paint and between successive paints should be carefully observed.
- The manufacturer's directions should be followed to assure the coating the chance it needs to develop the properties for which it was specified. Failure to follow the directions voids the manufacturer's responsibility for the failure of the coating to perform. If the manufacturer's directions are faithfully followed and the coating still does not meet specifications after application, the application should be stopped at once. The manufacturer's technical staff should be consulted immediately. These discussions should be as straightforward as possible.
- To avoid marring the film masking must be removed after the paint sets up but before it dries hard.
- Painting tools, including brushes, rollers, and spray guns, must be cleaned before the paint has dried or before the catalyzed coating is converted to an insoluble gel.

The right tools and their use   The right tools for applying any paint are those which bring out the properties of the paint for which it was specified and bring them out under the actual conditions of application. The manufacturer's directions should clearly state the following:

- The proper tools for applying the paint
- The proper additions, if any, to make the paint apply properly
- The proper rate of application, in square feet per gallon, dry film thickness, or the equivalent
- Any deviations in application technique from the normal use of the application tools such as spray pressure

The proper application tool depends upon the rheological (flow) properties of the paint being applied.

Brush. Paints applied by brush must have the following properties:

- Be thick enough to lay down an adequate film of wet paint.
- Be nonfluid enough (as manufactured) to stay in the brush when picked up on the brush for application.
- Remain fluid long enough after application to flow out the brush marks.
- Become nonfluid after the brush marks have flowed out and before the paint sags or drips.

Roller. Paint made for application by roller must be similar to brushing paint, with these differences:

- The paint must stay wet longer to flow out the roller pattern.
- The paint should be more nonfluid (as manufactured) to avoid dripping when picked up and applied by the roller.

- Paint for application with a stipple roller must be thicker and stickier and must not flow out after application.

Air-Spray Gun. Paint made for application by air spray must have these properties:

- It must be nonsticky when thinned to spraying viscosity so that it will break up easily into droplets.
- It must be stable when thinned to a viscosity low enough for spraying.
- The thinned paint, ready for application, must contain a solvent blend which is volatile enough to evaporate mainly in the spray cone before the paint reaches the surface but which will keep the paint wet long enough for the paint to flow out after being deposited on the surface.

Airless Spray. The paint is atomized by high-pressure pumping rather than being broken up by the large volume of air mixed with it. This procedure changes the required properties of the paint as follows:

- The paint can be sprayed at higher solids and higher viscosity, a thicker film being applied for each pass of the spray gun.
- The solvent blend need not evaporate so quickly because there is less solvent in the paint at application viscosity.

1.5.4 Correct drying cycle   The final properties of the dried or cured coating develop during the drying cycle. Unless conditions are correct, correct film properties will never develop. Only the manufacturers of the different coatings used in the paint system know the exact composition of their coatings. This composition determines the conditions for proper drying or curing (Fig. 1.1).

1.5.5 Protection against water   Water is the hidden enemy of paint. It is a pervasive element of deterioration. Water goes everywhere, either as

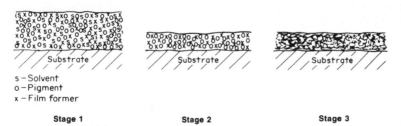

s – Solvent
o – Pigment
x – Film former

| Stage 1 | Stage 2 | Stage 3 |

Fig. 1.1 Drying stages of a typical paint film: Stage 1, wet film of paint represented by S (solvent), X (film former or resin), and O (pigment); Stage 2, in which the solvent has evaporated into the atmosphere, leaving only the pigment and the film former; and Stage 3, in which the film former becomes cross-linked or polymerized. A typical substrate is wood. NOTE: The figure shows volume relationships only. In reality pigment and film former are interspersed.

liquid water, as water vapor, or as a carrier for chemicals dissolved in water. It causes:

- Rusting and other corrosion
- Paint peeling
- Masonry efflorescence and spalling
- Wood rot
- Corrosive water solutions (staining, seawater)

Rusting and other corrosion   The corrosion of metals from exposure to water can be prevented if the proper paint system is correctly applied:

- Be sure that the surface is properly prepared to receive and to hold paint. All spots of metal corrosion, such as rust, and all loose material must be removed before priming the metal.
- Choose a primer that adheres to the metal being painted and itself has good water resistance.
- In some cases only a primer is needed.
- Be sure that the total paint system is water-resistant and is durable under the conditions of actual exposure.
- Coat *all* the metal surface that is exposed to water, not just the part that is easy to see and reach.

Paint peeling   This problem is usually caused by water getting behind the paint. If the paint system is not sufficiently resistant to water for the actual exposure, water can penetrate the surface in sufficient amount to cause the paint to peel. Usually, however, peeling is caused by water that has gotten behind the paint in other ways.

The most common cause of paint peeling is breaks in the paint surface due to cracks in the wood or open joints that are not properly caulked. (The newer latex caulks have proved to be far more durable than previous caulking materials.) Openings around doors and windows are a special source of trouble because of uncaulked or improperly caulked joints.

A second serious cause of paint peeling is excess moisture in the space behind the painted wall. Moisture condenses into liquid water and gets into the unpainted side of the wall behind the paint. Then, from behind, it pushes the paint off the wall. If the excess moisture comes from inside the building, it is usually caused by insufficient ventilation to the outside plus a wall paint system that is not waterproof. The cure is to provide more ventilation to the outside and, if necessary, to apply a waterproof paint system. If the excess moisture comes from the outside, it is usually caused by insufficient ventilation of the space between the inside and outside walls. The cure is to increase ventilation by adding ports in the outside wall. With energy savings as a motive, excessive caulking and sealing without ventilation ports can turn a house into a blister box, resulting in excessive peeling.

Another common source of water in painted walls is the presence of unpainted or unwaterproofed areas. Examples are the lower edges of wood siding and stucco walls, top and bottom ends of exterior doors that are not caulked and painted, and joints between siding that are too small to caulk but still are open. These joints should be waterproofed by repeated applications of a penetrating wood preservative applied with a full brush until the ends of the siding are saturated. The amount of penetrating wood preservative or wood primer required to protect bare wood is far greater than is generally realized. The ideal treatment is a 24-h cold soak in penetrating wood preservative to saturate the wood surface. The paint should be applied over the treated wood after drying.

When specifying penetrating wood preservatives or paint primers, it is important to understand how greatly the porosity of wood varies with the kind of grain exposed. Face grain, the least porous, requires the smallest amount of penetrating wood preservative or paint primer to satisfy the porosity of the wood. Edge grain, which is about twice as porous, requires about twice as much preservative or primer. End grain is about 4 times as porous as face grain and requires about 4 times as much preservative or primer. The ends of boards are all end grain. Slash grain is a combination of edge grain and end grain. The edges of boards are usually edge grain.

Because of the way plywood is constructed, its face is a mixture of face grain, edge grain, and end grain. All sawed edges of plywood contain layers of end grain, and when used outside they should be waterproofed *and* caulked. Although exterior-grade plywood is made with waterproof glues that will not delaminate, the plywood will delaminate if water penetrates it through the surface or the edge. The reason is that the layers of wood swell and split *between* the layers of waterproof glue. The end result is failure of both the decorative and the protective functions. For this reason, specifications should call for sealing the edges of plywood.

A third common cause of paint peeling is moisture or water creepage from moist ground under a house or other structure. Covering bare ground in the crawl space under the structure with tar paper is often used as a corrective measure. Caulking of breaks in floors and foundations can also help. In severe cases, removal of the source of excess water in the soil may become necessary.

Masonry efflorescence and spalling Efflorescence is the deposit of water-soluble salts on the outside of painted masonry surfaces such as stucco, concrete, volcanic-ash cinder block, and mortar between bricks. The water-soluble salts are leached out of the masonry by water that comes through the masonry from the inside. The water evaporates from the painted surface, leaving a deposit of salts. Although the salts can be washed off with a stream of water, they will form again unless the source of water behind the paint is stopped. Many times the source is a leaking roof.

If the efflorescence is only within about 16 in (406.4 mm) of the ground, it is probably due to the capillary action of the stucco sucking up groundwater. Efflorescence due to capillary action or groundwater will eventually dissolve enough soluble salts from the stucco to cause it to lose integrity and become powdery. Then the stucco will crumble or flake off at a touch, taking the paint with it. Capillary action can be stopped by keeping the ground around the base of the building dry. It can also be stopped by waterproofing the foundation with tar or waterproof paint for 2 ft (0.6 m) below ground level and a few inches above ground. If the building is coated with a paint such as an emulsion paint that readily transmits water vapor, efflorescence will occur. If the paint (e.g., an oil paint) is impervious to water vapor, it will probably peel off.

Spalling is the flaking off of masonry due to the freezing of water in it. The paint, of course, comes off with the masonry. The cause and the cure of spalling are the same as for efflorescence.

Wood rot   When water gets into painted wood, wood rot occurs. The water can come from leaks, from inadequately vented space between walls, or from excessive moisture inside an inadequately vented building. It can also come through porous wall paint. The water can come from groundwater sucked up into the wood behind the paint by the strong capillary action of the wood. Keeping wood at least 6 in (152.4 mm) above the ground eliminates this source of unwanted water. If the wood cannot be shortened, the earth should be dug away to at least 6 in below the wood.

Corrosive water solutions   These are water solutions containing dissolved material. Depending on the nature and amount of the material, they usually accelerate the deterioration of some function or property of the paint.

Dissolved compounds of iron will rapidly deteriorate the decorative function of paint. Rust stains form on the surface, spoiling its appearance. These stains are difficult or impossible to remove without damaging the paint film.

Seawater contains large amounts of dissolved salts. These salts accelerate water penetration of the paint film so that water comes into contact with the substrate.

1.6   Paint Ingredients   Ingredients of paint fall under four headings: vehicles, solvents, pigments, and additives.

The vehicle is that portion of a paint which gives it film continuity and also provides adhesion to the substrate. It is called the vehicle because it carries to the substrate the ingredients that will remain on the surface after the paint has dried. The vehicle contains the film former, which is the combination of resins, plasticizers, drying oils, etc., that gives film continuity and adhesion. In liquid paints the vehicle includes all the liquids in the paint and all the additives needed by those liquids.

The solvents are low-viscosity volatile liquids used in coatings to improve application properties.

The pigmentation, or combination of pigments, gives the coating properties that cannot be obtained from the vehicle alone. It contributes desirable properties in the following ways:

- For the decorative function, it contributes opacity, color, and gloss control.
- For the protective function, it contributes necessary specific properties such as hardness, resistance to corrosion, resistance to rapid weathering, abrasion resistance, and improved adhesion.
- It also serves such desired specialized functions as ease of sanding, flame retardance, and electrical conductivity.

Pigments are also used to fill space in paint films. This important function is often abused. The abuse arises from excessive use of "filler" or "inert" or "extender" pigments to reduce the raw-material cost of the paint.

Additives are ingredients formulated into a paint to modify the properties of either the vehicle or the pigmentation, or both. They give the wet paint or dried paint film properties not present in the vehicle and pigmentation system. Additives also improve certain other properties of the vehicle (such as speed of drying), the pigment (such as resistance to fading), or the entire paint (such as ease of application).

The ingredients of paint, commonly termed "paint raw materials," must be controlled to close specifications to produce coatings of consistent quality. Close inspection of incoming raw materials for conformance to preset quality standards is an important type of testing.

1.6.1 Paint vehicles   For simplicity in discussing how paint vehicles form films and how these films dry or cure, the vehicles are divided into six groups:

- *Solid thermoplastic film formers.* The solid resin is melted for application and solidifies after application.
- *Lacquer-type film formers.* The vehicle dries by solvent evaporation.
- *Oxidizing film formers.* Oxygen from the air enters the film and cross-links it to form a solid gel.
- *Room-temperature catalyzed film formers.* Chemical agents blended into the coating before application cause cross-linking into a solid polymer at room temperature.
- *Heat-cured film formers.* Heat causes cross-linking of the film former or activates a catalyst that is not active until heat has been applied.
- *Emulsion-type film formers.* The solvent evaporates, and the droplets of plastic film former floating in it flow together to form a film. The droplets of plastic are not soluble in the solvent, which is usually water.

Solid thermoplastic film formers   Hot-mop roof coatings are an old example of these vehicles. The tar is melted and resolidifies on cooling. A new application of this type of drying mechanism is the powder coating. The object to be coated is immersed hot in the powder coatings, and the resin in the coating melts, causing the coating to adhere to the object. Subsequent reheating of the object improves the flow-out of the coating.

Lacquer-type film formers   In describing the curing of a lacquer, the usual explanation is that the solvent evaporates and the film is then dry. Although the drying process is not that simple, the explanation is adequate. The most familiar type of lacquer is based on nitrocellulose. In addition to the nitrocellulose, which gives the fast and hard drying, the formula includes one or more softer resins to give adhesion. There are also one or more plasticizers to give flexibility. A complex solvent blend is normally used to give a controlled evaporation rate and to make sure that all the components stay in solution at all times until solvent evaporation is complete.

Oxidizing film formers   These film formers are based on drying oils, which react with the oxygen of the air to cross-link the molecules into a solid gel. The cross-links are oxygen bridges between the drying-oil molecules. Common drying oils include linseed, soybean, castor, safflower, and tung (china-wood) oils, fish oils, and tall oil (from the pine tree as a by-product of kraft-paper manufacture). Oxidizing film formers may be based on drying oils alone, on varnishes, or on oxidizing alkyds.

Varnishes.   These vehicles are made by cooking drying oils with hard resins. The properties of the varnish depend on the drying oil, the resin, the ratios of these to each other, and processing conditions. Among familiar resins used in varnishes are phenolic, ester-gum, maleic, and epoxy resins. Urethane varnishes are sometimes called urethane oils because of their low viscosity and great flexibility. Short-oil varnishes contain more resin and less oil, which makes them harder, more brittle, and faster-drying. Long-oil varnishes contain more oil and less resin, which makes them softer, more flexible, and slower-drying. Medium-oil varnishes are intermediate in composition and properties.

Alkyds.   These vehicles are synthetic drying oils in which a larger or smaller part of the drying-oil molecule has been replaced by a synthetic molecule. The result is a new molecule with superior properties. In a varnish the resin is dispersed in the oil gel; in an alkyd a totally new molecule is formed. For this reason alkyds were called synthetics for many years after their introduction. Air-drying alkyds dry at room temperature because of the amount of drying oil present in the molecule. Like varnishes, alkyds are classified as short-oil, long-oil, and medium-oil to describe differences in drying-oil content and the resulting differences in properties. Alkyds made with nondrying oils, such as coconut oil, are used in heat-cured film formers and as plasticizers.

Room-temperature catalyzed film formers The mechanism of catalyzed film formation and cross-linking to produce a solid gel differs from the formation of oxidizing film formers. In catalyzed coatings the oxygen bridge is not an important part of the cross-linking. Instead, chemical agents called catalysts cause a different type of direct chemical bonding. Some catalysts are effective in small amounts without becoming a significant portion of the film after curing. Others are used in large amounts and form a significant portion of the film after curing. Curing produces a new polymer in which the original reacting molecules are an integral part of the polymer and can no longer be identified. This direct linking without oxygen bridges possesses superior chemical resistance and better resists aging. Catalyzed coatings produce at room temperature the benefits of heat-cured film formers.

Heat-cured film formers Heat causes direct cross-linking between the molecules in the film former or activates a catalyst which is not active at room temperature. The superior properties of heat-cured films are similar to those of room-temperature catalyzed coatings. Since the application of heat under controlled conditions is necessary, these coatings normally are industrial finishes. The results obtained with these finishes are more consistent than those obtained with room-temperature catalyzed coatings because the conditions of application and curing can be more closely and more consistently controlled.

Emulsion-type film formers In its simplest form the emulsion-type film former consists of drops of plastic floating in water. As the water evaporates, the drops come together to form a film. The spherical drops of plastic flatten out in a form more like that of pancakes to overlap and stick together. Plasticizers are added to make the plastic film more flexible and to improve adhesion. Coalescing agents that act as (slow-evaporating) solvents for the plastic are added to improve film knitting. The other usual film ingredients, such as pigmentation, are also present.

1.6.2 Solvents Solvents have been defined as the low-viscosity volatile liquids used in coatings to improve application properties. Most liquid coatings cannot be applied without them, although many new solventless coatings are now available. Solvents will be discussed under four headings:

- Functions of solvents
- Types of solvents
- Selection of solvents
- Air pollution

Functions of solvents Solvents do their work in paints by performing simultaneously two or all three of the following important operations:

reactive) is limited to 5 percent by volume of the coatings solvent of which it is a part.

- Any combination of aromatic compounds (including aromatic hydrocarbons) with eight or more carbon atoms to the molecule (except ethyl benzene) is limited to 8 percent by volume.
- Any combination of ethylbenzene, ketones having branched hydrocarbon structures, trichloroethylene, or toluene is limited to 20 percent by volume.

The reader is referred to the particular set of air pollution control regulations for his or her area of operation. The manufacturers of coatings are responsible for the compliance of their paints with the regulations of the areas in which they are sold. The user is responsible for buying and using paints that the manufacturer stipulates are in compliance with local air pollution regulations. The user is also responsible for controlling the emission of solvent vapors at a rate not to exceed the maximum rate specified by the regulations. The reader should become familiar with the air pollution regulations for his or her district.

Recent air pollution laws restrict or prohibit sandblasting. Such laws may prove to have a drastic effect on surface-preparation costs.

1.6.3   Pigmentation   Paint pigments are solid grains or particles of uniform and controlled size that are permanently insoluble in the vehicle of the coating. This insolubility differentiates pigments from dyes. At some stage in their use dyes almost always are soluble in the carrier in which they are being used. Paint pigments must be unreactive at all times to do their job, while dyes must almost always be reactive at some stage in their use to do their job.

The properties of paint pigments, like those of paint vehicles, must be controlled to close tolerances to produce paints of consistent quality.

The discussion of pigmentation is grouped under the following four headings:

- Pigmentation and the decorative function
- Pigmentation and the protective function
- Pigmentation and specialized functions
- Pigments to fill space

Pigmentation and the decorative function   In a coating, pigments contribute opacity, color, and gloss control. The desired pigmentation properties are achieved by proper formulation, compounding procedures, and quality control on the part of the manufacturer. Three important aspects of pigmentation are opacity, color, and gloss control.

Opacity. Opacity is the ability of paint to hide or obscure the substrate. This in turn is due to the index of refraction of the pigment, which is the numerical measurement of the pigment's ability to bend the light rays that strike its surface. White, yellow, and orange pigments have the lowest indices of refraction, while black has the highest. "Clean" colors have the least opacity. "Dirty" colors, which are formed by the presence of a dark or muddying component, have greater opacity than the same colors would have as clean colors. "Dry hiding," sometimes called "high dry hide" or "dry high hiding," is the increase in white opacity caused by using an excess of extender or filler pigments in white or tinted paints.

Color. Color is due to the ability of pigmentation to absorb certain wavelengths of visible light and reflect the other wavelengths. Color is such a complex technical subject that only its general use in paints will be discussed here.

Colored pigments differ widely in properties and in cost because they are derived by different chemical processes or are quite different in chemical composition. Colored paints can vary widely in specific properties even when they appear to the unaided eye to be exactly the same color. The necessary specific properties of a paint for a specific end use must include the performance characteristics of the color. This is true because it is possible to make a color by means of many different pigment combinations. Yet pigment combinations can differ widely in resistance properties.

One important resistance property of paint colors is resistance to fading from exposure to light, chemicals, or heat. Another is resistance to bleeding, the migration of color from a coat of paint to a succeeding coat of paint. Bleeding is caused by solubility of a pigment or dark-colored resin in the solvent or the resin of the succeeding coat of paint. The specific resistance properties needed in the paint color for a given end use can be determined only by exposing the pigments in the color to similar conditions. The manufacturer of the pigment usually has made these exposures and compiled the test results. This information is readily available from the manufacturer, who should be able to estimate the durability of any blend of pigment color requested.

For these reasons the paint user is dependent on the advice and integrity of the paint manufacturer in choosing color. This is especially true for an exposure that is liable to cause premature color failure, such as exposure to full sunlight for most of the day, marine exposure, or exposure to corrosive fumes.

Gloss Control. Pigments control the gloss of paints by affecting the texture of the coating surface. If the surface desired is very fine and uniform, the formulator will choose pigments that will produce only minute roughness. Formulating with larger pigments will produce a rougher texture, and in this way the mirrorlike "specular reflectance" of a clear varnish or a gloss enamel becomes the more subdued "diffuse reflectance" of a satin varnish or a semigloss enamel. Still higher pigment loadings and/or larger-size pigments produce a flat finish with no shine at all, sometimes termed "dead flat."

Pigmentation and the protective function   Pigments contribute to the protective function of paint because a properly chosen pigment combination helps achieve the properties required for the specified end use. Pigments are normally blended as in color formulation, and the end properties are the result of the properties of the individual pigments in the blend. The pigments that contribute to the protective function usually appear either in primers or in topcoats with protective functions rather than in intermediate coats.

Pigments in primers contribute to adhesion and to protection of the substrate. By distributing the stresses of the drying or curing primer film, the pigment particles keep the stresses from causing the primer film to shrink excessively. Shrinkage tends to pull the paint film loose from the substrate. The pigment also provides tooth, or texture, to which the following coating can cling. The pigment in the primer also helps distribute the forces of shrinkage caused by the drying of the coating applied over it. If the shrinkage of successive coats of paint becomes cumulative, it may pull the entire paint system loose from the substrate. Pigments in primers contribute to the protection of the substrate by changing the relation of the film former to the substrate. For example, in metal primers the use of proper priming pigments increases the long-term adhesion of the primer by a process sometimes called passivation of the metal. Among typical metal primer pigments are red lead, zinc chromate, and red iron oxide. They are sometimes blended with each other and with other pigments in primer formulations.

Pigments in protective topcoats reinforce the properties of the film former. In exterior metallized coatings, the aluminum or other metal flakes significantly increase the weathering resistance of the film former by shielding it from ultraviolet degradation and other destructive forces. As a result, many aluminum finishes are formulated with inferior film formers to keep the cost down. The now-restricted lead pigments contributed to the durability of exterior finishes by forming compounds, called soaps, that reinforced the film. Their spiny balls caused gradual erosion of the paint film rather than some other less desirable type of failure.

Some pigments add to the protective feature of exterior coatings by mechanically reinforcing the film. Talc fibers and mica flakes are typical examples. Zinc oxide contributes film hardness and water resistance because of the nature of the zinc soaps formed with drying oils. Zinc oxide and mica also screen out deleterious ultraviolet rays of sunlight.

Pigmentation and specialized functions   Here, again, pigment blending is important. Ease of sanding is a good example of a specialized function of paint that is contributed by pigmentation. The use of a short-oil varnish with the proper pigmentation further enhances the ease of sanding. Texture is another specialized function made possible by pigmentation. Fire-retardant paints depend on specialized pigmentation that responds to heat by causing a chemical reaction that reduces the flammability of the coating.

Pigments to fill space   Most flat paints, such as wall paints and primers, contain a certain amount of pigment used to fill space in both the wet paint and the dry paint film. These pigments serve a useful purpose. In lower-cost paints of most kinds, filler pigments are present in sizable amounts to reduce cost, since the cost per unit of volume of filler pigment is much lower than that of either film former or hiding pigments.

The vehicles of quality paints, especially wall paints and primers, are sticky if used in high concentration because they are well-polymerized resins. This high degree of polymerization makes the product viscous and also improves adhesion, toughness, and chemical and weathering resistance. In addition, it prevents "striking in" by its improved resistance to the suction of the substrate. These well-polymerized film formers must be diluted with solvents for application. The lower solids resulting from this reduction of viscosity by thinning would leave too thin a film if applied alone or with a minimum of pigment. Minimum pigmentation is that which uses only prime pigment, that is, pigment necessary for decorative, protective, or specialized functions only.

Up to a certain volume ratio, the addition of selected filler pigment actually reinforces and improves film-former properties. Beyond that point, especially for exterior finishes, a greater amount of filler pigment (or prime pigment, for that matter) leads to a rapid deterioration of film properties. The volume relationship of pigment to total film-forming solids is called PVC (pigment-volume concentration) and is expressed in percentages. It passes through a critical value called the CPVC (critical pigment-volume concentration). When that critical volume relationship is exceeded (Fig. 1.2), the resistance properties of the paint rapidly deteriorate. The film former no longer bridges all the voids between the pigment particles. Striking in increases. The film becomes porous (Fig.

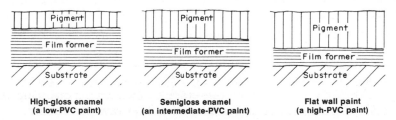

Fig. 1.2   Volume relationships between pigment and film former (e.g., resin) in dried films:

$$PVC = \frac{\text{volume of pigment}}{\text{volume of pigment} + \text{volume of film former}} \times 100.$$

1.3); as a result, it weathers much faster outside, is less washable, and loses some of its abrasion resistance, flexibility, and other desirable properties. Proper formulation allows for some migration of a coating into a porous surface. Striking in occurs when the system specification is wrong or a cheap paint is used.

Why, then, do manufacturers overpigment paint? First, extender pigment is much cheaper than film former. Using it in place of film former decreases the cost of the paint and increases the manufacturer's profit. Overpigmentation lowers the cost of the paint and raises the manufacturer's profit in another way: it reduces significantly the amount of white pigment (titanium dioxide) needed in the paint. This is dry hiding.

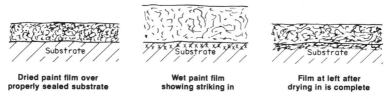

| Dried paint film over properly sealed substrate | Wet paint film showing striking in | Film at left after drying in is complete |

Fig. 1.3   Striking in. This condition weakens the paint film and causes porosity, blotchiness, and lower gloss.

1.6.4 Paint additives   Paint additives are ingredients added to the film former, pigmentation, and solvents. They impart needed paint properties not supplied by the other ingredients. Additives may also augment properties that are imparted to the paint by the other ingredients but still are not present in sufficient degree. The most familiar example is the addition of driers to oil-based paints to speed drying. Paint driers are catalysts for oxidizing film formers. They have other uses. Driers are soaps formed by the reaction of an organic acid with the oxide of a metal. The most common driers are formed by the reaction of compounds of cobalt, lead, calcium, zinc, zirconium, and other metals with tall-oil acids to form the tallates, with naphthenic acid from petroleum to form the naphthenates, etc.

Other classes of additives include wetting agents that help disperse pigments during the manufacturing process and defoamers that help break foam generated during agitation and application. Antisettling agents reduce settling of the pigmentation during storage of the paint. Preservatives and fungicides improve the storage stability of the paint by destroying organisms that cause putrefaction of the wet paint. Fungicides also kill organisms that grow on certain kinds of cured paint films and cause unsightly blotches of dark color.

Other additives include ultraviolet-screening agents that absorb ultraviolet rays of the sun in exposed coatings and reduce ultraviolet degradation of the film. Ultraviolet degradation drastically reduces the life expectancy of certain vehicles such as the vinyl chloride resins. Plasticizers increase the flexibility and adhesion of film formers. Many of the important film formers, such as the vinyl chloride resins, would not be usable without them. Acid acceptors accept and neutralize the acids formed by certain important resins, such as chlorinated rubber, during storage. Otherwise, the acids would cause a rapid deterioration of the film former. Antiskinning agents reduce or prevent skinning of the paint in the can, especially during long storage or after some of the paint has been used.

1.7 The Paint Formula   The formula lists the ingredients of the paint: vehicle, solvents, pigmentation, and additives. Amounts are normally stated in units of weight for accuracy. Accurate metering equipment permits measuring the liquids in units of volume. The significant relationships among the ingredients of the dried paint film are volume relationships, not weight relationships.

The film former may be present as drying oil, as varnish, as resin solution, as dry resin, as plasticizer, or as some combination of these. Solvent may be present as free solvent or as a component of varnishes or resin solutions. The pigments and the additives are usually listed separately.

Differences between the ratios of the principal ingredients (Fig. 1.4) is the most important factor in the differences between types of paints. The most important of these ratios is the volume of the pigmentation in the dried film compared with the total volume of dried film. The common types of paints in terms of the differences in the ratios of the ingredients they contain are clear finishes, stains, gloss enamels, semigloss (satin) enamels, flat paints, sealers and primers, house paints (for wood siding), stucco paints, and filling and caulking compounds.

1.7.1 Clear finishes   These materials are normally pure film former plus solvent and additives, such as a drier and an antiskinning agent. They may be oil and resin cooked into a varnish or a synthetic-resin solution such as an alkyd. Their color is normally transparent unless color is added in the form of a pigment or a dye. Clear finishes are glossy unless a flatting pigment is added. Properties depend on the oils and resins used and the conditions of processing the varnish.

1.7.2 Stains   Stains are low in both film former and pigmentation, especially if they are penetrating stains. They are made to soak into the surface to give color and some protection without forming a paint film on the surface. The so-called heavy-bodied stains or heavily pigmented stains are higher-solids versions of the same type of formulation. They are also called shake or shingle paints. The low ratio of film former to pigment in these heavier-bodied stains may limit durability and lead to earlier failure than is the case with a true paint.

1.7.3 Gloss enamels
These paints contain a high concentration of film former combined with enough opaque pigment to give both color and opacity. In drying, the high concentration of vehicle polymerizes and coats the pigment, giving a smooth, shiny surface. The film former in an enamel is chosen to dry to a hard, mar-resistant surface.

1.7.4 Semigloss enamels
Semigloss coatings contain less film former and more pigment than gloss finishes. Some of the pigment particles project through the surface to scatter the light and give the more diffuse reflectance needed for lower gloss.

1.7.5 Flat paints
These paints contain a great deal of pigment and a reduced amount of film former. The pigment particles project through the surface in enough places to give completely diffuse reflectance, which accounts for the complete lack of gloss or shininess. The film formers of quality flat paints, especially wall paints, are well polymerized to give good film continuity over surfaces of varying porosity.

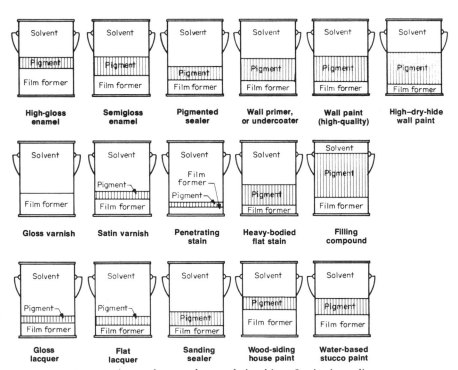

Fig. 1.4 Approximate volume relationships of paint ingredients.

1.7.6 Sealers and primers *Sealers* are designed to go over porous surfaces to seal off the suction caused by porosity. They contain more film former and less pigment than primers. *Primers* are designed to make the surface more paintable; e.g., a coat of flat paint is more paintable than old enamel, plastic, metal, etc. The pigment loading of sealers is more like that of semigloss enamels, while the pigment loading of primers is more like that of flat wall paints.

In recent years the terms "sealers" and "primers" have taken on a more general meaning because of the proliferation of new types of substrates to be coated. A sealer may now be any coating that keeps the following coats from soaking into the substrate. For example, a sealer over plastic may keep the solvent in following coats from attacking the plastic. A primer may now be any coating that provides improved adhesion for succeeding coatings. Either primer or sealer may be made with little or no pigment.

1.7.7 House paints House paints for wood siding are normally pigmented more heavily than exterior gloss enamels. Also, the film former is chosen to dry to a softer, more resilient film. Houses have large areas that will be repainted many times. The type of paint failure that leads to the need to repaint is important. A desirable type of failure is one which minimizes surface preparation for repainting. In house paints this is normally done by formulating a "chalking" paint. Chalking is the process of wearing away by turning to dust on the surface. The chalky paint can be removed by brushing to leave a finely textured surface suitable for painting without further preparation, except perhaps for washing the surface. "Self-cleaning" paints are those formulated so that the chalk formed does not cling to the surface but is washed away by the rain. There are both desirable and undesirable types of paint failure.

1.7.8 Stucco paints Stucco and other masonry paints are normally formulated with an emulsion type of film former pigmented about as heavily as a wood-siding paint. Stucco paints need not only to prevent the passage of liquid water (rain) through the film from the outside but also to permit the passage of water vapor through the film from the inside. Emulsion-type vehicles do this far more successfully than do other types.

This latter property is especially helpful in preventing the flaking off of stucco from the capillary action of groundwater. A word of warning: the same emulsion paint that resists pushing off from capillary groundwater is not waterproof below grade. Water under a static head will go right through it.

The ability of emulsion paint to permit water trapped behind the film to evaporate through the film will operate only if it is coated directly over the stucco or masonry. A coat of waterproof oil paint below the emulsion paint completely spoils this property of the emulsion paint.

Emulsion paints made with polyvinyl acetate and acrylic resins and some others also show two other important resistance properties: they resist deterioration from sunlight and from the alkali in the stucco better than oil paints.

1.7.9 Filling and caulking compounds   These materials contain a minimum amount of volatile solvents; otherwise they would shrink after application. Filling compounds are normally formulated with high pigment loading so that they dry hard and have good sanding properties. The film former must be tough and have excellent adhesion. Caulking compounds contain more film former than filling compounds do, and the film former is much more flexible because the dry product must have resiliency. The film former must adhere well to a variety of surfaces.

BIBLIOGRAPHY

American Paint and Coatings Journal, Feb. 12, 1979, pp. 52–55.

Flick, Ernest W.: *Solvent Based Paint Formulations,* Noyes Data Corp., Park Ridge, N.J., 1977.

Fontana, Mars G., and Roger W. Staehle (eds.): *Advances in Corrosion Science and Technology,* 3 vols., Plenum Publishing Corporation, New York, 1970–1973.

Fuller, Wayne R.: *Understanding Paint,* American Paint Journal Company, New York, 1965.

Martens, Charles R. (ed.): *Technology of Paints, Varnishes and Lacquers,* Robert E. Krieger Publishing Co., Inc., Huntington, N.Y., 1974.

Pierrehumbert, R. C.: "The Importance of PVC/CPVC Relationships in Interior Latex Coatings Formulations," *American Paint and Coatings Journal,* Apr. 2, 1979, pp. 53–59.

Chapter **2**

Paint Testing

TREVELLYAN V. WHITTINGTON
Aseptic-Thermo Indicator, Medical-Surgical Division, Parke, Davis & Co.

2.1 Purposes of Paint Testing   Paint testing is important to at least four different groups of people involved in the making and using of paint:

▪ The paint manufacturer, who should test paint as a routine part of paint manufacture and the development of new paints.

▪ The professional applicator and the large-scale user, who should evaluate the test results of the paint manufacturer. They should also do a somewhat different type of paint testing of their own.

▪ The specifier of paint, who should understand the purposes and principles of paint testing and the significance of the test results.

▪ The agencies and authorities that issue paint specifications.

Paint testing is conducted to generate definite information related to function, essential properties, specifications, and application procedures. For example, are the functions of paint (decoration, protection, specialized functions) adequately performed by the test paint? Does the test paint possess the essential properties of paint (adhesion, film integrity, ease of application, consistent quality) and the necessary specific properties for the intended end use? Does it possess them to the degree needed for the intended end use? Does the paint meet the specifications issued by the person or agency specifying the paint? Is the paint being applied so as to bring out the properties built into it by the manufacturer? Does the freshly dried paint film have the properties specified for the particular end use?

2.1.1 Quantitative and qualitative testing   Paint testing is both quantitative and qualitative. Quantitative tests give numerical answers. Color matching by eye is a qualitative test, while color matching by instrumental color measurement is a quantitative test.

2.1.2 Standard test methods   Standard test methods are those issued by an agency or an authority. An agency issues test methods and paint specifications for its own use in procuring paint. An authority issues test methods and specifications for use by anyone who wishes to employ standardized test methods and specifications.

The General Services Administration (GSA) of the United States federal government is a well-known agency that issues test methods. The American Society for Testing and Materials (ASTM) is a well-known authority that issues test methods. (Table 2.1, at the end of this chapter, cross-references the test methods described in the chapter with the closest ASTM and GSA test methods.)

2.1.3 Paint specifications   Paint specifications are of two principal types, performance and composition. Performance specifications specify the test methods that the paint must pass; this chapter is concerned primarily with this type of specification. Composition specifications give the required composition of the paint.

2.2   Classes of Paint Tests   Paint tests are conveniently grouped into three classes:

- Evaluation tests, which are conducted before a paint is adopted as a standard product
- Manufacturing control tests, which are conducted during the manufacturing process
- Application control tests, which are conducted during the application of the paint

2.2.1   Evaluation tests   Evaluation tests should be made by the paint manufacturer, the professional applicator, and the large user of paint. A summary of their differing reasons for making evaluation tests follows.

The manufacturer makes these tests for the following reasons:

- To gather needed information about an experimental coating during the process of development
- To determine whether an experimental coating is ready for adoption as a standard product
- To evaluate new raw materials, new ways to combine familiar raw materials, or new or different compounding procedures
- To evaluate the results of changes made in accordance with the rulings of a governmental regulatory agency

The applicator or the user makes these tests for the following reasons:

- To evaluate the suitability of the coating for the applicator's or the user's particular needs
- To verify or disprove the manufacturer's claims for any product when the claims seem unfamiliar or lavish
- To avail himself or herself of new advances in coatings technology by small-scale testing before risking large-scale application of an unfamiliar coating or application procedure
- To evaluate the quality control of the paint manufacturer

2.2.2   Manufacturing control tests   Manufacturing control tests are made by the paint manufacturer during the manufacturing process to acquire certain necessary information quickly:

- *Composition.* Whether the paint appears to contain the proper ingredients in the correct amounts
- *Procedure.* Whether the paint appears to have been compounded in accordance with the standard manufacturing procedure
- *Specifications.* Whether the paint meets established manufacturing control specifications, such as fineness of grind

NOTE: Experience shows that paints which meet these three sets of manufacturing control tests will normally meet all the specifications established by the manufacturer, such as those for exterior durability. The test results indicate whether the paint contains the same ingredients combined in the same way as in the paint tested in the evaluation tests before adoption.

2.2.3 Application control tests   Application control tests should be performed by both the applicator and the paint manufacturer:

- The paint manufacturer should check the paint before canning it to make sure that it has the required ease of application, which is an essential property of paint.
- The applicator should perform application control tests to make sure that the paint is being applied so as to bring out the properties built into it by the manufacturer.

2.3 Description of Typical Paint Tests   While space constraints limit the number of tests to be described, those chosen were selected to give the reader a comprehensive understanding of paint testing. The tests described are grouped to show effective ways to think about paint and its uses. The following groups are described:

1. Tests *primarily* for the *decorative* function
2. Test *primarily* for the *protective* function
3. Tests for *specialized* functions
4. Tests for *essential* properties
5. Tests for *necessary specific* properties
6. *Manufacturing* control tests
7. *Application* control tests

For easy reference this chapter uses the following general format to describe paint test methods:

- *Name of test.* The name or names in common use
- *What is being tested.* Property or properties of paint being tested
- *How test is performed.* Description of test procedure
- *Data yielded.* Measurements and observations made during test
- *Interpretation of data.* Conclusions about properties of the paint that can be drawn from the data

2.3.1 Tests primarily for the decorative function   These tests deal with the aesthetic properties, or pleasing appearance, of paint. The decorative function of paint can often be destroyed without seriously affecting the protective and specialized functions.

The following paragraphs describe six groups of tests that deal primarily with the decorative function:

1. Color
2. Gloss

3. Hiding power
4. Texture
5. Mar resistance
6. Resistance to chemicals

Color Five aspects of color testing are important to give a balanced view of the contribution of color to the decorative function of paint: (1) color matching by eye, (2) instrumental color measurement, (3) metamerism, (4) colorfastness, and (5) color bleeding.

Fig. 2.1 CG-6900 series Colorgard light booth. [*Courtesy of Gardner Laboratory, Inc., Bethesda, Maryland*]

Name of Test: Color Matching by Eye

What is being tested. How closely the paint color being tested matches the color standard.

How test is performed. A test swatch of the color being tested and a swatch of the color standard are viewed side by side under a standard light source by a standard procedure. The swatches should be dry. Since forced drying sometimes changes the color, swatches should be cured in a normal fashion.

Data yielded. The opinion of an experienced observer of how closely the test color matches the color standard and the ways in which it differs from the color standard. These ways include being lighter or darker, cleaner or muddier, having too much of a primary color (e.g., being too red), lacking in a primary color (e.g., needing red), etc.

Interpretation of data. Acceptance or rejection of the color match. In manufacturing the paint color, what should be done to correct a mismatch?

Name of Test: Instrumental Color Measurement. This test is often called Spectrophotometric Color Measurement or Color Difference Measurement.

What is being tested. Either (1) the reflectance or transmittance of light at all or selected wavelengths in the visible spectrum or (2) the difference between the reflectance of the color standard and the test paint.

How test is performed. In (1) above, a spectrophotometer is used (an abridged spectrophotometer takes measurements at selected wavelengths only). Light from the standard or test paint is passed through a prism or grating and broken up into its component wavelengths in the visible spectrum. The reflectance or transmittance can then be measured and plotted graphically. This is a slow process unless an abridged spectrophotometer is used. Even then a trained operator is required. In (2) above, a reflectometer or color-difference meter uses a combination of light filters and photocells to measure the difference in light reflectance between swatches of the test paint and the color standard.

Data yielded. In (1) above, a series of light reflectance or transmittance values; in (2) above, numerical differences between the reflectance of the test paint and a spectrally similar color standard.

Interpretation of data. In (1) above, the data give the spectral distribution of the components of the color, i.e., the percentage of the light that is absorbed or transmitted at each wavelength measured. How closely the test-paint color matches the standard can be seen by how closely the curve for the test paint lies on top of the curve for the color standard. In (2) above, the data give the difference in light reflectance. The smaller these differences, the closer the color match. By proper computer programming, spectrophotometric readings can be used to determine the amounts of carefully standardized colorants to be added to correct the color match of a batch of paint.

The spectrophotometer is considered the ultimate authority in color matching. It analyzes the composition of the color of both paint and standard in terms of wavelengths of light. Instrumental color measurement in general is less subject to difference between observers and to human fatigue than is color matching by eye. The ultimate criterion for the user, however, is how the paint looks to the eye of an experienced observer under actual on-the-job conditions.

Metamerism. This is the phenomenon of colors matching under one light source but not under another. It is due to differences in the blend of colored pigments used in making the test paint and in making the color standard. It is detected by comparing the colors under at least two different light sources, such as daylight and incandescent lamplight. Color matching should be done under a test light which provides the required light sources.

Name of Test: Colorfastness (Indoor). Outdoor colorfastness is usually tested as a part of the weathering test.

What is being tested. Resistance of paint color to fading.

How test is performed. Test panel is exposed to an arc light in the Fade-O-Meter with half of the panel masked.

Data yielded. Degree of aging of paint as measured by comparing the exposed portion with the masked portion or the number of hours of exposure to produce an observable change.

Interpretation of data. Fade resistance of a paint color compared with a standard paint color or fade resistance in hours of exposure to produce an observable degree of color fading. The test also measures the tendency of white and near-white paints to yellow or darken.

Name of Test: (Color) Bleed Resistance

What is being tested. Tendency of a colored pigment or resin to migrate into another coating applied over it because of solubility in the solvents of the topcoat.

How test is performed. There is no standard method of test. All tests in use involve coating a light-colored topcoat over the test paint which is suspected of bleeding. Migration of color into the topcoat is observed under the conditions of test. Baking aggravates bleeding because heat makes coatings more fluid and causes amalgamation between coats.

Data yielded. Whether bleeding occurs and how bad it is.

Interpretation of data. Whether the colored pigments (e.g., toluidine red) and dark resins (e.g., asphalt) in a test paint cause bleeding that cannot be tolerated under conditions of use.

Gloss Only one aspect of gloss will be considered: measurement of the light reflectance of the paint surface.

Name of Test: Gloss Measurement

What is being tested. Shininess of the paint (for gloss paint) or lack of it (for flat paint).

How test is performed. A beam of light is directed toward the test-paint surface at a certain angle from the horizontal. The percentage of the beam that is reflected at the same angle is measured by a photocell. Three standard angles are used: 60° for general gloss readings, 85° for high-gloss finishes, and 20° for flat finishes. The angle of incidence (e.g., 60° gloss) must be specified. Because visual comparison of the gloss of the test paint versus a gloss standard does not yield precise numerical values, it is less often used by the paint manufacturer, but it is a valuable test in the field.

Data yielded. Percentage of gloss of test paint. Completely specular light reflection (perfect gloss) would be 100 percent, and completely diffuse light reflection (mat or dead flat) would be 0 percent.

Interpretation of data. Whether the test paint falls within the specified range of gloss values. Visual comparison of the test paint with a gloss standard is less accurate except for so-called distinctness-of-image gloss.

Hiding power   The most important point is the ability of paint to hide the substrate in relation to the spreading rate of the paint. The test is of

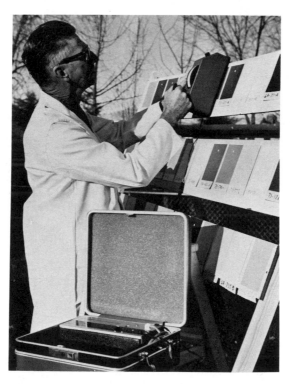

FIG. 2.2 GG-7260-A Glossgard gloss-meter system in use on a test fence. [*Courtesy of Gardner Laboratory, Inc., Bethesda, Maryland*]

Fig. 2.3 Opacity-display chart. [*The Leneta Company, Ho-Ho-Kus, New Jersey*]

most value for white and pastel colors and also for deeper colors based on such transparent pigments as clean yellows and oranges.

Name of Test: Hiding Power (Contrast Ratio)

What is being tested. The spreading rate of the paint at complete hiding, measured in square feet per gallon.

How test is performed. The paint is applied at different wet film thicknesses over black-and-white test charts, then dried. The light reflectance is read over the black and the white portions of the dried chart with a reflectometer. This process is repeated for each wet film thickness.

Data yielded. A series of light-reflectance readings.

Interpretation of data. The reflectance over the black area divided by the reflectance over the white area gives a number called the contrast ratio (CR). A CR of 0.98 is considered complete hiding. The important result for the user or specifier of the paint is the spreading rate in square feet per gallon at a CR of 0.98.

Texture   This is another property of paint associated with the decorative function for which there is no standard test method. Instead, approximately the following procedure is followed:

- The paint manufacturer checks the texture paint to see that the texture pattern can be controlled within certain limits.
- The applicator and the customer examine panels of possible texture patterns prepared by the applicator and agree on a certain texture pattern.
- The finished texture job is passed or rejected by the customer or the customer's inspector by comparing with the agreed-on standard panel.

Mar resistance   This is the ability of the *surface* of the coating to resist *moving* forces which deface surface appearance. The defacing is the result of fine scratches which may not interfere measurably with the protective function of the coating.

Name of Test: Mar Resistance.   A usual alternative is the name of the individual test method, such as BTL Balanced Beam Mar Tester.

What is being tested. The ability of the coating to resist defacing (breaking through the surface by scratching) under the test conditions.

How test is performed. Three types of testing methods are in use. Each emphasizes a slightly different aspect of marring.

- Some tests scratch the surface with a weighted sharp instrument such as a needle, making only one scratch or depression at a time.
- One test method causes particles of an abrasive such as Carborundum to fall on the surface in a controlled stream.
- Some tests rub a solid surface against the test paint in a controlled manner to scuff the surface. Abrasive wheels and wool blanket material are used in different test methods.

NOTE: Some of the tests used to evaluate mar resistance really evaluate the ability of the coating to resist surface streaking combined with surface marring. The Coin Mar Test and the Fingernail Mar Test do not break through the surface. Instead, they dent the surface and deposit a streak of foreign matter on it.

Data yielded. The data yielded depend on the test method. The single-scratch method gives the weight required to make a scratch that is visible and also to rupture the surface. The falling-abrasive method gives the change in light reflectance (gloss) for different amounts of abrasive. The scuffing methods also measure the change in gloss with the amount of scuffing.

Interpretation of data. How resistant the surface of the paint is to defacing by fine scratching under test conditions.

Resistance to chemicals   Lack of resistance to chemicals usually leads to rapid deterioration of the decorative function of paint. Tests for resistance to chemicals are usually either of the spot or of the immersion type.

Name of Test: (Chemical) Spot Test

What is being tested. Ability of the coating to resist staining by the chemical being tested when the chemical stands on the paint film.

How test is performed. The test chemical, usually liquid or in solution, is put on the test panel at a few locations, which are marked. The spots may be covered or left uncovered, as specified. At the end of the specified period of exposure the chemical is removed by blotting or sponging. Any residue is removed by a neutral solvent that will not affect the test. The test panel is then permitted to dry. After the specified recovery period, if any, the panel is examined and the test results reported.

The test chemical may be anything agreed on by the parties involved. Any special techniques should be spelled out in the test method specified.

Data yielded. Observation of all changes in appearance or other properties specified in the test method.

Interpretation of data. Resistance of the test paint to the chemical being tested when the chemical stands on the paint. (Flowing chemical fluids act differently.)

Name of Test: (Chemical) Immersion Test

What is being tested. Ability of the coating to resist staining by the chemical in which it is immersed.

How test is performed. The test panel of paint is immersed in the test chemical for a specified time. The test method must specify all special techniques, which include panel material, the method of application, the use of heat or other agent to accelerate the test, and special measurements.

Data yielded. Observation of all changes in appearance or other properties specified in the test method.

Interpretation of data. Resistance of the test paint to the chemical being tested when immersed.

These test methods also evaluate the deterioration of the protective function of the test paint.

Water, especially salt (sea) water, is the commonest corrosive chemical that paint encounters. Because of its importance, all test methods dealing specifically with the effect of water on paint are grouped together.

2.3.2 Tests primarily for the protective function
These tests deal with the ability of paint to protect the substrate from deterioration. To do this, the paint itself must be durable. The user should keep in mind that the protective function of paint may be preserved intact even when the decorative function is destroyed.

Discussion of tests related *primarily* to the *protective* function of paint will be limited to five groups of tests:

1. Resistance to weathering
2. Resistance to wear
3. Resistance to chemicals
4. Resistance to water
5. Flexibility (resistance to rupture on flexing)

Resistance to weathering   Weathering is the deterioration of paint due to outside exposure. The final test of any paint is the manner in which it holds up in actual use. However, testing paint weathering under actual use conditions is slow and expensive. The time and expense involved in such testing have led to the development of test methods designed to give answers more quickly and at less expense. The three aspects of resistance to weathering are (1) natural weathering, (2) accelerated outdoor weathering, and (3) artificial weathering.

The elements of weather that lead to paint deterioration are sunlight, temperature changes, and humidity (water and water vapor). Paint deterioration from water is so widespread that it is treated separately.

Name of Test: Natural Weathering (Test Fence)

What is being tested. How the test paint fails, and how rapidly it fails.

How test is performed. All steps of the procedure are carefully specified. The test-panel material and its cleaning, surface preparation, and any priming coats are done according to specification. Metal, wood, and masonry are the usual panel materials.

The test paint is applied in the specified manner and thickness, then dried or cured as specified. The dried or cured test panel is then exposed on an outside rack. Observations of the weathered panel are made in a specified manner after specified intervals of weathering. A standard reference paint of known performance is usually included, frequently on a part of the test panel.

Data yielded. The form recommended jointly by the Federation of Societies for Coatings Technology (FSCT) and the ASTM lists 16 aspects of paint deterioration for observation:

Appearance	Peeling
Gloss	Dirt
Chalking	Mildew
Erosion	Rusting
Checking	Fading
Cracking	Darkening
Flaking	Yellowing
Scaling	Blistering

Ratings are made by experienced observers, using a scale of 0 to 10. The rating for an unexposed specimen, which, of course, shows no degree of failure, is 10; the rating for complete failure or complete loss of a property is 0. The reader is referred to the original specification for exact definitions of the 16 terms listed above and for information on weathering-test methods.

It is obvious that this is a qualitative test, in which the data yielded are the opinions of experienced observers. It is also quantitative, in that these opinions are given numerical values. A group of experienced observers is used whenever possible, and their opinions are averaged.

Interpretation of data. The averaged opinions of experienced observers, given numerical scores, rate the durability of the test paint. The number of the 16 terms to be included in the rating should be spelled out in the original specification.

The racks used to hold the test panels of paint during natural-weathering tests are called test fences (see Fig. 2.2). The panels on the test fence are exposed at various angles. In the United States the commonest angles are vertical south and 45° south. This means that the painted side of the test panel is toward the south, where it receives more sun. It is either vertical or 45° from the vertical, with the test paint on the upper side. Usually a portion of the panel is covered so that the paint in its unweathered state can be seen as a means of comparison.

Analytical kits for field-testing weathered paints are now available. They can be used by the average field technician to determine the type of coating on the surface, the cause of premature coating failure, and other dry film characteristics (Fig. 2.4).

Accelerated Outdoor Weathering. In an effort to shorten the time required on the test fence, many variations of exposure conditions have been tried. Exposing the panels at the horizontal or at 5° from the horizontal accelerates weathering. Racks that rotate to follow the sun are sometimes used. One type of test rack follows the sun, has mirrors to concentrate the sun's rays on the test panels, and also sprays water on the panels at intervals.

All accelerated outdoor-exposure methods, even the common 45° south exposure, must be correlated with the vertical south-exposure method by comparing actual exposure data. Accelerated exposure methods usually accelerate the deterioration of some paint properties more rapidly than others. Differences in test-panel material (for example, differences in wood) substantially affect the results. Each step of the test procedure therefore must be carefully controlled.

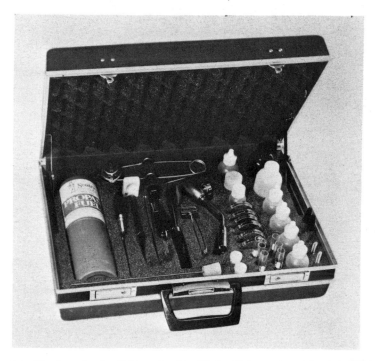

Fig. 2.4 Weathered-paint field testing kit. [*Tinker & Rasor, San Gabriel, California*]

Name of Test: Artificial Weathering (Accelerated Weathering)

What is being tested: How the paint fails and how rapidly it fails, as in natural weathering.

How test is performed. Steps are similar to those of natural weathering on the test fence, with three important differences. The test panels are exposed inside a laboratory in an artificial-weathering machine, most commonly the Atlas Weather-Ometer. The elements of natural weathering (sunlight, temperature changes, humidity) are replaced by a combination of artificially produced elements. The light is supplied by a carbon arc, a xenon arc lamp, fluorescent lighting, or some combination of them. Temperature changes, if used, are ordinarily supplied by the cycles of the carbon arc light. Humidity is supplied by a periodic water spray. Sometimes extra elements of weathering are introduced to accelerate the testing. An example is the periodic injection of sulfur dioxide gas to simulate the corrosive conditions caused by sulfur gases.

Data yielded. About the same as for natural weathering on the test fence. Artificial weathering may change the relative rates at which different paint properties fail, however, and results must be checked against those of test-fence weathering or against actual in-service testing. A useful check on accelerated-weathering data for new paints is the inclusion of panels of paints for which both natural-weathering and artificial-weathering data have been collected.

Interpretation of data. About the same as for natural weathering on the test fence. However, results are to be interpreted as an indication only of what the paint will *probably* do on the test fence.

Resistance to wear   Paint films wear away by contact with abrasive materials, usually in solid form. Wear resistance is the result of a combination of other properties of the coating, such as adhesion, cohesive strength, and toughness (which itself is due to a combination of properties). Wear resistance cannot be reliably predicted by measuring the other properties. It must be measured directly. The most important direct test of wear resistance is the test for abrasion resistance.

Name of Test: Abrasion Resistance

What is being tested. The ability of the test paint to resist wearing away by contact with abrasive material under controlled test conditions.

How test is performed. The choice of abrasive medium and its application vary with the test method. The choice depends on what the developer of the test considers to make the greatest contribution to wear resistance (e.g., cohesive strength).

Some tests rotate abrasive wheels over the test paint. Others rub it with sandpaper. Some let abrasive particles such as sand, emery, or silicon carbide fall against the panel while it is supported at an angle, usually 45°. Others blow the abrasive particles against the test panel with compressed air. Some tests use pebbles, gravel, or steel shot as the abrasive agent, the paint being coated on the inside of a box containing the agent. In other tests water is directed against the panel at high velocity to simulate rain erosion on aircraft coatings. In still another test method a variety of abrasive materials can be used in wet or dry form.

Interior paints are often tested for wet-abrasion resistance on a wet-abrasion machine. A wet sponge is rubbed with a cake of grit soap, weighted, and then drawn back and forth over the test paint until the paint wears through. The same machine is used to check ease of soil removal from the paint and, indirectly, to measure resistance to water.

Data yielded. Weight or film-thickness loss per number of cycles of abrasion; or the amount of abrasive material, time of application, or cycles of abrasion required to wear through the coating to the substrate.

Interpretation of data. How resistant the coating is to abrasion under the test conditions.

Traffic marking paints constitute a large and important group of coatings with their own special group of wear-resistance tests. These tests attempt to simulate closely the actual use conditions of the paints. Some of them correlate closely with results obtained in actual use of the paints.

Resistance to chemicals   The test methods for the resistance of paint to a chemical or chemicals are usually the same as those used primarily for decorative properties. (Special tests can be devised for any particular chemical.) The properties examined by these methods, which are related primarily to the protective function, deal with film integrity. After exposure to the test chemical, the film is observed for film continuity, film erosion, adhesion, softening, and swelling. Loss in any of these properties weakens or destroys the protective function. Salt water is one of the most common corrosive chemicals.

Resistance to water   Water deserves special treatment because it causes so many paint failures. It deteriorates paint when it is present as liquid water, as water vapor, or as a solution of chemicals in water. The following tests describe the deleterious effects of water in each of these three forms.

Name of Test: Water Immersion

What is being tested. The ability of the coating to protect the substrate (usually steel) from deterioration during immersion in water.

How test is performed. Test panels prepared according to specified procedures are immersed in water under specified conditions and examined after specified periods of exposure.

Data yielded. The degree of attack of water on the substrate by penetrating through the film (e.g., rusting) and also the degree to which the coating itself loses desirable properties (e.g., loss of adhesion, an essential property of paint, a loss which produces blistering).

Interpretation of data. Test methods use photographic standards of rusting or blistering, which provide a rating scale from 10 (unaffected) to 0 (badly affected).

Name of Test: Humidity Test (Humidity Cabinet; Water Fog Testing; Condensation Tester)

What is being tested. The ability of the coating to resist damage from the transmission of water through the film when it is exposed to water vapor which condenses on the surface.

How test is performed. Test panels prepared according to specified procedures are exposed in a specially constructed cabinet and examined after specified periods of exposure. The temperature and relative humidity in the cabinet are maintained within specified limits.

Data yielded. Degree of blistering and rusting as in the Water Immersion Test.

Interpretation of data. The ability of the coating to resist damage to the substrate and to itself from the transmission of water through the film from the condensation of water vapor on the surface.

NOTE: There is also a group of tests for measuring the rate at which paint films will transmit water vapor. However, the humidity tests described above seem more in keeping with the basic concepts of this chapter.

Name of Test: Blister Resistance

What is being tested. The ability of the paint film to resist blistering from water that is condensed in the substrate behind the paint film and tries to pass through the film to a colder outside atmosphere.

How test is performed. Test panels (usually wood) are coated on one side only. The uncoated side is exposed to the inside of a box in which high humidity is maintained under controlled conditions, and the coated side is exposed to the colder temperature of the room in which the box is placed.

Data yielded. The tendency of test paint to blister (lose adhesion) under test conditions.

Interpretation of data. The results depend on the ability of the coating to adhere to wet wood and also to transmit water through the film. The relative importance of these two properties has not been determined. The significant finding is whether or not the paint blisters.

Name of Test: Marine Exposure Tests.

There are three types of tests in common use: atmospheric, tide-range, and submerged.

What is being tested. The ability of the test paint to protect the substrate (usually metal) from corrosion by seawater or from the accumulation of organisms, such as barnacles growing in seawater. Seawater is the most common of the corrosive water solutions because of the salts dissolved in it. These dissolved salts accelerate the rate at which water goes through the paint film and in the process degrade the film itself.

How test is performed. The substrate material, the application of the test coating, and the exposure procedure are specified. In *atmospheric* marine exposure tests, the panels are exposed on test racks near enough to the sea so that they are subjected to salt spray. In *tide-range* marine exposure tests, the test-panel rack is located in tidewater so that panels are submerged at high tide and completely out of seawater at low tide. In *submerged* marine tests, the test panels are kept submerged. In ship-bottom patch tests, a portion of the ship below the waterline is coated (usually two separate portions of the bottom are coated). The test panels are observed after specified times of exposure.

Data yielded. How the test paint fails and how rapidly it fails under test conditions. Since the protective function is emphasized, the important test data are blistering (loss of adhesion), rusting or other corrosion, and the growth of undesirable marine organisms (on antifouling paints).

Interpretation of data. Whether and how well the test coating resists penetration by seawater (as shown by blistering and substrate corrosion) and/or how well it resists the growth of marine organisms on it.

Name of Test: Salt Spray (Fog) Testing. This test method is based on the same corrosiveness of salt spray that forms the basis of the atmospheric marine exposure test. It is not a substitute for atmospheric marine exposure testing, but it obtains results much more quickly and has a broader application to corrosion problems. The test was developed originally to test the corrosion resistance of metals, especially that of plated metals.

What is being tested. Resistance of the coating to corrosion from salt fog under test conditions (there are variations of the test to accelerate the process).

How test is performed. The substrate material, the application of the test coating, the exact composition of the salt spray, and the operating parameters of the salt-spray cabinet are all specified. The test panels are scratched with an X through the test paint to the metal substrate to measure rust creepage under the edge of the paint film.

Data yielded. How rapidly the substrate rusts from water penetrating the paint film and how rapidly rust creeps under the film from the X scratch.

Interpretation of data. Resistance of the coating to water transported through it by salt water and to corrosion creepage under the edge of the paint film. Corrosion is an electrochemical phenomenon which is delayed or prevented by the electrical resistance of the coat of paint. Electrical resistance is proportional to the thickness of the paint film: thin spots become corrosion sites. Here again is a reminder of the importance of applying paint properly so as to get a uniform film thickness uniformly cured.

Moisture Meters. Electric moisture meters are used to check the moisture content of wood or plaster. The level of water in the substrate determines whether or not it is too wet for good painting. Determination of the moisture content can also be useful in finding the causes of paint problems.

Flexibility Flexibility is a measure of how much a paint film can expand without rupture when the substrate on which it is coated expands, as by bending. This is a complex property in which adhesion and film extensibility both play a part. Elasticity is not a necessary part of flexibility. There are two established tests for flexibility. The first involves a slower rate of film expansion, and the second an impact or almost-instantaneous rate of film expansion.

Fig. 2.5   MG-1412 mandrel set. [*Courtesy of Gardner Laboratory, Inc., Bethesda, Maryland*]

Name of Test: Flexibility by Mandrel Test

What is being tested. How much the paint film can be extended without rupturing by bending the panel on which it is coated.

How test is performed. The test paint is applied by specified procedures to a metal test panel of specified size, thickness, type of metal, and surface preparation. After the specified drying period, the test panel is bent over a conical mandrel (equipped with a drawbar for bending the panel) or a cylindrical mandrel (available in a series of diameters).

Data yielded. How much the film expands at the point of rupture when the panel is bent over the mandrel.

Interpretation of data. The percentage of extension which the paint film will undergo without rupture when the substrate expands at the rate of bending used in the test.

Name of Test: Impact Test

What is being tested. The ability of the paint film to resist rupture from expansion when the metal substrate is deformed abruptly by impact.

How test is performed. The test panel is prepared much as in the Flexibility by Mandrel Test. A hemispherical piece of hard metal is impacted against the panel by a falling weight. The test is continued by letting the weight fall farther or by using a heavier weight until the test film cracks.

Data yielded. The weight and distance of fall required to cause the test paint to crack. The weight is permitted to fall onto the painted side (direct impact) or the unpainted side (reverse impact) of the panel.

Interpretation of data. The amount of impact (usually in units of inch-pounds) and the resulting film extension that the test paint can undergo without rupture when tested by this procedure.

2.3.3 Tests for specialized functions   Some specialized functions of paint have standardized test methods; others do not. Examples of both types will be described to round out this description of test methods. The tests described include those for (1) fire retardancy, (2) sandability, (3) biological deterioration, and (4) strippability.

Fire retardancy   This is the property of delaying the spread of flames, sometimes by the sacrifice of the paint film itself. Fire retardancy in any given fire depends on a combination of factors, only one of which is the fire retardancy of the coating. This circumstance makes testing for fire retardancy difficult and limits the usefulness of the test data.

Name of Test: Tunnel Test. Test tunnels vary in length from 25 down to 2 ft (from 7.6 down to 0.6 m). The longest tunnel is used in the Underwriters Laboratories (UL) test method. The great expense of this method has prevented its universal adoption and led to the development of shorter and less expensive test tunnels to encourage cooperative testing.

What is being tested. Rate of flame spread under conditions of test.

How test is performed. A test panel of specified length is put into place under the ceiling of the tunnel. The airflow through the tunnel is regulated to a specified rate. The end of the test panel is set afire by gas burners, and the flame spreads down the test panel to the vent end of the tunnel. A value of 100 is given to the time required for the flame to spread to the vent end on bare red-oak flooring, select grade. Thermocouples check temperature and a light-source–photocell combination measures smoke density.

Data yielded. Rate of flame spread, temperature, and smoke generated.

Interpretation of data. How much the test coating retards the rate of flame spread or reduces temperature and smoke generation.

Other Tests of Fire Retardancy. Several other techniques are in use. All are simpler than the UL tunnel test. Some tests apply heat to the test panel with a gas or other flame. The rate of flame spread or the weight loss or amount of charring is measured. Other methods apply flame from a controlled source to a test panel of specified dimensions (usually 3 by 3 ft, or 0.9 by 0.9 m, or larger) and measure flame spread, temperature rise, charring, afterglow, or some combination of these.

One test uses a cabinet to contain the test panel. Another places the test panel or panels in an open-front box. A group of related tests evaluates the resistance of the test paint to high heat with flame. The test methods vary. One includes a weathering test of the heated test panel. Results from this test are rated by comparison with a paint of known performance properties.

Sandability   This is an important property of many coatings but one which often is not tested. Sandability is a prime requirement of sanding sealers, undercoats, and filling compounds. These must all sand easily, but the ingredients facilitating easy sanding must not degrade the durability of the coating system. Ease of sanding is caused by additives or by standard paint raw materials used in an uncommon way. The sanding coat requires a properly balanced formula, for a poor-quality coat will degrade the entire coating system.

Name of Test: Sanding Characteristics

What is being tested. The quality of surface produced by the sanding and the ease of removal of what is being sanded off.

How test is performed. The coating being tested is applied to a standardized panel material which has been sanded smooth. The coating is applied and dried under controlled conditions. Multiple coats of different materials can be specified, each coat being dried and sanded before the next is applied. Both the technique of sanding and the sandpaper to be used are specified.

Data yielded. Observations of the smoothness and uniformity of the surface produced by sanding, the nature of the paint removed by sanding (powdery or gummy), and the gumming of the sandpaper.

Interpretation of data. The quality of surface produced and the ease and cost of producing it.

Biological Deterioration.   Both the decorative and the protective functions of paint can be destroyed by the growth of microorganisms on the dried paint film. The decorative function is affected more quickly and more often. Mildewed paint is one of the most unsightly examples of failure of the decorative function. Latex paint in the can is also subject to spoilage from the growth of fungus; paint thus contaminated becomes unusable.

Some paints serve as food for microorganisms whose growth is promoted by warm and humid conditions. Their growth is discouraged by paints that do not serve as food and by paint additives called mildewcides that are poisonous to microorganisms.

Name of Test: Resistance to Fungus Growth. Usually, however, each test has its own name.

What is being tested. The ability of the test paint to resist the growth of fungus in a can of wet paint or on the film of dried paint.

How test is performed. The wet paint is inoculated with a specimen of "spoiled" paint containing the test organism (fungus) and then, after specified periods of aging, examined for living organisms. The resistance of a dried film of paint to fungus growth is checked by coating it onto a porous substrate such as filter paper to make test panels. A small test panel is laid on a bed of agar or broth (fungus food) which is inoculated with the test organism or organisms. The inoculated test specimen is then cultured for a specified number of days at a specified temperature and relative humidity.

Data yielded. Whether the wet paint or the dried paint film will support the organism under test conditions.

Interpretation of data. Whether the paint in the can or the dried paint film will be subject to fungus growth under normal conditions of manufacture, storage, and use.

Strippability   Strippable coatings illustrate a specialized function of paint that is infrequently used but is important to the user. Coatings perform many such specialized functions for which there are no standardized test methods because use conditions vary widely. Strippability is also an example of the need for the paint-system approach to enable paint to perform a function.

Strippable coatings are normally designed for use in spray booths, studio scenery sets, and other places requiring quick removal of coatings to the substrate so that the process can start over. A strippable coating works best when applied over a base coat that gives it only limited adhesion. The strong cohesion of the strippable coating thus exceeds the limited adhesion to the base coat. The strip coat is readily peeled off together with all coats of paint that have been applied over it.

2.3.4 Tests for essential properties   Essential properties of paints are those which all paints require. They are especially important if the paints are to be used as engineering materials. These four essential properties are (1) adhesion, (2) ease of application, (3) film integrity, and (4) consistent quality.

Adhesion   Adhesion is the paint property of sticking to a surface, whether the surface is a bare substrate or another coat of paint. It is the most important property of paint. Tests to measure adhesion usually give quantitative measurements, which are measurements of the force required to remove the film from the surface under testing conditions.

Because of the importance of adhesion, there are a large number of tests to measure it. No one method yields all the desired information.

The tests discussed below are grouped according to the methods used to remove the coating from the substrate.

Name of Test: Adhesion (by Cutting the Paint Film from the Substrate). Each individual test method has its own name, such as Arco Microknife, Adherometer, or Penknife Method.

What is being tested. The force required to cut or push the coating away from the substrate under testing conditions.

How test is performed. A sharp cutting edge (harder than the paint) is forced against the paint film at an angle so as to cut or push the film away from the test panel. The force required is measured.

The original form of this test, still used, is the Penknife Adhesion Test. A sharp penknife is employed to cut a ribbon of paint from the test panel. Although this is a qualitative test, an experienced observer can compare coatings to get reproducible results. All the other test methods in this group may be considered attempts to quantify the penknife method.

Data yielded. Measurements of the force required to cut or push the coating away from the substrate.

Interpretation of data. The force required for removal is proportional to the adhesion, so that the greater the force required for removal, the greater the adhesion.

Name of Test: Adhesion (by Scratching or Scraping the Paint Film). Each test method has its own name, such as Crosscut Adhesion, Scrape Adhesion, or Scratch Tester.

What is being tested. In the Crosscut Adhesion Test, the percentage of the film that resists removal by the test procedure; in the Scrape Adhesion Test, either the weight required on the stylus or the hardness of the stylus (or pencil) required to penetrate the film to the substrate.

How test is performed. The film is scratched or scraped under controlled conditions, and some quantity related to adhesion is measured:

1. *Crosscut adhesion.* A series of closely spaced parallel scratches is made through the film; then a second series is made at right angles to the first series. The number of small squares remaining is a measure of adhesion. Alternatively, adhesive or masking tape with known pull strength can be pressed down over the crosscut area and jerked away before counting the squares remaining.

2. *Scrape adhesion.* A loaded stylus is moved at an angle along the surface of the coating. The load is increased until the stylus cuts through the film to the substrate. The weight required on the stylus is measured, or the hardness of the stylus required is observed. In the second form of the test, the hardness of the pencil lead needed to scratch through the

film is used as a measure of film adhesion. This method is very suitable for field testing.

Data yielded. In the Crosscut Adhesion Test, the squares remaining of the total number formed by the cutting process; in the Scrape Adhesion Test, either the weight that must be applied to the stylus or the hardness of the stylus (or pencil) required to penetrate the film to the substrate.

Interpretation of data. The adhesion of the paint film as proportional to the weight that must be applied to a stylus or to the hardness of a stylus (or pencil) required to penetrate the film, or the percentage of the film resisting removal with tape after crosshatching.

Name of Test: Adhesion (by Direct Pull). The paint is tested as an adhesive rather than as a coating. Each test method has its own name, such as Tensile Method for Adhesion or Spring-Scale Pull-Off Test. Reproducibility may be poor.

What is being tested. The resistance of the coating to removal by a direct pull on the coating.

How test is performed. The test coating is pulled away from the substrate by a force perpendicular to the surface. Another surface is cemented to the test coating and pulled away while the force required is measured.

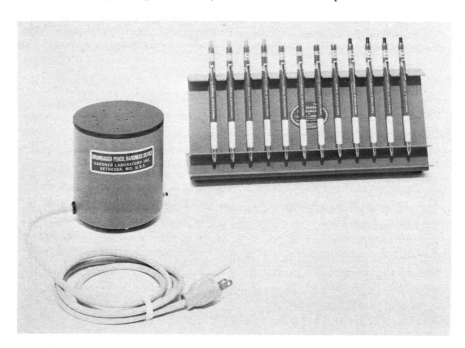

Fig. 2.6 Hg-3000 and HG-3001 Brumbaugh pencil hardness device with holder and lead guides. [*Courtesy of Gardner Laboratory, Inc., Bethesda, Maryland*]

Data yielded. A measurement of the force required to remove a certain area of the test film.

Interpretation of data. How good an adhesive the coating is under testing conditions. The author would include in this group both wet-tape and dry-tape adhesion tests made without scratching the film. These two types of adhesion tests are of considerable value to the applicator in on-the-job testing.

Testing for Adhesion by Other Methods. Two other test methods are important: peel tests and impact tests. *Peel tests* for adhesion are based on the resistance of paint film to peeling away from a break in the film when the film is pulled back from the break. The most common test, the Tape Test Method, is performed as follows. An X is scratched through the film with a sharp blade, and masking tape of a controlled pull strength is pressed down over it. The film is then jerked away abruptly, and the resistance of the film to peeling away is observed. This test method is of most value to users of paint. *Impact tests* measure adhesion when force is applied abruptly and the film is expanded in size.

Ease of application   There is no widely accepted group of standardized test methods to evaluate ease of application, probably because the technique of the individual applicator plays too large a part except in mechanized application. Still, if paint lacks this property, it does not have all the properties essential for use as an engineering material.

Users of paint are therefore on their own to evaluate the ease of application of paint. Before any paint is specified, the coating should be evaluated under actual use conditions. In addition, the manufacturer should check paint for ease of application before packaging it. Unfortunately, this is not universal practice.

Film integrity   The dried paint film should have film continuity without weak spots. Although it is not always practical to check the dried film on the job for all the properties that it is supposed to have, the user should prepare test panels to assure that the batch meets the required specification, especially for large jobs or jobs on which problems could result in expensive repainting costs or in legal battles.

On the job the test for film integrity is to observe the paint film for evidence that it is being applied properly and is knitting properly. Such observations must be made on each coat as it is being applied, after it has been applied, and before the next coat is applied.

Consistent quality   Uniformity in paint properties is the result of proper formulation, manufacturing, and testing procedures. Consistent quality is one of the surest signs of a paint manufacturer's integrity. The manufacturer should keep complete records of all paint tests performed and refer to them. The specifier should, on occasion, ask to examine these records.

A paint is uniform in quality if it meets the same specifications whether it is freshly made or aged. This criterion makes certain demands on the manufacturer:

- All batches of the paint must be manufactured to the same specifications.
- Each property that is tested must have a definite range (e.g., viscosity must be between stated values). This range must not be so broad that paints at its upper and lower limits will be different enough to cause the user problems.
- The properties of the paint must remain stable for at least the longest period in which the paint is liable to be sold for use. If the paint should not be used after a certain date, the label should say so.

2.3.5 Tests for necessary specific properties (NSP)   Necessary specific properties are those needed for certain specific end uses. (The paint must also have the four essential properties.) Two criteria should be kept in mind in choosing NSP:

- The paint has all the NSP needed for the specific end use and has them to the required degree.
- The paint will retain the NSP to an adequate degree for its required life-span.

The best assurance that paint will meet these criteria is passing of the proper group of tests. These tests are chosen to evaluate the NSP for the specified end use, and their selection is most important. Tests are required for each coat of paint in the paint system selected.

2.3.6 Manufacturing control tests   These tests are performed by the paint manufacturer during the manufacturing process. They are designed to determine quickly whether the paint (1) appears to have the correct composition, (2) appears to have been compounded according to the correct procedure, (3) meets the manufacturing control specifications established for the paint, and (4) meets additional specifications set up by the customer.

Some common manufacturing control tests have already been described. These tests determine what adjustments in composition are needed before a batch of paint is released. They include Color Matching by Eye, Gloss, and Contrast Ratio, for decorative functions; and tests for adhesion and ease of application, for essential properties.

Another group of manufacturing control tests is aimed more specifically at determining whether the paint has been put together properly. These tests also help determine what adjustments are needed before the paint is released. Typical tests evaluate (1) density, (2) dispersion, (3) viscosity, and (4) rate of drying.

Density   Tests for density determine quickly whether a batch of paint falls within the range set by the manufacturer. Too high a density indicates that some liquid has probably been omitted or that extra pigment has been added. Pigment usually has 3 to 5 times the density of varnish or solvent. Too low a density indicates that pigment has been omitted or that extra varnish or solvent has been added.

Name of Test: Density. This test is better known as the Pounds-per-Gallon Test when expressed in United States customary units.

What is being tested. The density of the completed paint.

How test is performed. A small container of accurately known volume and weight is filled with the completed paint at a standard temperature, usually 77°F (25°C). The filled container is weighed, and the weight of paint found by subtracting the tare weight of the empty container. The density is found by dividing the weight of paint by the volume of paint.

Fig. 2.7 Containers used to test density. [*Courtesy of Gardner Laboratory, Inc., Bethesda, Maryland*]

The test is simplified by using a weight-per-gallon cup. There are two sizes of cups. The larger size holds 83.3 g of water at 77°F. It can be used satisfactorily on a triple-beam balance and is suitable for use in the field to check for overthinning of paint. The net weight of the paint in grams divided by 10 is the density in pounds per gallon. The smaller-size cup, which is only one-tenth as large, is designed for use on a laboratory analytical balance.

Data yielded. The density of the completed paint.

Interpretation of data. Whether the paint meets the specification for density. To the paint manufacturer or the user, this means that the ratio of ingredients is correct. To the applicator on the job, it means that the paint has been thinned correctly before use.

Dispersion Dispersion is the measure of the approximate diameter of the coarsest group of particles (usually pigment) in the paint. It does not measure the particle-size distribution, which gives the percentage of particles in each of a group of particle-size ranges, from the finest to the coarsest.

Name of Test: Fineness of Grind (Grind Gauge Test)

What is being tested. The fineness of the coarsest group of particles in the batch of paint.

How test is performed. The test procedure described here is based on use of the ASTM gauge. It is employed as a manufacturing control test during the pigment-dispersion stage of paint manufacturing (see Sec. 1.6.3). The paint paste is placed into grooves in a carefully machined steel block, and a carefully machined scraper is drawn down over the block in the direction of the grooves. The grooves (about 5 in, or 127 mm, long) are graduated uniformly in depth from 0 at the end to 0.004 in (0.1 mm) at the other end. The reading is taken at the point at which the specks (large particles) in the film appear to predominate (the film looks more specky than smooth).

Data yielded. A scale reading in either United States customary units (mils, or thousandths of an inch) or in metric units.

Interpretation of data. The scale reading tells whether the largest particle size is in the proper size range. If the particle size is too large, the gloss either will be too low or will have a specky or seedy appearance. If the particle size is too fine, the gloss will be too high. Thus, an overground flat paint will be a sort of poor semigloss, and an overground semigloss will be a sort of poor gloss. Other properties may also be adversely affected by overgrinding. The grinding gauge is a very helpful and easy tool to use in the field.

Viscosity   Viscosity measures the fluidity of the paint. Tests for viscosity are quantitative, but the method of measurement must be chosen according to the way in which the paint is to be applied. The reason is that paints designed for different methods of application require different kinds of fluidity. For example, paint for application by paint roller has a viscosity quite different from that of paint for application by spraying. For this reason, several different methods of measuring viscosity are in use. Each emphasizes a different aspect of viscosity. The following paragraphs describe four different methods of measuring viscosity.

Name of Test: Viscosity by Rising Bubble or by Falling Ball or Plunger.   Each test method has its own name.

What is being tested. Fluidity of the coating, measured by the rate of rise of an air bubble or the rate of fall of a ball or plunger going through the liquid.

How test is performed

1. *Rate of rise of an air bubble (bubble viscometer).* The viscosity of a clear oil, varnish, lacquer, or resin solution is measured by the rate at which an

air bubble rises through a transparent tube of the substance. The tube is of a standard length and diameter, and the bubble is of a standard size. The viscosity is determined by comparing the rate of rise with that of a liquid of known viscosity or by measuring the time required for the bubble to rise a certain distance in the tube.

2. *Rate of fall of ball (falling-ball viscometer).* Like the first test, it also measures the viscosity of clear oils, varnishes, lacquers, or resin solutions. The length and diameter of the tube and the size and density of the ball are closely controlled. A series of balls of differing sizes and densities can be used to cover a wide range of viscosities.

3. *Rate of fall of a plunger (mobilometer).* A plunger in the form of a closed cylinder fits loosely inside a tube. In other forms the plunger has a perforated disk or cone at the bottom through which the coating passes. This type of test is suitable for both clear and opaque coatings. The dimensions of all parts and the weight of the plunger are closely controlled. The time required for the plunger to sink through a certain distance is proportional to the viscosity of the coating.

Data yielded. The viscosity of the coating in the units used by the particular test method. The bubble viscometer gives values in letters corresponding to the particular standard tube matched by the coating being tested or in seconds for a certain distance of travel. The falling-ball viscometer gives an answer in poises or centipoises. The mobilometer gives an answer in seconds required for the plunger to fall a certain distance.

Interpretation of data. An estimation of the fluidity of the coating. The bubble viscometer and the falling-ball viscometer are limited to transpar-

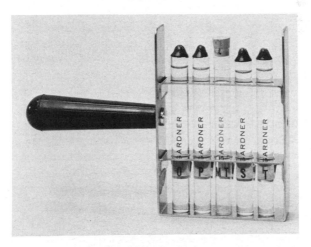

Fig. 2.8 VG-7420 five-tube viscosity-tube holder. [*Courtesy of Gardner Laboratory, Inc., Bethesda, Maryland*]

ent coatings with little or no false body. The mobilometer is also suitable for opaque coatings and for coatings with false body (which must be broken down by agitation before the coating becomes truly fluid).

Name of Test: Viscosity by Efflux Viscometer. Each test method has its own name.

What is being tested. The viscosity of the coating, measured in time units, usually seconds.

How test is performed. The test instrument consists of a cup with an orifice (hole or tube) in the bottom. Both cup and orifice are of controlled volume, diameter, and length. The cup is filled at a standard temperature, and the time required for the coating to flow through the orifice is measured. The more closely the orifice resembles a capillary tube (its length is great compared with its diameter), the more accurate the viscometer is. Two efflux viscometers are described:

1. The Ford cup is widely used for testing coatings. It has a conical bottom, and a range of orifice sizes is available. The Ford cup is usually considered a laboratory type of instrument because it must be properly set up in a stand and is not portable while in use (Fig. 2.9).

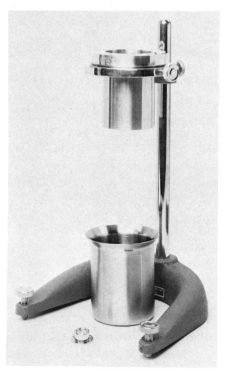

Fig. 2.9 VG-7204 No. 4 Ford cup mounted on a VG-7206-A Ford-cup stand. [*Courtesy of Gardner Laboratory, Inc., Bethesda, Maryland*]

2. The Zahn cup, which is portable, is suitable for field use in checking viscosity in the shop or at the application site. The orifice is a hole in the rounded bottom of the cup. A range of orifice sizes is available. Both the Zahn cup and the Ford cup are more suitable for thin viscosities and are of limited value for false-bodied coatings.

Data yielded. The time in seconds required for the standard volume of coating held by the cup to flow through the orifice.

Interpretation of data. The time in seconds is proportional to the viscosity of the coating.

Name of Test: Viscosity by Paddle Viscometer. Only the Krebs-Stormer viscometer will be discussed.

What is being tested. The viscosity of the coating in Krebs units (KUs), an arbitrary scale of viscosity devised for this instrument but widely used throughout the coatings industry.

How test is performed. A sample of the paint in a pint paint can is cooled or heated to a specified temperature and then placed on the movable shelf of the instrument. The shelf is raised until the paddle is immersed to the calibration mark on the shaft. The clutch is then released, causing the slotted weight on the cord that drives the paddle to descend. The time required for 100 revolutions of the paddle is checked with a stopwatch, and the corresponding KUs are read off a chart on which weight in grams is shown against seconds. The time required for 100 revolutions should be between 27 and 33 s.

Additions to and modifications of the Krebs-Stormer viscometer have increased its usefulness. The most widely used modification is a stroboscopic timer, which helps the operator quickly select the proper weight for 200 rpm (100 revolutions in 30 s).

Data yielded. Seconds for 100 revolutions with a given number of grams, which gives a KU value from the chart. With the stroboscopic attachment, only the weight needed for 200 rpm is yielded. Both methods use the same chart for conversion to KUs.

Interpretation of data. 100 KUs was chosen as good brushing viscosity of puffy, flat wall paint in the original development of this test method. Thicker paints rate over 100 KUs; thinner paints, under 100 KUs.

This test method is of greatest value when the manufacturer has determined the proper range of KU values for the paints to be tested. The Krebs-Stormer test data are then used to adjust the viscosity as needed to meet manufacturing specifications.

The three types of methods for testing viscosity that have been described thus far have limited value for testing full-bodied paint. A more versatile instrument, one that will measure the yield points at which false-bodied coatings break down and become more fluid, is needed. The ability to measure a wider range of viscosity is also desirable. Moreover, the ability to measure viscosity at high shear rates as in brushing paints is needed. A fourth type of test method, described below, provides the needed flexibility.

Name of Test: Rotational Calibrated-Spring Viscometer. Of several such instruments in use two will be described. The particular test method is named after the viscometer used.

What is being tested. The viscosity of the coating or its brushability.

How test is performed. In both test methods a cylindrical bob is rotated in the coating. The shaft, or spindle, of the bob is connected through a calibrated spring to a dial and is driven by an electric motor.

The Brookfield viscometer uses a series of solid bobs and rotates them at a series of speeds. The dial readings are converted into centipoises by a series of conversion factors for the different speeds and spindles. The viscosity range covered can be from 0 to 8 million centipoises (0 to 8000 Pa · s) on the most sophisticated model of this instrument.

The Brushometer has one cylindrical bob run at one speed. The bob is hollow and has a perforation on each side through which the paint is

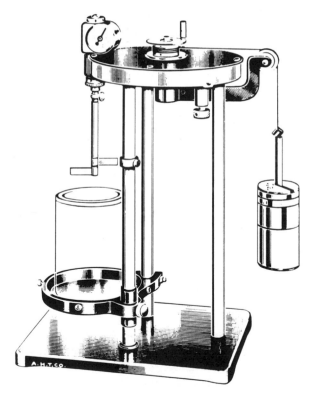

Fig. 2.10   A paddle viscometer. [*Courtesy of Gardner Laboratory, Inc., Bethesda, Maryland*]

pumped during rotation. The dial can be calibrated as desired. This instrument is especially helpful with brushing paints, in which the rate of shear is high.

Data yielded. With the Brookfield viscometer, viscosity in centipoises; with the Brushometer, brushing viscosity of the paint in the units chosen for calibrating the instrument.

Interpretation of data. With the Brookfield viscometer, viscosity is converted to absolute units of viscosity, which require no interpretation. With the Brushometer, brushing viscosity is proportional to the scale reading observed.

Rate of drying   Testing the rate of drying is a quantitative test which measures the time required for the paint to reach certain stages of drying. These stages are defined in each test method.

Two test methods are generally accepted. The first is the time-honored finger-test method. The second is one of many instrumental methods in use. Only five stages of drying will be included in the descriptions, although as many as eight stages are recognized by some authorities. The number of stages is restricted to show a close parallel between the test data yielded by the finger-test method and those yielded by the instrumental method.

Name of Test: Drying Rate by Finger-Test Method

What is being tested. The time required for the drying film of paint to reach certain stages of drying as defined in this test method.

How test is performed. A film of the test paint of specified thickness is applied to the test panel by the specified method. The drawdown method is preferred. The drying film of paint is tested at intervals by the techniques described below.

First stage: set-to-touch. The paint feels as if it is sticking to the finger when touched lightly, but none actually does stick. To be sure that none sticks, the finger is touched against a clean glass plate.

Second stage: tack-free. The finger is pressed firmly against the drying film of paint, then lifted off. The paint does not feel sticky.

Third stage: dry-through or dry-to-handle. The panel of test paint is placed on a table, paint side up. The table is of such a height that the paint tester's thumb just reaches the panel when his or her arm is hanging straight down. The tester puts a thumb on the test paint with maximum pressure, then twists the thumb through a quarter turn. The film is dry-through when the paint film is not deformed by this technique. "Deformed" means that the twisting of the thumb causes a detectable movement of the surface of the film.

Fourth stage: dry-to-recoat. The test film shows no irregularities from applying a second coat over itself or from applying over it another coating designed to go over it. The drying time of the second coat, when applied as described, must not exceed its normal specification for drying time. The dry-to-recoat stage can be reached before the dry-through stage if the paint is a soft-drying one, such as a linseed-oil house paint.

Fifth stage: dry print-free. The test film shows no thumbprint at all when tested by the dry-through test method described under the third stage. This last test stage may be almost meaningless for soft-drying paints. For example, a linseed-oil house paint may require 30 days or more to reach this stage.

Data yielded. The time (in minutes or hours) required to reach the different stages of drying defined in the test method.

Interpretation of data. The *approximate* times required for the paint to reach the different stages of drying under actual use conditions.

Name of Test: Gardner Circular Drying-Time Recorder

What is being tested. The time required to reach certain stages of drying as defined in this test method.

How test is performed. A film of the test paint of specified thickness is applied to the test panel by the specified method. The drawdown method is preferred. The Gardner circular drying-time recorder (Fig. 2.11) is promptly set in place on the wet paint film and the test started.

A Teflon ball at one end of a balanced arm is placed on the wet paint film. A weight of 12 g is then placed on top of the ball. The ball travels at a constant speed in a circular path and leaves a pattern to show the stages of drying. Recorders are available in a range of speeds to measure a

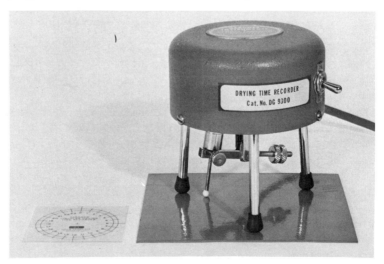

Fig. 2.11 Gardner circular drying-time recorder. [*Courtesy of Gardner Laboratory, Inc., Bethesda, Maryland*]

range of rates of drying. A series of four types of patterns is made by the plastic ball or stylus in the drying paint film.

First stage: set. The paint no longer flows back into the groove made by the stylus. At first the paint may flow back together, and later it may leave a groove; the pattern is smooth in both cases.

Second stage: surface-dry. The paint forms a skin, and the skin is ruptured by the plastic stylus, making a rough pattern.

Third stage: dry-through. The paint is now firm all the way through. The stylus no longer ruptures the drying film but leaves only a trace on the surface.

Fourth stage: dry-to-recoat. As in the finger-test method, this stage of drying is determined by actually recoating the test paint with the specified topcoat.

Fifth stage: dry print-free. The plastic stylus leaves no trace on the drying paint film.

Data yielded. The time required to reach the different stages of drying defined in the test method.

Interpretation of data. The *approximate* times required for the paint to reach the different stages of drying under actual use conditions.

Comparison of Finger-Test and Circular-Recorder Methods

1. In both methods the earlier stages of drying can be more positively identified than the later stages.

2. Neither method nor any other method gives a reliable indication of dry-to-recoat time. This must be determined by actually applying a second coat.

3. The interpretation of the data yielded by different types of paint is quite different. This is true in both methods. A certain amount of experience is required to interpret the data, particularly in the finger-test method because of the difference in size, weight, texture, oiliness, etc., of different observers' fingers and thumbs.

2.3.7 Application control tests   This group of tests is performed by the user during application to make sure that the paint is being applied properly. Proper application is application according to specification, and it should bring out the properties built into the paint by the manufacturer.

The manufacturer should perform the same or a similar group of tests on the paint before canning it. This assures that the paint has the required ease of application.

The tests used by the applicator must be few in number and quick and easy to perform. Otherwise they may delay application, causing an unjustified increase in application costs.

What should these tests tell the applicator about the paint? All the following are important:

1. Whether the paint *viscosity* is in the proper range
2. Whether the paint has the proper *percentage of solids*
3. Whether the paint has the proper *ease of application*
4. Whether the paint is being applied to the proper *wet film thickness*
5. Whether the paint is being applied to a *uniform wet film thickness*
6. Whether the paint is being applied to the proper *dry film thickness*
7. Whether the paint is *drying* properly
8. Whether the paint is *adhering* properly
9. Whether *special requirements* are being satisfied

Viscosity   Is the viscosity within the proper range? Correct viscosity may be critical for spray and automated applications. Because paints applied by these methods are normally thinned before application, the thinned or reduced viscosity is what is important. The Zahn cup is the easiest to use in the field.

Paints for brushing and rolling require a heavier false body, which is not suitable for measurement with the Zahn or other efflux types of viscometer. The Brushometer is more suitable. These paints can often be judged adequately for viscosity by checking ease of application.

Percentage of solids   Is the percentage within the proper range? If it is not, the applied film thickness may not be. Solids are not readily checked by the user, but density can be checked by employing the pounds-per-gallon cup. This cup is also an easy means of learning whether or not the thinned paint has been properly thinned. A pounds-per-gallon cup and a small triple-beam balance are excellent investments for checking whether paint has been thinned properly on the job. This combination, plus the Zahn cup, offers an excellent low-cost means of checking proper thinning in the field.

Ease of application   Is application as easy as it should be? If it is not, time will be wasted in "fighting" the coating, application results will be nonuniform, results will be unacceptable, or the coating process will not function at all.

The surest test of ease of application is a trial application under actual-use conditions. If this is not possible, the test method which most closely simulates actual conditions will give the most reliable results. Test equipment which simulates use conditions may be quite expensive and suitable only for the paint manufacturer's laboratory. If time permits, on-the-job testing is invaluable to the paint specifier.

Wet film thickness   Is the paint being applied to the proper wet film thickness? If it is and the percentage of solids is correct, then two of the most important properties of the applied film of paint—dry film thickness and film integrity—will probably be correct. Film integrity is defined as the total properties of the paint that can develop only if the film is properly applied to the correct thickness so that the film can knit properly.

Two quick and easy methods of measuring wet film thickness in the field are in common use. Each is named for the instrument used.

Nordson or Elcometer Wet-Film Gauge. This instrument has a small square of metal or plastic with notches or teeth of graduated depth on all four edges. One edge is set down in the wet film, pulled out, and the notches observed. The process is continued until the paint wets the top inside of one notch and not the next-deeper notch. The wet film thickness is between the two notch depths. Squares are available with different ranges of notch depth.

Fig. 2.12 Elcometer wet-film gauge. [*Courtesy of Gardner Laboratory, Inc., Bethesda, Maryland*]

Inmont (Interchemical) Gauge. This instrument has a pair of parallel wheels with an off-center wheel mounted rigidly between them. One outside wheel has a scale showing the clearance between the circumference of the outside wheels and the off-center wheel. The gauge is rolled across the wet paint film, starting at the deepest point, the point of maximum distance between the circumference of the inner and outer wheels. The point at which the wet paint first touches the off-center wheel is read on the outside calibrated wheel as the wet film thickness. The Inmont Gauge is available in a number of ranges.

Uniform wet film thickness   The uniformity of wet film thickness is best gauged by visual observation of the wet film as it is being applied and as it begins to dry. Mechanized application often lends itself to automatic metering or gauging of the wet paint so that a wet film of uniform thickness can be applied. Metering is not possible in manual application; what is desired is a film which is uniformly wet when applied and which maintains a uniform appearance as the wet look disappears.

High–dry-hide paints dry lighter, producing a blotchy appearance after drying begins. The blotchiness disappears after the solvent has evaporated. Many emulsion paints dry darker, also causing a blotchy appearance after drying begins. This blotchiness also disappears when drying is complete. In both cases, uniformity of application must be judged by the wet film.

Dry film thickness   Dry film thickness is measured by many methods, each named for the instrument used. It is strongly recommended that the applicator try to apply the film properly by using the preceding five application control tests. It is expensive to learn by the dry-film test method that the film has been improperly applied.

Two types of methods for checking dry film thickness will be described. The first is nondestructive (i.e., it does not damage the film in any way) but is limited to testing film thickness over metal. The second is destructive (i.e., it damages the film by making a gash in it) but is suitable for testing paint film thickness over any type of substrate.

Magnetic- or Electric-Current Gauge. These gauges depend on measuring the intensity of magnetic attraction or magnetic flux or electrical eddy currents produced when the test instrument is brought near the metal substrate. The instrumental reading varies with the nearness of the instrument to the metal substrate. The dial of the test instrument is calibrated according to the film thickness, which determines how closely the instrument's probe is to the metal surface. The Elcometer is a familiar test instrument based on magnetism but is limited to ferrous-metal substrates. The Permascope Thickness Gauge depends on electrical eddy currents and is suitable for nonferrous-metal substrates.

Destructive Film-Thickness–Testing Gauges. These gauges make a hole or gash in the film to measure its thickness. Those that make a tiny hole with a needle or a tiny drill depend on making electrical contact with a metal substrate. Those that cut a gash in the film down to the substrate with a tiny chisel can be used on any type of substrate. The film thickness is determined by measuring under magnification the exposed edge of the film where the groove is cut.

Drying   In the field, drying can be adequately judged by the finger-test method. The applicator must have become familiar in advance with the drying characteristics of the particular paint. He or she should know the rate of drying from the paint manufacturer's specifications or from his or her own tests with an instrument such as the Gardner Circular Drying Time Recorder.

Adhesion   Adhesion of the dried coating in the field is tested by a direct-pull method. The paint passes the test if the paint film is not removed. Two suitable methods follow:

Tape Adhesion. This is the quicker and less accurate of the two methods, but it is usually adequate. Adhesive tape of specified pull strength is used. A test patch of the tape of standard test length, usually about 1 in (25.4 mm), is pressed down firmly against the dried paint film with the thumb. The strip is left attached to the roll of tape, and the test patch is removed by an abrupt motion (almost a jerk) of the roll perpendicular to the painted surface.

The adhesion test is made more severe by covering the test area with water for a standard period of time before testing. (Water reduces adhesion by migrating into the paint film.) This can be done by taping to the surface a patch of gauze or other absorbent material saturated with water.

Adhesion Tester. This instrument, made by Zormco, gives more accurate data. It includes an aluminum button that is attached to the coating with a suitable adhesive. This adhesive must make a bond to the paint surface that is stronger than the required bond of the paint to the substrate. After the adhesive has dried or been cured, the claw of the Adhesion Tester is attached to it and the test started. The knurled wheel is turned to increase the pull until the specified pounds of pull is reached or the aluminum button pulls the paint film loose. The button can be left on the film and painted over if the paint passes the test.

Special requirements   Texture, multicolor effects, etc. (in fact, all properties possible) should be judged while the painting crew is still on the scene. Touching up and repainting are much cheaper if the crew does not have to be brought back to do the work.

2.4 Pointers for Specifying Paint   The emphasis in this chapter is on paint performance rather than on paint composition. Therefore, the emphasis in specifying paint should be on specifying performance.

2.4.1 Performance versus economics   The paint specifier must take into account not only paint performance but paint economics, balancing performance against total cost. In considering paint performance, it is important to remember that one function can fail while other functions are still good for a long time. This is especially true of paint color (decorative function).

Cost must include all elements that affect the applicator of the paint and the customer:

- Cost of surface preparation
- Cost of the paint system
- Cost of paint application
- Cost of paint maintenance and repainting
- Cost of staying in business

The inclusion of the last two items of cost may be challenged by some readers. If applicators plan to continue in business, they must consider these costs. Their ability to get future business depends upon the durability of the paint jobs they turn out and also on their ability to operate at a profit. A customer employing a paint applicator should remember that if the applicator is not a good businessman or businesswoman, turning out quality work at a profit, sooner or later he or she must short-change some or all customers to stay in business. If the applicator goes out of business, the customer has no recourse for correcting shoddy work.

2.4.2 Specifying necessary specific properties Necessary specific properties are paint properties which are needed for a specific application but which may not be needed for another specific application. The following points are recommended in specifying NSP:

1. List the conditions to which each paint will be exposed. Include especially elements of exposure which may cause premature failure of the paint. Examples are direct sunlight for long hours, unusual water conditions, exposure to chemicals (including fumes), abrasion, impact, grease, heat, frequent cleaning, and flexing.
2. Determine the lifetime of the coating. If necessary, discuss this matter with the customer.
3. Learn the types of failure that the paint undergoes in the conditions of exposure, the factors involved in paint maintenance (touch-up, etc.), and the surface preparation needed for repainting.
4. Choose specifications for the paint system, including surface preparation and method of application, which will meet the performance specifications set forth in the preceding three steps. If you are unsure, consult the paint manufacturer, especially the manufacturer's technical staff.
5. Remember that in coatings the paint-system approach is necessary. The best paint system is the system with the best combination of NSP for the specified end use.

2.4.3 General pointers The following advice applies to all phases of specifying and using paint. The pointers are merely good business and good engineering practice applied to the process of specifying paint and getting a good paint job.

Strive for clarity and completeness in all phases of communication. Let all persons involved know exactly what is expected of them, their products, or their workers. Paint specifications should be complete. Use a specification, and also specify the paint tests to be used and the complete paint system. Avoid use of the phrase "or equal to." This phrase is an admission of ignorance and a failure to accept responsibility for specifying the proper paints and their performance. Paint application specifications should either be standard application specifications or be spelled out in detail.

Inspect each stage of the work before the next stage begins. This means inspection after surface preparation and after the application of each coat of paint. Inspection must be performed by someone with enough experience to recognize the quality of work being done. Inspection of the final job before acceptance is a must. The most common failure of management at all levels is failure to check to see that orders have been followed. Checking is the purpose of inspection: to see that specifications have been met at each stage.

2.4.4 Paint-system thinking   To ensure the durability of a paint job a specifier should remember the following points:

- Old paint already on the surface becomes a part of the cured paint system.
- Adhesion of any coat of paint and of the coats above it depends on the adhesion of the coats beneath it.
- Properties of the finished paint job are the sum of the properties of the components of the paint system acting together.

In addition, a specifier should consider limiting his or her sources of supply of paint in order to assign responsibility for product performance of each coat of paint. Assignment of responsibility is more difficult with paint (coatings) than with most other engineering systems, especially when several manufacturers are involved. The reason is that all the coats of paint must blend into each other to provide intercoat adhesion. This blending blurs the distinction between coats, and the malfunctioning of one coat can easily be confused with that of an adjacent coat. The coats cannot be cleanly separated after drying to determine the precise cause of the malfunction, such as marginal drying or slow hardening. A machine can be taken apart so that each part can be examined separately to pinpoint failure, but a paint system cannot be examined in this way.

A partial solution to this problem is to specify a painting system from one manufacturer or dealer. Another is to make sure that inspection is performed on each coat of paint before approval is given to apply the next coat. Any paint malfunction should be promptly called to the manufacturer's attention, and the manufacturer's representative should inspect it before another coat of paint is applied.

2.4.5 Referee sample   The specifier of paint should always stipulate that a referee sample of each batch of paint be available for delivery to the jobsite when needed because of a paint problem. This sample should be taken when the batch is canned and after all final batch adjustments have been made. The sample is to be used by the applicator and the supplier in an objective attempt to resolve differences of opinion on the causes of the problem.

Cross-reference of test methods   Table 2.1 refers to ASTM and GSA test methods that most nearly fit the test methods described in this chapter. The ASTM test methods may be found in Part 27 of the *Annual ASTM Standards,* which also cross-references GSA test methods (see ASTM D-2833). The GSA test methods may be found in *Federal Test Method Standard No. 141,* which also cross-references ASTM test methods.

TABLE 2.1

Name of test*	ASTM method	GSA method
Color Matching by Eye	D-1729	4249.1, 4250
Instrumental Color Measurement	D-2244	6123
Colorfastness (Indoor)	D-2620	4561.1
(Color) Bleed Resistance	D-868, D-969	4571.1
Gloss Measurement	D-1471	6101
Hiding Power (Contrast Ratio)	D-344, D-2805	4121, 4122.1
(Chemical) Spot Test	D-1308	6081
Natural Weathering (Test Fence)	D-1006	6161.1
Accelerated Outdoor Weathering		6160
Artificial Weathering (Accelerated Weathering)	D-822	6151, 6152
Abrasion Resistance	D-658, D-968	6191, 6192, 6193
Water Immersion	D-870	6011
Humidity Test	D-1735	6201
Blister Resistance	D-714	6461
Salt Spray (Fog) Testing	B-117	6061
Flexibility by Mandrel Test	D-522, D-1737	6221, 6222
Impact Test	G-14	6226
Tunnel Test	D-1360, D-1361	
Sanding Characteristics		6321
Resistance to Fungus Growth	D-2574, D-3273	6271.1
Adhesion (by Cutting the Paint Film from the Substrate)		6304.1
Adhesion (by Scratching or Scraping the Paint Film)	D-2197	6302.1, 6303.1
Adhesion (by Direct Pull)		6301.1
Ease of Application	D-2931	2112, 2131, 2141.1
Film Integrity	D-3258	6261
Consistent Quality	D-869	3011.1, 3022, 3027, 4208
Density (Pounds-per-Gallon Test)	D-1475	4184.1
Fineness of Grind	D-1210	4411.1
Viscosity by Rising Bubble or by Falling Ball or Plunger	D-1545	4271, 4272.1
Viscosity by Efflux Viscometer	D-1200	4282
Viscosity by Paddle Viscometer	D-562	4281
Rotational Calibrated-Spring Viscometer	D-2196	4287
Drying Rate by Finger-Test method	D-1640	4061.1
Percentage of Solids	D-2832	4041.1, 4042
Inmot (Interchemical) Gauge	D-1212	
Magnetic- or Electric-Current Gauge	E-376	6181
Destructive Film-Thickness–Testing Gauges	D-2691	6183

NOTE: Some widely used methods, such as that employing the Sward Hardness Rocker, are not covered by a test-method specification because different laboratories do not duplicate each other's methods well enough.

* Tests are listed in the order of their description in this chapter.

BIBLIOGRAPHY

Annual Book of ASTM Standards, part 27: "Paint: Tests for Formulated Products and Applied Coatings," American Society for Testing and Materials, Philadelphia, 1978.

Annual Book of ASTM Standards, part 28: "Paint: Pigments, Resins and Polymers," American Society for Testing and Materials, Philadelphia, 1978.

Annual Book of ASTM Standards, part 29: "Paint: Fatty Oils and Acids, Solvents, Miscellaneous, Aromatic Hydrocarbons, Naval Stores," American Society for Testing and Materials, Philadelphia, 1978.

Berger, Dean M.: "Detecting Film Flaws in Coatings," *Chemical Engineering,* Mar. 17, 1975, pp. 79–83.

_____: "Inspecting Coatings for Film Thickness," *Chemical Engineering,* Feb. 17, 1975, pp. 106–110.

Champion, Frederick A.: *Corrosion Testing Procedures,* 2d ed., John Wiley & Sons, Inc., New York, 1965.

Corrosion Engineers Buyer's Guide, 1978–80, National Association of Corrosion Engineers, Katy, Tex., 1978.

Federal Test Method Standard No. 141b, Government Printing Office, Feb. 1, 1979.

"Federation Training Series Test Methods," cassette, Federation of Societies for Coatings Technology, Philadelphia, 1975.

Gardner Measurement and Testing Catalog, Gardner Laboratories, Inc., Bethesda, Md., 1979.

Gaynes, Norman I.: *Testing of Organic Coatings,* Noyes Data Corp., Park Ridge, N.J., 1977.

Guevin, Paul R.: "Review of Skid and Slip Resistance Standards Related to Coatings," *Journal of Coatings Technology,* August 1978, pp. 33–38.

Holtzen, Dwight A., "An Improved Drawdown Blade," *Journal of Coatings Technology,* March 1980, pp. 43–46.

National Research Council: *Deterioration of Materials,* Reinhold Publishing Corporation, New York, 1954.

Paint Standards and Tests from International Organization for Standards (IOS), nos. R842, R1512–1524, R2409, and R2431. Available from American National Standards Institute, New York.

Payne, Henry F.: *Organic Coating Technology,* 2 vols., John Wiley & Sons, Inc., New York, 1954–1961.

Speller, F. N.: *Cause and Prevention of Corrosion,* McGraw-Hill Book Company, New York, 1951.

Vesce, Vincent C.: *Exposure Studies of Organic Pigments in Paint Systems,* Allied Chemical Corporation, New York, 1959.

Wilson, C., and J. A. Oates: *Corrosion and the Maintenance Engineer,* Hart Publishing Company, New York, 1968.

Chapter **3**

Raw Materials

ELIAS SINGER
Troy Chemical Company

3.1 Introduction   Paint is a mechanical mixture or dispersion of pigments or powders, at least some of which are normally opaque, with a liquid or medium known as the vehicle. It must be able to be applied properly, and it must adhere to the surface on which it will be applied and form the type of film desired. Paint must also perform the function for which it is being used: protection, decoration, or a special functional job.

The vehicle portion of the paint will normally consist of a nonvolatile portion which will remain as part of the paint film and a volatile portion which will evaporate, thus leaving the film. The dried paint film will therefore consist of pigment and nonvolatile vehicle. The volatile portion of the vehicle, which leaves the paint film, is normally used for proper application properties.

The proportion of pigment to nonvolatile vehicle will normally determine the type of gloss that the dried film will have. If this proportion is small (e.g., less than 25 percent of the total nonvolatile volume), the result probably would be a glossy film, since there would be more than enough nonvolatile vehicle to cover the pigment completely. As the per-

TABLE 3.1 Typical Components of Paint or Lacquer

I. Vehicles
   A. Nonvolatile
      1. Solvent-based
         a. Oils
         b. Resins
         c. Driers
         d. Additives
      2. Lacquers
         a. Cellulosics
         b. Resins
         c. Plasticizers
         d. Additives
      3. Water-based
         a. Styrene-butadiene
         b. Polyvinyl acetate
         c. Acrylic
         d. Other polymers and emulsions
         e. Copolymers
         f. Additives
   B. Volatile solvents
      1. Trade sales and maintenance aliphatic solvents and, in some cases, aromatics
      2. Chemical and industrial solvents including some aromatics
      3. Lacquer solvents such as ketones, esters, and acetates
II. Pigments
   A. Opaque
   B. Transparent
   C. Special-purpose types

centage of pigment volume goes up, the gloss goes down. At a 45 percent pigment-volume concentration (PVC) the paint would probably be a semigloss, and at a 70 percent PVC the sheen is likely to be dull or flat. Paint or lacquer components are shown in Table 3.1.

Normally we think of two types of coatings: those that are solvent-based, that is, those that are reducible by an organic solvent; and those that are water-based, those that may be thinned or reduced by water. The specific properties of a coating will depend almost wholly on the specific properties of the pigments and vehicles being used and on the proportions of one to the other.

There are, of course, many coatings that contain little or no pigmentation. These are the clear coatings, including clear lacquers and varnishes. They are usually used over wood when the beauty of the substrate is not to be hidden or obliterated. Clear coatings normally dry to a high gloss, but pigmented clear coatings dry to a dull finish. Special flatting types of pigments that give no color and have no obliterating properties are normally used in these dull-finish clear coatings.

Following is a discussion of some of the basic raw materials in paint. The list is not meant to be all-inclusive.

3.2 Oils   Oils are used in coatings either by themselves, as a portion of the nonvolatile vehicle, or as an integral part of a varnish, when combined with resin, or of a synthetic liquid, when combined with the resinous portion of the synthetic. Among the more important properties of oil are the following:

- Oil improves the flexibility of the paint film: eliminating oil from certain formulations would cause the film to crack.
- In exterior finishes, oil gives durability.
- As part of the nonvolatile vehicle, oil improves gloss.
- Some oils give moderate resistance to water, soap, chemicals, and other corrosive products.
- Some oils give specialty properties such as wrinkling (for wrinkle finishes).
- With special treatments, oils can be used to improve leveling and the flow, nonpenetration, and wetting properties of the vehicle. They also have other desirable characteristics.

Composition   Most of the oils are triglycerides of fatty acids. Glycerin, $C_3H_5(OH)_3$, has three OH groups, each of which can react with the carboxyl group of a fatty acid. Such a reaction will result in water being split off and a triglyceride being formed. This is the oil as it is found in nature.

Properties   The properties of the specific oil depend largely on the type of fatty acids in the oil molecule. Thus, highly unsaturated fatty acids will give improved drying properties but have a greater tendency toward yellowing. Drying is especially improved if the double bonds are in a conjugate system, in which two double bonds are separated by a single bond. Such oils also have a faster bodying rate when heated and somewhat better water and chemical resistance.

Oil Treatments   Many of the oils cannot be used in the raw state, as they are produced by the crushing of seeds, nuts, fish, etc., and must be treated to make them usable. Others can be used in the raw state but are often treated to give them special properties. Among these treatments are the following:

■ *Alkali refining.* The oil is treated with alkali, which lowers its acidity and makes it less reactive and also improves its color.

■ *Kettle bodying.* The oil, usually refined, is heated to a high temperature for several hours to polymerize it. This increases its viscosity and improves its dry, color retention, flow, gloss, wetting properties, and nonpenetration. However, the process impairs brushability.

■ *Blowing.* Air or oxygen is passed through the oil at elevated temperatures. The resultant oil has improved wetting, flow, gloss, drying, and setting properties, but brushability and, often, color and color retention are impaired. In addition, paints containing blown oils have a greater tendency toward pigment settling.

Among the more important paint oils are the following.

3.2.1   Linseed oil   This is the largest-volume oil used by the coatings industry. It is very durable, yellows in interior finishes but bleaches in exterior paints, and has good nonsagging properties, easy brushing, good drying, fair water resistance, medium gloss, a medium bodying rate, and poor resistance to acids and alkalies. It is used largely in house paints, trim paints, and color-in-oil pastes. Alkali-refined and kettle-bodied linseed oil is used in varnishes and interior paints. Linseed oil is an important modifying oil in synthetic alkyds.

3.2.2   Soybean oil   This is a semidrying oil that can be used only with modifying oils and resins to improve its drying properties. The refined oil has excellent color and color retention. Soybean oil is one of the most important modifying oils in alkyds and is used in nonyellowing types of paint.

3.2.3   Tung oil (china-wood oil)   This oil contains conjugated double bonds and cannot be used in its raw state since it would dry to a soft, cheesy type of film. In its kettle-bodied state it gives the best-drying and most resistant film of any of the common paint oils. It has a good gloss and good durability and is used in finishes for which dry and resistance are important: spar varnishes, quick-drying enamels, floor, porch, and deck paints, concrete paints, etc.

3.2.4 Oiticica oil   This oil is similar to tung oil in its properties, but its drying, flexibility, and resistance characteristics are not quite as good. It also has somewhat poorer color and color retention. However, it has better gloss and better leveling qualities than tung oil. Oiticica oil is normally used as a substitute for tung oil when there is a large price difference between them.

3.2.5 Fish oil   This is a poor-drying oil that cannot be used in its raw state because of its odor. In its kettle-bodied state it has relatively easy-brushing and good nonsagging properties. It also has fairly good heat resistance. Fish oil is used in paints when poor flow is necessary, as in stipple finishes. It is also used in low-cost paints since it is usually lower-priced than the other oils.

3.2.6 Dehydrated castor oil   Raw castor oil is a nondrying oil that is used in lacquers as a plasticizing agent to make them more flexible. When it is treated chemically to remove water from the molecule, additional double bonds are formed; this makes it a drying oil. The dehydrated oil dries better than linseed oil, although paints made with it sometimes have a residual tack that is difficult to remove. Dehydrated castor oil has very good water and alkali resistance—almost as good as that of tung oil. It also has excellent color and color retention, on a par with that of soybean oil. The oil is used in finishes for which color and dry are important: alkyds, varnishes, and quick-drying paints.

3.2.7 Safflower oil   This oil, a relative newcomer to the coatings industry, has some of the good properties of both soybean oil and linseed oil. It has the excellent nonyellowing features of soybean oil and dries almost as well as linseed oil. Safflower oil can therefore be used as a substitute for linseed oil in many white formulations for which color retention is important, especially kitchen and bathroom enamels.

3.2.8 Tall oil   This is not really an oil, but it is often used as an oil or as a combination of an oil and a resin. Tall oil is a combination of fatty acids and rosin. Normally it is separated into its separate ingredients, which are sold and used as such. The rosin is sold and used for the rosin properties, and the tall-oil fatty acids are used for the fatty-acid properties. As a component in alkyds, the fatty acids give vehicles similar to those made with soybean fatty acids. When limed, tall oil gives a liquid that is low in cost and high in gloss, has poor flexibility, and tends to yellow very badly on aging.

3.3 Resins   If coatings were made with oil as the only nonvolatile component with the exception of driers, the result would be a relatively soft, slow-drying film. Such a film would be satisfactory for house paints, ceiling paints, or other surfaces for which hardness and fast dry are not important but totally unsatisfactory for many trade sales and maintenance coatings and for most industrial or chemical coatings. In addition to improving hardness and speeding drying time, specific resins give other important properties. Thus, they often improve gloss and gloss retention, and they also usually improve adhesion to the substrate. Resistance to all types of agents such as chemicals, water, alkalies, and acids would not be obtained without the use of different types of resins. Low-cost resins are used to reduce the raw-material cost of a coating. Following are properties of the more popular resins.

3.3.1 Rosin   This low-cost natural resin, derived from the sap of trees, is essentially abietic acid, $C_{20}H_{30}O_2$. It must be largely neutralized before it can be used. This is normally done by reacting the rosin with lime, in which case it is known as limed rosin, with glycerin, which gives ester gum, or with pentaerythritol, which yields pentaresin. Liming rosin gives a resin with a high gloss, excellent gloss retention, and fine adhesion. However, the resin is relatively poor in drying time and in resistance to water and chemicals. Since it tolerates large quantities of water, it is popular for low-cost finishes. A solution of limed rosin in mineral spirits, called gloss oil, is popular in low-cost floor paints, barn paints, and general-utility varnishes.

3.3.2 Ester gum   This resin, made by reacting rosin with glycerol, $C_3H_5(OH)_3$, which neutralizes or esterifies the abietic acid, might be considered the first synthetic resin. Ester gum dries somewhat more slowly than limed rosin but has much-improved color-retention and resistance characteristics. It gives a very high gloss and has excellent adhesion. The higher–acid-number ester gums are compatible with nitrocellulose and therefore are used in lower-cost gloss lacquers.

3.3.3 Pentaresin   When pentaerythritol, $C(CH_2OH)_4$, is the alcohol used to react with rosin, the result is a resin with a higher melting point that has good heat stability, color, and color retention and gives a high gloss. When the resin is cooked into varnishes with different oils, good drying properties and a moderate degree of water and alkali resistance are obtained. Similar to the other resin esters, pentaresin has good adhesion to all types of surfaces.

3.3.4 Coumarone-indene (Cumar) resins   These resins, derived from coal tar, are essentially high polymers of the complex cyclic and ring compounds of coumarone and indene. They are completely neutral and thus are ideal for leafing types of aluminum paints. In addition, they have good alcohol and electrical breakdown properties. They also are resistant to corrosive agents such as brine, dilute acids, and water. On the negative side, they have poor color retention and only fair drying properties and gloss. Their cost is normally quite low.

3.3.5 Pure phenolic resins   These are pure synthetic resins made by reacting phenol with formaldehyde. There are two essential types: a type that is cooked into oil and is used largely in trade sales and marine paints and a type that is sold dissolved in a solvent and is applied in that form and baked. The first type has excellent water resistance and durability, making it ideal for exterior, floor, porch, deck, and marine paints or varnishes. Since it also has fine chemical, alkali, and alcohol resistance, it can be used for furniture, bars, patios, and similar applications. In some instances, adhesion is rather poor. The solvent type is heat-reactive and becomes extremely hard and resistant to chemicals when properly cured. It is used for can linings, linings for the interior of tanks, and similar applications. All phenolics tend to yellow.

3.3.6 Modified phenolic resins   Combinations of ester gum and pure phenolics, these resins have properties between those of their components. They have very good water, alkali, and chemical resistance, and the ester-gum portion gives them good adhesion. They offer a good dry and a high gloss. These resins are fine for floors, porches, and decks, in sealers, for spar varnishes, and for any other uses for which a combination of good resistance, a hard film, and fast drying is desirable and for which yellowing can be tolerated.

3.3.7 Maleic resins   These resins are made by reacting maleic acid or anhydride with a polyhydric alcohol such as glycerin in the presence of rosin or ester gum. They have very fast solvent release, good compatibility with nitrocellulose, and good sanding properties. This combination makes them ideal resins for sanding lacquers. Maleic resins also have a fast dry and good color retention so that they can be used in quick-drying white coatings. They should be used only in shorter oil lengths, since in longer oil lengths they have some tendency to lose dry as they age.

3.3.8 Alkyd resins   These resins, which are made by reacting a polybasic acid such as phthalic acid or anhydride with a polyhydric alcohol such as glycerin and pentaerythritol and which are further modified with drying or nondrying oils, are probably the most important resins used in solvent-based trade sales paints and in many industrial coatings. Those that are modified with large percentages of drying oils are normally used in trade sales paints; they are known as long-oil or medium-oil alkyds. Those that are modified with smaller percentages of oil or with nondrying oils are used in industrials, baking finishes, and lacquers; they are known as short-oil or nondrying alkyds. Normally, the larger the percentage of glyceryl phthalate, or resinous portion, the faster the dry, the more brittle the finish, and the better the baking properties. Other properties depend on the type of modifying oil and the type of polybasic acid used.

Generally, alkyds have excellent drying properties combined with good flexibility and resultant excellent durability. Color retention, when modified with nondrying oils or with oils having good retention such as soybean or safflower oils, is very good. Gloss and gloss retention in alkyd paints are unusually good. In baking finishes, alkyds are normally combined with other resins such as urea and melamine to obtain top-grade films. The resistance characteristics of alkyds, though good, do not compare with those of pure phenolics and are not equal to those of modified phenolics. If high-resistance characteristics are not required, however, alkyds are second to none in good overall properties. Thus they are ideal for all types of interior, exterior, and marine paints and for a large percentage of industrial coatings.

3.3.9 Urea resins   The short-oil, high-phthalic alkyds previously mentioned are combined with ureas and melamines in baking finishes. Urea resins can be used only in baking types of coatings since they convert from a liquid to a solid form under the influence of heat, in a type of polymerization often called curing. The ureas, a product obtained from the reaction of urea and formaldehyde, give a film that is hard, fairly brittle, and colorless. This brittleness and rather poor adhesion can be corrected by combining them with alkyd resins or plasticizers. The ureas have excellent color retention and fine resistance to alcohol, grease, oils, and many corrosive agents. They make excellent finishes for many metallic surfaces such as those of refrigerators, metal furniture, automobiles, and toys.

3.3.10 Melamine resins   These resins, synthesized from melamine, a ring compound, and formaldehyde, act much as urea resins do. However, they cure more quickly or at lower temperatures and give a somewhat harder, more durable film with higher gloss and better heat stability. Although they are more expensive, they are to be preferred for high-quality white finishes because their shorter baking cycle produces a film that is whiter and has the best color retention.

3.3.11 Vinyl resins   Solvent-based vinyl resins are normally copolymers of polyvinyl chloride and polyvinyl acetate, though they are available as polymers of either one. They are usually sold as white powders to be dissolved in strong solvents such as esters or ketones, but may be sold already dissolved in such solvents. They are then plasticized to make an acceptable film. The chloride is very difficult to dissolve but has extreme resistance to chemicals, acids, alkalies, and solvents. The acetate is not as resistant but is much more soluble. The more practical copolymer still exhibits exceptional resistance to corrosive agents, chemicals, water, alcohol, acids, and alkalies. Vinyl resins do an exceptionally fine job in coatings for cables, swimming pools, cans, masonry, or any surface requiring very high resistance.

3.3.12 Petroleum resins These completely neutral, rather low-cost resins are obtained by removing the monomers during the cracking of gasoline and polymerizing them. They have good resistance to water, alkalies, alcohol, and heat. Some have good initial color, but they all tend to yellow on aging. Petroleum resins are very good for aluminum paints, and they make good finishes for bars, concrete, and floors when cooked into tung or oiticica oil.

3.3.13 Epoxy resins These resins, more correctly called epichlorohydrin bisphenol resins, are chain-structure compounds composed of aromatic groups and glycerol, joined by ether linkages. Various modifying agents are used to give epoxies of different properties, but all such resins generally have excellent durability, hardness, and chemical resistance. They can be employed for high-quality air-drying and baking coatings, and some can even be used with nitrocellulose in lacquers.

3.3.14 Polyester resins In addition to the alkyd resins, which are polyesters modified with oil, there are other types of polyesters, such as polyester polymers, that have a light color and good color retention, excellent hardness combined with good flexibility, and very good adhesion to metals. They are useful in many industrial-type coatings for which such properties are important.

3.3.15 Polystyrene resins Resins of this group, made by the polymerization of styrene, are available with a variety of melting points that depend on the degree of polymerization. They are thermoplastic. The higher–melting-point resins are incompatible with drying oils, but the lower polymers are compatible to some degree. Polystyrene resins have high electrical resistance, good film strength, high resistance to moisture, and good flexibility when combined with oils or plasticizers. They are useful in insulating varnishes, waterproofing paper, and similar applications.

3.3.16 Acrylic resins These thermoplastic resins, obtained by the polymerization or copolymerization of acrylic and methacrylic esters, may be combined with melamine, epoxy, alkyd, acrylamide, etc., to give systems that bake to a film with excellent resistance to water, acids, alkalies, chemicals, and other corrosives. They find use in such applications as coatings for all types of appliances, cans, and automotive parts and for all types of metals.

3.3.17 Silicone resins These polymerized resins of organic polysiloxanes combine excellent chemical-resistance properties with high heat resistance. They are expensive and therefore are not usually used for their chemical-resistance properties, which can be obtained from lower-priced resins, but for their very important heat- and electrical-resistance properties, which are superior to those of other resins. At a lower cost, they can be copolymerized with alkyds and still retain some of their important properties.

3.3.18 Rubber-based resins   These resins, based on synthetic rubber, give a film, when properly plasticized, that has high resistance to water, chemicals, and alkalies. They are excellent for use in swimming-pool paints, concrete-floor finishes, exterior stucco and asbestos-shingle paints, and other coatings requiring a high degree of flexibility and resistance to corrosion.

3.3.19 Chlorinated resins   Paraffin can be chlorinated at any level from 42 percent, which gives a liquid resin, to 70 percent, which gives a solid resin. Chlorinated resins are popularly used in fire-retardant paints. The 70 percent resin is also used in house paints and in synthetic nonyellowing enamels for improved color and gloss retention. Chlorinated biphenyls with high-resistance characteristics can also be made; they are often combined with rubber-based resins for coatings requiring a high degree of alkali resistance. Rubber also is chlorinated and is sold as a white granular powder containing about 67 percent chlorine. It is quite compatible with alkyds, oils, and other resins such as phenolics or cumars. It has high resistance to acids, alkalies, and chemicals and is useful for alkaline surfaces such as concrete, stucco, plaster, and swimming pools.

3.3.20 Urethanes   Three general classes of urethane resins or vehicles are available today: amine-catalyzed two-container systems, moisture-cured urethane, and urethane oils and alkyds. The first and second types contain unreacted isocyanate groups which are available to achieve final cure in the coating. In the first case, an amine is used to catalyze a cross-linking reaction that results in a hard, insoluble film; in the second, the moisture in the air acts as a cross-linking agent.

Urethane oils and alkyds, on the other hand, are cured by oxidation, in the same way as alkyds and oils, and require driers or drying catalysts. However, cure occurs more quickly, and the resultant film is very hard and abrasion-resistant and has greatly improved resistance to water and alkalies. However, color retention is somewhat poorer. Because of their advantages, urethane oils and alkyds are widely used in premium floor finishes and for exterior clear finishes on wood. The hardness of the film tends to impair intercoat adhesion, and care must be exercised to sand the surface lightly between coats to provide tooth.

3.4 Driers   The basic difference between lacquer and solvent-based paint is that lacquer dries by evaporation of the solvent and paint by a combination of oxidation and polymerization. To speed the drying action of a paint, driers are required. Without them paint would dry in days instead of in hours, and in many cases the film would be softer and have poorer resistance properties.

Most driers are metallic soaps that act as polymerization or oxidation agents, or both. The soaps must be in such form that they are soluble in the vehicle. Everything being equal, the more soluble the soaps are, the more effective they are as driers. Tall-oil driers, based on tall-oil fatty acids, are somewhat less soluble than naphthenates based on naphthenic acid. Synthetic-acid driers based on octoic, neodecanoic, and similar acids are now the most popular. In addition, the metal portion of the more active driers is normally oxidizable. One theory is that these driers, especially the oxidation catalysts, act in their reduced form by taking oxygen from the air, become oxidized, pass the oxygen on to the oil or other oxidizable molecule, become reduced again, and are therefore in a position to take on additional oxygen to pass on to the oxidizable vehicle. This process is repeated until the film is completely oxidized.

Soaps of the following metals are normally used as driers.

3.4.1 Cobalt   The cobalt drier, sold containing 6 or 12 percent cobalt as metal, is the most powerful drier used by the coatings industry. It acts as an oxidation catalyst and is known as a top drier, drying the top of the film. Excessive amounts of cobalt drier will set up stresses and strains in the paint film that can result in wrinkling. Though purple in color, cobalt has low tinting strength and will not discolor a paint.

3.4.2 Lead   This drier is normally sold in strengths containing 24 or 36 percent lead as metal. It is very light in color and thus will not discolor a paint. Lead is a polymerization catalyst and therefore makes an ideal combination with cobalt, since it tends to harden or dry the bottom of the film. Because of lead laws, this type of drier is gradually being replaced by calcium, zirconium, or both.

3.4.3 Manganese   This drier, sold normally in strengths of 6, 9, or 12 percent metal, is what is known as a through drier, acting on both the top and the bottom of the film. Mainly, however, it is an oxidation rather than a polymerization catalyst and can therefore cause wrinkling if employed in excessive amounts. It is often used in combination with cobalt and lead to cut the cobalt content and reduce skinning. At other times, it is used with lead as a manganese-lead drier combination. It is brownish in color and tends to discolor paints if used in large amounts.

3.4.4 Calcium   This very light-colored drier, which has no tendency to discolor paints, acts as a polymerization agent similar to lead. It also tends to improve the solubility of lead if used in combination with it and thus makes lead more effective as a drier. It is sold in metal contents of 4, 5, and 6 percent. Calcium is becoming increasingly popular for use as a substitute for lead in lead-free paints.

3.4.5 Zirconium   Like calcium, zirconium is light in color and acts usually as a polymerization catalyst. In lead-free paints it is often used with cobalt or in combination with cobalt and calcium. Zirconium is light in color and sold in concentrations of 6, 12, and 18 percent metal content.

3.4.6 Nonmetallic Driers   The elimination of lead has focused attention on nonmetallic driers. The most popular of these is orthophenathroline, which often gives excellent drying properties, sometimes superior to those of standard combinations, when used with manganese and sometimes with cobalt.

3.4.7 Other metals   Other metals are sometimes used as driers. Among the most popular are iron, useful in colored baking finishes, and zinc, useful as a wetting and hardening agent. Zinc is also used to reduce skinning tendencies in a paint. Sometimes cerium is used as a drier.

3.5 Additives   This group of raw materials is used in relatively small amounts to give coatings certain necessary properties. (Driers really belong in this category.) Since additive compositions are not normally revealed by manufacturers, the following discussion refers to trade names. On occasion additives are used on the jobsite if problems arise. In such cases there should be close coordination and supervision by the paint manufacturer to avoid even bigger problems.

3.5.1 Antisettling agents   This group of agents is used to prevent the separation or settling of the pigment from the vehicle. Most commonly this is done by using additives that set up a gel structure with the vehicle, trapping the pigment within the gel and preventing it from settling to the bottom.

3.5.2 Antiskinning agents   These are essentially volatile antioxidants that prevent oxidation, drying, or skinning of the paint while it is in the can but volatilize and leave the paint film, allowing it to dry properly once it has been applied. The most common antiskinning agents are methyl ethyl ketoxime, very effective in alkyds, and butyraldoxime, effective in oleoresinous liquids. Phenolics are sometimes used, but they can slow the drying time of the coating.

3.5.3 Bodying and puffing agents   These products are used to increase the viscosity of a paint. Without them paint is often too thin to be sold or used. In solvent-based paints, gelling or thixotropic agents such as those mentioned in the subsection "Antisettling Agents" may be used. There are also liquid bodying agents that are based largely on overpolymerized oils. In water-based paints, the most common bodying agents are methyl cellulose, hydroxyethyl cellulose, the acrylates, and the bentonites. These agents also tend to improve the stability of the emulsion.

3.5.4 Antifloating agents   Most colors used in the paint industry are a blend of colors. Thus, to form a gray some black is added to a white paint. It is important that one color does not separate from the other, and antifloating agents are used for this purpose. Silicones are sometimes used, but they pose serious bubbling and recoatability problems. Special antifloating agents are sold under various trade names (Fig. 3.1).

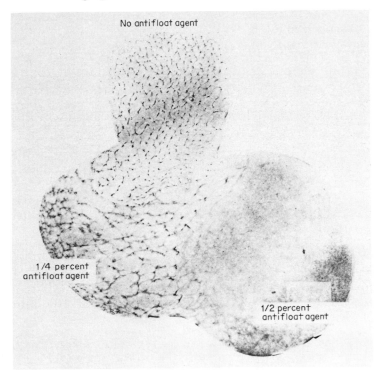

Fig. 3.1 Antifloating agents prevent colors from separating. [*Troy Chemical Company*]

3.5.5 Loss of dry inhibitors   Certain colors such as blacks, organic reds, and even titanium dioxide tend to inactivate the drier, and the paint loses drying on aging. Agents are therefore introduced to react slowly with the vehicle and feed additional drier to replace what was lost. In the past most of the agents have been lead compounds such as litharge, but these are now being replaced by agents based on cobalt.

3.5.6 Leveling agents   Sometimes a paint does not flow properly and shows brush or roller marks. These can often be corrected by special wetting agents that cause the vehicle to wet the pigment better.

3.5.7 Foaming (bubbling)   This is much more of a problem in water-based than in solvent-based paints. The presence of bubbles not only makes for an unsightly paint when applied (Fig. 3.2) but results in a partially filled paint can when the bubbles leave the paint while it is in the can.

3.5.8 Grinding of pigments   Unless a pigment is properly ground, the result is a coarse film of poorer opacity and, in a gloss-finish type of paint, usually in a poorer gloss. Certain types of wetting agents tend to improve the ability of the disperser or mill to separate these pigment particles more easily and thus to obtain a better grind.

3.5.9 Preservatives   Almost every formulation based on water must have a preservative for can stability. Until recently, most preservatives

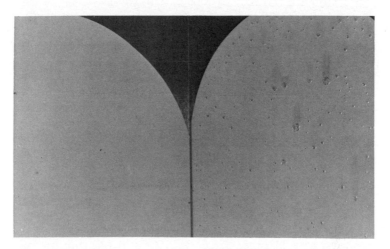

Fig. 3.2   Effect of a solvent-based debubbling agent. [*Troy Chemical Company*]

have been mercurials, but these are being partially replaced by complex organics.

3.5.10 Mildewcides   Most exterior paints will suffer a blackish-greenish discoloration due to the growth of fungi or mildew on the surface. Until now this condition has been prevented by the inclusion of a mercurial in the paint, often in combination with zinc oxide. Now non-mercurials also are available.

3.5.11 Antisagging agents   When applied, a paint sometimes flows excessively so that it causes what are known as curtains, runs, or sags. Most bodying or antisettling agents prevent this tendency. Some of them prevent sag without increasing paint body (Fig. 3.3).

3.5.12 Glossing agents   Sometimes the gloss in solvent-thinned gloss-type formulations is low. Though it can usually be increased by changing

Paint without antisagging agent   Paint with Anti-Sag

Fig. 3.3 Effect of an antisagging agent. [*Troy Chemical Company*]

vehicles or pigmentation or by increasing the ratio of nonvolatile vehicle to pigment, the use of an additive may be a simpler step.

3.5.13 Flatting agents   Just as gloss is desirable in gloss finishes, flatness is needed in flat finishes. Flatness is easy to obtain in regular flat paints, but in clear coatings such as flat varnishes or lacquers this goal is much more of a problem. It can be accomplished by the use of special flatting agents such as amorphous silica.

3.5.14 Penetration   In some systems the paint is supposed to penetrate the surface. Penetration is important in stains and in paint that will be applied to a poor surface. Most paints, however, require good nonpenetration for improved sealing properties and good color and sheen uniformity. This goal is accomplished mainly by agents that set up a gel structure in the paint.

3.5.15 Wetting agents for water-based paints   Many different types of wetting agents are necessary in water-based paints. Some are used for improved pigment dispersions, while others are employed to improve adhesion to a poor surface such as a slick surface (Fig. 3.4).

3.5.16 Freeze-thaw stabilizers   These are necessary in water-based paints to prevent them from coagulating or flocculating when the paints are subjected to freezing temperatures. The stabilizers, such as ethylene or propylene glycol, lower the temperature at which the paint will freeze.

Another way of accomplishing this goal is to use an additive that improves the stability of the emulsion.

3.5.17 Coalescing agents   The purpose of these agents in water-based paints is to soften and solvate partially the latex particles in order to help them flow together and form a more nearly continuous film, particularly at low temperatures. This can be done with ether alcohols such as butyl cellosolve and butyl Carbitol.

3.6 Lacquers   Lacquers dry essentially by evaporation of the solvent, and they are dry as soon as the solvent is gone. Raw materials consist of substances that form a dry film, or that can become part of a dry film, without the necessity of going through oxidation or polymerization steps, and of the solvents in which these film formers are dissolved.

The basic film formers of lacquers are the cellulosics. In addition, most lacquers also contain resin for improved adhesion, build, and gloss and plasticizers for improved flexibility. Each of these three types of lacquer film formers is briefly examined.

3.6.1 Cellulosics   By far the most important cellulosic is nitrocellulose; second is ethyl cellulose. Cellulose acetate is also of some importance.

Nitrocellulose, made by nitrating cotton linters, comes in two grades: RS (regular soluble types) and SS (spirit- or alcohol-soluble types). Both are available in a variety of viscosities and form a film that is hard, tough, clear, and almost colorless.

Ethyl cellulose, made by reacting alkali cellulose with ethyl chloride, also comes in different viscosities. It has greater compatibility with waxes,

Fig. 3.4   Some paint additives prevent crawling on a slick substrate. [*Troy Chemical Company*]

better flexibility, better chemical resistance, less flammability, and a higher dielectric constant. It is also somewhat softer, tends to become brittle when exposed to sunlight and heat, and is more expensive. These disadvantages can be partially overcome by the use of proper modifying agents and solvents.

Cellulose acetate lacquers are tough and stable to light and heat. They also have good resistance to oils and greases and are durable. However, they have poor solubility and compatibility, and this defect partially limits their usefulness.

3.6.2 Resins   In most instances, the lacquer film will contain a larger percentage of resin than of the cellulosic. The reason is that resins add many important properties to lacquer films and usually are lower in cost. The most valuable property they add is adhesion; this is of particular importance, since nitrocellulose by itself has rather poor adhesion. In addition, resins give higher solids and therefore a thicker film, improve gloss, reduce shrinkage, and improve heat-seal properties.

In choosing a resin, make certain that it is compatible with the cellulosic being used. It must also be soluble in a mixture of esters, alcohols, and hydrocarbons so as to give a clear, transparent film.

Among the resins in common use are rosin esters such as ester gum, used for its low cost; maleic resin, used in wood finishes for its good sanding properties; and alkyds, employed for their good resistance and durability. Alkyds modified with coconut oil are often used; they may be further modified with other resins such as terpenes for good heat-seal properties and phenolics for good water resistance.

3.6.3 Plasticizers   Without plasticizers, most lacquers would be much too brittle, would tend to crack, and therefore would not be durable. In addition to giving flexibility, plasticizers increase the solids content so as to produce films of practical thickness, and they also tend to improve gloss, especially of pigmented lacquers. Another plus feature, especially of chemical plasticizers, is that they act as a solvent for the cellulosic and thus enable more of this cellulosic to be used. In addition, they help slow the setting time of the lacquer, enabling it to level out satisfactorily.

Plasticizers must be completely nonvolatile so that they remain in the film permanently. There are some exceptions to this requirement, in lacquers such as nail polish which do not remain on a surface permanently.

Since most plasticizers are lower in cost on a solids basis than cellulosics, there might be a tendency to use excessive amounts. This would be dangerous, since the result would be a tacky, soft film with poor chemical and water resistance and poor abrasion resistance.

Two types of plasticizers, the oil type and the chemical type, are generally used in lacquers. A good example of the nonsolvent oil type is raw and blown castor oil, which gives perpetual flexibility, is low in cost, has good color and color retention, and is insensitive to temperature change. Excessive amounts tend, however, to spew from the film. Solvent-type chemical plasticizers such as dibutyl phthalate, triphenyl phosphate, and dioctyl phthalate have excellent compatibility and good heat-seal properties. The chlorinated polyphenyls have good resistance characteristics. All tend to produce a good, tight film.

3.7   Water-Based Polymers and Emulsions   Manufacture of these types of coatings is the fastest-growing part of the coatings industry. Most of the trade sales and architectural paints are now water-based. Even in the industrial field, more and more water-based or water-thinnable paints are being manufactured. The major advantages of these coatings are that they can be thinned with water and, in the case of trade sales paints, that there is little odor, a fast dry, better nonpenetration and holdout, very good alkali resistance, excellent stain resistance, and easy cleanup with water. In all cases, they practically eliminate the release of solvent fumes into the atmosphere—a big plus in view of environmental restrictions.

3.7.1   Styrene-butadiene   This is the oldest and initially was the only polymer available for latex paints. It is a copolymer of polystyrene, a hard, colorless resin, and butadiene, a soft, tacky, rubberlike polymer. Paints based on the polymers of styrene-butadiene have some disadvantages in their tendency toward poor freeze-thaw stability and low critical PVC. There is also a greater tendency toward efflorescence, the appearance of a white crystalline deposit on a painted surface. The use of styrene-butadiene polymer is now very small.

3.7.2   Polyvinyl acetate   This is one of the most popular polymers used in the manufacture of latex paints. The polymer itself is a thermoplastic, hard, resinous, colorless product having good water resistance. Normally it is bought as a water emulsion containing surface-active agents, protective colloids, and a catalyst. It is much more stable and easier to use than styrene-butadiene and therefore has largely replaced it in latex paints. The film is clear, colorless, and odorless and has very good water and alkali resistance. The polymer gives a breathing type of film which prevents blisters if applied over somewhat moist surfaces. Since by itself the film would be too brittle, it must be plasticized, either internally or in the paint formulation. Polyvinyl acetate (PVA) types have advantages over styrene-butadiene types in durability, stability to light aging, and nonblistering properties. The emulsion tends to be acidic, and formulating with it requires some caution.

3.7.3   Acrylic   Acrylic polymers are probably the best in quality of the emulsions popularly used in the manufacture of latex paints. They are made essentially by polymerization or copolymerization of acrylic acid, methacrylic acid, acrylonitrile, and the esterification of them. The properties of acrylic polymers depend to a large degree on the type of alcohol from which the esters are prepared. Normally alcohols of lower molecular weight produce harder polymers. The acrylates are generally softer than the methacrylates.

The acrylics differ from the PVAs in being basic (i.e., nonacid). The danger of their causing containers to rust is thus reduced. Moreover, since the acrylics are almost completely polymerized prior to application as a paint film, there is practically no embrittlement or yellowing on aging. This factor improves the durability of acrylic paints; in fact, durability is a special feature of acrylics. They are the most stable of the polymers and require a minimum of such stabilizers as protective colloids, dispersing agents, and thickeners. They will also withstand extremes of temperature to a high degree.

The acrylics have excellent resistance to both scrubbing and wet abrasion. Moreover, the extreme insolubility of the dried paint film gives it excellent resistance to oil and grease. As a result, oil stains and other dirt marks can easily be removed without injuring the film.

The major disadvantage of acrylics is cost, which is higher than that of other latices. In partial compensation, acrylics will take higher pigmentation, and more low-cost extenders may therefore be used.

3.7.4 Other polymers and emulsions

Though most trade sales paint is water-based, this is not true of industrials. Because of the special requirements of industrial coatings, satisfactory water-based polymers with the required properties have not yet been developed. Nevertheless, much progress has been made, and satisfactory water-reducible coatings have been made for many industrial applications.

The most popular general type of aqueous industrial vehicles is the so-called water-soluble resin. The basic approach is to prepare the resin at a relatively high acid number and then to neutralize it with an amine such as ammonia or dimethylaminoethanol. A wide variety of resins, including alkyds, maleinized oils, epoxy esters, oil-free polyesters, and acrylics, is produced in this manner. These resins may be either air-dried or baked vehicles. Driers such as cobalt, manganese, calcium, or zirconium may be added as cross-linkers to the baking vehicles. Coatings made with these vehicles generally are competitive with solvent-based industrials in terms of gloss, film properties, and overall resistance. However, there is a problem with air-drying efficiency on aging because of the complexing of the driers with the amines used.

Emulsion vehicles, particularly acrylic and styrene-acrylic types, are also being promoted for baking industrial finishes. These cure by cross-linking mechanisms, generally through the use of melamine or urea resins. It is more difficult to obtain high gloss with emulsions as compared with water-soluble resins, but because of their higher molecular weight emulsions may offer advantages in film strength and resistance properties.

3.7.5 Copolymers   Some types of polymers can be copolymerized. The types of acrylics are the acrylates, methacrylates, and acrylonitriles. To obtain special properties, polymers are frequently blended or copolymerized.

3.8 Solvents   There are essentially three types of volatile solvents: a true solvent, which tends to dissolve the basic film former; a latent solvent, which acts as though it were a true solvent when used with a true solvent; and a diluent, a nonsolvent that is tolerated by the coating. Thus, in a lacquer, ethyl acetate is the true solvent, ethyl alcohol is the latent solvent, and petroleum hydrocarbon is the diluent. In a latex paint water might be considered the true solvent, but in an alkyd enamel it would be a diluent.

The question could very well be asked: why use an item that does not remain a part of the film? The answer is, of course, that this item gives coatings essential application and film properties. With the exception of the newer 100 percent solids coatings such as powder coatings, paint simply could not be applied without a solvent, since in most instances the result would be a semisolid mass. It can therefore be said that the most important property of a solvent is to reduce viscosity sufficiently so that the coating can be applied, whether by brush, roller, dipping, or spraying. Besides this most important property, the solvent has other significant features. It controls the setting time of the paint film, which in turn controls the ability of one panel of paint to blend with another panel applied later. In addition, it controls important properties such as leveling or flow, gloss, drying time, durability, sagging tendencies, and other good or bad features in the wet paint or paint film.

3.8.1 Petroleum solvents   These constitute by far the most popular group of solvents used in the coatings industry. They consist of a blend of hydrocarbons obtained by the distillation and refining of crude petroleum oil. The faster-evaporating types, which come off first, are used as diluents in lacquers or as solvents in special industrials. Solvents of the intermediate group are used in trade sales paints. Members of the slowest group, beginning with kerosine and going into fuel oils, are used for heating, lubrication, and other applications.

The most important group used in trade sales paints and varnishes consists of mineral spirits and heavy mineral spirits. Mineral spirits are petroleum solvents with a distillation range of 300 to 400°F (149 to 204°C). They are sometimes considered a turpentine substitute because the distillation ranges are approximately the same. Because of their low price, proper solvency, and correct evaporation rate, mineral spirits are probably the most popular solvents used by the coatings industry. Normally they are the sole solvents in all interior and exterior paints with the exception of flat finishes. Special grades that pass antipollution regulations are now being sold. Heavy mineral spirits are a slower-evaporating petroleum hydrocarbon and an ideal solvent for flat-type finishes. During cold winter weather, the formulator might use a combination of regular and heavy mineral spirits.

The U.S. Environmental Protection Agency has set new guidelines, based on regulations already adopted in California, that severely limit the amount of solvent in architectural coatings. The recommended limit is 250 g of volatile organic material per liter of paint. This limit also affects water-based paints containing organic freeze-thaw agents and additives. Architects switching to new high-solids coatings should work closely with the manufacturer to assure proper performance and be certain that application personnel are properly trained to handle the more complex systems.

A faster-evaporating petroleum solvent with a distillation range of 200 to 300°F (93 to 149°C), known as VM&P naphtha, is sometimes used by painters as an all-purpose thinner. Its fast evaporation rate might cause the paint to set too quickly. It is also used by some manufacturers in traffic paints, for which a fast setting time and dry are desirable.

In some industrials and lacquers, a still faster-evaporating type, having a distillation range of 200 to 270°F (93 to 132°C), is desired. In many coatings it gives satisfactory spraying and dipping properties. An even faster-evaporating type, with a distillation range of 130 to 200°F (54 to 93°C), is sometimes used when very fast evaporation and drying are desired, but it might cause blushing or flatting of the paint or lacquer film.

Because of regulations regarding air pollution, the straight types of hydrocarbon solvents that hitherto have been the backbone of the coatings industry are being phased out and replaced by mixtures that will pass the stringent regulations of various states including California, Illinois, and New York.

3.8.2 Aromatic solvents

This group of cyclic hydrocarbons is obtained normally from coal-tar distillation or from the distillation of special petroleum fractions. These hydrocarbons are almost pure chemical compounds and are much stronger solvents than petroleum hydrocarbons. With the exception of high-flash naphtha, they are rarely used in trade sales coatings but are employed in industrial and chemical coatings for which vehicles having weak solvent requirements are not normally used. Since aromatic solvents are pure chemicals, they have regular boiling points rather than distillation ranges. Naturally, those with the lowest boiling points will evaporate more quickly and thus give a faster dry. The most popular of these are as follows:

▪ Benzol C_6H_6; boiling point, 175°F (79°C). Quite toxic, it is used in paint and varnish removers. It can cause blushing or whitening of a clear film.

▪ Toluol, $C_6H_5(CH_3)$; boiling point, 230°F (110°C). It is very popular in fast-drying industrials and in lacquers.

▪ Xylol, $C_6H_4(CH_3)_2$; boiling point, 280°F (138°C). It is popular in industrials and lacquers for which slower evaporation is acceptable.

▪ High-flash naphtha, a blend of slower-evaporating aromatics. The distillation range is 300 to 350°F (149 to 177°C) for brushing-type industrials and lacquers.

These products also are slowly being replaced by others that can pass stringent air pollution requirements.

3.8.3 Alcohols, esters, and ketones   A great many of these types of solvents are used in industrials and, especially, in lacquers. Among the more popular solvents of this type are the following:

▪ *Acetone, CH_3COCH_3*. Very strong and very fast-evaporating, it can cause blushing. It is used in paint and varnish removers.

▪ *Ethyl acetate, $CH_3COOC_2H_5$*. This is a standard fast-evaporating solvent for lacquers. It is relatively low in cost.

▪ *Butyl acetate, $CH_3COOC_4H_9$*. This is a very good medium-boiling solvent for lacquers. It has good blush resistance.

▪ *Ethyl alcohol, C_2H_5OH*. Used only in a denatured form, it is a good latent solvent for lacquers and also is used to dissolve shellac. It is relatively low in cost.

▪ *Butyl alcohol, C_4H_9OH*. This is a medium-boiling popular latent solvent for lacquers.

Other popular ketones used in lacquers are methyl ethyl ketone and the slower-evaporating methyl isobutyl ketone. They are very strong and relatively low in cost.

Very slow-evaporating solvents are sometimes used in lacquers to prevent blushing or for brushing application. Among popular products are the lactates, cellosolve, and Carbitol.

3.9 Pigments   All the raw materials discussed thus far form portions of the vehicle. In nonpigmented clear coatings these raw materials are all that would be used. In pigmented coatings, or paints, it would be necessary to add a pigment or pigments to obtain the essential important properties of the paint that differentiate it from the clear coating. Paints may contain both a hiding, or obliterating, type of pigment and a nonhiding or, as it is sometimes known, an extender type of pigment.

One of the most important properties of pigments is to obliterate the surface being painted. This property is often known as hiding power, coverage, or opacity. We frequently hear such terms as "one-coat hiding power." This simply means that one coat of paint, normally applied, will completely cover the substrate or surface which is being painted. Sometimes, however, especially if a radical change in color is made, two or even three coats of paint may be required to do so, especially if the paint lacks good hiding power.

Another important reason for using pigments is their decorative effect. This means giving the desired color to the surface being painted. Usually when paint is applied, great care is taken about the color scheme so as to make the surface as attractive as possible. Pigments are also used because they protect the surface being painted. Everyone will recognize red lead as a pigment used to protect steel from rusting. Not so well known are zinc chromate, zinc dust, and lead suboxide.

Still other pigments are used to give a paint special properties. For example, cuprous oxide and tributyl tin oxide are used in ship-bottom paints to kill barnacles, and antimony oxide is used to give fire retardance to paint. Pigments may also give the desired degree of gloss in a paint. Everything being equal, the higher the pigmentation, the lower the gloss. In addition, pigments are used to give other desirable properties. Thus they can be employed to give a coating the desired viscosity, to control the degree of flow or leveling, to improve brushability by enabling the use of additional easy-brushing solvent, and to give very specific properties such as fire retardance, fluorescence and phosphorescence, and electrical conductance or insulation.

3.9.1 White hiding pigments   White is important not only as a color in its own right but because it forms the basis for a great many shades and tints in which it constitutes a large or small percentage of the color. The number of important white pigments being used by the paint industry has been dwindling (Table 3.2). Thus pigments such as lithopone, basic

TABLE 3.2 Relative Opacity or Hiding Power of White Pigments

Pigment	Hiding units
Basic lead carbonates	15
Basic lead sulfate	15
Zinc oxide	20
Lithopone	25
Antimony oxide	28
Anatase titanium dioxide	100
Rutile titanium dioxide	125–135

lead sulfate, titanium-barium pigment, titanium-calcium pigment, zinc sulfide, and many leaded zinc oxides have practically disappeared. Of the white pigments now being used, the most important by far is titanium dioxide.

Titanium dioxide, TiO₂   This pigment comes in two crystalline forms. The older anatase form has about 75 percent of the opacity, or hiding power, of the present rutile form. Both forms are excellent for both interior and exterior use. Titanium dioxide is used in both trade sales and chemical coatings. Very little anatase is now being used except in some specialty coatings. The rutile comes in types designed for use in enamels and flats, for solvent-based and water-based coatings. Normally 2 to 3 lb/gal (240 to 359 kg/m³) of rutile titanium dioxide will give adequate coverage in most formulations. Anatase is less chalk-resistant.

Zinc oxide, ZnO   Despite its rather poor hiding power (only about 15 percent of that of TiO₂), zinc oxide still maintains its importance in the coatings industry. This is due to unusually good properties which more than offset the relatively high cost of the pigment per unit of hiding power. Zinc oxide's most important use is in exterior finishes; it tends to reduce chalking and the growth of mildew in house paints. In enamels it tends to improve the color retention of the film on aging. Zinc oxide also is sometimes used to improve the hardness of a film.

3.9.2 Extender pigments   These pigments, though they have practically no hiding power, are used in large quantities with both white and colored hiding-power pigments. An important property of some extender pigments is to lower the raw-material cost (RMC) of the paint. Most of these pigments are so-called nonhiding pigments such as whiting, talc, and clay. If prime, or hiding, pigments had to be used to lower the gloss in order to obtain a flat finish, the RMC would be extremely high in most instances. Instead, extender pigments are used to accomplish this task at a small fraction of the cost.

Whiting (calcium carbonate)   This is probably the most important extender pigment in use. It comes in a variety of particle sizes and surface treatments, and it can be dry-ground, water-ground, or chemically precipitated. Normally quite low in cost, it can be used to control such properties as sheen, nonpenetration, degree of flow, degree of flatting, tint retention, and RMC.

Talc (magnesium silicate)   Though used widely as an extender in interior finishes, this pigment finds its greatest use in exterior solvent-based coatings, especially house paints. This is due largely to a combination of durability and low cost. Most grades of talc tend to have good nonsettling properties and give a rather low sheen.

China clay (aluminum silicate)   This extender, though used to some degree in solvent-based coatings, finds its greatest use in water-based paints. It disperses readily with high-speed dispersers, in the normal method of manufacturing latex finishes, and does not impair the flow characteristics of the paint. Some grades will improve the dry hiding power of water-thinnable or solvent-based paint.

Other extenders   Among extenders that are sometimes used are diatomaceous silica, used to reduce sheen and gloss; regular silica, which gives a rough surface; barites, used to minimize the effect of the extender; and mica, which because of its platelike structure is used to prevent the bleeding of colors.

3.9.3 Black pigments

Next to whites, blacks are probably the most important colors used in the coatings industry. The reason for their wide use is twofold. First, black is a very popular color and is often used in industrial finishes, trim paints, toy enamels, quick-drying enamels, etc. Second, it is also very popular as a tinting color, particularly for all shades of gray, which are made by adding black to white.

The two most popular blacks in use consist of finely divided forms of carbon; they are known as carbon black and lampblack. Carbon black, the most widely used of the blacks, is sometimes called furnace black. It is made by the incomplete combustion of oil injected into the combustion zone of a furnace. Lampblack, or channel black, is made by the impingement of gas on the channel irons of burner houses. Both types of black come in a variety of pigment sizes and jetness. Practically all black colors are made with carbon black. They have tremendous opacity: only 2 to 4 oz/gal (15 to 30 kg/m³) of paint is necessary in most instances for proper coverage. They also have excellent durability, resistance to all types of chemicals, and lightfastness. Even the most expensive, darkest jet blacks are inexpensive to use because only a small amount is needed.

Whereas carbon black is used principally as a straight color, lampblack, a coarse furnace black made from oil, is used mainly as a tinting color for grays, olive shades, etc. Largely because of its coarseness, lampblack has little tendency to separate from the TiO_2 or other pigments with which it is used and to float up to the surface, as do the carbon blacks with their much finer particle size. Floating, a partial color float to the surface of the film, and flooding, a more nearly complete and uniform color float, are, of course, undesirable, and for this reason carbon black is rarely used as a tinting color. Lampblack has very poor jetness but gives a nice bluish shade of gray. It also has excellent heat and chemical resistance.

Other blacks that are sometimes used are black iron oxide, used as a tinting black having brown tones and in primers, and mineral and thermal blacks, used as low-cost black extenders.

3.9.4 Red pigments

In discussing white or black colors, everyone knows what colors are meant and what they look like. Other colors, however, come in different shades. Thus, there are a great variety of reds, some of which are briefly mentioned below.

Red iron oxides   These are good representatives of a series of metallic oxides that have very important properties. Though relatively low in cost, they have such fine opacity that 2 lb/gal (240 kg/m³) is normally adequate, and they also possess high tinting strength. In addition, they have good chemical resistance and colorfastness, and they disperse easily in both water and oil so that high-speed dispersers can be used in manufacturing paints based on iron oxide pigments. Red iron oxides give a series of rather dull colors having excellent heat resistance. These colors are used popularly in floor paints, marine paints, barn paints, and metal primers and as popular tinting colors.

Toluidine reds   These popular, very bright azo pigments come in colors ranging from a light to a deep red. They have excellent opacity, so that ¾ to 1 lb/gal (90 to 120 kg/m³) of paint normally gives adequate hiding power. Since they also have fine durability and lightfastness, they are used in such finishes as storefront enamels, pump enamels, automotive enamels, bulletin paints, and similar types of finishes. The toluidines tend to be somewhat soluble in aromatics, which should therefore be kept to a minimum. They are also not the best pigments for baking finishes since they sometimes bronze, or for tinting colors, since they are somewhat fugitive in very low concentrations. They also bleed.

Para red   This azo pigment is deeper in color than toluidine and not quite as bright. It has very good coverage, about 1 lb/gal (120 kg/m³) giving adequate coverage. Para red is not as lightfast as toluidine and tends to bleed in oil to a greater degree. Moreover, it has poor heat resistance and cannot be used in baked coatings. Its lower cost makes it attractive for bright interior finishes and some exterior finishes.

Rubine reds   These bright reds, sometimes known as BON (β-oxynaphthoic acid) reds, are available in both resinated and nonresinated forms. They have good bleed resistance but only fair alkali resistance.

Lithol red   This complex organic red has very good coverage, 1 lb/gal (120 kg/m³) giving adequate coverage in most instances. It is bright and has a bluish cast. Lithol red is relatively nonbleeding in oil but tends to bleed in water, and its durability and lightfastness are only fair. Since it is relatively low in cost, it is used in such applications as toy and novelty enamels.

Naphthol reds   These arylide pigments have excellent alkali resistance and are relatively low in cost. They bleed in organic solvents and are more useful in emulsion than in oil-based paints.

Quinacridone reds   These pigments come in a variety of shades, ranging from light reds to deep maroons and even violets. They have good durability and lightfastness and high resistance to alkalies. They also tend to be nonbleeding and show good resistance to heat.

Other reds   Among reds sometimes used are alizarine (madder lake) red for deep, transparent finishes, pyrazolone reds for high heat and alkali resistance, and a larger series of vat colors.

3.9.5 Violets   The demand for violets is small because they are expensive and often have poor opacity. However, several violets may be mentioned.

Quinacridone violets   These pigments are durable and have good resistance to alkalies and to heat.

Carbazole violets   These pigments have very good heat resistance and lightfastness. They also are nonbleeding, and their high tinting strength makes them useful for violet shades.

Other violets   Violets in use include tungstate and molybdate violets for brilliant colors and violanthrone violet for high resistance and good lightfastness.

3.9.6 Blue pigments   Blues not only are important as straight and tinting colors but also are popular for use in combination with other colors to produce different shades and colors.

Iron blue   This popular blue, a complex iron compound also known as prussian blue, milori blue, and chinese blue, is one of the most widely used blue pigments in the coatings industry. It combines low cost, good opacity, high tinting strength, good durability, and good heat resistance. However, it has very poor resistance to alkalies and cannot be used in water paints or in any paints that require alkali resistance.

Ultramarine blue   This color, sometimes known as cobalt blue, is popularly used as a tinting color. It gives an attractive reddish cast when added to whites. Ultramarine blue has poor opacity, high heat resistance, and good alkali resistance. While it can be used in latex paints, special grades low in water-soluble salts must be obtained. It is often used for whites to give extra opacity and make them look whiter by lending them a bluish cast.

Phthalocyanine blue   This blue is increasingly popular because of its excellent properties. It gives a bright blue color and has excellent opacity, durability, and lightfastness. In addition, it is relatively nonbleeding and gives a greenish blue shade when used as a tinting color. Its high chemical and alkali resistance makes it satisfactory for water-based coatings as well as for all types of interior and exterior finishes. The price, though high, is not so high as to prohibit the use of this blue in most finishes.

Other blues   Sometimes used are indanthrone blue, which has a reddish cast and high resistance; and molybdate blue, which is used when a very brilliant blue is desired.

3.9.7 Yellow pigments   Various yellow pigments are discussed below.

Yellow iron oxide   Although yellow iron oxide pigments give a series of rather dull colors, they have excellent properties. They are relatively easy to disperse, are nonbleeding, and have good opacity despite low cost. They also have fine heat resistance. Since their chemical and alkali resistance is excellent, they may be used in both water- and solvent-based paints. Excellent durability makes them useful for all types of exterior coatings. They are also popular shading colors, for when added to white they give such popular shades as ivory, cream, and buff.

Chrome yellow   This once-popular bright yellow comes in a variety of shades, from a very light greenish yellow to dark reddish yellow. Chrome yellow paints have good opacity and are easy to disperse, but they tend to darken under sunlight. Because they are lead pigments, they are gradually being phased out of use.

Cadmium yellow   Largely a combination of cadmium and zinc sulfides plus barites, cadmium yellow pigments are sold in a variety of shades. They have good hiding and lightfastness if used as straight colors. They also are bright and nonbleeding, bake well, and have good resistance except to acids. Since they are toxic, however, they are being phased out of use.

Hansa yellow   With the elimination of chrome yellow and cadmium yellow, hansa yellow pigments are becoming increasingly important as bright yellows. They come in several shades, from a light to a reddish yellow. Hansa yellow pigments have excellent lightfastness when used straight but are somewhat deficient in tints. Although their hiding power is only fair, they have excellent tinting strength, which makes them good tinting pigments, especially in water-based coatings, for which they have excellent alkali resistance. However, they bleed in solvents and do not bake well.

Benzidine yellow   Along with hansa yellow pigments, benzidine yellow pigments are finding increasing usage as the use of lead-containing yellows becomes illegal. They are stronger than hansa yellows and have good alkali and heat resistance. Their resistance to bleeding is also better. Since their lightfastness is poorer, however, they are unsatisfactory for exterior coatings.

Other yellows   Among other yellows in use are nickel yellows, which have good resistance and make greenish yellow colors; monarch gold and yellow lakes, which are used for transparent metallic gold colors; and vat yellow, which has extremely good lightfastness and good resistance to heat and to bleeding.

3.9.8 Orange pigments   A number of orange pigments are in use.

Molybdate orange   This very popular bright orange, with its reasonable cost, hiding power, brightness, and colorfastness, is being phased out because of its lead content.

Chrome orange   This lead pigment is also being phased out. In money value it is inferior to molybdate orange.

Benzidine orange   Benzidine orange pigments are bright and have good alkali resistance and high hiding power. They also have good heat resistance and resistance to bleeding and can be used in both water- and solvent-based paints. Since their lightfastness is only fair, they are not the best pigments for outside use.

Dinitroaniline orange   This bright orange has very good lightfastness and good alkali resistance, making it a good exterior pigment for aqueous systems. It tends to bleed in paint solvents.

Other oranges   Among oranges sometimes used are orthonitroaniline orange, which is lower in cost but inferior in most properties to dinitroaniline orange; transparent orange lakes, which are used for brilliant transparents and metallics; and vat orange, which is high in overall properties but also high in price.

3.9.9 Green pigments   Four green pigments are widely used.

Chrome green   Until recently the most popular of all greens for its brightness, durability, hiding power, and low cost, chrome green is gradually being replaced by other greens because of its lead content. It comes in a combination of shades from a yellowish light green to a bluish dark green. Chrome green has poor alkali resistance and cannot be used in latex paints.

Phthalocyanine green   This is fast becoming the most important green pigment of the coatings industry. A complex copper compound of bluish green cast, it has excellent opacity, chemical resistance, and lightfastness. It also is nonbleeding and can be used in both solvent- and water-based coatings, both as a straight color and for tints. It is rather expensive.

Chromium oxide green   This rather dull green pigment has excellent durability and resistance characteristics and can be used for both water and oil, in both interior and exterior paints. It has moderate hiding power and is easy to emulsify. Its high infrared reflection makes it an important green in camouflage paints.

Pigment green B   This pigment is used mainly in water-based paints because of its excellent alkali resistance, but it can also be used in solvent-based paints. Its lightfastness is only fair, so that it is not satisfactory for exterior paint use. It does not give a clean shade of green but is satisfactory in most instances.

3.9.10 Brown pigments   Two brown pigments are discussed below.

Brown iron oxide   Most of the browns used by the coatings industry are iron oxide colors. Essentially combinations of red and black iron oxides, they have very good coverage, excellent durability, good light resistance, and good resistance to alkalies. They are suitable for both water- and solvent-based paints and for both interior and exterior finishes.

Van Dyke brown   This essentially organic brown gives a purplish brown color. Lightfast and nonbleeding, it is used largely in glazes and stains.

3.9.11 Metallic pigments   Aluminum, bronze, zinc, and lead pigments are in use.

Aluminum   By far the most important of the metallic pigments, aluminum is platelike in structure and silvery in color and comes in a variety of meshes and in leafing and nonleafing grades. The coarser grades are more durable and brighter, while the finer grades are more chromelike in appearance. Aluminum powder has high opacity, excellent durability, and high heat resistance. The nonleafing grade is used when a metallic luster is wanted by itself or with other pigments. The leafing grade is used when a silvery color is desired. This grade is highly reflective, making it ideal for storage tanks, since it tends to keep the contents cooler. It is also very popular for structural steel, automobiles, radiators, and other products with metallic surfaces. The nonleafing grade is used for so-called hammertone finishes.

Bronze   Gold-colored bronze powders consist mainly of mixtures of copper, zinc, antimony, and tin. They come in a variety of colors, from a bright yellowish gold to a dark brown antique type of gold. Bronze powders are used mainly for decorative purposes. Their opacity is poorer and their price higher than those of aluminum.

Zinc   Zinc dust is assuming increasing importance as a protective pigment for metal, especially since lead is gradually being eliminated. It is used in primers for the prevention of corrosion on steel when employed as the sole pigment in so-called zinc-rich paints, and it is used in combination with zinc oxide in zinc dust–zinc oxide primers. Zinc dust–zinc oxide paints are satisfactory for both regular and galvanized iron surfaces. Zinc-rich paints are used with both inorganic vehicles such as sodium silicate and organic vehicles such as epoxies and chlorinated rubber. Both types have excellent rust inhibition and show good resistance to weather.

Lead   Lead flake has found useful application in exterior primers, in which it exhibits excellent durability and rust inhibition.

3.9.12 Specialty and special-purpose pigments   Some pigments are used, not for their color or opacity, but for the special properties that they give a coating. Two of these have been mentioned in the metallic-pigment category: zinc dust and lead flake, which are used primarily for rust inhibition. Others are mentioned below.

Red lead   This bright orange pigment is used almost exclusively for corrosion-inhibiting metal primers, especially on large structures such as bridges, steel tanks, and structural steel. Since it has poor opacity, it is sometimes combined with red iron oxide for improved opacity and low cost. Because of restrictions on the use of lead, its employment is being phased out.

Basic lead silicochromate   This also is a bright orange pigment that is used primarily as a rust-inhibiting pigment for steel structures. Because of its low opacity it can be combined with other pigments to give topcoats of different colors that still have rust-inhibiting properties.

Lead silicate   This pigment is used mainly in water-based primers for wood, in which it reacts with tannates and prevents them from coming through and discoloring succeeding coats of paint. It may be eliminated from home use because of restrictions on the use of lead.

Zinc yellow   This hydrated double salt of zinc and potassium chromate is used principally in corrosion-inhibiting metal primers. It is becoming one of the few permissible pigments to use on steel connected with houses or apartments. It is greenish yellow in color and has poor opacity.

Basic zinc chromate   This pigment has properties somewhat similar to those of zinc yellow. It is used in metal pretreatments, especially in the well-known "wash primer" government specification for conditioning metals, in which capacity it promotes adhesion and corrosion resistance for steel and aluminum.

Cuprous oxide   This red pigment is used almost exclusively in antifouling ship-bottom paints to kill barnacles that would normally attach themselves to a ship below the waterline.

Antimony oxide   This white pigment is used almost entirely in fire-retardant paints, in which it has been very effective, especially in combination with whiting and chlorinated paraffin.

BIBLIOGRAPHY

Bentley, Kenneth W.: *The Natural Pigments,* Interscience Publishers, Inc., New York, 1960.

Mellan, Ibert: *Industrial Solvents Handbook,* Noyes Data Corp., Park Ridge, N.J., 1977.

Patton, Temple C.: *Pigment Handbook,* John Wiley & Sons, Inc., New York, 1973.

Preuss, Harold P.: *Pigments in Paint,* Noyes Data Corp., Park Ridge, N.J., 1974.

Rothenberg, G. B.: *Paint Additives: Recent Developments,* Noyes Data Corp., Park Ridge, N.J., 1978.

Stevens, V. L., and R. H. Lalk: "Solvent Option for Air Quality Compliance," Water-Borne and Higher Solids Coatings Symposium, sponsored by University of Southern Mississippi and Southern Society for Coatings Technology, New Orleans, La., Mar. 10–12, 1980.

Chapter 4

Selecting the Paint System

PAUL WEAVER
Consulting Engineer

4.1 Introduction   Because of the very large number and varieties of coatings available and the many different trade names, the selection of protective coatings is unduly complicated. A specification writer must consider proper surface preparation, use of a suitable priming system when required to ensure adequate adhesion and inhibit underfilm corrosion, careful workmanship, and proper selection of the protective coating itself.

In many cases, clients devote considerable research to the coating field because of specific requirements. The main reasons for painting plants, buildings, and structures are public relations, corrosion protection, standards requirements, and employee relations. Other important reasons are product purity, safety, antistick properties, abrasion resistance, emissivity, electrical insulation, noise attenuation, poison (barnacles on boats), provision of a nonskid surface, light reflection, fire protection, resistance to radiation, and impact resistance.

Through laboratory tests, which fairly promptly screen out obviously unsuitable coatings, and through extensive field tests, which give further useful data about the performance of coatings, the client often establishes a list of manufacturers whose products must be specified. Because of the very rapid rate at which technology is growing, coatings should also be selected on the following bases:

▪ When feasible, obtain coatings from a reputable company with facilities close to the area where the coatings are to be used, so that service for assistance in proper application will be available when needed.

▪ Obtain complete formulation data from the coatings manufacturer for each type of coating being considered. These data, of course, should be kept confidential.

▪ Obtain comparative exposure test data from industrial sites or from qualified independent testing organizations.

From the foregoing data, it is possible to choose acceptable coatings without extensive testing simply by comparing data on a given generic type of coating with those of another coating of the same generic type. It is good practice to evaluate new coatings or coatings from vendors not on the approved list for at least 2 or 3 years. After this time, coatings may be added to the approved list with some assurance that they will perform satisfactorily. Restriction to a specific list of manufacturers reduces the number of products to be considered and simplifies the task, but the necessity of selecting the proper coating is not eliminated.

When alternative materials are being selected, their relative costs do not entirely reflect the cost of the finished system because materials account for only a percentage of the total cost. In many cases the material cost may be less than 15 percent of the total applied cost. Therefore, using inferior coatings is poor economy. On the other hand, expensive coatings should not be used if they are not needed. For a checklist of properties desired in a coating, see Table 4.1.

TABLE 4.1   Checklist to Determine the Properties Desired in a Paint Job*

Purpose of painting	Yes	No	Not applicable
Is the paint to be applied for:			
Public relations			
Employee relations			
Corrosion protection			
Product purity			
Antistick properties			
Impact resistance			
Abrasion resistance			
Emissivity			
Electrical insulation			
Protection required by standards			
Safety and identification			
Good weathering properties			
Resistance to radiation			
Is the paint resistant to the temperature to which it will be exposed?			

* A specification writer can add items to this table, then rate the "yes" items in the order of priority to help determine the paint to be specified.

Labor costs are reduced by minimizing the number of coats, but they may also be reduced by other factors, such as modern methods of surface preparation with automated centrifugal abrasive-blast machines, automated abrasive-blast equipment, and pickling. In addition, many modern application techniques, such as airless spray, electrodeposition, fluidized bed, and dip coatings, lower the labor cost of coatings. Some of these coating methods are applicable to shop coating only. For photographs of powder-coated pipe, see Figs. 4.1, 4.2, 4.3, and 4.4.

4.2   Paint-System Selection by Customer Requirements

4.2.1   Public relations   Studies show that an attractive building or plant has a marked influence on a company's image and in some cases can

Fig. 4.1   Powder-coated pipe.

improve sales. This is particularly true of highly visible plants and of firms dealing directly with the public. Coating selection is affected by the client's desire either to recoat frequently or to apply the most durable system initially.

When corrosion is not a problem and appearance is of prime importance, there may be a minimum of painting, such as a primer and one color coat over a properly prepared substrate. The coating may be an

Fig. 4.2   Small-diameter pipe being powder-coated.

Fig. 4.3 Pipe being powder-coated.

alkyd, which has very good color and gloss retention when exposed to the weather. At intervals of, say, 2 years or more, depending on climatic and other exposure conditions, an additional coat of paint may be applied. Thus the facility looks good at all times.

On the other hand, a silicone-alkyd enamel may be used. This is slightly more expensive than a straight alkyd, but it sheds dust very well and has excellent weather resistance, producing a good-looking job for a longer time than a straight alkyd.

Fig. 4.4 Pipe being powder-coated.

Fig. 4.5 Sandblasted tanks to NACE No. 1 finish primed with inorganic zinc and painted with three coats of vinyl.

In other cases, two color coats of alkyd or silicone-alkyd are applied over a properly primed and prepared substrate for a performance of 8 to 10 years or longer. If the paint is in good condition, dust and other deposits on large surfaces, such as tanks, may be removed with clean water and detergent to recapture the appearance of a newly painted surface. Depending on the material on the surface, washing should cost

Fig. 4.6 Sandblasted metal with four coats of chlorinated rubber, now corroded.

less than half of the cost of a coat of paint. One reason for this saving is that washing may be carried out by laborers instead of painters.

If corrosion is a problem or tall process towers are to be painted, for economic reasons a different approach is taken. Proper surface preparation, such as sandblasting to white metal, National Association of Corrosion Engineers (NACE) No. 1 surface finish, or Steel Structures Painting Council (SSPC) Specification SP-5, is followed by a zinc-rich primer and a proper topcoat for the exposure. The tanks in Fig. 4.5 were sandblasted to white metal (NACE No. 1 finish), primed with inorganic zinc, and painted with three coats of vinyl in 1957. When this picture was taken in 1973, there was no rust, although the environment was very corrosive. Because of weathering and color fading, two color coats were applied in 1962, one in 1965, and another in 1970. In each instance there was no surface preparation because there was no rust.

Other steel in the same area, which was not primed with inorganic zinc but was sandblasted to white metal (NACE No. 1 finish) and painted with a four-coat chlorinated-rubber system in 1957, was repainted four times. In the interim the steel corroded severely (Fig. 4.6). Besides being unsightly, the steel had to be replaced in part because of severe corrosion.

Although the original cost of the inorganic zinc–vinyl system was higher than that of the chlorinated-rubber system, the yearly cost ("equal annual cost"),[1] including depreciation and return on investment, was lower for the inorganic zinc–vinyl system. An important bonus was the fact that this system looked good at all times. An economic evaluation of the inorganic zinc–vinyl system and the chlorinated-rubber system is given below.

Cost of painting the four tanks by using the inorganic zinc–vinyl system:

Per square foot

Original cost, 1957	$0.55
Two coats, 1962	0.15
One coat, 1965	0.10
One coat, 1970	0.12
Total	$0.92

Cost of painting the other steel in the area with chlorinated rubber, including the 1973 painting:

Per square foot

Original cost, 1957	$0.38
Maintenance painting, 1960	0.25
Maintenance painting, 1963	0.25
Maintenance painting, 1967	0.29
Maintenance painting, 1970	0.33
Maintenance painting, 1973	0.37
Total	$1.87

[1] H. G. Thuesen, *Engineering Economy*, 2d ed., Prentice-Hall, Inc., Englewood Cliffs, N.J., 1957.

No additional painting of the tanks was required before 1975. To obtain a common cost basis, all costs are reduced to 1957 present worth. We then have:

Per square foot

For 1957	$0.5500
For 1962 $= \dfrac{1}{(1+i)^n} = 15 = (0.4019)(15) =$	0.0603
For 1965 $= \dfrac{1}{(1+i)^n} = 10 = (0.2326)(10) =$	0.0233
For 1970 $= \dfrac{1}{(1+i)^n} = 12 = (0.0935)(12) =$	0.0112
Total present worth, 1957	$0.6448

Where $\dfrac{1}{(1+i)^n}$ = present worth[2]

i = interest rate (time value of money) of 20 percent per year

n = number of years

For an equal annual cost from 1957 through 1975 we have:

$$\text{Equal annual cost} = \$0.6448\,(\text{CRF}) = \$0.6448 \left[\frac{i(1+i)^n}{(1+i)^n - 1} \right]$$

$$= \$0.6448(0.20781) = \$0.1340 \text{ per square foot per year}$$

Where CRF = capital recovery factor, or $\dfrac{i(1+i)^n}{(1+i)^n - 1}$

i = 20 percent per year

n = 18 years

Since the steel coated with chlorinated rubber was not repainted before 1975, to have a common basis for cost comparison all costs were reduced to 1957 present-worth costs. Costs for the chlorinated-rubber system for 1957 present worth follow:

Per square foot

For 1957	$0.3800
For 1960 $= \left[\dfrac{1}{(1+i)^n}\right] (25) = (0.5787)(25) =$	0.1447
For 1963 $= \left[\dfrac{1}{(1+i)^n}\right] (25) = (0.3349)(25) =$	0.0837
For 1967 $= \left[\dfrac{1}{(1+i)^n}\right] (29) = (0.1615)(29) =$	0.0468
For 1970 $= \left[\dfrac{1}{(1+i)^n}\right] (33) = (0.0935)(33) =$	0.0309
For 1973 $= \left[\dfrac{1}{(1+i)^n}\right] (37) = (0.0541)(37) =$	0.0200
Total present worth, 1957	$0.7061

[2] Ibid.

For an equal annual cost from 1957 through 1975 we have:

Equal annual cost = $0.7061 (CRF) = $0.7061(0.20781)
$$= \$0.1467 \text{ per square foot per year}$$

This is 9 percent more ($0.1467 versus $0.1340) in cost per square foot per year for the steel not coated with inorganic zinc. If the time value of money is ignored, which does not give a true economic comparison but often is done, the average cost per square foot per year of painting cost is as follows:

Tanks: $0.92 per square foot for 18 years = $0.0511 per square foot per year for 18 years

Other steel: $1.87 per square foot for 18 years = $0.1039 per square foot per year for 18 years

In this case, with the time value of money ignored, the average annual painting cost of the steel not coated with inorganic zinc is more than twice the cost of the steel coated with inorganic zinc.

In time, the use of inorganic zinc primers will reduce maintenance painting costs as a percentage of capital invested. In fact, when sandblasting is required on maintenance work, inorganic zinc primers are an excellent choice. Obviously, on items for temporary use, such as pilot plants and facilities planned for a short duration, say, less than 5 years, an expensive coating system is not needed.

If the surface temperature of a substrate is above the temperature limits of organic coatings, say, above 250 or 300°F (121 or 149°C), it is good practice to sandblast the surface to white metal, prime with inorganic zinc, and apply a suitable topcoat, such as a silicone or modified silicone coating. The extra initial cost will be repaid many times because of the number of years during which this system will perform.

The same analysis is applicable to pure architectural coatings used on homes, buildings, or commercial establishments. The principles are the same. On a stucco surface we might compare a paint based on a 100 percent acrylic emulsion with one made from a polyvinyl chloride emulsion or with other coatings. A similar economic analysis would indicate the best value based on average annual cost.

Let us concentrate for the moment on in-plant maintenance painting as opposed to normal architectural surfaces. In painting process towers, good surface preparation and good coatings should be used because scaffolding and labor costs are high. Towers 25 ft (7.6 m) in diameter by 200 ft (61 m) tall are not uncommon. Since scaffolding is a major cost in repainting, it is good practice to sandblast to white metal, then use an inorganic zinc primer and a topcoat with an appropriate coating to fit the environment so that the paint job will last several years. Labor is not nearly so productive 200 ft in the air as it is on the ground. As a rule of thumb, the labor production rate diminishes by a factor of around 25 to

30 percent for each 75 ft (23 m) of height. The rate, of course, will vary. It is obvious that the cost of paint material is insignificant in such cases. The same thinking holds true when painting multistory buildings, water tanks, and similar structures.

4.2.2 Employee relations   Studies show that an attractively painted plant has fewer accidents, less employee absenteeism, and better housekeeping. These factors cannot be expressed in exact dollar values. The insignificant extra amount of capital cost for painting for employee benefit is doubtless regained in a very short time. Painting items with pleasing colors for highlighting, in contrasting colors for safety, or with lighter colors for better visibility pays off.

Proper painting may promote good employee relations by

- Providing better light, thus resulting in less eyestrain
- Providing better contrast, by highlighting certain parts of a machine or structure, thus resulting in increased safety
- Providing pleasing colors

All these reasons concern color, whereas for corrosion protection alone color is not necessary. Black is satisfactory and in some cases is more corrosion- and weather-resistant then are colored materials.

Some specification writers avoid color because they think that color costs too much. This is not necessarily so. Some colors cost about the same as black, provided, of course, only one color is used in an area. Others may cost as much as 2 percent more than black (this 2 percent is the percentage increase of an entire sandblast and paint job and is not the difference in cost of the material alone). It would be safe to assume that the finish or color coat of a two-color scheme would not have an overall increase in cost of more than 7 to 14 percent over black. Actually the final coat is usually the only color coat and therefore the only one that involves more than one color. Labor costs on this one coat should not exceed 10 percent more than those of a single finish color, provided surfaces close to each other are painted the same color.[3]

Lighting costs are reduced and vision is better with white or other light colors. No amount of light can make visibility good when walls, equipment, and fixtures are coated with deep, dull colors.

Colors present a pleasing contrast when one item, say, pipes are one color and vessels another. On machine tools, buff and medium gray have been found to give excellent results. Extensive tests showed that a combination of buff and medium gray resulted in 33 percent more efficiency when mercury vapor lights were used and 31 percent more efficiency when incandescent lights were used.

[3] R. H. Bacon and V. B. Volkening, "Color Can Revolutionize Your Plant," *Petroleum Refinery,* January 1957.

Aesthetic value should not be underestimated. It is practically impossible to put a money value on this quality, but it is real. Not only the quantity but the quality of work increases, and, most important of all, attitudes are altered by colors. This is true in office buildings and commercial buildings as well as in industrial facilities.

Color triads go well together, provided they are chosen from equidistant spacings on the color wheel or the color cone (Figs. 4.7 and 4.8). Red, orange, and yellow suggest warmth, whereas green, blue, and the blue greens suggest coldness. Red colors suggest gay and cheerful moods. Black, when used alone, is depressing and suggests gloom.

Changing the color of a package increased its sale by more than 1000 percent. It is not inconceivable that pleasing colors in a plant or office can

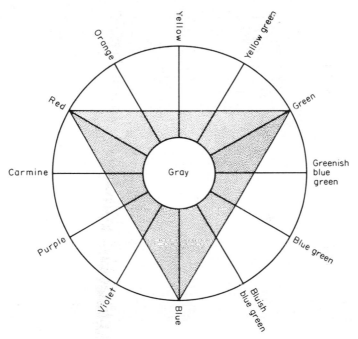

Fig. 4.7 Color wheel. Colors diametrically opposite on the color wheel are complementary; that is, they go well together. Color triads go well together, but they should be 120° apart on the color wheel.

greatly increase employee morale and, therefore, efficiency, which would be enhanced by greater interest and better attitudes.

A note about pigments: the specification writer should ensure that the pigment used fits the environment. For example, aluminum should not be used in strong acid or alkali areas. Copper or lampblack pigments should not be used in the first coat on steel, especially under moist conditions. Copper and carbon are more noble than steel and would therefore promote galvanic corrosion of the steel, resulting in severe pitting.

4.2.3 Corrosion protection   For very mild corrosive conditions, in which relative humidity is rarely above 50 percent and in which the corrosion rate is less than 25 mdd (milligrams per square decimeter per

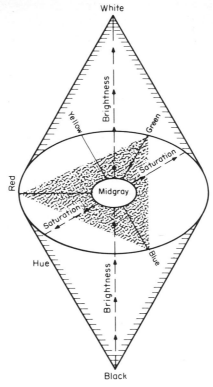

Fig. 4.8 Color cone. Not all the colors are shown here, but relative locations would be the same as on the color wheel, Fig. 4.7. Complementary colors are diametrically opposite each other, and triads are 120° apart, but they should be in the same horizontal plane. Near the top of the cone are the pastels. They all go well together. All colors on the axis are gray except the top and bottom points, which are white and black respectively. Going down the axis, we pass from white, through all shades of gray, to black at the bottom.

day, or 0.19 lb of steel per square foot per year), it would be difficult to justify painting from the standpoint of corrosion. However, such mild corrosive conditions rarely are found in industrial plants.

Table 4.2 gives different rates of corrosion and the cost per year based on 50 cents per pound for erected unpainted structural steel. Severe corrosion rates of 300 mdd are not uncommon in industrial marine exposures. Carbon steel and many other metals should be protected from such exposures.

TABLE 4.2   Cost of Steel Corrosion at Various Corrosion Rates

Corrosiveness	mdd*	Pounds of steel lost per square foot per year	Cost of corrosion (cents per square foot per year)†
Mild	50	0.37	18.5
Medium	100	0.74	37.0
	150	1.11	55.5
	200	1.48	74.0
Severe	250	1.85	92.5
	300	2.22	111.0

* To convert from mdd to ipy (average penetration in inches per year) for any metal or alloy use the following equation:

$$mdd \times \frac{(0.001437)}{d} = ipy$$

where d = density of metal in grams per cubic centimeter.
† In 1977 dollars.

Many chemicals, among them chlorinated solvents, hydrocarbon solvents, ketones, sodium hydroxide (concentrations not in excess of 50%), and phenol, are not considered corrosive to carbon steel at ambient temperatures. However, spills of these materials may promote corrosion by removing the protective coating from the steel, thus permitting the steel to be corroded by atmospheric exposure. Even if the atmosphere is not corrosive, the spills produce an unsightly appearance in the tank, structural steel, or other item from which some of the coating has been stripped.

When spills do occur, it is good practice to recoat the stripped area as soon as feasible, and at least within 1 week if the atmospheric corrosion rate is 100 mdd or more or within 4 weeks if the corrosion rate is less than 100 mdd. Prompt recoating eliminates expensive surface preparation. Spill damage is unsightly (Fig. 4.9) and causes maintenance problems, but when no spills occur, tanks look very good (Fig. 4.10).

Coating or lining a vessel may involve a choice of metallic, inorganic, or organic materials. For example, in a 73% caustic soda environment catalyzed epoxies perform well if temperatures are correct, as do neoprene sheet, nickel-clad steel, pure nickel, some stainless steels, and other metals. Table 4.3 gives applicable data.

As an example, a 10,000-gal (37.85-m³) tank car can be lined with a catalyzed-epoxy coating or, as an alternative, clad inside with nickel. If the car is loaded at all times and the average temperature is 300°F (149°C), the clad metal will last at least 10 years; if it is loaded 40 percent of the time and the average temperature is not over 200°F (93°C), it will

Fig. 4.9 Spill damage caused by corrosive chemicals.

last about 30 years. In the following calculation the life of coatings in comparable service is estimated at 18 months. Using a capital recovery factor

$$i(1 + i)^n/[(1 + i)^n - 1]$$

when i = interest rate per year, or 20 percent
        n = number of years

the cost of the nickel cladding is \$2700 per year on a 30-year basis, while

Fig. 4.10 Well-maintained tank farm on which no spills occur. The tanks were last painted 5 years before they were photographed.

TABLE 4.3   Economics of Lining a Tank Car (10,000 Gal; Interior Surface, 825 Ft^2)

Protection	Years	Total cost	Cost per year*	Cost per year†
Organic coating	1.5	$1,000	$667	$927
	2.0	$1,000	$500	$655
Nickel-clad	30	$13,500	$450	$2,700

NOTE: Costs are given in 1977 dollars.
* Cost of money is not included.
† Cost of money is calculated at 20 percent a year.

the coating costs $927 per year on an 18-month basis, or $655 per year on a 2-year basis.

In this calculation the following factors have not been considered: taxes, premature obsolescence, changes in interest rates or the cost of money, effect of sustained high temperatures, cost of outage while the car is being recoated, etc. All these factors must be considered when a complete economic evaluation is made.

Many people line carbon steel tanks that are used for steam condensate. The author has long advocated not lining condensate tanks. The carbon steel piping which carries the condensate to and from the boilers is not lined, and corrosion is not severe. There certainly cannot be much oxygen in the condensate. If there were, the boilers and pipelines would corrode severely.

Under normal conditions, unlined condensate tanks last more than 15 years, and some are still in very good condition after that time. As a rule, the author does not advocate lining for protection from corrosion any tank that will perform satisfactorily 10 years or more without a lining.

An economic analysis follows. Assume that an unlined 40-ft by 40-ft (12.2-m by 12.2-m) condensate tank, installed and painted, costs $12,000. Then assume that lining the tank with a baked phenolic costs $10,500, making a total of $22,500. With an equal annual cost at 20 percent time value of money for 10 years, for the unlined tank we have:

$$i(1 + i)^n/[(1 + i)^n - 1](12,000) = (0.23852)(12,000) = \$2862 \text{ per year}$$

and for the lined tank:

$$i(1 + i)^n/[(1 + i)^n - 1](22,500) = (0.23852)(22,500) = \$5367 \text{ per year}$$

In fact, the equal annual cost of the lined tank is almost twice that of the unlined tank. Actually, the difference is much greater, because unlined tanks last considerably longer than 10 years. They will probably last as long as the piping system.

The importance of proper surface preparation and priming in corrosive environments cannot be overemphasized (Figs. 4.5, 4.11, 4.12, 4.13,

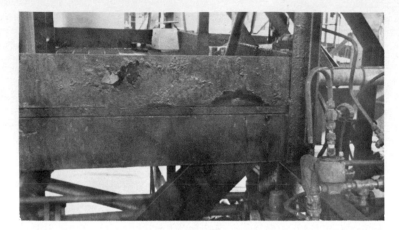

Fig. 4.11   There is heavy rust on the kickplate. While the other parts look new after 2 years, the kickplate was not properly sandblasted or primed.

and 4.14). The steel in Fig. 4.11 had been properly sandblasted, primed with inorganic zinc, and painted except for the kickplate, which had not been sandblasted. This photograph was made only 2 years after the steel was painted. Note the nice appearance of all the steel except the kickplate, which is severely corroded.

Fig. 4.12   Structural steel in good condition after 7 years of service.

Fig. 4.13 Chlorinated-rubber coating over a properly prepared and primed surface looks good after 7 years of service.

The structural steel shown in Figs. 4.12 and 4.13 had been sandblasted, primed with inorganic zinc, and coated with chlorinated rubber 7 years before these photographs were made. There was no visible rust on the steel. On the other hand, a jib crane only a few feet away (Fig. 4.14)

Fig. 4.14 There is rust on this jib crane. The vendor's painting system did not include sandblasting and priming.

began to peel after 1 year's service. The crane had been coated with the vendor's standard paint system, which for competitive reasons did not include sandblasting or zinc priming. The crane had to be sandblasted, primed with inorganic zinc, coated with chlorinated rubber, and reinstalled. It would have been much more economical to order the crane properly cleaned and painted in the first place. Surface preparation is an extremely important part of the painting system (Figs. 4.15 and 4.16).

4.2.4 Product purity   Many products do not corrode carbon steel, but for certain uses, such as containing pharmaceuticals, coatings are applied to avoid contamination by iron.

Some chemicals polymerize when only a small amount of iron is picked up; others darken upon reacting with iron. Carbon steel vessels are lined to preserve the purity of food products, glycols, phenol, sodium hydroxide, styrene, and other substances. At moderate temperatures, say, not over 150°F (66°C), thin film coatings perform satisfactorily as linings for carbon steel vessels for these materials. The linings must be resistant to the product. Furthermore, for foods and pharmaceuticals the linings should meet U.S. Department of Agriculture (USDA) or U.S. Food and Drug Administration (FDA) standards, or both. The FDA materials list was promulgated under the Food, Drug, and Cosmetics Act (Food Additives Act) of 1958. Copies are available from the FDA. The Meat Inspection Division of the USDA also lists coatings and pigments that may be used in contact with food. Specifications may be obtained from the Meat Inspection Division.

Linings for product purity need not be pinhole-free if the product is not corrosive to the substrate. Naturally, some restrictions must be put in the specifications so that the lining will not be left full of holes like a sieve. The following information should be included in the lining specifications:

▪ After curing has been completed, the owner's representative will check the lining for thickness and for pinholes.

▪ Visible discontinuities in the finished lining will not be permitted, and the minimum film thickness specified is the minimum acceptable as determined by measurement with a Mikrotest film-thickness gauge, as manufactured by the Nordson Corporation, or with an owner-approved equivalent. Areas which are determined by measurement to be less than the minimum dry thickness specified shall have sufficient paint added to attain the specified dry thickness.

▪ The maximum allowable number of pinholes in the finished lining after curing, as determined by a Tinker-Rasor MI holiday detector, shall not exceed 5 per 100 ft² of given contiguous area, with no more than 2 pinholes within any one square foot.

Fig. 4.15 The white tank shown here and in Fig. 4.16 is a classic example of the importance of surface preparation. Whereas the tank is severely rusted, there is no rust on the tank saddle, which was sandblasted, primed with inorganic zinc, and coated with epoxy. The tank itself was not sandblasted. The pictures were made 3 years after the tank had been installed and painted.

Fig. 4.16

Insertion of these provisions in lining specifications has reduced prices by as much as 20 percent below the cost of a pinhole-free lining. Naturally, when a product is highly corrosive to the substrate (steel, for example), the lining must be pinhole-free.

4.2.5   Antistick coatings   Here again the product may or may not be corrosive to the substrate. Polypropylene, polyethylene, tetrafluoroethylene (TFE), fluorinated ethylene propylene (FEP), baked phenolics, and other materials are used for their antistick properties. Polypropylene and polyethylene are usually applied as a sheet lining, and TFE and FEP as high-baked coatings. However, if corrosion resistance is desired, sheet material should be used. Baked phenolic is applied as a coating.

4.2.6   Emissivity of coatings   The emissivity of a body is the same as its absorptivity, so that a good radiator is a good absorber and a good reflector is a poor radiator. Surfaces such as bright aluminum sheet have very low emissivities and are therefore very good reflectors. White or aluminum-colored coatings are used to paint items to retard the pickup of radiant-heat energy. For example, the state of New Jersey requires that tanks of more than 10,000 gal (37.85 m^3) be painted white if the vessel contains hydrocarbon liquids whose vapor pressure is greater than 0.02 lb/in^2 (137.9 Pa). This is done to avoid losses by evaporative emissions.

To prevent radiant-heat pickup, rubber-lined tanks are painted white. This keeps the temperature somewhat lower inside the tanks, especially in the vapor zone.

Care should be used in specifying paints for radiant-heat gain or loss. When it is desirable to radiate internal heat *away* from a body, dull black should be used. For instance, insulation of high-voltage electrical lines *should* be black. A white or aluminum finish could cause insulation to retain enough heat to overload and burn a line in two.

4.2.7   Electrical insulation   Coatings with high electrical resistance are used to encapsulate items in the electrical field. Epoxies, silicones, polyethylenes, rubber, vinyls, and other types of coatings are used for electrical insulation. Coatings with good electrical resistance are needed for such items as buried pipelines. Otherwise galvanic corrosion can perforate the pipe if it is not cathodically protected. If the line is cathodically protected, excessive current will be used if the pipe coating does not have good electrical resistance.

4.2.8   Basis for selection of coating systems   There are numerous criteria for selecting coating systems. For coatings that will be exposed to the weather, climatic conditions are of prime importance. If the relative humidity does not exceed 50 percent and there are no chemical or other industrial exposures, steel need not be sandblasted. In such areas cleaning with power tools or a wire brush, followed by an oleoresinous primer and a topcoat, is ample.

Steel exposed to moderately severe environments should be sandblasted to white metal or to near white metal, primed with a zinc-rich coating, and topcoated with a coating resistant to the environment. Steel exposed to severely corrosive environments, such as marine, humid, coastal, or chemical exposures, should be sandblasted to white metal, primed with a zinc-rich coating, and topcoated with a coating suitable for the exposure.

One of the most serious problems of the chemical industry is corrosion. Oil-based paints are unsatisfactory in most chemical-industry environments. Some of the reasons are given below.

1. *Slow drying time.* Chemical dust and fumes would be deposited between coats because each would have to dry 24 h or longer before a subsequent coat could be applied. This sandwiching of chemicals between the layers not only causes the paint to deteriorate rapidly owing to chemical action but also permits the chemicals to attack the steel through thin prime coats.

2. *Lack of resistance to chemicals.* Many chemicals quickly destroy pigments and vehicles in oil paints, leaving the surface vulnerable to chemical attack.

3. *Oil solubility.* Ordinary cleaning fluids and lubricating oils dissolve oil-based paints, making equipment covered with these paints look unsightly and destroying the protective qualities of the coatings.

Although there are a number of special coatings for specific end uses, there are also many good coatings for general use. Among the best and most widely used are acrylic latex, vinyl, vinyl ester, phenolic epoxies, chlorinated rubber, epoxies, phenolics, alkyds, asphalts and coal tars, inorganic and organic zinc, and silicone alkyds.

CRITERIA FOR SELECTING COATINGS

 1. Abrasion resistance
 2. Adhesion
 3. Impact resistance
 4. Flexural qualities
 5. Resistance to a given medium
 6. Resistance to sunlight
 7. Temperature resistance
 8. Drying time
 9. Appearance
10. Wetting time
11. Applied cost
12. Antistick properties

The coating or coatings having the best properties required for a given set of conditions should be selected, provided the cost is not prohibitive. Because more than a dozen generic coating materials are available commercially and each has its own characteristics, it is important to select the best for the anticipated service. Obviously, this can be done only if the generic nature of the product is known.

Rating the relative resistance of typical commercial coatings to sunlight and weather, stress and impact, abrasion, heat, water, salts, solvents, alkalies, acids, and oxidation by using a value of 10 as the ultimate for each property gives a relative comparison of coating materials (Table 4.4).

Top-quality coatings should be compared generically, but identification by generic name per se is no guarantee of quality. Coatings should be purchased on specifications from reliable coatings manufacturers. *It is false economy to purchase a protective coating without knowing its solids content and the resin content of the solids.*

Economics plays the major role in selecting a coating system. Specify the system that performs best for the least amount of money over a given period of time. This does not necessarily mean using the least expensive or the most expensive system. Because of repainting costs and corrosion, a "cheap" system may very well cost more per year than a more expensive system. Don't gild the lily by using very expensive coatings. If, for example, a coating system performs adequately for 5 years and costs only $0.75 per square foot, a coating costing $1.50 per square foot cannot be justified economically for the same service even if the more expensive coating performs indefinitely. Money does have a time value even if it is invested in a paint system.

If the time value of money is assumed to be 20 percent per year, the equal annual cost of the two systems would be as follows for a 5-year period:

$i(1 + i)^n/[(1 + i)^n - 1](0.75) = (0.33438)(\$0.75) = \$0.251$ per square foot per year

$i(1 + i)^n/[(1 + i)^n - 1](1.50) = (0.33438)(\$1.50) = \$0.502$ per square foot per year

By redoing the moderately priced system every 5 years, the equal annual cost would be approximately $0.251 per square foot per year, whereas the expensive coating over a 15-year period (no touch-up is assumed) would cost:

$i(1 + i)^n/[(1 + i)^n - 1](1.50) = (0.21388)(\$1.50) = \$0.321$ per square foot per year

Once a suitable system has been chosen, investigate ways to lower the cost without sacrificing quality. For example, centrifugal blasting with

TABLE 4.4 Comparative Resistance Value of Typical Commercial Coating Formulations

Condition	Generic type										
	Neoprene	Vinyl	Saran	Epoxy	Chlorinated rubber	Styrene copolymer blends	Furan	Phenolic	Alkyd	Asphalt	Oil-based
Sunlight and water	8	10	7	9	7	6	8	9	10	7	10
Stress and impact	10	8	7	3	7	6	1	2	4	5	4
Abrasion	10	7	7	6	7	7	5	5	6	3	4
Heat	10	7	7	9	5	6	9	10	8	4	7
Water	10	10	10	10	10	10	10	10	8	10	7
Salts	10	10	10	10	10	10	10	10	8	10	6
Solvents	4	5	5	8	3	4	10	10	4	2	2
Alkalies	10	10	8	9	10	10	10	2	6	7	1
Acids	10	10	10	6	10	10	10	10	6	10	1
Oxidation	6	10	10	6	9	8	2	7	3	2	1
Total	88	87	81	80	78	77	75	75	63	60	43

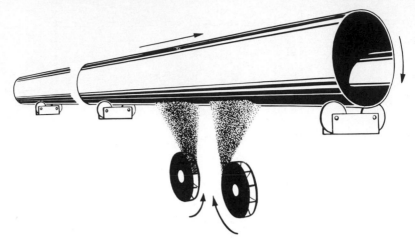

Fig. 4.17   Schematic drawing of a pipe being abrasive-blast–cleaned with centrifugal blast wheels. [*Courtesy of Wheelabrator-Frye, Inc., Mishawaka, Indiana*]

metal abrasives rather than sandblasting saves money; the same degree of cleanliness is attained at only a fraction of the cost (Figs. 4.17, 4.18, 4.19, and 4.20).

Sometimes it is possible to use two-coat high-build systems rather than

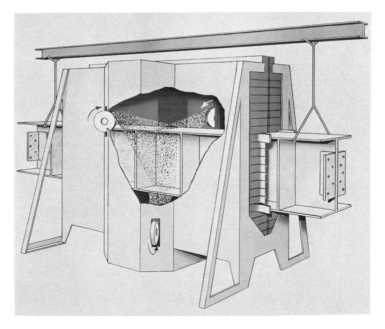

Fig. 4.18   Schematic drawing of structural steel being abrasive-blast–cleaned with centrifugal blast wheels. [*Courtesy of Wheelabrator-Frye, Inc., Mishawaka, Indiana*]

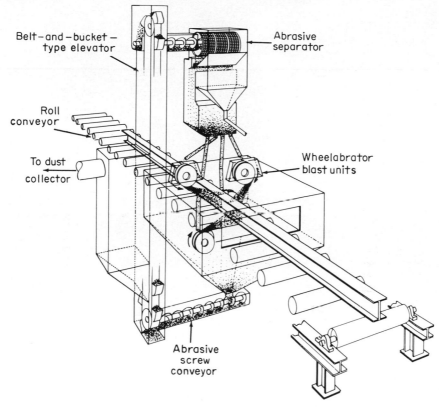

Belt-and-bucket-
type elevator

Abrasive
separator

Roll
conveyor

To dust
collector

Wheelabrator
blast units

Abrasive
screw
conveyor

Fig. 4.19   Schematic drawing of a typical four-Wheelabrator-unit structural cleaner. [*Courtesy of Wheelabrator-Frye, Inc., Mishawaka, Indiana*]

three or more thin coats, thus reducing labor costs per unit area. Shop painting and even field painting before erection can save money because painting an item after it has been erected entails lowered labor efficiency.

4.2.9 Protection required by standards   The nuclear industry has sophisticated requirements set forth in *ANSI Standard N101.2* for coatings evaluation, testing, and quality control for the interior surfaces of nuclear containments. This standard gives specific coating types having good radiation resistance and also lists coatings that are easily decontaminated.

The FDA and the USDA have regulations for coatings that will be in contact with food or drugs. For example, no lead or other lethal materials are permitted in coatings that will come in contact with food or drugs or in places where infants could ingest the peelings. When specifying coatings for food service, include a certification clause by the coating manufacturer that the coating conforms to FDA and USDA regulations.

Fig. 4.20 Large structural member being abrasive-blast–cleaned by Wheelabrator unit. [*Courtesy of Wheelabrator-Frye, Inc., Mishawaka, Indiana*]

4.3 Service Requirements

4.3.1 Systems selection for refineries, chemical plants, and sewage and water treatment plants The following painting information primarily concerns carbon steel, cast steel, and cast iron. The systems listed (Table 4.5) are designed for longtime trouble-free service.

Epoxy There are several distinct types of epoxy coatings including amine-cured and polyamide-cured epoxies for conventional and airless sprays and coal-tar epoxies, both amines and polyamide-cured. Typical epoxy coatings cover as follows: conventional, 320 ft²/gal (7.85 m²/l), 2 mils (0.0508 mm) thick; high-build, 60 to 70 ft²/gal (1.47 to 1.72 m²/l), 15 mils (0.381 mm) thick; and coal-tar epoxies, 70 ft²/gal (1.72 m²/l), 16 mils (0.4064 mm) thick.

Generic primers, rather than inorganic zinc primer, may be used for the coatings listed in the table, but for a performance of 15 years or longer inorganic zinc primer is recommended. If more than one system is shown in the table for a given exposure, all those listed may be used successfully. Furthermore, it is acceptable to specify chlorinated rubbers in lieu of epoxies in exposure to alkalies.

Epoxies are not noted for retaining good color and gloss over a long time. The color and gloss retention of an epoxy system may be improved

TABLE 4.5 Systems Selection for Refineries, Chemical Plants, and Sewage and Water Treatment Plants*

Item	Exposure	Surface preparation	Generic type Primer	Topcoats (temperature limits)	Dry film thickness (mils)		Remarks
Structural steel, including equipment supports	Fumes and spills of acids and acid salts	Near-white metal blast (NACE No. 2)†; (SSPC SP-10)‡	One coat, inorganic zinc§	Two coats, high-build vinyl (up to 150°F)	Primer / Topcoats / Total	3.0 / 6.0 / 9.0	The first coat of vinyl should be applied over the primer before exposure to acid fumes or spills
		Commercial blast (NACE No. 3 finish); (SSPC SP-6-63)	One coat, coal-tar epoxy	One coat, coal-tar epoxy (up to 200°F)	Primer / Topcoat / Total	8.0 / 8.0 / 16.0	This material is black. It is a good heavy-duty coating when looks are not important
	Alkalies and alkaline salts spills	Near-white metal blast (NACE No. 2); (SSPC SP-10)	One coat, inorganic zinc	Two coats, polyamide-cured epoxy (up to 200°F)	Primer / Topcoats / Total	3.0 / 4.0 / 7.0	The first coat of epoxy should be applied over the primer before exposure to alkaline spills
		Commercial blast (NACE No. 3 finish); (SSPC SP-6-63)	One coat, coal-tar epoxy	One coat, coal-tar epoxy (up to 200°F)	Primer / Topcoat / Total	8.0 / 8.0 / 16.0	This material is black. It is a good heavy-duty coating when looks are not important
	Solvent spills	Near-white metal blast (NACE No. 2); (SSPC SP-10)	One coat, inorganic zinc	Two coats, polyamide cured-epoxy (up to 200°F)	Primer / Topcoats / Total	3.00 / 4.00 / 7.00	This system performs well when exposed to spillage of aliphatic solvents, aromatic solvents, and some chlorinated solvents

TABLE 4.5 Systems Selection for Refineries, Chemical Plants, and Sewage and Water Treatment Plants* (Continued)

Item	Exposure	Generic type			Dry film thickness (mils)	Remarks
		Surface preparation	Primer	Topcoats (temperature limits)		
Structural steel, including equipment supports (continued)		Near-white metal blast (NACE No. 2); (SSPC SP-10)	One coat, inorganic zinc	One coat, inorganic topcoat (up to 750°F)	Primer 3.0 / Topcoat 6.0 / Total 9.0	This system is resistant to spills of strong chlorinated solvents, provided they are acid-free. It is also resistant to phenol spills
	Calcium and sodium hypochlorite spills	White metal blast (NACE No. 1 finish); (SSPC SP-5-63)	None	One coat, vinyl ester (210°F, all concentrations of calcium hypochlorite)	Total 20.0	This material has very good resistance to calcium and sodium hypochlorite (up to 15% sodium hypochlorite, up to 180°F)
Steel piping, aboveground, uninsulated	Fumes and spills of acids and acid salts	Near-white metal blast (NACE No. 2); (SSPC SP-10)	One coat, inorganic zinc	Two coats, high-build vinyl (up to 150°F)	Primer 3.0 / Topcoat 6.0 / Total 9.0	The first coat of vinyl should be applied over the primer before exposure to acid fumes or spills
		Commercial blast (NACE No. 3 finish); (SSPC SP-6-63)	One coat, coal-tar epoxy	One coat, coal-tar epoxy (up to 200°F)	Primer 8.0 / Topcoat 8.0 / Total 16.0	This material is black. It is a good heavy-duty coating when looks are not important

Exposure	Surface preparation	Primer	Topcoat	Coats	Thickness, mils	Remarks
Alkalies and alkaline salts spills	Near-white metal blast (NACE No. 2); (SSPC SP-10)	One coat, inorganic zinc	Two coats, polyamide-cured epoxy (up to 200°F)	Primer Topcoats Total	3.0 4.0 7.0	The first coat of epoxy should be applied over the primer before exposure to alkaline spills
	Commercial blast (NACE No. 3 finish); (SSPC SP-6-63)	One coat, coal-tar epoxy	One coat, coal-tar epoxy (up to 200°F)	Primer Topcoat Total	8.0 8.0 16.0	This material is black. It is a good heavy-duty coating when looks are not important
Solvent spills	Near-white metal blast (NACE No. 2); (SSPC SP-10)	One coat, inorganic zinc	Two coats, polyamide-cured epoxy (up to 200°F)	Primer Topcoats Total	3.00 4.00 7.00	This system performs well when exposed to spillage of aliphatic solvents, aromatic solvents, and some chlorinated solvents
	Near-white metal blast (NACE No. 2); (SSPC SP-10)	One coat, inorganic zinc	One coat, inorganic topcoat (up to 750°F)	Primer Topcoat Total	3.0 6.0 9.0	This system is resistant to spills of strong chlorinated solvents, provided they are acid-free. It is also resistant to phenol spills
Calcium and sodium hypochlorite spills	White metal blast (NACE No. 1 finish); (SSPC SP-5-63)	None	One coat, vinyl ester	Total	20.0	See remarks on structural steel
Hot piping, 201 to 750°F, all areas	Near-white metal blast (NACE No. 2); (SSPC SP-10)	One coat, inorganic zinc	One coat, modified silicone	Primer Topcoat Total	3.0 2.0 5.0	Apply topcoat before exposing to acid or alkaline fumes or spills

TABLE 4.5 Systems Selection for Refineries, Chemical Plants, and Sewage and Water Treatment Plants* (Continued)

Item	Exposure	Surface preparation	Generic type		Dry film thickness (mils)		Remarks
			Primer	Topcoats (temperature limits)			
Steel piping, aboveground, uninsulated (continued)	Hot piping, 751 to 1200°F, all areas	White metal blast (NACE No. 1 finish); (SSPC SP-5-63)	One coat, silicone for hot surface	One coat, silicone for hot surfaces	Primer / Topcoat / Total	1.0 / 1.0 / 2.0	Do not apply thick coats. They will spall off
Steel piping, aboveground, insulated	Fumes and spills of acids, acid salts, alkalies, and alkaline salts; process temperature, 0 to 200°F	Near-white metal blast (NACE No. 2); (SSPC SP-10)	One coat, inorganic zinc	One coat, polyamide-cured epoxy	Primer / Topcoat / Total	3.0 / 2.0 / 5.0	No coating is needed on insulated surfaces with continuous operating temperatures below 0°F or above 200°F Omit coat of epoxy when not subject to chemical spills
Buried steel piping, ambient temperature	Coating with hot tar or asphalt, and wrapping with tar or asphalt felt	Machine-blast commercial finish (NACE No. 3); (SSPC-SP-6-63)	Coal tar for coal-tar system; asphalt system	Hot-melt coal tar Hot-melt asphalt	Total Total	94 94	For various specifications refer to National Association of Pipe Coaters
	Extruded polyethylene or polypropylene	Machine-blast commercial finish (NACE	Asphalt	Polyethylene or polypropylene	Total, 6 in and up / 3 and 4 in / 2 in	40 / 35 / 30	These materials in long runs are competitive with coating and wrapping. They have

Application	Surface preparation	Primer	Topcoat	Film thickness, mils	Remarks
	No. 3; (SSPC-SP-6-63)			Under 2 in / 25	excellent electrical resistance
Powdered coating	Near-white metal blast (NACE No. 2); (SSPC SP-10)	Phenolic or epoxy	Phenolic or epoxy	Primer 1 to 1.5; Topcoat 10 to 11; Total (minimum) 12	These materials are competitive, have good electrical resistance, and have good resistance to sunlight
Fumes and spills of acids and acid salts	Near-white metal blast (NACE No. 2); (SSPC SP-10)	One coat, inorganic zinc	Two coats, high-build vinyl (up to 150°F)	Primer 3.0; Topcoats 6.0; Total 9.0	The first coat of vinyl should be applied over the primer before exposure to acids, fumes, or spills
	Commercial blast (NACE No. 3 finish); (SSPC SP-6-63)	One coat, coal-tar epoxy	One coat, coal-tar epoxy (up to 200°F)	Primer 8.0; Topcoat 8.0; Total 16.0	This material is black. It is a good heavy-duty coating when looks are not important
Alkalies and alkaline salts	Near-white metal blast (NACE No. 2); (SSPC SP-10)	One coat, inorganic zinc	Two coats, polyamide-cured epoxy (up to 200°F)	Primer 3.0; Topcoats 4.0; Total 7.0	The first coat of epoxy should be applied over the primer before exposure to alkaline spills
	Commercial blast (NACE No. 3 finish); (SSPC SP6-63)	One coat, coal-tar epoxy	One coat, coal-tar epoxy (up to 200°F)	Primer 8.0; Topcoat 8.0; Total 16.0	This material is black. It is a good heavy-duty coating when looks are not important
Solvent spills	Near-white metal blast (NACE No. 2); (SSPC SP-10)	One coat, inorganic zinc	Two coats, polyamide-cured epoxy (up to 200° F)	Primer 3.0; Topcoats 4.0; Total 7.0	This system performs well when exposed to spillage of aliphatic solvents, aromatic solvents, and some chlorinated solvents

Fabricated-steel tanks, vessels, and heat exchangers, uninsulated

TABLE 4.5 Systems Selection for Refineries, Chemical Plants, and Sewage and Water Treatment Plants* (*Continued*)

Item	Exposure	Surface preparation	Generic type		Dry film thickness (mils)		Remarks
			Primer	Topcoats (temperature limits)			
Fabricated-steel tanks, vessels, and heat exchangers, uninsulated (*continued*)		Near-white metal blast (NACE No. 2); (SSPC SP-10)	One coat, inorganic zinc	One-coat, inorganic topcoat (up to 750°F)	Primer Topcoat Total	3.0 6.0 9.0	This system is resistant to spills of strong chlorinated solvents, provided they are acid-free. It is also resistant to phenol spills
	Calcium and sodium hypochlorite spills	White metal blast (NACE No. 1 finish); (SSPC SP-5-63)	None	One coat, vinyl ester	Total	20.0	See remarks on structural steel
	High temperature, 201 to 750°F, all areas	Near-white metal blast (NACE No. 2); (SSPC SP-10)	One coat, inorganic zinc	One coat, modified silicone	Primer Topcoat Total	3.0 2.0 5.0	Apply topcoat before exposing to acid or alkaline fumes or spills
	High temperature, 751 to 1200°F	White metal blast (NACE No. 1 finish); (SSPC SP-5-63)	One coat, silicone for hot surfaces	One coat, silicone for hot surfaces	Primer Topcoat Total	1.0 1.0 2.0	Do not apply thick coats. They will spall off
Fabricated-steel tanks, vessels, and heat ex-	Process temperatures, 0 to 200°F, subject to fumes	Near-white metal blast (NACE No. 2); (SSPC SP-10)	One coat, inorganic zinc	One coat, polyamide-cured epoxy	Primer Topcoat Total	3.0 2.0 5.0	Omit coat of epoxy when surface is not subject to chemical spills. No coating is

	Exposure	Surface preparation	Priming	Finish coat	Coats and dry-film thickness, mils	Remarks
changers, insulated	and spills of acids, acid salts, alkalies, and alkaline salts					needed on insulated surfaces with continuous operating temperatures below 0°F or above 200°F
Blowers, compressors, conveyors, cranes, extruders, fans, hoists, pumps, and miscellaneous mechanical equipment:						
Cast-iron and cast-steel shelf items	Fumes and spills of acids and acid salts	Dust, dirt, grease, and other foreign matter removed	Vendor's standard paint	Two coats, vinyl	Primer 1.5 / Topcoats 6.0 / Total 7.5	Primer must be compatible with vinyl topcoat
Fabricated-steel items	Fumes and spills of acids and acid salts	Near-white metal blast (NACE No. 2); (SSPC SP-10)	One coat, inorganic zinc	Two coats, high-build vinyl (up to 150°F)	Primer 3.0 / Topcoats 6.0 / Total 9.0	The first coat of vinyl should be applied over the primer before exposure to acids, fumes, or spills
		Commercial blast (NACE No. 3 finish); (SSPC SP-6-63)	One coat, coal-tar epoxy	One coat, coal-tar epoxy (up to 200°F)	Primer 8.0 / Topcoat 8.0 / Total 16.0	This material is black. It is a good heavy-duty coating when looks are not important

TABLE 4.5 Systems Selection for Refineries, Chemical Plants, and Sewage and Water Treatment Plants* (Continued)

Item	Exposure	Surface preparation	Generic type		Dry film thickness (mils)		Remarks
			Primer	Topcoats (temperature limits)			
Mechanical equipment: Cast iron, cast steel, and shelf items	Spills of alkalies and alkaline salts; vendor's standard paint	Dust, dirt, grease, and other foreign matter removed	One coat, barrier primer	Two coats, polyamide-cured epoxy	Primer Topcoats Total	1.5 4.0 ―― 5.5	Primer must be compatible with epoxy topcoat
Fabricated-steel items		Near-white metal blast (NACE No. 2 finish); (SSPC SP-10)	One coat, inorganic zinc	Two coats, polyamide-cured epoxy (up to 200°F)	Primer Topcoats Total	3.0 4.0 ―― 7.0	The first coat of epoxy should be applied over the primer before exposure to alkaline spills
		Commercial blast (NACE No. 3 finish); (SSPC SP-6-63)	One coat, coal-tar epoxy	One coat, coal-tar epoxy (up to 200°F)	Primer Topcoat Total	8.0 8.0 ―― 16.0	This material is black. It is a good heavy-duty coating when looks are not important
Cast iron, cast steel, and shelf items	Solvent spills; vendor's standard paint	Dust, dirt, grease, and other foreign matter removed	One coat, barrier primer	Two coats, polyamide-cured epoxy	Primer Topcoats Total	1.5 4.0 ―― 5.5	This system performs well when exposed to spillage of aliphatic solvents, aromatic solvents, and some chlorinated solvents
Fabricated-steel items		Near-white metal blast (NACE No. 2 finish); (SSPC SP-10)	One coat, inorganic zinc	Two coats, polyamide-cured epoxy (up to 200°F)	Primer Topcoats Total	3.0 4.0 ―― 7.0	This system performs well when exposed to spillage of aliphatic solvents, aromatic solvents, and some chlorinated solvents

Item	Service	Surface preparation	Primer coat	Topcoat		DFT (mils)	Remarks
Mechanical equipment (*continued*): Cast iron, cast steel, and shelf items	Calcium and sodium hypochlorite spills; vendor's standard paint	Near-white metal blast (NACE No. 2 finish); (SSPC SP-10)	One coat, inorganic zinc	One coat, inorganic topcoat (up to 750°F)	Primer Topcoat Total	3.0 6.0 9.0	This system is resistant to spills of strong chlorinated solvents, provided they are acid-free. It is also resistant to phenol spills
		Dust, dirt, grease, and other foreign matter removed	One coat, barrier primer	One coat, vinyl ester	Primer Topcoat Total	1.5 20.0 21.5	Primer must be compatible with vinyl ester topcoat
Fabricated-steel items		White metal blast (NACE No. 1 finish); (SSPC SP-5-63)	None	One coat, vinyl ester	Total	20.0	See remarks on structural steel
Mechanical equipment, uninsulated, hot surfaces							See remarks on fabricated-steel tanks, uninsulated, hot surfaces
Mechanical equipment, insulated							See remarks on fabricated-steel tanks, insulated

* The coatings referred to in this table are all solvent-type (not water-thinnable), except that the inorganic zinc primer may be of either type. The epoxies referred to are catalyzed epoxies, not epoxy esters.
† National Association of Corrosion Engineers.
‡ Steel Structures Painting Council.
§ Organic zinc primer may also be used.

by applying a thin coat (1.5 mils, or 0.381 mm, dry thickness) of a urethane or vinyl acrylic enamel over the epoxy. The urethane coatings have good chemical resistance. An acrylic-epoxy may also be used. This coating has very good color and gloss retention.

Offshore drilling rigs are coated with inorganic zinc and epoxy.

Vinyls   Among the most widely used resins for protective coatings are vinyls. The basic materials for vinyl resins are available as several different copolymers, but the most frequently used is the copolymer containing approximately 87 parts by weight of vinyl chloride and 13 parts by weight of vinyl acetate. These materials may also be called vinyl chloroacetates.

When properly applied, vinyls yield a general-purpose coating resistant to splashes or fumes from water solutions of most acids, salts, and alkalies at temperatures below 150°F (66°C). They are also available commercially as emulsions.

A typical specification for a vinyl coating is vinyl resin, 22 to 27.5 percent; specific gravity, 0.9 to 0.95; viscosity, 200 to 500 centipoises (0.2 to 0.5 Pa · s). Pigments must be ground suitably so that they are completely dispersed, and the resin must be compounded so that it has adequate resistance to heat and sunlight. One gallon (3.785 liters) of a material of this type should cover a minimum of 225 ft² (20.9 m²) of surface to give a film approximately 1.7 mils (0.04318 mm) thick. Vinyl-resin coatings may contain either liquid or solid plasticizers. The latter are sometimes superior because they do not migrate from the surface over a long period.

High-build vinyl coatings specifically designed for application by airless-spray equipment are now available. In comparison with conventional vinyl coatings, fewer coats give adequate protection, thus saving labor. Coverage for a typical high-build vinyl is 220 ft²/gal (5.4 m²/l), 1.8 mils (0.04572 mm) thick. Some high-build vinyls may produce satisfactory coatings up to 5 mils (0.127 mm) thick at a coverage rate of 90 ft²/gal (2.2 m²/l).

Vinyl esters   Coatings and linings formulated from vinyl esters have excellent resistance to acids and strong oxidizing agents such as sodium hypochlorite. They also have good resistance to sodium hydroxide. These coatings are catalyzed and applied in much the same way as polyester coatings.

Inorganic zincs   These coatings give excellent protection to steel. For the best long-term protection a top coating of acrylic latex, epoxy, vinyl, or other suitable resin, depending on the exposure, should be applied over the inorganic zinc after it has been properly cured. The two basic types are postcured and self-cured coatings. There are also hydrocarbon-solvent and water-miscible types. Very good formulations are available in all types.

Inorganic zinc coatings are excellent shop primers. They are good on cold surfaces and perform excellently on hot surfaces to well above 600°F (316°C) by themselves or when used with high-heat paints. These coatings cathodically protect the steel substrate and effectively prevent underfilm corrosion. They are as close to a permanent primer as is now available. The coatings are very resistant to abrasion of cables and chains during the handling of steel. They are not, however, resistant to strong acids and alkalies, and when they are used in such exposures, they must be protected by suitable topcoats. Typical coverages of inorganic zinc coatings are 230 to 330 ft²/gal (8.1 m²/l), 3 mils (0.0762 mm) thick.

Organic zincs   Organic zinc coatings based on catalyzed epoxies, acrylics, polystyrene, and chlorinated rubbers are now available.

Silicone-alkyds   These materials have the general properties of alkyds with the following notable exceptions. The silicone-alkyds have exceptional resistance to sunlight and excellent color retention, and they are more or less self-cleaning.

4.3.2 Systems selection for nuclear and conventional power plants   If unpainted galvanized steel is not objectionable from an aesthetic standpoint, it is a very good material to use when it is not exposed to spills or fumes of acids and alkalies. For example, galvanized structural steel will perform satisfactorily for from 15 to 20 years, even in areas of high humidity, without being painted.

The equal annual cost for galvanized steel at $140 per ton for galvanizing, with an annual interest rate of 20 percent and a 15-year life for the zinc coating, would be $i(1 + i)^n/[(1 + i)^n - 1](140) = (0.21388)(140) = 29.94 per year per ton of steel. Medium construction has an average of about 250 ft² (23.2 m²) of surface per ton of structural steel. This would be equivalent to $29.94/250, or $0.12 per year per square foot of steel.

Coating systems for conventional power plants are shown in Table 4.6.

Coatings for a nuclear power plant should be the same as for a conventional power plant (Table 4.6), except for items susceptible to exposure to radioactive materials. West and Watson[4] found epoxies and vinyl to have good resistance to gamma radiation. These materials were also easily decontaminated.

For specific exposures, tests should be made in accordance with *ANSI Standard N101.2*. The reason for the tests is that some coatings of a specific generic type might perform well when exposed to nuclear radiation, whereas other coatings of the same generic type might not. Another

---

[4] G. A. West and C. D. Watson, "Gamma Radiation Damage and Decontamination Evaluation of Protective Coatings and Other Materials for Hot Laboratory and Fuel Processing Facilities," ORNL-3589, U.S. Department of Commerce, National Bureau of Standards, Clearinghouse for Federal Scientific and Technical Information, Springfield, Va.

TABLE 4.6 Systems Selection for Conventional Power Plants

Item	Exposure	Surface preparation	Generic type		Dry film thickness (mils)		Remarks
			Primer	Topcoats (temperature limits)			
Structural steel, including equipment supports	Mild; very little sulfur in the fuel; relative humidity rarely above 50 percent	Power-tool cleaning or sand brush-off blast (NACE No. 4 finish);* (SSPC SP-7-63)†	One coat, oleo-resinous primer	Two coats, silicone-alkyd (250°F)	Primer Topcoats Total	1.5 4.0 5.5	For very mild conditions, sandblasting other than NACE No. 4 finish cannot normally be justified for structural steel
	Mild; very little sulfur in the fuel; relative humidity above 50 percent most of the time	Pickling or centrifugal blast to white metal (NACE No. 1 finish); (SSPC SP-5-63)		Hot-dip-galvanized (600°F)	Total	3.5	If suitable galvanizing facilities are available and painting is not required for aesthetics, galvanizing is a very economical coating
		Near-white metal blast (NACE No. 2); (SSPC SP-10)	One coat, inorganic zinc	Two coats, chlorinated rubber (200°F)	Primer Topcoats Total	3 4 7	A silicone-alkyd system is recommended when the primary reason for painting is appearance
	Oxides of sulfur from fuels high in sulfur content (above 2 per-	Near-white metal blast (NACE No. 2); (SSPC SP-10)	One coat, inorganic zinc	Two coats, polyamide-cured epoxy (200°F)	Primer Topcoats Total	3 4 7	Sulfur oxides form sulfurous and sulfuric acids, which are very corrosive to carbon steel and other metals

Location	Environment	Surface preparation	Priming coat	Finish coats	Component	Dry film thickness, mils	Remarks
	cent); relative humidity above 50 percent most of the time						
Steel piping, aboveground, uninsulated	Mild; very little sulfur in the fuel; relative humidity rarely above 50 percent	Power-tool cleaning or sand brush-off blast (NACE No. 4 finish); (SSPC SP-7-63)	One coat, oleo-resinous primer	Two coats, silicone-alkyd (250°F)	Primer Topcoats Total	1.5 4.0 5.5	Lighter pigments will darken in temperatures above 250°F
	Mild; very little sulfur in the fuel; relative humidity above 50 percent most of the time	Pickling or centrifugal blast to white metal (NACE No. 1 finish); (SSPC SP-5-63)		Hot-dip galvanized (600°F)	Total	3.5	If suitable galvanizing facilities are available and painting for aesthetics is not required, galvanizing is a very economical coating system
		Near-white metal blast (NACE No. 2); (SSPC SP-10)	One coat, inorganic zinc	Two coats, chlorinated rubber (200°F)	Primer Topcoats Total	3 4½ 7½	A silicone-alkyd system is recommended when the primary reason for painting is appearance
	Oxides of sulfur from fuels high in sulfur content (above 2 percent); relative humidity above 50 percent most of the time	Near-white metal blast (NACE No. 2); (SSPC SP-10)	One coat, inorganic zinc	Two coats, polyamide-cured epoxy (200°F)	Primer Topcoats Total	5 4½ 9½	Sulfur oxides form sulfurous and sulfuric acids, which are very corrosive to carbon steel and other metals

TABLE 4.6 Systems Selection for Conventional Power Plants (Continued)

Item	Exposure	Surface preparation	Generic type — Primer	Generic type — Topcoats (temperature limits)	Dry film thickness (mils)	Remarks
Steel piping, aboveground, uninsulated (continued)	Hot piping, 201 to 750°F, all areas	Near-white metal blast (NACE No. 2); (SSPC SP-10)	One coat, inorganic zinc	One coat, modified silicone (750°F)	Primer 3.0 Topcoat 2.0 Total 5.0	
	Hot piping, 751 to 1200°F, all areas	White metal blast (NACE No. 1 finish); (SSPC SP-5-63)	One coat, silicone for hot surfaces	One coat, silicone for hot surfaces (1200°F)	Primer 1.0 Topcoat 1.0 Total 2.0	Do not apply thick coats. They will spall off
Steel piping, aboveground, insulated	Process temperature, 0 to 200°F	Near-white metal blast (NACE No. 2); (SSPC SP-10)	One coat, inorganic zinc or hot-dip-galvanized	None	Total 3 (3.5 if galvanized)	No coating is needed on insulated surfaces with continuous operating temperatures below 0°F or above 200°F. No coating is needed for 80 to 200°F if relative humidity is below 50 percent
Buried steel piping, ambient temperature	All areas					See Table 4.5
Fabricated steel tanks, vessels, and	Mild; very little sulfur in the fuel; relative	Power-tool cleaning or sand brush-off blast	One coat, oleo-resinous	Two coats, silicone-alkyd	Primer 1.5 Topcoats 4.0 Total 5.5	Lighter pigments will darken in temperatures above 250°F

Equipment	Environment	Surface preparation	Primer coat	Topcoat	Component	Thickness, mils	Remarks
	humidity rarely above 50 percent	(NACE No. 4 finish); (SSPC SP-7-63)	primer	(250°F)	Total	3.5	If suitable galvanizing facilities are available and painting for aesthetics is not required, galvanizing is a very economical coating system
heat exchangers, uninsulated	Mild; very little sulfur in the fuel; relative humidity above 50 percent most of the time	Pickling or centrifugal blast to white metal (NACE No. 1 finish); (SSPC SP-5-63)		Hot-dip-galvanized (600°F)			
		Near-white metal blast (NACE No. 2); (SSPC SP-10)	One coat, inorganic zinc	Two coats, chlorinated rubber (200°F)	Primer Topcoats Total	3 4 7	A silicone-alkyd system is recommended when the primary reason for painting is appearance
	Oxides of sulfur from fuels high in sulfur content (above 2 percent); relative humidity above 50 percent most of the time	Near-white metal blast (NACE No. 2); (SSPC SP-10)	One coat, inorganic zinc	Two coats, polyamide-cured epoxy (200°F)	Primer Topcoats Total	3 4 7	Sulfur oxides form sulfurous and sulfuric acids, which are very corrosive to carbon steel and other metals
Hot surfaces, 201 to 750°F, all areas		Near-white metal blast (NACE No. 2); (SSPC SP-10)	One coat, inorganic zinc	One coat, modified silicone (750°F)	Primer Topcoat Total	3.0 2.0 5.0	

TABLE 4.6 Systems Selection for Conventional Power Plants (Continued)

Item	Exposure	Surface preparation	Generic type		Dry film thickness (mils)		Remarks
			Primer	Topcoats (temperature limits)			
Fabricated steel tanks, vessels, and heat exchangers, uninsulated (*continued*)	Hot surfaces, 751 to 1200°F, all areas	White metal blast (NACE No. 1 finish); (SSPC SP-5-63)	One coat, silicone for hot surfaces	One coat silicone for hot surfaces (1200°F)	Primer Topcoat Total	1.0 $\underline{1.0}$ 2.0	Do not apply thick coats. They will spall off
Fabricated steel tanks, vessels, and heat exchangers, insulated	Process temperatures, 0 to 200°F	Near-white metal blast (NACE No. 2); (SSPC SP-10)	One coat, inorganic zinc or galvanized	None	Total (3.5 if galvanized)	3	No coating is needed on insulated surfaces with continuous operating temperatures below 0°F or above 200°F. No coating is needed for 80 to 200°F if relative humidity is below 50 percent
Stacks	Mild; very little sulfur (less than 1 percent in fuel); relative humidity rarely above 50 percent	For stack temperatures below 250°F; power-tool cleaning or sand brush-off blast (NACE No. 4 finish); (SSPC SP-7-63)	One coat, oleo-resinous primer	Two coats, silicone-alkyd (250°F)	Primer Topcoat Total	1.5 $\underline{4.0}$ 5.5	
		For temperatures from 250 to 750°F; white	One coat, inorganic zinc	One coat, modified silicone	Primer Topcoat Total	3.0 $\underline{2.0}$ 5.0	

Environment	Surface preparation	Primer	Topcoat		Mils	Remarks
	metal blast (NACE No. 1 finish); (SSPC SP-5-63)					
	For temperatures from 715 to 1200°F; white metal blast (NACE No. 1 finish); (SSPC SP-5-63)	One coat, silicone for hot surfaces	One coat, silicone for hot surfaces	Primer Topcoat Total	1.0 <u>1.0</u> 2.0	Do not apply thick coats. They will spall off
Mild; very little sulfur (less than 1 percent in fuel); relative humidity above 50 percent most of the time	For stack temperatures from 200 to 750°F; white metal blast (NACE No. 1 finish); (SSPC SP-5-63)	One coat, inorganic zinc	One coat, modified silicone	Primer Topcoat Total	3.0 <u>2.0</u> 5.0	
	For temperatures from 751 to 1200°F; white metal blast (NACE No. 1 finish); (SSPC SP-5-63)	One coat, silicone for hot surfaces	One coat, silicone for hot surfaces	Primer Topcoat Total	1.0 <u>1.0</u> 2.0	Do not apply thick coats. They will spall off

TABLE 4.6 Systems Selection for Conventional Power Plants (Continued)

| Item | Exposure | Surface preparation | Generic type | | | Remarks |
			Primer	Topcoats (tempera-ture limits)	Dry film thickness (mils)	
	Oxides of sulfur from fuels high in sulfur content (above 2 percent); relative humidity above 50 percent most of the time	Same as above; see "Remarks" column.	Same as above	Same as above	Same as above	If the stack temperature is below 300°F, sulfurous and/or sulfuric acid will condense on the inside wall of the stack, causing severe corrosion unless stack is properly lined
Blowers, boilers, compressors, conveyors, cranes, fans, hoists, pumps, and miscellaneous mechanical equipment:						
Cast-iron and cast-steel shelf items	Vendor's standard paint, all areas					No additional paint is required
Fabricated-steel items	All areas; see "Remarks" column					Clean and paint as specified for fabricated steel tanks, vessels, and heat exchangers

* National Association of Corrosion Engineers.
† Steel Structures Painting Council.

good reference is "Protective Coatings (Paints) for the Nuclear Industry," *ANSI Standard N5.9-1967*. Detailed nuclear specifications are beyond the scope of this chapter.

4.3.3 Systems selection for mining facilities   Mine waters are often dilute sulfuric acid solutions containing sulfates. Coal-mine waters contain ferric and ferrous sulfate and are very corrosive to carbon steel and to copper-base alloys. Some copper and gold mines contain sulfuric acid in the mine water and therefore are very corrosive to carbon steel.

Although stainless steel is resistant to dilute sulfur compounds, it is much more expensive than carbon steel protected by a corrosion-resistant coating. The following items may be made of carbon steel and coated as specified for mechanical equipment exposed to fumes and spills of acids and acid salts (Table 4.5): cars, hoists, skips, cages, loaders, tipplers, washers, agitators, safety ladders, chutes, dust collectors, and conveyors.

4.3.4 Systems selection for pipeline and marine applications   There are numerous coating systems for pipelines. Three of the best are listed in Table 4.5 for buried steel piping; these systems are competitive in cost. If the pipe will be exposed to sunlight for long periods of time, as in storage, or if sections are installed aboveground, a powdered coating is recommended. This coating has much better resistance to sunlight than does polyethylene. It also costs considerably less than field-sandblasted and -coated pipe.

For underground service, all three systems mentioned are good. The polyethylene and the powdered coating both have very good dielectric properties, and thus fewer amperes could be used to protect the pipe cathodically.

For marine-systems selection see Table 4.7.

4.3.5 Systems selection for interior surfaces of tanks, pipes, and miscellaneous equipment   The three principal reasons for lining or coating the interior of tanks, pipe, and equipment are

- Product purity
- Nonstick properties
- Corrosion protection

In some cases the lining may perform all three functions. If the lining or coating is applied for product purity and/or nonstick properties, it is not necessary to obtain a pinhole-free lining. Such a lining can cost as much as 20 percent more than a lining with a few pinholes. Pinholes should not be left in a lining when the product is corrosive to the substrate. For coatings and lining systems for carbon steel tanks, pipes, and miscellaneous equipment see Table 4.8.

TABLE 4.7 Marine-Systems Selection

Item	Exposure	Surface preparation	Generic type		Dry film thickness (mils)		Remarks
			Primer	Topcoats			
Steel, dock piling	Seawater	Near-white metal blast (NACE No. 2 finish); (SSPC SP-10)	One coat, coal-tar epoxy	One coat, coal-tar epoxy	Primer 8.0 Topcoat 8.0 Total 16.0		
Barges (exterior)	River water and/or seawater	White metal blast (NACE No. 1 finish); (SSPC SP-5-63)	One coat, high-build epoxy primer	One coat, high-build polyamide-cured epoxy	Primer 10.0 Topcoat 10.0 Total 20.0		
Offshore oil rigs, structural steel, piping, fabricated-steel items	Seawater	White metal blast (NACE No. 1 finish); (SSPC SP-5-63)	One coat, inorganic zinc	Two coats, high-build polyamide-cured epoxy	Primer 3.0 Topcoat 8.0 Total 11.0		From the splash zone down, one of the systems listed should perform satisfactorily. In any case, steel from the splash zone down should be cathodically protected in addition to coating

TABLE 4.8 Coatings and Linings for Carbon Steel Tanks, Pipes, and Miscellaneous Equipment

Product	Purpose of lining			Lining systems[a]								
	Product purity	Non-stick	Corrosion resistance	No. 1: epoxy	No. 2: epoxy phenolic	No. 3: odorless vinyl	No. 4: vinyl	No. 5: baked phenolic	No. 6: coaltar epoxy	No. 7: neoprene	No. 8: inorganic	No. 9: sheet rubber
Acetaldehyde					2			2			1	
Acetic acid, 10%									1		1	
Acetone	X										1	
Aliphatic esters	X										1	
Alkyl benzene	X										1	
Alum, 15%					2				1			2
Aluminum chloride, nitrate, and sulfate, 10%					1				2			
Ammonia water, 10%	X				2				1			
Ammonium chloride, nitrate, and sulfate, 10%			X	1	2		2	2	2	2		2
Aniline	X										1	
Benzene	X										1	
Benzoic acid, 10%	X			1	2		2	2		2		
Boric acid, 10%	X			1	2		2	2		2		
Butadiene	X										1	
Butane	X										1	
Butanol (butyl alcohol)	X										1	
Butyl cellosolve	X										1	
Butyl glycidol ether	X										1	
Calcium chloride, saturated			X						1	2		2
Calcium hypochlorite			X									1
Carbitol	X										1	

TABLE 4.8 Coatings and Linings for Carbon Steel Tanks, Pipes, and Miscellaneous Equipment (Continued)

Product	Purpose of lining			Lining systems[a]								
	Product purity	Non-stick	Corrosion resistance	No. 1: epoxy	No. 2: epoxy phenolic	No. 3: odorless vinyl	No. 4: vinyl	No. 5: baked phenolic	No. 6: coal-tar epoxy	No. 7: neoprene	No. 8: inorganic	No. 9: sheet rubber
Carbolic acid (phenol)	X										1	
Carbon tetrachloride	X										1	
Carbonic acid	X							2				
Castor oil	X			1	1							
Caustic soda, 10 and 20%	X			1	2		2		2	2		2
Caustic soda, 50%	X			1	2		2		2	2		2
Caustic soda, 73%[b]			X		2							
Cellosolve	X										1	
Chlorine (dry)[c]												
Chlorine (wet)[d]			X									1
Chlorobenzene (dry)	X							2			1	
Chloroform (all grades)	X							2			1	
Citric acid	X			1	2	2	2	2			1	
Crude oils, sour			X	2	2		2		1	2		
Crude oils, sweet			X	2	2		2		1	2		
Cumene	X										1	
Cyclohexane	X										1	
Cyclohexanone	X										1	
Diacetone alcohol	X										1	
Diethylene glycol	X			2	2	1	2	2			1	
Diethylene glycol ethyl ether	X										1	
Diisobutylene	X										1	
Dipropylene glycol	X			2	2	1	2	2			1	
Dow-Per and Dow-Per CS	X				3			3			3	

Chemical	C1	C2	C3	C4	C5	C6	C7	C8	C9	C10	C11
Dowanol (all grades)							3			3	
Dowclene EC							3			3	
Dowclene 10			3				3		3		
Dowfrost				3			3				
Dowfume (all grades)							3				
Dowicide 2S							1				
Dowtherm A							3			3	
Dowtherm SR-1				3			3			1	
Ethyl acetate	X							3			
Ethyl acrylate	X									1	
Ethyl alcohol	X						3			1	
Ethyl cellosolve										3	
Ethyl dichloride				3				3			
Ethyl glycol	X		2	2		2	2			2	
Ethylbenzene	X			2			1			1	
Ferric chloride		X						2			1
Fuel oil	X		2	2		2			2	1	
Gasoline	X			2		2	2			1	
Glycerin (glycerol)	X			2	2		2			1	
Heptane				3						3	
Hexane				3						3	
Hydrochloric acid (all concentrations)		X									1
Hydrofluoric acid, 20%		X									1
Hypochlorite bleach (sodium hypochlorite, 5%[e])		X									1
Isobutyl alcohol							3			3	
Isobutyl ketone										3	
Isophorone										3	
Isopropyl alcohol										3	
Isopropyl benzene										3	
JP-4 fuel										3	

TABLE 4.8 Coatings and Linings for Carbon Steel Tanks, Pipes, and Miscellaneous Equipment (Continued)

Product	Purpose of lining			Lining systems[a]								
	Product purity	Non-stick	Corrosion resistance	No. 1: epoxy	No. 2: epoxy phenolic	No. 3: odorless vinyl	No. 4: vinyl	No. 5: baked phenolic	No. 6: coal-tar epoxy	No. 7: neoprene	No. 8: inorganic	No. 9: sheet rubber
Kerosine				3							3	
Lube oil					3			3		3	3	
Magnesium oxide					3			3			3	
Mesityl oxide												
Methyl acrylate								3			3	
Methyl alcohol											3	
Methyl cellosolve											3	
Methyl chloroform					3						3	
Methyl ethyl ketone (MEK)											3	
Methyl isobutyl carbinol											3	
Methyl isobutyl ketone (MIBK)											3	
α-Methyl styrene					3			3			3	
Methylene chloride								3			3	
Methylene dichloride (dry)											3	
Milk	X					1						
Mineral oil											3	
Monobromobenzene											3	
Monochlorobenzene											3	
Monoethanolamine											3	
Monoisopropanolamine											3	
Morpholine											3	
Motor oils				3	3		3	3			3	
Naphtha							3				3	
Naphthalene											3	

Nitric acid^f								3
Octyl alcohol								3
Perchloroethylene					3			3
Petroleum								3
Petroleum ethers								1
Phenol (carbolic acid)	X	2		2				
Phosphoric acid, 85%	X	2				3		1
Pine oil								3
Polyethylene glycol^g		2		2				3
Polyethylene pellets (dry)^h	X			2				3
Polystyrene (dry)^h	X							
Polyvinyl chloride (dry)^h		1						
Propane								3
Propyl alcohol								3
Propylene dichloride^i						3	1	
Sodium chloride	X							2
Sodium hydroxide, 10 and 20%		1			1			
Sodium hydroxide, 50%	X	1	2	2	2	2	2	2
Sodium hydroxide, 73%^b	X	1	2		2		2	2
Sodium hypochlorite (see HYPOCHLORITE BLEACH)								
Styrene monomer	X							1
Sulfuric acid, 10 to 75%	X							
Sulfuric acid, 95 to 100% (dry)^j								
Trichloroethylene						2		3
Triethanolamine								1
Triethylene glycol	X	2	2	1		2	2	2
Tripropylene glycol		2	2	1		2	2	2
Turpentine								3
Vinylidene chloride			3					

TABLE 4.8 Coatings and Linings for Carbon Steel Tanks, Pipes, and Miscellaneous Equipment (*Continued*)

NOTE: An X in one or more of the first three columns indicates the purpose of the lining. A 1 in a lining-systems column indicates that this is the recommended lining. A 2 indicates that the lining is suitable for the product but is not necessarily the recommended lining. For example, the lining may be more expensive than is necessary for a given product. A 3 indicates that a lining is not required but that the subject lining will not be damaged by the product. This can be valuable information for barge and truck owners who transport one type of cargo in one direction and another type on the return trip. A blank space does not necessarily indicate that a given lining is unsuitable; it means only that the system was not considered for the product or products in question. Unless otherwise noted, the lining performance is based on a product temperature not exceeding 100°F (38°C), although many linings will perform well up to 200°F (93°C) or higher. Each case should be checked or, preferably, tested for performance above 100°F.

a Descriptions of the numbering lining systems follow:

Number	Description	Dry film thickness (mils)
1	Two-coat, 100 percent solids epoxy	20
2	Five-coat epoxy phenolic, low-baked	10
3	Five-coat odorless food-grade vinyl	10
4	Seven-coat chemical-resistant vinyl	6
5	Five-coat baked phenolic, high-baked (400°F)	4.3
6	Two-coat coal-tar epoxy	16
7	Four-coat neoprene	15.5
8	Two-coat system; one coat, lead-free inorganic zinc, and one coat, inorganic topcoat	4
9	Sheet rubber (various types; some pressure-cured, some exhaust-steam-cured; consult manufacturer for specific rubber for a given product)	3/16 in

NOTE: A white metal blast (NACE No. 1 finish) should be obtained for each lining system.

b Use nickel for long-term performance.

c No lining is required.

d Use vinyl ester FRP tanks and pipes as an alternative.

e Vinyl ester FRP should be considered in lieu of steel.

f Carbon steel is not recommended.

g For pharmaceuticals use only food-grade coating.

h Food-grade epoxy; two coats; 5 mils.

i If dry, No. 8 system is satisfactory.

j No lining is required as long as no moisture enters.

In addition to the linings in Table 4.8, pipes and pumps may be lined with Kynar,[5] Penton,[6] polypropylene, Saran, and Teflon.[7] These materials have good resistance to a large number of products. Chemical-resistance charts are available from resin manufacturers.

Prime coating   For industrial exposures the conventional shop prime coat is inadequate and often represents money wasted. Because the primed surfaces are not sandblasted and red lead and chromate generally are used, they will not stand up in chemical plants or in any other aggressive environment even if the steel is assembled and painted at once. What is worse, the steel usually is allowed to lie exposed to the weather for months and often for more than a year before being erected and painted. When this is the case, all the primer has been destroyed and corrosion and pitting have started. Although the primer has deteriorated and the steel shows considerable rust, the finish painting usually proceeds according to specifications, with one or two coats of finish over the shop coat. It is not necessary to know much about corrosion to realize that the cost of the entire paint job, including the original shop coat, is money wasted. Furthermore, holdups in construction for repainting can cost hundreds of thousands of dollars. For example, lengthy delays in the construction of a San Francisco skyscraper were caused by improper painting procedures. Architects should make sure that inspection takes place before and after steel is erected.

The author highly recommends that structural steel, steel tanks, vessels, pump and compressor bases, and other fabricated items be sandblasted to at least a near-white metal (NACE No. 2 finish), preferably at the site of the installation, and given a coat of inorganic zinc. Items subject to strong acid or alkali spills should be coated with a primer appropriate to the exposure. For very mild corrosion conditions, when topcoats will be applied within 2 months, oleoresinous primers are satisfactory.

An inorganic zinc primer provides years of corrosion-free protection to steel. It offers several advantages over other primers:

1. *Simplicity.* When properly cured, it is compatible with vinyls, epoxies, acrylic latices, chlorinated rubber, silicones, and other coatings. It is suitable for surfaces that are to operate below ambient temperatures, and some formulations can operate at up to 1200°F (649°C).

2. *Excellent weathering properties.* Inorganic zinc does not deteriorate rapidly on exposure to the weather, as do most other primers; so only a

---

[5] Trademark for polyvinylidene fluoride; Pennwalt Plastics Department, Pennwalt Corporation.

[6] Trademark for a chlorinated polyether; Hercules Powder Company.

[7] Trademark for tetrafluoroethylene (TFE) fluorocarbon resins and for fluorinated ethylene propylene (FEP) resins; E. I. du Pont de Nemours & Co.

minimum of touch-up priming is necessary before topcoating. Items coated with inorganic zinc should not be exposed in acid or alkali areas without a suitable topcoat to protect the zinc.

3. *Good heat resistance.* A single coat of high-heat paint over the inorganic zinc primer will give excellent results on surfaces that operate at 600°F (316°C). Thus this material is suitable for vessels that operate at, say, 400°F (204°C) on the bottom and −30°F (−34°C) or colder at the top.

4. *Additional advantages.* Inorganic zinc primers protect steel under insulation in the critical range (32 to 150°F, or 0 to 66°C) without topcoats, greatly extend the service life of surface preparation,[8] and cathodically protect the steel substrate. When properly used, they are more economical than other conventional primers.

However, there are several disadvantages. The first cost is greater than that of standard primers such as epoxies, vinyls, chlorinated rubbers, and wash primers; the material is more difficult to apply than standard or conventional primers; and it is not resistant to strong acids and alkalies and therefore requires topcoating before being placed in such services.

Economic comparison   When inorganic zinc is specified for structural steel and miscellaneous steel items, it is good practice to allow the vendor to galvanize and phosphatize part or all of the items because galvanizing is cheaper on items that have a large number of square feet per ton than is inorganic zinc–coated steel. One study shows that the economic break-even point between galvanizing steel and coating it with inorganic zinc is about 167 ft^2 (15.5 m^2)/ton; that is, steel with more than 167 ft^2/ton would be cheaper to galvanize, and steel with less than 167 ft^2/ton could be inorganic zinc–coated more cheaply. The break-even point varies with the actual relative costs of galvanizing and inorganic zinc–coating.

After erection, steel should be washed free of mud and other dirt and spot-blasted to remove rust, and all bare spots and scratches should be spot-primed. If regular sandblasting equipment cannot be used because of space requirements or possible damage from flying sand and dust, special equipment should be used for spot cleaning. This equipment sucks the sand and dust back into the system. It is usually good practice to apply an additional prime coat and follow it as quickly as possible with one or two finish coats, depending on the material used and the job specifications.

Sandblasting or centrifugal blasting and priming of new steel may

---

[8] Actual field use proved that piping, vessels, and structural steel primed with inorganic zinc and topcoated with three coats of vinyl could last 15 years in a very corrosive area (other steel in the same area that had not been primed with inorganic zinc was repainted five times). Although subject to frequent spills of sodium chloride brine, thus far there was not a speck of rust on the items primed with inorganic zinc.

seem expensive, but when the cost is compared with the cost of the steel and amortized over 10 to 15 years (or more with some preventive maintenance), the first painting becomes an inexpensive form of insurance against corrosion.

4.4 Nature of Substrate   Much of the data discussed thus far relates to carbon steel. Other substrates are discussed below.

4.4.1 Metals

Aluminum   This metal is prepared for painting by solvent-cleaning the surface to remove any traces of oil or grease. It is then treated with a 10% solution of phosphoric acid and rinsed with clean water. There are also several proprietary surface treatments to prepare aluminum for painting. Another method of preparing aluminum for painting is anodizing.

Once the surface has been prepared, the aluminum can be painted in much the same way as steel, except that an inhibited primer should be used rather than inorganic zinc.

Aluminum is used increasingly as a structural material and for other purposes, including piping and tanks, in modern industrial plants. Ordinarily it does not require coating for atmospheric protection, but it does need protection from most acids and alkalies. When, however, aluminum must be coated, as in jacketing over insulation on a piping system in a highly corrosive area, the correct coating should be specified. Alkyds and polyamide-cured epoxies perform excellently on aluminum surfaces. Aluminum also is now available precoated in colors with vinyls, epoxies, and phenolics when color is necessary or when the environment demands it.

Aluminum and copper are used extensively in industrial plants for electrical busbars. Buses coated with epoxies maintain a high current-carrying capacity because the high emissivity of the epoxy coating increases the radiation of heat from the bars and also protects them from corrosion.

Because of an unfavorable galvanic relationship, copper-, arsenic-, or mercury-bearing paints (i.e., antifouling coatings) should not be applied to aluminum.

Copper   Copper is prepared for painting by treating the surface with a solution consisting of 5% ferric chloride and 5% muriatic acid at commercial concentration in water. (Caution: Wear protective goggles and clothing. Muriatic acid, i.e., hydrochloric acid, is hazardous.) After the material is allowed to react for a few minutes, the surface should be washed with fresh water.

An inhibited primer rather than inorganic zinc should be used for the various systems discussed heretofore.

Galvanized surfaces   These surfaces are prepared for painting in the manner specified for aluminum.

Magnesium   Magnesium-alloy surfaces are prepared for painting by chemical or anodic treatments; the latter are the more protective.[9]

4.4.2 Concrete and masonry

Concrete   Surfaces must be clean, dry, and free of previously applied coatings and of disintegrated or chalky material. All imperfections such as water and air pits in poured concrete surfaces should be corrected by filling with cement grout as follows:

- Smooth the concrete surface, breaking down all rough protrusions.
- Apply cement grout (1 part fine sand and 2 parts cement) by sacking or by working the grout into the surface with a handstone.
- Cure for 3 days, keeping the surface damp at all times.
- Lightly stone the concrete-grout surface with a Carborundum brick to remove any rough areas.

This procedure is necessary to provide a pore-free coating over the surfaces. Rough areas in the concrete or pinholes allow the penetration of corrosive chemical reagents. In breweries the smallest pinhole is a haven for organisms that could upset the manufacturing process. Inspect pinholes carefully after concrete (especially prestressed concrete) has been sacked. Use a knife or other tool to assure that a large cavity does not exist behind the pinhole. Employ experienced personnel to do testing on jobsites.

Concrete surfaces, whether original or cement-grouted, should be acid-etched to remove glaze and concrete laitance as follows:

- Etch with a solution of 1 part commercial concentrated hydrochloric acid and 2 parts clear water.
- Apply the acid solution to concrete by a brush or a garden spray until the solution runs.
- The concrete must be well wet with acid. It will bubble for ¼ to ½ min and then stop, at which point the acid is neutralized.
- When etching is complete, wash the surface with clean water, using a garden hose. Brush during washing with a stiff brush to remove concrete salts. A properly etched surface should be slightly granular and free of glaze. On very dense machine-troweled surfaces several applications of acid may be necessary to get proper tooth or slightly granular surface.
- Dry the surface thoroughly. A moist surface will not allow proper adhesion of the coating.

---

[9] Refer to ASTM D-1732 for a method of preparing magnesium-alloy surfaces for painting.

Concrete surfaces may also be suitably prepared for painting by light sandblasting.

Polyamide-cured–epoxy coatings and urethane coatings are good for laboratory floors subject to chemical spills. Both these materials have good chemical resistance as well as impact and abrasion resistance. The epoxy coatings have good resistance to nuclear radiation and are used for floor toppings when concrete is susceptible to exposure to radioactive materials. The topping (epoxy grout) should be at least ¼ in (6.35 mm) thick with a seal coat of epoxy at least 10 mils (0.254 mm) thick without aggregates.

When concrete floors are subject to spills of strong acids and alkalies, such as sulfuric acid, hydrochloric acid, and sodium hydroxide, expensive floor toppings are not recommended because of the difficulty in maintaining a continuous topping without cracks. The acids or alkalies will reach the concrete through cracks in the topping and destroy it. This can be dangerous. In the case of sodium hydroxide, it is usually more economical to replace or patch the floor at intervals rather than to apply epoxy grout.

Let us assume that the concrete floor costs \$0.60 per square foot in place and that an epoxy-grout topping costs \$2.50 per square foot. If the concrete alone will last 5 years, as indicated below, the topping can hardly be justified if the time value of money is 20 percent.

Concrete-floor equal annual cost:

$$(\$0.60) \left[\frac{i(1 + i)^n}{(1 + i^n - 1)} \right]$$
$$= (0.60)(0.33438) = \$0.201 \text{ per square foot per year}$$

Concrete-floor-plus-topping equal annual cost:

$$(\$0.60 + \$2.50)(0.33438) = \$1.04 \text{ per square foot per year}$$

Even if the topping performed indefinitely, it would still have a higher equal annual cost than the concrete alone.

Concrete floor plus topping, 20 percent, 100 years; equal annual cost:

$$(\$0.60 + \$2.50)(0.20) = \$0.62 \text{ per square foot per year}$$

Thus the concrete floor plus topping has an equal annual cost more than 3 times as high as that of concrete alone.

When spills of acids can corrode concrete foundations, the concrete may be protected by acidproof brick, which is beyond the scope of this book.

Masonry   Surfaces are prepared for painting by removing dust, dirt, and other foreign matter. Porous surfaces such as lightweight concrete blocks should have a filler applied before painting if they are exposed to the weather. Inside walls do not require filling except in cafeterias and washrooms, where it is desired to seal or smooth wall surfaces.

4.4.3 Wood surfaces   By comparison, very little preparation is required for wood surfaces that are to be painted. Exterior wood surfaces should be clean and dry. An oleoresinous undercoat should be applied before a finish coat of silicone-alkyd. If wood is exposed to chemicals, it should be painted similarly to steel in the same area, except, of course, that no sandblasting or priming with inorganic zinc is necessary. Epoxies, chlorinated rubbers, and vinyls may be applied directly over clean, dry wood. As a result of pollution regulations that limit hydrocarbon emissions, some water-based primers are on the market.

Certain woods, such as redwood, are resistant to acids and therefore do not need protection. However, all woods need protection from strong alkalies such as sodium hydroxide. Treating wood with pentachlorophenol will extend its life by retarding bacterial growth.

Interior wood surfaces should be clean and dry. Nails should be set and holes puttied before an enamel undercoater is applied. This is followed by a coat of enamel.

4.5 Basic Functions of Coatings Systems

4.5.1 Corrosion protection   Alkyds and silicone-alkyds have excellent durability but only fair chemical resistance. They should not be exposed to strong solvents, acids, or alkalies. Coal-tar epoxies, being black, are used principally for corrosion protection from acids, alkalies, and salts. Chlorinated-rubber coatings have good acid and alkali resistance and also retain color and gloss well. Epoxies have excellent resistance to alkalies, many acids, salts, and solvents.

Epoxy phenolics have excellent alkali resistance and are used as linings for tank cars containing caustic soda of both 50 and 73% concentrations at temperatures as high as 260 to 270°F (127 to 132°C). Many cars are coated with phenolic epoxies in this service. The same coatings have excellent solvent resistance and often are used as linings for tanks or tank cars when there is a combination of exposure to solvents and to caustics or other chemicals.

Though not resistant to alkalies and impact, baked-phenolic coatings have been used for many years for drums and containers holding mildly corrosive materials such as aqueous formaldehyde and food products. This coating prevents iron contamination of the stored liquid. Phenolic coatings are applied as liquid resins dissolved in alcohol and are dried and baked at temperatures around 300 to 400°F (149 to 204°C).

Terpene-phenolic resins have color retention and chemical resistance superior to those of straight phenolics. Oil-soluble modified phenolic resins with suitable pigments are used to protect underwater structures and to afford atmospheric protection of steel. Although these coatings are not very chemical-resistant, they do have good solvent resistance compared with oleoresinous coatings.

Urethanes have good resistance to acids and alkalies. They also have good gloss and color retention, but they tend to be more expensive per mil-square foot applied than epoxies.

Vinyls have excellent resistance to many acids and alkalies, as well as to some solvents, but not to chlorinated solvents. Vinyl ester is one of the very few coatings resistant to hypochlorites.

4.5.2 Selection for appearance   Some coatings used for corrosion protection, such as silicone-alkyds, chlorinated rubbers, epoxies, acrylic-epoxies, urethanes, and vinyls, look good. However, epoxies and vinyls tend to lose their gloss and color. A color coat of vinyl acrylic or urethane is often applied over an epoxy system to give good corrosion resistance while maintaining gloss and color. This combination is less expensive than a straight urethane system. An acrylic-epoxy system has very good color and gloss retention. The appearance of coal-tar epoxy can be improved by applying a coat of polyamide-cured epoxy over it.

4.5.3 Selection for safety and identification   This objective requires a selection of colors rather than a particular type of paint. However, there are fluorescent paints that stand out and are easily seen. These materials are very expensive and do not last long when exposed to the weather or to chemicals.

Section 1910.44 of the Department of Labor's *Occupational Safety and Health Standards* sets rules and regulations for safety color codes for marking physical hazards. The regulations specify the exact colors to be used. These colors are defined in *ANSI Standard Z53.1-1971*.

4.5.4 Selection for specific cases   Coatings containing cuprous oxide or an organic tin pigment are used for barge and boat hulls to help keep barnacles off. These materials are poisonous.

An abrasive such as pumice, clean sand, or alumina thrown in almost any coating while it is still wet will provide a nonskid surface. Alkyd, chlorinated-rubber, and epoxy coatings are available with abrasives already in them.

4.6 Application Limitations and Costs

4.6.1 Application conditions   Although a few coatings can be applied over damp surfaces, they are exceptional. Surfaces normally must be dry and free from contamination. As a rule, catalyzed epoxies should not be applied when the air or the substrate is below 50°F (10°C) unless the coating has been specifically formulated for curing at low temperatures.

Painting should not be permitted when the temperature is above the autoignition point of the paint solvent being applied or below the coalescing point of emulsions in water-based paints. Caution should be used in specifying inorganic zincs that do not cure readily when the relative humidity is less than 50 percent for extended periods.

4.6.2 Material limitations   Catalyzed materials (epoxies, neoprenes, vinyl esters, urethanes, etc.) have a limited pot life. The higher the temperature, the shorter the pot life. Placing the spray pot or other container in cold water or cracked ice can extend pot life for several hours. Some coatings such as coal-tar epoxies can be applied in thick coats, but others cannot. Coatings such as chlorinated rubbers, epoxies, and vinyls with thixotropic agents can be applied in thicker coats on vertical surfaces than would otherwise be possible. Chlorinated rubbers tend to become brittle on aging.

4.6.3 Specific application problems   Some vinyls, especially high-build materials, give problems such as bubbling when applied over inorganic zinc. This can be remedied by applying a tie coat or by first applying a mist coat of the vinyl, followed promptly by a full coat. In working with a specific product, it is a good practice to use the coating manufacturer's printed application instructions. Most coating companies are ready to provide assistance when needed.

Tanks that are to be field-lined with a high-baked phenolic (400°F, or 204°C) should be placed on insulating concrete, such as a 1 : 2 : 4 mix of Lumnite, Haydite, and vermiculite. Otherwise it will be extremely difficult to cure the phenolic coating on the floor of the tank.

4.6.4 Applied cost of coatings   In evaluating protective coatings on an economic basis, all costs including the time value of money must be considered. There are three generally used methods: capital recovery with a return, discounted cash-flow rate of return, and the capitalized-cost method. All these methods give essentially the same results.

Although case histories are useful in providing costs, they do not tell the whole story. For example, let us assume that a coating system costs $1 per square foot and will last 10 years and that another system costs $0.50 per square foot and will last 5 years. It appears that protection by either system would cost $0.10 per square foot per year, but this is true only if money has no time value, which it certainly does have.

Let us assume that the time value of money in this case is 15 percent per year. Now let us compare the two systems.

System 1:

$$(\$1.00) \left[\frac{i(1 + i)^n}{(1 + i)^n - 1} \right] = \text{equal annual cost}$$

$$(\$1.00)(0.19925) = \$0.19925 \text{ per square foot per year}$$

where $\dfrac{i(1 + i)^n}{(1 + i)^n - 1}$ = capital recovery factor

$$i = 15 \text{ percent}$$
$$n = 10 \text{ years}$$

System 2:

$$\left[(\$0.50) + (\$0.50)\dfrac{1}{(1 + i)^n} \right]\dfrac{i(1 + i)^n}{(1 + i)^n - 1} = \text{equal annual cost}$$

$[(\$0.50) + (\$0.50)(0.4972)](0.19925) = \$0.14$ per square foot per year

where $\dfrac{1}{(1 + i)^n}$ = present worth of paint job 5 years hence

Same system ($0.50 per square foot)
where $i = 15$ percent
$\quad\quad n = 5$ years

and

$$\dfrac{i(1 + i)^n}{(1 + i)^n - 1} = \text{capital recovery factor}$$

where $i = 15$ percent
$\quad\quad n = 10$ years

It is now seen that System 1 costs 33.6 percent per year more than System 2. In this case, therefore, System 2 is more economical than System 1.

If the protective properties of coatings of the same generic type are the same and the percentage of solids by volume varies, the following calculation is helpful to get a comparable cost:

$$C = \dfrac{P}{5 \times 16.04}$$

where C = cost per square foot, 1 mil thick, in dollars
$\quad\quad P$ = dollars per gallon
$\quad\quad S$ = percentage of nonvolatile solids by volume

The approximate applied costs of some coatings are given in Table 4.9. These costs include surface preparation.

4.7 The Coating Specification

4.7.1 Basis for specification   Reliable technical information is required to write the specification and performance history before selecting any coating or protective system. The specifier must know the chemical and physical properties of the generic system as well as materials limitations.

Field application procedures, including detailed instructions to the applicator, are a necessary part of the specification.

4.7.2 Understanding of conditions Before the specification is written, the specifier must understand clearly what is to be painted, the service environment, and the desired duration (temporary or permanent installation).

TABLE 4.9 Applied Cost of Coatings

Coating type	Dry film thickness (mils)	Cost per square foot
Alkyd and silicone-alkyd	5.5	$0.66
Chlorinated rubber and inorganic zinc	7	$0.83
Epoxy and inorganic zinc	7	$0.83
Coal-tar epoxy	16	$0.91
Vinyl and inorganic zinc	9	$0.94

NOTE: Costs will vary with time and with location. The costs shown are for the Gulf Coast area for the year 1972. The reader may make adjustments to these costs by consulting McGraw-Hill's *Engineering News Record,* which gives labor and other costs for the current year and previous years. Also refer to the 1980 *Dodge Manual for Building Construction Pricing and Scheduling.*

SPECIFICATION DETAILS

Specifications should be specific. They should include at least the following items:

1. *Surface preparation.* State whether items are to be hand-cleaned, power-tool–cleaned, sandblasted, or prepared by some other means. If sandblasting is to be done, specify the surface finish required, such as NACE No. 1 sandblasting to white metal, NACE No. 2 sandblasting to near-white metal, NACE No. 3 sandblasting to a commercial finish, or NACE No. 4 brush-off blast-cleaned finish. Visual standards for abrasive-blast–cleaned steel surfaces may be obtained from the National Association of Corrosion Engineers, Houston, Texas, and other organizations.

2. *Material.* Specify the generic types of coatings to be used, and give names and manufacturer's numbers of acceptable materials.

3. *Application.* State whether the material is to be brushed, sprayed, or rolled. If the choice is optional with the applicator, say so.

4. *Number of coats.* Specify the minimum number of coats desired.

5. *Film thickness.* Specify the minimum dry film thickness per coat and the total minimum dry film thickness. Also, specify the type and make of film-thickness gauges and holiday detectors that will be used for inspecting the coating.

6. *Drying time.* If the matter is critical, specify the minimum or maximum drying time between coats.

7. *Minimum temperature.* Specify the minimum temperature at which painting can be done or the dew-point requirement.

8. *Manufacturer's application instructions.* State that the paint manufacturer's printed application instructions shall be followed unless additional coats, film thicknesses, etc., are to be applied. Mixing, thinning, application, and curing, if required, shall be in accordance with the manufacturer's printed instructions.

9. *Scope.* The scope of the work should be very clearly defined. What is to be painted, and what is not to be painted? What is to be furnished by the contractor or vendor? What will the owner furnish? The quality of work desired should also be mentioned.

10. *Special conditions.* It is much easier for contractors or vendors to give sensible bids when they are completely familiar with all details. For example, if there are restrictions due to plant operations, be sure contractors or vendors know about them in advance. This is especially important for safety considerations including fire hazards.

11. *Comments on specifications.* Normally, specifications should not tell how a job should be done unless this is necessary for clarification. They should specify the end results desired. The following statement tells *how* to perform the work and for the most part is meaningless: "The workmanship shall be of the best quality." This sounds good, but what does it mean? "Best" as compared with what? Does the writer really intend to compare this work with all other work to see if it is the best? If not, why so specify? Even if a comparison were made, the owner would have one idea about what was best and the vendor, contractor, or fabricator would have another idea. Both, in their own opinions, would be right. As clearly as possible use measurable engineering quantities (mil thickness, etc.) to assure compliance.

12. *Surface preparation.* The writer needs to specify the end results desired, such as NACE No. 1 white metal finish. He or she also needs to specify the maximum anchor pattern, or surface-profile depth, if this is critical. Be ready actually to measure this depth or to prepare and cite NACE or other standards that will produce the desired anchor pattern on a given substrate by using specific abrasives, air pressure, nozzles, etc. For example, a specification may call for a machined-steel surface with a maximum roughness of 63 μin (1.6 μm). This surface is then specified to be sandblasted and coated with a plastic coating without degradation of the surface finish. This requirement is not feasible. Don't contradict yourself.

13. *Inspection.* Specify the type and make of instruments that will be used to inspect the coating. Mention the calibration of the instru-

ments. Different instruments measure to different depths into the surface profile. Also set timetable milestones to ensure that tests are done at the proper points in the process of surface preparation and application (e.g., when the film is wet, when it is dry, when the first coat is applied, etc.). Do not permit the job to continue until after each checkpoint, or milestone, has been approved.

4.7.3 Types of specification   There are three basic types of specification: open, restricted, and closed.

Open specification   This is a performance specification in which conditions are stipulated and must be met as a minimum according to coverage, surface preparation, and generic type. All coating suppliers can offer their materials. Often the cheapest coating material is selected by the coating applicator. This consequence may not be the intent of the specifier, and it could result in a sacrifice in quality.

Restricted specification   This specification normally requires a material or a system by generic type, specific standard, or military specification. A few suppliers are listed with their product names to pin down the desired system. Coating contractors bid on the same limited number of materials and therefore depend on their efficiency to apply a system economically.

Closed specification   Only one coating system and one supplier are written into the specification. This specification is often a client requirement. It tends to increase costs because of the absence of competition. If application difficulties are encountered, there is little choice but to remedy them rather than switch to competitive materials.

EXAMPLE OF SPECIFICATION

1. Bid information
   1.1 The bid shall be a lump-sum proposal for furnishing all supervision, material, labor, tools, and equipment to perform the painting as specified herein.
2. Special conditions
   2.1 Bidders' attention is directed to the fact that this work will take place in an operating plant. Safety rules of the XYZ plant will apply at all times. These rules include the wearing of approved safety helmets, monogoggles, and safety glasses. Smoking or carrying matches is not permitted in Area 21.
   2.2 Rubber boots and rubber gloves must be worn in the XYZ building.
   2.3 Spark-producing tools may be used on all this work except the following: Area 21.

2.3.1 A daily safe work permit will be required in this area. Safe work permits will not be required daily in other areas.

2.4 Chemical respirators are required in all areas and will be supplied by the contractor.

2.5 At the completion of this work, signs shall be stenciled with ¾-in-high letters and numbers on each tank and vessel and on each structure and pipeway, in locations designated by the paint inspector, giving the following information:

2.5.1 Painted: month, year

2.5.2 Area No.

2.5.3 Work Order No.

2.5.4 System No.

2.6 The item number shall be stenciled on each tank, vessel, and piece of equipment, whether or not the number was on the item before it was painted. These numbers shall be of the same size and color as those on tanks and vessels in the area.

2.7 All painted signs and numbers on items that are to be painted shall be marked and tagged securely before being painted and shall be restenciled after the items have been painted.

2.8 The contractor shall provide storage and change-room facilities, including a portable sanitary toilet.

2.9 In areas where gratings must be removed, the openings shall be roped off during the day, and the contractor shall place the gratings back in position before leaving at the end of the workday.

2.10 Drawing No. ABC-1, issued with these specifications, shows area numbers referred to herein.

3. Scope of work

3.1 This work shall include cleaning and painting the following:

3.1.1 Area 21 XYZ building interior.

3.1.1.1 Include all insulation, hose racks, piping, structural steel, conduit, instruments, and all other exposed metal surfaces.

3.2 Omit all gratings in all areas, but gratings shall be removed to paint supporting steel and shall be replaced when painting has been completed.

4. Painting

4.1 Surface preparation of steel items

4.1.1 Sandblast to white metal (NACE No. 1 finish).

4.1.2 All valve stems, glass surfaces, and bolt threads shall be lubricated before painting is done so that paint will not adhere to them. After painting has been completed, the lubricant shall be removed from glass surfaces.

4.2 Application

4.2.1 All coats of paint on cable trays shall be brushed. Care shall be exercised to keep paint off electric cables.

4.2.2 Insulation shall be painted as follows:

4.2.2.1 Two coats of PVA emulsion shall be applied in accordance with Specification 123.

4.2.3 On all items except insulation, apply one coat of Z primer in accordance with Specification 134 and two coats of epoxy in accordance with Specification 156.

4.2.4 After all pipe and nozzle flanges have been cleaned and painted, PVC tape of proper width shall be applied around the flanges, sealing the space between them and covering the edges of the flanges. Two color coats of paint shall be applied over the tape.

4.2.5 Zinc-rich primers may be spray-applied. All other first-coat primers shall be brushed. All other coats may be brushed, rolled, or sprayed if overspray can be controlled.

4.2.6 Care and protection of special surfaces

4.2.6.1 Valve stems shall not be painted.

4.2.6.2 All glass surfaces such as gauge and sight glasses, etc., shall be protected from paint and shall be left clean and kept free of paint.

4.2.6.3 All identification nameplates shall be kept free of paint.

4.2.7 Minimum film thicknesses specified are the minimum acceptable dry film thicknesses of each paint or paint system as determined by measurement of a flat surface with XYZ film-thickness gauge, as manufactured by the ABC Corporation, or owner-approved equivalent. Areas in any coat of paint which are determined by measurement to be less than the minimum dry thickness specified shall have sufficient paint added to attain the specified dry thickness.

4.3 Color schedule

4.3.1 The color schedule shall be the same as now exists.

BIBLIOGRAPHY

Architectural Specification Manual, sec. 09900: "Painting," Painting and Decorating Contractors of America, Specification Services, Falls Church, Va., 1975.

Bacon, R. H., and V. B. Volkening: "Color Can Revolutionize Your Plant," *Petroleum Refiner,* January 1957.

Johnson, Keith: *Polyurethane Coatings,* Noyes Data Corp., Park Ridge, N.J., 1972.

Kline Guide to the Paint Industry, 1980, Charles H. Kline & Company, Inc., Fairfield, N.J., 1980.

Protective Coatings for Atmospheric Use: Their Surface Preparation and Application, Physical Characteristics, and Resistances, National Association of Corrosion Engineers, Katy, Tex., 1957.

Report 6A256 on Epoxy Resins, National Association of Corrosion Engineers, Katy, Tex., 1956.

Roberts, Aaron Gene: *Organic Coatings, Properties, Selection, and Use,* U.S. Department of Commerce, National Bureau of Standards, Government Printing Office, Washington, 1968.

Steel Structures Painting Manual, vol. 2: *Systems and Specifications,* Steel Structures Painting Council, Pittsburgh, Pa., 1964.

Theusen, H. G.: *Engineering Economy,* 2d ed., Prentice-Hall, Inc., Englewood Cliffs, N.J., 1957.

Treseder, R. S.: *NACE Corrosion Engineers Reference Book,* National Association of Corrosion Engineers, Katy, Tex., March 1980.

Weaver, Paul E.: *Industrial Maintenance Painting,* 4th ed., National Association of Corrosion Engineers, Houston, Tex., 1973.

West, G. A., and C. D. Watson: *Gamma Radiation Damage and Decontamination Evaluation of Protective Coatings and Other Materials for Hot Laboratory and Fuel Processing Facilities,* ORNL-3589, Clearing House for Federal Scientific and Technical Information, National Bureau of Standards, Springfield, Va., 1965.

Chapter **5**

Surface Preparation: Part I

GUY E. WEISMANTEL
Chemical Engineering

5.1 Introduction   The quickest way to achieve paint failure is improper surface preparation. It is just as important to qualify a surface-preparation contractor and applicator as it is to specify a painting system. In fact, with labor costs rising, the cost of paint becomes a minor factor in total costs (Table 5.1), especially when service life and maintenance are considered. The specification writer must give serious thought to surface preparation and inspection procedures to assure a properly prepared surface. Selection criteria for a contractor should follow strict qualification guidelines, which might include interviews and approvals of the project manager and painting foremen, with assurances that specific individuals will run the job. The quality of surface preparation and surface repair on a new surface significantly affects the amount of preparatory work that will be required for all subsequent repaints during the useful life of the surface. Adequate surface preparation and surface repair are the most important requirements to get maximum economy and durability from a paint system.

TABLE 5.1 Painting Costs

Portion of job	Percentage of total
Paint application	25–60
Surface preparation	15–40
Cleanup	5–10
Accessory products	3–6
Paint	10–20

Because the results of surface preparation and repair are quickly concealed by the first coat of paint, the effects of a poor specification or poor workmanship are not usually evident until premature paint failure develops. Some appearance defects such as lapping and poor sheen uniformity, however, may be apparent as soon as the job has been completed.

The surface preparation and surface repair required on a structure, whether new or repaint, are best determined by a paint inspector or a reliable painting contractor. Construction defects or designs that minimize the effectiveness of a paint system should be corrected or the system be modified to minimize adverse effects. Quite often paint failure is attributable to poor design practice (Fig. 5.1). A good painting contractor examines all drawings and specifications for the project and advises the architect, owner, and project engineer of any conflict between the contractor's work and that of other trades as well as of errors, omissions, or impractical design details that may contribute to paint failure. This examination and report can be part of prebid conferences or be integrated as a preliminary step in the construction or painting project.

Specifications for surface preparation and surface repair should take into consideration health, safety, and hazards as required by the Occupational Safety and Health Act (OSHA). The requirements of OSHA will sometimes be the final hurdle in choosing procedures for surface preparation and repair. This is especially true of sandblasting, which is outlawed in some areas and is suspected of being a serious health hazard.

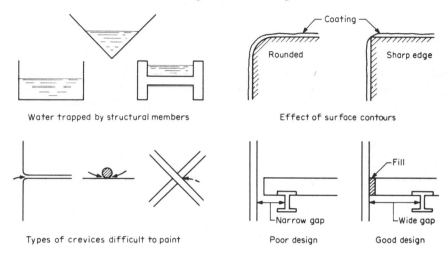

Fig. 5.1 Design pitfalls that may cause paint failure.

5.2 General Specifications   To present the proper information to all parties concerned, be as specific as possible. The contractor needs to know what is expected. Do not expect an estimator to be a mind reader. Do not assume that details will automatically be covered by some all-inclusive phrase. Most contractors will employ the most economical method of surface preparation, and their inspection procedures to qualify the surface as satisfactory for painting may be quite different from what the specification writer has in mind or even from what is called for. Paint specifications frequently are lengthy, and the same details should appear in the surface-preparation portion of the specifications.

This chapter covers specific surface-preparation procedures, but the following subsections give a general overview of surface-preparation considerations.

5.2.1 Surface examination   No exterior painting or interior finishing shall be done under conditions which jeopardize the quality or appearance of painting or finishing.

5.2.2   Preparation   All surfaces shall be in proper condition to receive the finish. Woodwork shall be hand-sandpapered and dusted clean. All knotholes, pitch pockets, or sappy portions shall be shellacked or sealed with knot sealer. Nail holes, cracks, or defects shall be carefully puttied after the first coat with putty matching the color of the stain or paint.

5.2.3   Interior woodwork   Finishes shall be sandpapered between coats. Cracks, holes, or imperfections in plaster shall be filled with patching plaster and smoothed off to match adjoining surfaces.

5.2.4   Plaster or masonry   Plaster or masonry shall be dry before any sealer or paint is applied. After the primer-sealer coat is dry, all visible suction spots shall be touched up before succeeding coats are applied. Work is not to proceed until all such spots have been sealed. In the presence of high-alkali conditions, surfaces should be washed to neutralize the alkali. If the paint to be applied is of the latex-emulsion type, wash with a 4% solution of tetrapotassium pyrophosphate (TTPP) in water: 5 oz (0.14 kg) dissolved in 1 gal (3.785 l) of water. Be sure to rinse the surface to remove all TTPP. If the paint is the conventional oil type, wash with a zinc sulfate solution: 3 lb (1.36 kg) of zinc sulfate to 1 gal of water. Then rinse.

5.2.5   Metals   Metals shall be clean, dry, and free from mill scale and rust. Remove all grease and oil from surfaces. Wash unprimed galvanized metal with a solution of chemical phosphoric metal etch and allow it to dry.

5.2.6   Concrete and brick surfaces   These surfaces shall be wire-brushed clean. Surfaces which are glazed or have traces of parting compound on them shall be sandblasted or acid-etched. The acid etch, if specified, shall be composed of an acid detergent and dilute muriatic acid. For this treatment, mix 1 part concentrated muriatic acid and 4 parts water and add 1 part detergent. Then, remove the acid with a water rinse. Upon completion of the sandblasting or acid-etching the architect shall be notified in writing by the painting contractor that all glaze or parting compound has been completely removed so that inspection can be carried out.

Concrete stains resulting from the weathering of corroded metals can be removed with a solution of 2 oz (0.057 kg) of sodium metasilicate in 1 gal (3.785 l) of water. Stained areas on weathered surfaces should be thoroughly wet with water before the solution is applied. Gentle rubbing with a cloth wet with the solution will generally remove the stain, but in severe cases two applications may be required.

5.3 Cleaning Methods   Current procedures include hand or power wire-brushing, flame cleaning, power tools, acid cleaning, water blasting, and sandblasting (Table 5.2). A commercial blast removes most of the paint from a previously painted surface except that which adheres tightly. A brush-off blast removes only loose paint and other material and generally is not recommended for complete repair jobs. The chosen cleaning method affects system-maintenance painting costs over the years, and the specification should call for the correct abrasive for the job requirements (Tables 5.3 and 5.4).

TABLE 5.2   Surface Preparation

Type of service	Service	Surface preparation (Order of preference; 1 is best)				
		Water blast	Sand-blast	Acid-etching	Sack	Hand
Poured concrete or precast slab	Light-duty maintenance	3	5	4	1	2
	Heavy-duty maintenance	NR^a	1	NR^a	3^b	2^c
	Architectural	2	NN	3	1	2
Concrete block	Chemical or architectural	2	NN	NN^d	. . .	1
Gunited surface	Chemical or architectural	2	4	3	. . .	1
Floors	Chemical or architectural	NR	2	1	. . .	3^{e,f}

SOURCE: National Association of Corrosion Engineers, *Surface Preparation Handbook*, copyright 1977 and used by permission.

NOTE: NR = not recommended; NN = not needed.

^{a} Not recommended for dense concrete, as it cannot open voids, but may be used in some instances.

^{b} Generally not recommended for use with typical tank-lining materials because the sack coat is not as strong as the base concrete, but may be used with flexible coatings such as vinyls.

^{c} Hand impact tools will do an adequate job but are not recommended, as they are too slow.

^{d} May be required if chemically contaminated with oil or grease. These must be removed before coating. Several strong alkaline cleaners which are adequate for this purpose are available.

^{e} May be used if concrete is hard and irregular such as a broom-finish concrete.

^{f} Power scarification may also be used on smooth floors.

TABLE 5.3 Abrasives Used for Blasting

Abrasive	Location	Natural or manufactured	Major chemical component	Generic type	Shape	Bulk specific gravity	CAB* abrading number (mils), 45°	CAB* breakdown percent of sample	CAB* breakdown etching (mils), 90°	CAB hardness
Steel shot	Cleveland, Ohio	Manufactured	Iron	Metallic	Spherical	7.28	46	0	19	100
Steel grit	Cleveland, Ohio	Manufactured	Iron	Metallic	Angular	7.65	115	0	45	100
Chilled iron grit	Cleveland, Ohio	Manufactured	Iron	Metallic	Angular	7.40	154	8	116	92
Pure aluminum oxide	Worcester, Mass.	Manufactured	Aluminum	Oxide	Cubical	3.80	101	24	75	76
Reclaimed aluminum oxide	Worcester, Mass.	Manufactured	Aluminum	Oxide	Cubical	3.76	92	34	81	66
Garnet	Fernwood, Idaho	Natural	Iron-silica	Oxide	Cubical	4.09	102	46	80	54
Mineral slag	Hammond, Ind.	Manufactured	Silica-alumina iron	Slag	Cubical	2.79	99	61	83	39
Mineral shot	Birmingham, Ala.	Manufactured	Silica-alumina iron	Slag	Round	2.78	53	71	54	29
Flintbrasive	Joplin, Mo.	Natural	Silica	Silica	Very angular	2.61	103	67	75	33
Flint-Missouri	Joplin, Mo.	Natural	Silica	Silica	Very angular	2.61	107	67	77	33
Walnut shell	Los Angeles, Calif.	Natural	Cellulose	Vegetable	Cubical	1.30	0	14	0	86
Aplite	Piney River, Va.	Natural	Silica alumina	Oxide	Irregular	2.72	36	78	6	22
Flint shot	Ottawa, Ill.	Natural	Silica	Silica	Round	2.63	46	57	30	43
Wedron	Wedron, Ill.	Natural	Silica	Silica	Round	2.63	42	57	25	43
Yellow jacket	Bay City, Wis.	Natural	Silica	Silica	Round	2.62	56	62	36	38
Portage	Phalanx, Ohio	Natural	Silica	Silica	Cubical	2.62	77	74	50	26
Silica sand	Valley Park, Mo.	Natural	Silica	Silica	Cubical	2.61	63	77	37	23
All-purpose sand	Bridgeton, Mo.	Natural	Silica	Silica	Angular	2.64	53	90	28	10
Albast	Birmingham, Ala.	Natural	Silica	Silica	Angular	2.63	76	85	48	15
Silica sand	Mauricetown, N.J.	Natural	Silica	Silica	Angular	2.63	75	90	51	10
Texblast R	Romayor, Tex.	Natural	Silica	Silica	Cubical	2.63	61	76	41	24
Texblast C	Corrigan, Tex.	Natural	Silica	Silica	Angular	2.61	80	88	65	12
Clemtax No. 3	Eagle Lake, Tex.	Natural	Silica	Silica	Angular	2.62	89	78	55	22
Silica sand	San Jacinto, Tex.	Natural	Silica	Silica	Angular	2.63	65	84	45	16
Silica sand	Pearl River, La.	Natural	Silica	Silica	Angular	2.64	58	79	35	21
Silicon carbide	Worcester, Mass.	Manufactured	Silicon carbide	Carbide	Blocky	3.81	114	57	84	43
Glass beads	Jackson, Miss.	Manufactured	Silica	Silica	Round	2.78	44	35	42	65
Mullite	Dillwyn, Va.	Natural	Alumina silicate	Oxide	Angular	2.68	37	88	24	12
Brady sand	Brady, Texas	Natural	Silica	Silica	Round	2.62	64	89	36	11
Crystal amber	Monterey, Calif.	Natural	Silica	Silica	Angular	2.61	83	85	61	15
Clemco Monterey	Monterey, Calif.	Natural	Silica	Silica	Angular	2.61	83	85	61	15
Corncob grit	Chicago, Ill.	Natural	Cellulose	Vegetable	Angular	1.26	0	12	0	88
Standard sand	Davenport, Fla.	Natural	Silica	Silica	Angular	2.62	65	84	45	16
Staurolite	Starks, Fla.	Natural	Staurolite	Silica	Round	3.60	...	...	...	...
Calcium carbonate	Joplin, Mo.	Natural	$Al_2O_3 \cdot CaCO_3$	Calcium	Angular	2.10	...	...	...	...

SOURCE: Clemtex, Ltd., Houston, Texas; National Association of Corrosion Engineers, Houston, Texas.

* CAB (cabinet for abrasive breakdown) values relate breakdown and abrading characteristics of various abrasives as obtained from a standard testing cabinet.

TABLE 5.4 Recommended Abrasives as Used in Abrasive Blast Cleaning

Recommended service	Silica abrasives	Slag sand	Slag shot	Flint abrasives	Natural mineral abrasives	Synthetic abrasives	Special abrasives	Vegetable abrasives	Glass abrasives	Chilled iron grit	Chilled iron shot	Annealed abrasives	Steel grit abrasives	Steel shot	Staurolite	Calcium carbonate
General blast cleaning when abrasives can be recycled and reused economically	X	X	X	X	X	Y	X	X		Y	X	X				
General blast cleaning when abrasives cannot be economically reclaimed	Y	X	X	X											X	X
Premetallizing blasting	X	X		Y	X	X				Y		X	X			
Blasting when metal tolerances cannot be changed							X	Y								X
Blasting in rooms and cabinets	X	X	X	X	X	Y	X	X	X	Y	X	X			X	
Blasting when elimination of food contamination of nonmagnetic abrasives are required								Y								
Blasting to obtain a high luster on aluminum, brass, etc.							X	X	Y							
Liquid Hone-Hydro Hone-Wet hone blasting	X				X	Y	Y	Y	Y							
Centrifugal wheel blasting												X	Y	Y*		
Maintenance work at critical parts, seals, joints, and threads																X

SOURCE: Clemtex, Ltd., Houston, Texas.

NOTE: X = most commonly used abrasives; Y = preferred abrasives to use.

* Round shot alone can peen a surface, so it is good practice to combine shot with some grit.

PAINT REMOVERS

Federal Specification TT-R-251: *Paint Remover* (*Organic Solvent Type*) lists most of the special kinds of paint removers generally available to the painting contractor as follows:

Type I. Flammable mixture with paintable retardant
Type II. Flammable mixture with nonpaintable retardant
Type III. Nonflammable mixture with paintable retardant

In addition to specification-type removers, the "flush-off," or "water-rinsable," type is widely available on the market. The flush-off type is based on nonflammable methylene chloride as the solvent surfactant (surface-active agent) and emulsifier which make it water-flushable, a cellulosic thickener to give it body or viscosity, and usually some paraffin wax to retard evaporation. Such removers are safe and very effective but, like all paint removers, are messy to use. The flammable type of paint remover is based on combinations of benzol and acetone with a paraffin-wax evaporation retardant. Such removers are hazardous from both a health and a flammability standpoint.

5.3.1 Sandblasting   There are three general methods of sandblasting:

1. *Conventional dry sandblasting.* The sand is not recycled. Dust respirators and other safety precautions are taken. For environmental reasons restrictions on dry blasting are becoming increasingly severe.

2. *Vacuum sandblasting.* This method reduces health hazards and recovers the sand. It is more costly and less efficient than dry blasting, but its efficiency can be increased by holding the vacuum cone at a slight distance from the surface. The vacuum method is useful inside shops and in areas where dust might damage machinery.

3. *Wet sandblasting.* This method reduces the dust hazard and may be required by legal restrictions. The wet sand and paint residues accumulate on ledges and other flat areas, necessitating a rinsing operation.

5.3.2 Wire-brushing and scraping   Power and hand wire-brushing are used mainly on small jobs, in cleaning small areas after sandblasting, and on surfaces for which sandblasting is not feasible. Hand scraping is used on small areas, in places where access is difficult, and for final cleanup after other methods have been employed.

5.3.3 Power tools   Power tools such as rotary wire and disc tools, rotary impact chippers, and needle scalers may be used if sandblasting is not feasible.

5.3.4 Water blasting

Water blasting is a clean and effective method for the removal of old paint from masonry surfaces that is becoming more widely used. It is generally acceptable for health and environmental requirements. Water blasting is the most efficient method of cleaning large masonry surfaces now in use, and it is often the preferred blast method for underwater or marine work.

5.3.5 Chemical methods

Acid-etching Use a 5 to 10% solution of muriatic acid, with or without a detergent, to roughen dense, glazed surfaces. This technique is also suitable for parting and curing compounds. It is important to rinse thoroughly acid-etched surfaces to remove the residual soluble reaction products of calcium and magnesium chloride, which affect the adhesion and stability of latex paints in particular.

Paint removers Both conventional solvent-based and water-rinsable types may be used to remove old paint. Paint removers, particularly of the water-rinsable type, are useful if contamination from sand or paint residues must be avoided. Most paint removers contain some wax, which must be removed completely before repainting. Wax inhibits the drying of paints and destroys adhesion.

Steam cleaning Steam cleaning with or without detergents is frequently used in food-packing plants. A mildewcide such as sodium hypochlorite is usually added. Low-pressure steam cleaners are available for use on walls in homes and offices.

Alkali cleaning Alkali cleaners are usually based on trisodium phosphate (TSP) and other detergents. Their effectiveness is increased by using them in temperatures ranging from 150 to 200°F (66 to 93°C). Alkali cleaners should not be used on masonry surfaces adjacent to aluminum, stainless steel, or galvanized metal. A thoroughly clean water rinse is essential, for residual alkali and detergents can cause greater damage than the original soil if they are not removed completely.

5.4 Masonry Surfaces

"Masonry" is a rather loose term designating a great variety of nonwood and nonmetallic surfaces. One definition is: "Masonry refers to any construction material either made with or held together by cement. It includes concrete, concrete block, cinder block, brick, stone, stucco and cement-asbestos board.[1]

5.4.1 Repairing masonry

Surface preparation and repair procedures for new surfaces are determined by the type of construction and workmanship. Surface preparation and repair of previously painted surfaces

---

[1] Federation of Societies for Coatings Technology, *Recent Developments in Architectural and Maintenance Painting*, Philadelphia, 1968, p. 7.

are determined by structural failures and by the quality of the original paint system and the degree of deterioration.

Unpainted masonry   The surface of exterior masonry shall be free of all oil, grease, loose paint, or other foreign matter. Defective or improper previous coatings must be removed by scraping or sandblasting or be properly prepared with a suitable surface conditioner. (Surfaces previously painted with water-emulsion or water-thinned finishes should be cleaned out and patch-filled with mortar similar to the original surface and uniformly textured.)

NOTE: A waiting period of from 60 to 90 days shall be allowed for new brick, stucco, and masonry surfaces to dry before painting. All suction spots or "hot spots" in plaster shall be touched up before successive coats are added.

When the topcoat is a conventional alkyd-based floor paint, new concrete floors shall stand for 90 days before painting and then be etched with a solution of muriatic acid formed by adding 1 pt (0.473 l) of acid to 1 gal (3.785 l) of water. Care should be taken to etch the surface uniformly. After the etching, the surface should be flushed with clean water, allowed to dry for 3 days, and then given the first coat of paint thinned with 1 pt of thinner per gallon. When water-based floor paints are specified, surfaces should be etched with the same solution of muriatic acid as for alkyd-based paint. Care should be taken to etch the surface uniformly. After the etching, the surface should be flushed with clean water, allowed to dry until damp, and then given a coat of latex paint.

Painted masonry surfaces   If the existing finish is sound and shows only normal chalking, sand lightly and scrub with a stiff brush, using a solution of trisodium phosphate in hot water (2.5 lb/5 gal, or 59.9 kg/m³), which will properly clean most surfaces. Rinse surfaces well, allow to dry, and apply one coat of a penetrating masonry conditioner. If blistering, cracking, or peeling exists, repair or replace all loose or defective cement and stucco. Cut open hairline cracks, and clean and fill them with exterior patching compound. If required, texture the repaired areas to match the adjoining cement or stucco. Apply one coat of surface conditioner or penetrating masonry primer to the repaired areas, allow it to dry, then coat with the finish specified.

5.4.2   Repair materials

Concrete patch   This is essentially a cement-sand mix and may contain hardening accelerators. Latex may be added to improve adhesion and flexibility.

Portland-cement grout   This is basically a mixture of cement and fine sand. Its properties may be modified by adding hardeners or retardants to increase the working time.

Fill coats for masonry   There are two generally used types:

- Latex type based on vinyl-acrylic or acrylic latices. This is the commonly used type.
- Epoxy type based on epoxy-polyamide resin. It provides maximum chemical resistance and hardness.

Caulking compounds   Caulking compounds are intermediate in flexibility and adhesion between putty and glazing compounds and sealants. There are several types:

1. *Oil-based.* These compounds continue to harden because of the oxidation of the oil and have a limited service life.
2. *Butyl rubber.* These compounds possess good flexibility and adhesion but are subject to degradation in sunlight.
3. *Acrylic latex.* These compounds have good flexibility and a long service life. While relatively new, they are rapidly expanding in use.
4. *Silicone.* Silicone compounds are very durable and have a very low temperature coefficient.

Sealants   These materials are usually made from polysulfide or polyurethane resins. Sealants based on these resins are the best available materials from the standpoint of adhesion, flexibility, and durability. They are commonly used in certain types of wall construction.

Patching plaster   This is gypsum plaster modified with lime or other materials to increase working time.

Joint cements   These materials are used primarily for filling drywall nailheads and joints with a tape reinforcement. Most joint cements are made with polyvinyl acetate latex. If the joint cement is a dry powder, spray-dried polyvinyl acetate latex is used. If it is a liquid ready to use, polyvinyl acetate latex is used.

Glazing compounds   Glazing compounds are used mainly for setting glass in a wood or metal sash. Formulations may be used on oil-modified resins or latices.

Putty   Made with oil or oil-modified resins, putty is used to set glass in a wood sash and to fill small holes such as nailheads in wood after setting. If latex-based topcoats are used, the putty should be thoroughly dry to prevent glossy spots caused by the bleeding through of uncured oil.

Plastic wood   Plastic wood is made from wood dust and a nitrocellulose vehicle. Dust from various types of wood is used to facilitate color matching.

Epoxy-resin compounds   Patching compounds based on epoxy-polyamide resins are useful when maximum chemical resistance and hardness are required, as on concrete floors.

5.4.3 Physical and chemical properties of masonry surfaces   The physical and chemical properties of masonry surfaces are important in determining the type of surface preparation to be used as well as the paint system to be specified. All masonry surfaces that contain cement or lime mixed with cement are alkaline. Surface alkalinity gradually decreases with aging through the reaction of carbon dioxide in the air with calcium hydroxide to form calcium carbonate. Newly formed cracks or other breaks in the surface require a new aging period of 6 months or more to reduce alkalinity. While not often used in today's formulations, chrome green and chrome yellow pigments burn out when in contact with alkalies. With their disuse and the widespread use of alkali-resistant latex and chlorinated-rubber vehicles, aging masonry surfaces longer than 30 days or pretreating them with a wash of muriatic acid is not always necessary. Still, some masonry surfaces such as concrete blocks with mortars high in lime content or cinder blocks like those common in Arizona construction might require alkali-resistant primers.

The porosity of masonry surfaces varies greatly. Cinder blocks, concrete blocks, and similar unit masonry usually require a filler to provide waterproofness and a good, paintable surface. The smoothness or profile of masonry surfaces also varies widely, particularly with poured concrete and tilt-up concrete. Concrete floors are usually very smooth and may have a hard-glazed surface, depending on the concrete mix and the troweling of the surface. Extremely smooth, hard surfaces may require light sandblasting or etching with muriatic acid.

The best concrete contains a minimum amount of water. A 30-day waiting period is desirable to permit the curing of concrete and the evaporation of unreacted surface water before repairing for painting.

5.4.4 Problem masonry   There are usually one or more contaminants on new concrete that require treatment or removal. They include the following:

Laitance   Laitance is a powdery mixture of cement and sand (an incompletely hydrated cement scum) that is usually caused by poor curing conditions, too much water, or poor troweling. Laitance is similar to a chalky paint surface and should be removed, particularly on floors, before painting. Since the bonding and strength of laitance are poor, any coating applied over it will eventually peel. For the best coating performance, laitance should be removed by whip sandblasting, dry abrasion, vacuum blasting, abrasive blasting with water injection, high-pressure water blasting (with or without an abrasive), or acid-etching. Concrete that is to be coated should be finished with a wooden float rather than steel trowels, which induce glazing. In general, acid-etching is most effective in preparing horizontal surfaces such as floors, while sandblasting is better suited to walls.

Efflorescence   Efflorescence represents deposits of soluble calcium and magnesium salts that are usually found around cracks which permit the entrance of water. The hydration of these salts represents forces of the order of 100,000 lb/in² (690 million Pa) and must be removed before painting, and the source of moisture eliminated. Even after concrete has been coated, if water continues to enter it, efflorescence can develop and disrupt a paint film. Efflorescence on concrete or brickwork retaining walls can be removed with muriatic acid. Its recurrence can be prevented with sound waterproofing.

Surface voids and bug holes   Voids and bug holes should be treated as part of surface repair and filled with a concrete patch.

Surface glaze   Surface glaze affects paint adhesion, particularly on floors. It is due to the particular quality of concrete used or to finishing by steel troweling. The glaze can be removed by sandblasting or by acid-etching and rinsing.

Curing and parting compounds   Most curing compounds that are sprayed on concrete to retain moisture are poor anchor coats for paint and should be removed. In the tilt-up concrete construction common in the Pacific area, a combination curing and parting compound which can contain wax is used so that a slab poured on top of another slab will not stick to it. The compound, which may be as thick as ⅛ in (3.175 mm) in low areas, is sprayed on the first flat slab. While curing and parting compounds deteriorate fairly rapidly outdoors, thick spots are likely to remain more than 6 months on the north side of a building. Laitance may be combined with a parting and curing compound. Clear sealers are sometimes used to bond the compound to the concrete, but the best procedure is removal by sandblasting or by detergent acid-etching and rinsing.

Form oils   Form oils are usually mineral oils and generally are present in thin films that weather rapidly. If necessary, remove form oils with painter's thinner or a detergent wash and thorough rinsing.

In recent years, a major source of failure in masonry coatings has been nonadherence caused by a film of form grease or oil that has been transferred from the forms to the concrete surface. Such films are used as freeing agents to facilitate the removal of forms. Because even acid-etching does not clean form films satisfactorily, they should not be matched with surfaces that are to be given protective coatings for chemical immersion.

Form oils can contain large amounts of wax or soap, which is used because it strips easily. A residuum of these compounds remains on the concrete surface, however, and can be removed only by drastic measures such as sandblasting, detergent scrubbing, or flame volatilizing. The effect of waxes or soaps on coatings is the same as on forms: the finished coating peels away readily from the surface (waxes used as curing compounds produce the same effect). If concrete is to be coated for corrosion protection, the form should preferably be coated with a lacquer or a hard epoxy that remains on the form when it is dismantled. Moreover, curing compounds should not be applied to tank bottoms, floors, or other areas that later are to be coated. Evidence of form-oil problems shows up quickly as cratering or "fisheyes" in the finished coating. Because the coating cannot wet the oil, it flows away from the oil-saturated area. Recent use of nonmigrating epoxy-polyamide form-release agents and curing compounds has helped to eliminate this difficulty, yielding concrete surfaces for which no further preparation is necessary. Epoxy topcoats may be applied immediately after curing.

Mildew   All mildewed surfaces should be scrubbed clean. One procedure is to blend thoroughly three heaping teaspoons of trisodium phosphate, 1½ cups of bleach, and 1 gal (3.785 l) of water. Use a scrub brush to apply the mixture, and allow the scrubbed area to sit for 15 min. *Rinse thoroughly,* allow to dry, and paint as soon as possible to prevent reinfection.

Cement work   The quickest way to ruin a concrete surface is to permit workers to apply a cement wash just prior to painting. Cement washes bond poorly to concrete because of different hydration rates, and they eventually powder or flake off, taking top paint with them. Do not specify or allow this procedure.

Concrete hardeners   Concrete hardeners are sometimes used to increase surface hardness, decrease permeability to liquids, and provide added resistance to mild chemical attack. They are made from sodium silicate or from magnesium, zinc, or lead fluorosilicates. Hardeners should not be used when coatings are to be applied, because they migrate or float to the pour surface, giving it a glossy, grayish brown appearance. Organic coatings will not adhere to inorganic hardeners, and the latter cannot be removed by acid-etching. The only way to prepare a hardened concrete surface properly to receive paint is to sandblast it.

5.4.5 Special problems with concrete   Air pockets or bubbles persist in the surface of all concrete, although good vibration and working techniques minimize the number of resulting surface imperfections. Small surface holes and pits can be covered with a coating applied by spray, brush, or roller, but many of them remain uncoated because air or solvent in the holes expands to cause blistering. If at first the coating bridges the void space, the film dries from both sides and shrinks, leaving a hole. Subsequent coats meet the same fate. When there are air pockets beneath a thin layer of cement that has little strength, coatings applied over this layer can fail from mechanical damage or from a loss of adhesion. These pockets should be opened by mechanical means or by whip sandblasting. The open pockets are then filled with cement plaster, thick organic coatings, or cement-latex coatings applied by brush, trowel, or squeegee.

"Sacking" a surface (dressing it with a Carborundum stone and cement grout to fill voids) is useful if the procedure is performed soon after the forms have been removed; timing is crucial to ensure that the pour surface and grout have roughly equal hydration times. The proper procedure is first to brush on a coarse mortar and then to rub with a clean wet sack to remove surface scum and film. Sacking can produce a very smooth surface that, depending on the coating system, can require the application of one·or more very thin wash coats of the initial paint. This procedure facilitates penetration and adhesion. Before any surfacing material is applied, however, holes and voids should be opened by blasting or by power-sanding.

Sacking is not recommended for tank linings. For tank linings or for immersion service, high-solid organic mastics can be squeegeed into voids; the lining should be a coat of the same material. Although this type of sealing is more expensive, it is the best for chemical immersion service, which requires a nonporous, nonflaking surface finish.

Many concrete slabs encounter ground moisture, and during severe rainfalls water vapor can penetrate exposed walls. If it is suspected that an underground or below grade concrete surface is subject to excessive capillary moisture or hydrostatic pressure, a simple check should be made. The condition can be spotted by placing a rubber mat or a plastic sheet over the finished concrete surface for 10 h. If moisture collects under the cover, the surface should be sealed to prevent paint peeling as well as loose floor tiles, musty odors, mildew, rusting metal, and other problems common to dampness. Air conditioning can aggravate the moisture problem by lowering the inside temperature below the dew point.

Never paint incompletely dried concrete with a nonbreathing film coating. Moisture trapped within the coating will ultimately find its way out and, in so doing, dislodge the impermeable film. Ideally, one coating surface should be a breather film if the other is to be impermeable.

Before a solvent-based coating can be applied to a concrete surface, the surface should be dry throughout at least the top $1/16$ in (1.588 mm) of depth. Dry concrete appears light gray in color. If there is any question about dryness, use a commercial probe-type moisture meter to check the moisture content.

For concrete floors subject to hydrostatic pressure use a vapor barrier. Choose the material on the basis of moisture conditions and building-use requirements rather than capital cost. The permeability rating is a measure of the rate of vapor transmission (Table 5.5); low readings provide the best vapor barrier. However, using a vapor barrier does not eliminate the need for proper drainage. The American Concrete Insti-

tute publishes details on service-life comparisons and installation recommendations.

Concrete itself is subject to numerous failure mechanisms, all of which
contribute to coating failure if the surface is painted. Typical problems
include the following:

- Cracking (inadequate cure, freeze-thaw action, improper control
joints, improper reinforcement, premature troweling, etc.)
- Curling (too thin a slab, improper joint distance or curing)
- Shrinkage (high winds, low humidity, improper mix, etc.)
- Dusting (excessive clay, improper troweling or finishing, etc.)
- Blotchiness (batch contamination, variations in water-cement ratio,
etc.)
- Crazing (improper troweling or curing, overuse of vibration, etc.)
- Scaling (freezing, inadequate air content, etc.)
- Blistering (air, too high a sand or cement content, lean mix, etc.)

Although complete concrete specifications and troubleshooting are
beyond the scope of this handbook, the specification writer must consider potential concrete problems in drawing up paint specifications.

TABLE 5.5   Vapor Barrier Permeability Ratings

Material	Permeability rating
⅛-in (3.175-mm) asphalt panels	0.000
Butyl rubber sheeting	0.002
55-lb roofing felt	0.03–0.08
4–6 mil (0.1016–0.1524 mm)	
polyethylene sheeting	0.10
15-lb roofing felt	0.6–2.0

5.4.6 Inspection   Qualified inspectors should check surface conditions
prior to painting. Specifications should cover surface-preparation approvals in great detail (Table 5.6).

To ensure the proper surface preparation and application of coatings,
the American Society for Testing and Materials Committee D01.43 has
set forth guidelines in its *Proposed Manual of Coating Work for Nuclear
Power Plant Primary Containment Facilities.* It has written sections on
concrete-surface preparation, on the qualification of journeyman painters for containment coatings, and on the inspection of concrete coatings.

Each inspector should have a checklist for the various hold points
during concrete-surface preparation. The inspector should identify the
curing compound, hardeners, form oil, spackling compounds, and
grouts and caulking compounds while recording the condition of the

concrete, noting the following:

Honeycomb or porosity
Roughness or smoothness
Cracking
Efflorescence
Laitance
Looseness
Contaminants
Temperature and humidity
Excessive moisture
Type of cleaning method and abrasive used (sweep blasting, sandblasting, acid-etching, etc.)

It is possible to monitor coating adherence by using an adhesion tester such as the commercially available Elcometer. Epoxy glue is used to hold the test dolly in place; when the dolly is pulled free, the type of failure and the tensile strength of the adhesion are recorded.

The performance of all coating-application work should comply with the Occupational Safety and Health Act (29 C.F.R. Parts 1910 and 1926). The Occupational Safety and Health Administration will furnish material safety data sheets, or they may be requested from the coating supplier. These sheets are required if certain materials are applied in confined work areas.

5.5 Plaster Surfaces   The most common types of plaster are gypsum plaster, acoustical plaster, and putty-coat or high-lime plasters. Before painting, allow about 30 days for proper curing and the reduction of moisture content. Roughen high-lime plasters that are very dense and smooth by sanding them lightly to improve the adhesion of prime coats.

Wash previously painted surfaces in kitchens and other areas where oil and grease accumulate with TSP or another alkaline solution, adding detergents as required. After cleaning with this solution, thoroughly rinse the surface with fresh water to remove residual TSP and detergents, which would adversely affect paint adhesion.

5.6 Drywall Surfaces   Drywall, commonly referred to as Sheetrock, wallboard, and gypsum board, consists essentially of a gypsum (hydrated calcium sulfate) core with a paper laminate. It is now the most generally used type of wall construction. Nail holes and joint open spaces are filled with a joint compound in which tape is embedded to increase its strength.

Repaint surfaces are usually dusted and, if necessary, repaired with a patching compound. Do not wash drywall surfaces with TSP or other detergents, which could damage the paper laminate.

TABLE 5.6 Surface Preparation of Masonry and Related Products

Structural material and use	Surface preparation	Surface repair
Concrete walls and concrete unit masonry, poured concrete	For new exterior walls, remove form oils and parting and curing compounds by sandblasting, water blasting, power wire-brushing, impact tools, or detergent muriatic wash. Sandblast or water-blast tilt-up concrete to remove curing and parting compounds. If oiled forms are used, remove residual oil with painter's thinner or detergent wash or age 6 months. Blow dust off with compressed air or hose down with water	Fill large voids and nonmoving cracks with a concrete patch. Caulk joints and moving cracks, including fiber-glass cloth. Texture patches with sand to match the original surface. Repair mortar. A silicone water repellant may be used if the paint system is of the latex type. If old paint is flaking or chalking severely, sandblast or water-blast. If the flaking is spotty, use a power wire brush, rotary disc tool, rotary impact tool, or power sander. Blow dust off with compressed air or hose down with water. Fill cracks with a concrete patch. Repair joints with joint filler
Cement-asbestos board (Transite) walls	Aging 6 months to 1 year is recommended before painting. Dust off or use a detergent wash. Use the same treatment for repaint surfaces up to mild chalk. Wire-brush or blast lightly peeling and flaking old paint surfaces	Repair with cement-based patching material if required. Do not apply coatings to a glazed surface
Concrete floors	Sandblast or treat first with a 5 to 10% muriatic acid solution. Follow this with a clean-water rinse. Use a vapor barrier when required	For voids and cracks, use a concrete patch or an epoxy-resin patching material. If the existing finish is bad, remove by sandblasting or by a chemical remover. If the existing finish is sound, clean with a ¼-lb (0.1134-kg) TSP solution and rinse thoroughly. For other than hairline cracks, clean and fill
Stucco walls	For new walls, age 30 days or longer. For repaints, flaking and peeling paint should be lightly sandblasted. Cement-based paints and lime washes should	Fill voids and cracks with a stucco patch. Caulk joints with an oil- or latex-based caulk. If oil paint is chalking or flaking severely, sandblast lightly. Blow dust

Surface		
	be bonded with a surface conditioner. Dust off with compressed air or hose down with water. Fill joints with joint cement	off with compressed air or hose down with water. Repair cracks with a stucco patch. Moderate to heavily chalking surfaces in good condition may be prepared with a surface conditioner only. If the old surface is chalking lightly to moderately, blow dust off with compressed air or hose down with water. An adhesion additive and chalk wetter may be added to a first coat of the latex type. Repair joints with joint filler
Plaster interior walls and ceilings	For new plaster walls, age 30 days or longer before painting. For repaints, grease and dirt, particularly in kitchen areas, should be washed with a detergent solution. Glossy enamels should be dulled with a solvent etch or sandpaper	Fill cracks with a patching compound, sand smooth, and dust off. On moving cracks, use a joint compound with a fiber-glass or nylon cloth inner layer. Wash all glossy, greasy, or grimy paint surfaces with a mild alkali and rinse thoroughly. Remove all loose, blistered, or otherwise-defective paint; smooth and feather edges by sanding. Plaster on which existing paint is loose or peeling, shows poor adhesion or checking, or is otherwise unsuitable for repainting is to be stripped to the bare plaster. Cut out and fill all plaster cracks with a patching compound. Spot-prime all spackled patches. Remove calcimine by washing and rinsing with hot water
Drywall, gypsum-board, or Sheetrock interior wall surfaces	Dust off new walls only. For dirty repaints, wash with a detergent solution	For new walls, use tape imbedded in joint cement. For repaints, set nails if necessary and fill with a patching compound. Sand and dust off. Spot-seal with a primer or first coat
Canvas and paintable wall coverings	Clean surface and coat with a suitable prime coat or topcoat, or both	Repair and patch as indicated, or remove existing covering and prepare walls properly

5.7 Plastic Surfaces   Plastics are now common in building construction, fixtures, and furniture. Among the many types in use are polystyrene, acrylonitrile, butadiene, fabrics impregnated with vinyl acetate-acrylic copolymers, polyethylene, and polypropylene. Some plastics such as polystyrene may craze if they are cleaned or painted with materials containing strong solvents.

Clean and paint plastics with caution, especially if their identity is unknown. If necessary, use patch tests to test adhesion or other desired qualities of the finished system.

5.8 Hardboard, Particle Board, and Compressed Board   New hardboards and particle boards usually have hard, dense surfaces and should be roughened slightly by sanding to improve adhesion. Compressed board made of compressed paper for inside construction should be treated as drywall surfaces are. For exterior construction, treatment is similar to that for wood siding. Because these surfaces are very absorbent, their spongelike properties should be considered in writing coating specifications.

5.9 Wood Surfaces   Wood is a universal building material, and many varieties are used (Table 5.7). Surface preparation, surface repair, and painting involve numerous special problems and techniques (Table 5.8). The growing scarcity and cost of natural wood have increased the need to use lower grades of lumber, and this trend has broadened painting problems.

5.9.1 New wood   For unpainted exterior wood, the surface should be free of dirt and of loose or peeling paint. Fill nail holes and cracks with putty after the primer has been applied. Determine the cause of paint peeling on previously painted surfaces, and correct the condition before repainting. Nails used in new wood for siding and similar structural purposes should be countersunk, spot-primed, and filled with putty or plastic wood. Cracks and similar defective areas should also be spot-primed and filled with putty or plastic wood. Some poor-quality flat-grain pieces of lumber may require sanding.

The choice of sandpaper is largely determined by the required quality of the job. Table 5.9 relates mesh numbers to grit sizes. Sandpaper grits are made of natural minerals including flint (quartz), emery, and garnet, as well as synthetic minerals including aluminum oxide and silicon carbide.

Clean knots, sap streaks, and pitchy areas of softwood with painter's thinner and then seal with shellac, MIL-S-12935, polyvinyl butyral sealer, aluminum paint, or a suitable water-based primer. Allow sufficient drying time before applying succeeding coats of paint. Open-grain

TABLE 5.7   Wood Quality and Coating Systems

Type of wood	Soft closed grain	Hard open grain	Hard closed grain	Surface preparation and painting consideration
Alder	X			Accepts stain
Ash		X		Requires filler; accepts stain
Aspen			X	Normal surface preparation
Basswood			X	Normal surface preparation before painting
Beech			X	Poor for painting but all right for varnish
Birch			X	Accepts paint or varnish
Cedar	X			Accepts paints or varnish; bleeds
Cherry			X	Accepts varnish well
Chestnut		X		Needs filling; paints poorly
Cottonwood			X	Normal surface preparation before painting
Cypress			X	Accepts paint or varnish
Elm		X		Requires a filler; paints poorly
Fir	X			Surface and paint to be prepared carefully
Gum			X	Varnishes well
Hemlock	X			Fair for painting
Hickory		X		Requires a filler
Mahogany		X		Requires a filler; Philippine grade can be low-quality
Maple			X	Varnishes well
Oak		X		Requires a filler; satisfactory for stain
Pine	X			Slash grain hard to paint
Redwood	X			Paints well; bleeds
Teak		X		Requires a filler; excellent rubbing wood
Walnut		X		Requires a filler; stains well

hardwoods such as oak, walnut, and mahogany normally used in paneling and doors require filling and sanding before painting.

5.9.2   Previously painted wood   The surface preparation and repair of previously painted exterior wood is determined largely by the quality of the original paint job and paint as well as by the effects of weathering. Weathering effects vary appreciably between northern and southern exposures and under eaves. On exterior siding, nails should be reset, spot-primed, and filled with putty. Flaking, peeling, cracking, and alligatoring areas should be sanded and scraped, then dusted or given a clean-water rinse. Refill joints with joint compound as required. Clean moderately to severely chalked surfaces with a detergent wash. If mildew is present, remove it with a sodium hypochlorite–detergent wash.

For repaint work around a sash, remove loose or shrunken putty or glazing compound which has cracked open. Thoroughly clean the rabbet, and apply one coat of primer. When the surface is dry, reglaze with fresh glazing compound on metal or with putty for a wood sash.

For painted wood such as sidings, sash, trim, and doors, sand lightly if

TABLE 5.8 Surface Preparation of Wood and Related Materials

Structural material and use	Surface preparation	Surface repair and pretreatment
Factory-primed siding and factory–preservative-treated wood for exterior walls	Dust off. Remove pitch spots with painter's thinner. Seal knots and pitchy areas with shellac or poly-vinyl butyral type of sealer or aluminum paint	If steel nails are used, countersink, prime, and putty. Fill cracks and damaged areas with putty or plastic wood. Fill joints with a joint compound
New siding for exterior walls, plywood, hardboard, particle board, and clapboard	Check moisture content with a moisture meter. A moisture content of 9 to 14 percent is best for painting. Remove any mildew. Sand rough areas. Clean knots and pitchy areas with painter's thinner and seal with shellac or polyvinyl butyral sealer (Mil-S-12935)	If steel nails are used, countersink, prime, and putty. Fill cracks and damaged areas with putty or plastic wood. Fill joints with a joint compound
Exterior wood walls, fascia, and trim previously painted	If chalk is mild, dust off or wash with water or a mild detergent solution. If chalk is moderate to heavy, dust off or wash and select a suitable primer. Dull glossy surfaces, under overhangs, with solvent etch or sandpapering. Peeling and flaking paint areas should be scraped, sanded, or wire-brushed carefully, or a combination of methods used. Flame cleaning may be used on old oil paints but not on latex paints	Caulk joints with a caulking compound. Set nails, spot-prime, and putty. Fill small cracks and gouges with putty or plastic wood
New interior wood walls, beams, trim, door casings, and moldings	Sand to smooth the surface. Fill open-grain hardwoods such as oak and mahogany. Remove oily spots and pitch with painter's thinner and seal with shellac or polyvinyl butyral sealer	Fill cracks and damaged areas with putty, filler, or plastic wood
Interior wood walls, beams, trim, door casings, and moldings previously painted or varnished	If necessary, remove old finish with paint remover or sanding. Glossy surfaces should be dulled by using solvent etch or sanding	Fill damaged areas with putty, plastic wood, or wood filler
Wood floors	Machine-sand. Use an electric edger with abrasive discs on edges. If the floor is fir or pine, clean knots and pitchy areas with painter's thinner and seal with shellac or polyvinyl butyral sealer. Fill open-pore hardwoods such as oak. On repaint jobs, remove the old finish by machine-sanding or using paint remover	Fill cracks and damaged areas with putty, plastic wood, or wood filler

TABLE 5.9   Grit Sizes

Mesh number	Symbol	Mesh number	Symbol
600	None	100	2/0
500	None	80	1/0
400	10/0	60	½
360	None	50	1
320	9/0	40	1½
280	8/0	36	2
240	7/0	30	2½
220	6/0	24	3
180	5/0	20	3½
150	4/0	16	4
120	3/0	12	4½

the existing finish is sound and free of defects and shows normal chalking. Wherever the existing finish is badly checked, alligatored, peeling, or in generally poor condition, completely remove existing finish with a power sander or a burning torch as conditions require. Remove all grease and dust before applying the prime coat.

For shingles and stained wood siding, wire-brush or sand off all loose scale, dust surfaces, and touch up bare spots. For clear finishes, if the existing finish is sound, wash surfaces with a mild detergent, rinse, and sand when dry to assure the adhesion of specified coatings. If the existing finish is cracked and peeling, remove the coating to the bare wood by power-sanding to a smooth finish; wipe the surface dust-free. After the removal of the old coating, touch up discolored areas with a penetrating stain to assure the uniformity of the final appearance of finished surfaces. For floors, decks, and surfaces to be clear-finished, remove the existing finish to the bare wood, free of marks and discoloration. Renail loose flooring, and fill nail holes with filler tinted to match the selected color.

For other painted wood, wash all glossy, greasy, or grimy paint surfaces with a mild alkaline such as TSP and rinse thoroughly. Remove all loose, blistered, or otherwise-defective paint. Putty or fill in all cracks and other minor irregularities. Sand down thoroughly, and prime all bare wood. If the existing finish is checked, cracked, loose, or alligatored, strip to the bare wood, fill holes and minor irregularities, and sand lightly. Dull glossy old enamels on woodwork with a solvent etch or a TSP wash, or sand to assure good adhesion.

5.9.3 Manufactured-wood surfaces

Plywood Plywood is made from many kinds of wood for both interior and exterior use. Some plywoods are specially treated with mildew and decay preventives such as sodium pentachlorophenol. These additives offset surface preparation and aging time before painting. A key factor in the successful use of exterior plywood is the proper and absolute sealing of all edges of the panel to prevent moisture from entering.

Surface-laminated wood   Some building panels are surface-laminated with phenolic-resin–impregnated paper. They usually require little surface preparation.

Hardboards and particle board   Hardboards may be tempered with oils or be untempered. The smooth, dense surfaces should be roughened by sanding to improve adhesion. They are extremely porous and may require sealing.

5.9.4   Structure design and quality of work in fabrication   The design of a structure may affect the choice of surface-preparation methods on the basis of accessibility and similar considerations. Even more important is the quality of workmanship in fabrication. Defects caused by poor workmanship require repair before painting.

The type of fasteners used in construction may also affect the choice of surface-preparation methods. For example, if steel nails are used in place of galvanized, stainless-steel, or aluminum nails and a latex paint system is specified, the nailheads must be spot-primed to prevent rusting. The quality of the joints determines the need for a joint compound. Overlapping or adjoining construction materials sometimes determine the best procedure for surface preparation.

5.9.5   Surface preparation and choice of paint system   The moisture content of wood is very important in both original construction and painting. If green lumber high in moisture is used in construction, it will shrink on drying. The drying process will open joints, which then will need puttying or joint compounds. The best moisture range is 9 to 14 percent for exterior wood, as measured by a moisture meter, for painting operations. For repaint work, the moisture content usually is stabilized and is not a problem.

Both slash-grain or flat-grain and edge-grain woods are used in construction, particularly for siding. Paint durability is best over edge-grain wood. Slash-grain or flat-grain wood may require more sanding to smooth the surface for painting, and even then these grades of wood are prone to peeling at the hard lignin.

The microbiological condition of wood, especially previously painted wood, is important in surface-preparation procedures. Mildew is commonly found in warm, humid climates, as in the Southern states. In addition, it is always a serious problem in the interior of food and beverage plants. Mildew is best removed by detergent–steam cleaning. The cleaning solution should contain a mildewcide such as sodium hypochlorite to kill the mildew spores. On exterior surfaces a detergent–sodium hypochlorite mixture is suitable. After the surface has been cleaned, a solvent solution of a mildewcide and a sanitizing solution are applied prior to painting. The sanitizing-solution film is not washed off.

Some woods such as redwood and cedar contain natural dyes or stains that are soluble in certain paint vehicles. The dyes in redwood and cedar are quite soluble in water. These dyes require sealing or bleaching before topcoating. Latex primers containing a dye reactant are now available for water-based coatings systems.

The degree of chalking of previously painted wood determines the type and amount of surface preparation. If chalking is mild or very mild, simple dusting or a clean-water rinse is all that is necessary even for most latex topcoats. If the surface is chalking moderately, a primer (or an adhesion additive added to the first coat) may be necessary to bind the chalk. Remove heavy chalk before painting. Use a vigorous detergent wash followed by several rinses to remove all traces of detergent film. Paint soon after surface cleaning to ensure that the surface does not rechalk, thus necessitating repeated surface preparation.

Film cracking, peeling, flaking, or alligatoring requires sanding, wire-brushing, scraping, or the use of paint remover. Flame removal or burning now is not often used on old paints because of cost, hazards, and environmental effects. It also is not a very effective removal method for thermoplastic types of latex paints.

BIBLIOGRAPHY

Berger, Dean M.: "Preparing Concrete Surfaces for Painting," *Chemical Engineering*, Oct. 25, 1976, pp. 141–143.

———: "Preparing Concrete Surfaces for Painting," *Concrete Construction*, September 1977, pp. 481–484.

———: "Preparing for Painting," *Chemical Engineering*, Oct. 28, 1974, pp. 130–132.

Painting, Construction Specifications Institute, Washington, 1970.

Surface Preparation Abrasives for Industrial Maintenance Painting, National Association of Corrosion Engineers, Katy, Tex., 1964.

Surface Preparation of Concrete for Coating, National Association of Corrosion Engineers, Katy, Tex., 1966.

Williams, Alec: *Paint and Varnish Removers*, Noyes Data Corp., Park Ridge, N.J., 1972.

Wood Treatment and Stabilization, Application and Use Specification No. 7304-01, Flecto Company, Oakland, Calif., February 1980.

Chapter 6

Surface Preparation: Part II

KENNETH B. TATOR

KTA-Tator Associates, Inc.

6.1 Introduction   Surface preparation is perhaps the most important single factor in the longevity of a specified coating system. Deficient or improper surface preparation is directly responsible for a greater number of premature coating failures than anything else.

There are many surface-preparation methods. Acid or alkali chemical cleaning, solvent washing, grinding, scarifying, sandblasting, high-pressure water blasting, scraping, and wire-brushing all are commonly used to prepare various surfaces for painting. Choosing the right cleaning method is an integral part of the paint specification. Choice depends on the type of substrate being prepared (steel, galvanized metal, wood, concrete, etc.). Another factor that influences the choice of a surface-preparation method is the coating system itself. Generally, synthetic-resin coating systems such as vinyls, epoxies, chlorinated rubbers, and phenolics require more stringent surface preparation (for example, blast cleaning or pickling) than do oil-based coatings such as alkyds and epoxy esters.

Economics is certainly a factor. For new construction, surface preparation and painting in the fabricating shop or manufacturing facility are almost always less expensive than painting after field erection or installation. This is true because pickling, rotary-wheel blast cleaning, and other cost-effective surface-preparation methods cannot be carried out in the field. Personnel safety and hazards to adjacent equipment or operations are also limiting factors. Many jobs do not allow open blast cleaning because of the danger of sparking and, ultimately, of explosion. Abrasive-blast cleaning is also discouraged if electric motors, hydraulic equipment, and the like are in close proximity. Even government, through restrictive legislation, has in some areas prohibited open sandblasting or other surface-preparation methods that contribute to particulate emissions into the atmosphere or expose workers to what is considered an unhealthy environment (by virtue of airborne contaminants, excessive noise, etc.).

From the foregoing, it is apparent that specifying proper surface preparation not only is extremely important but also can be complicated. The purpose of this chapter is to provide the specifier with a basic understanding of some of the types of surface preparation and to present certain "standard" specifications and useful information to facilitate proper writing of surface-preparation specifications. Typical surface-preparation specifications are presented for guidance at the end of the chapter.

Surface preparation is simply doing something to a surface so that paint will adhere properly to it. The most important aspect is cleanliness. The surface should be cleaned of oil, grease, dirt, dust, etc., so that the coating will be in direct contact with the surface being painted. Simple washing, with or without a detergent, in many cases is sufficient to provide this cleanliness. On the other hand, old deteriorated paint, rust or rust scale, mill scale, or other adherent surface contaminants must be removed prior to painting. Abrasive cleaning or chemical cleaning is often required.

The second function of surface preparation is to provide an anchor pattern, etch, or tooth. Such roughness to the substrate aids paint adhesion in two ways:

▪ It increases the surface area, allowing a greater number of chemically polar groups within the coating to come in contact with the substrate.

- It provides a mechanical anchor, or tooth, to facilitate physical adhesion.

Some substrates, among them wood, concrete, and cast iron, are sufficiently porous or have suitable roughness without further surface preparation for good coating adhesion.

Chemical cleaners and etchants are preferred for soft metals like aluminum, zinc, copper, and lead. For most carbon steels and higher-alloy steels, however, blast cleaning is by far the most effective means of both cleaning and roughening. The specifier should be aware of problem surfaces such as those mentioned below.

6.1.1 Mill scale   New hot-rolled steel has a bluish black scale called "mill scale" on its surface (Fig. 6.1). This scale results from normal cool-

Fig. 6.1 Mill scale. [*KTA Tator Associates, Inc., Coraopolis, Pennsylvania*]

ing and oxide formations during rolling. Initially it is intact and protective, but because of its brittleness and a coexpansion rate different from that of steel, it will ultimately crack. Moisture will penetrate the cracks, and rust will form beneath the scale, spalling it from the steel. Paint applied over the scale will retard but not eliminate this process. Paint failure over mill scale is perhaps one of the most common problems. Mill scale must be removed from steel prior to painting in all but the mildest service environments. This is best done by blast-cleaning or pickling the surface.

6.1.2 Rust scale   Rust scale (Fig. 6.2) is a heavy iron oxide corrosion deposit. In humid environments it can build up into thick, voluminous layers. All such scale must be removed prior to painting. This is best done with hand or power chipping hammers or other impacting tools.

6.1.3 Oil and grease   Grease and oil should be removed from metal by wiping it with solvents. Even new, protected, unrusted metal may have an invisible oil film that has been deposited during processing. Unless this film is removed, the bond between paint and metal is obstructed. The use of turpentine and mineral spirits is not recommended since these materials leave an oily deposit of their own.

Fig. 6.2 Rust scale. [*KTA-Tator Associates, Inc., Coraopolis, Pennsylvania*]

6.1.4 Design-problem-area treatment   Sharp edges of metal and threads on pipes are difficult to cover adequately with primer (Fig. 6.3). Frequently these are the first areas to rust after painting. Particular care should be taken to get a continuous coating on all metal surfaces. If protrusions such as weld spatter and laminations can be power-ground flat and smooth, the extra work pays off in total paint performance. Sharp edges such as the edges of structural members and rough flame- or saw-cut edges should be rounded by grinding, preferably to a ⅛-in- (3.175-mm)-minimum radius. Crevices, sharp corrosion pits, and deep gouges in the metal should be filled with weld metal. Corrosion design problems such as tack welds, back-to-back angles, and lap joints should be eliminated in the design if possible or be sealed by welding or caulking. In general, the greater number of surface irregularities that are eliminated prior to painting, the longer paint can be expected to protect these problem areas.

Welding leaves a hard-glaze deposit that will later crack and disbond. The glaze must be removed by mechanical abrasion or blast cleaning.

6.1.5 Galvanized metal   Galvanized metal is iron or steel with a coating of zinc metal, approximately ½ to 3 mils (0.0127 to 0.0762 mm) in

thickness, that is applied by a hot-dipping or strip-coating process. Although galvanizing itself is a protection against corrosion, time and weather will destroy this protection and rusting will occur. If sufficient weathering has taken place, galvanized metal can be cleaned to remove all zinc corrosion products and surface contaminants and then primed without etching the surface. New galvanizing, however, will not hold

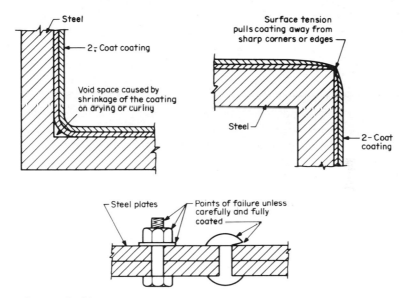

Fig. 6.3 Problem areas. [*KTA-Tator Associates, Inc., Coraopolis, Pennsylvania*]

most conventional metal primers and must be chemically etched or, instead, primed with chemical-conversion pretreatments. Some latex-emulsion primers formulated specifically for metal and galvanized surfaces, primers containing zinc dust, and portland-cement primers will adhere to new, clean galvanized surfaces without special etching.

6.2 Surface-Preparation Methods   All surface-preparation methods can be divided essentially into two categories, mechanical and chemical. Mechanical surface-preparation methods are those in which mechanical energy in the form a propelled abrasive or agitated liquid vapor is used to remove contaminants physically. Chemical methods are those in which a chemical action, generally dissolution or emulsion, is used to remove surface contaminants.

6.2.1 Mechanical cleaning methods

Hand cleaning   This method covers the manual use of hand tools, including wire brushes, scrapers, chipping hammers, knives, and chisels, as well as of emery or sandpaper. Any grease is removed prior to mechanical cleaning.

Hand-tool cleaning is used when a job is inaccessible to power tools or other cleaning methods, when other cleaning methods are not available, or when a job is too small to warrant bringing in power tools. In general, however, hand-tool cleaning is suitable only for removing loose rust, mill scale, paint, and other poorly adhering contaminants. Intact mill scale cannot be removed by this means. Tightly adhering paint, rust scale, or weld or cement spatter is extremely difficult to remove with hand tools.

Hand tools provide effective cleaning only on small jobs on which a worker does not lose interest because of the slow cleaning rate. For large or production jobs, it is best to use another cleaning method.

Power-tool cleaning   The rotary wire brush and grinder, powered by electricity or air, are perhaps the most commonly used power tools. They have a rotary shaft with various head attachments. Like hand-tool cleaning, power wire-brushing will not remove intact mill scale or very tightly adhering contaminants. However, the cleaning rate is faster, and the degree of cleaning is much better. Rotary discs in a variety of sizes, shapes, and abrasives are also available. These clean by grinding into the base metal and removing all surface contaminants or by buffing to remove loose surface contaminants without damaging the metal surface. Abrasive disc grinders can remove intact mill scale and tightly adhering contaminants.

Chippers and scalers are impact tools that drive a chisellike chipping hammer or a series of pointed descaling "needles," impacting the metal surface being cleaned and thereby loosening and removing contaminants. Impacting tools work well on hard or brittle surface contaminants such as rust, mill scale, and weld flux. The rate of cleaning is quite slow, however, and there is danger that a surface contaminant may be driven into the surface of the metal by the impact action.

There are available rotary impact scalers which include cutting tips that impact the surface at a rate much faster than that of the piston-driven type. Moreover, the power of impact is smaller. Residual surface contaminants must be removed by wiping or wire-brushing when using any impact cleaning method. There is also danger that the metal being descaled will be cut and gouged, leaving sharp burrs where paint may fail prematurely.

Wire-brushing will not remove intact mill scale, but it can remove loosely adhering material and all paint. Too long a use of wire-brushing in one spot can result in burnishing, leaving a polished, glossy surface to which paint will adhere poorly. Wire brushes are useful over rivets and bolt heads and irregular surfaces.

Danger from sparks and flying particulates should always be considered when using power tools. The operator and adjacent workers should use goggles and protective clothing while these tools are in use.

Flame cleaning   Flame cleaning is a method of preparing a metal surface by passing a high-velocity oxyacetylene flame over the surface being cleaned. This method depends upon the high heat generated at the surface, which removes some of the mill scale either by the thermal difference between the coefficients of expansion of the underlying steel and the mill scale on the surface, which will spall off the mill scale, or by the explosive action of water vapor generated underneath the scale, or by a combination of the two methods. Temperature is not a determining criterion for success, because it has been shown that the flame must have high velocity to obtain the intense, rapid transfer of heat to the mill scale. In some cases flame cleaning has advantages over water blasting and power-tool cleaning, but it always falls far short of pickling and blast cleaning as a method of surface preparation.

Flame-cleaning equipment utilizes the same principle as the oxyacetylene cutting torch; the surface is heated and a jet of oxygen turned on it, rapidly burning the paint and blowing the char from the surface. For best results, the flame temperature should be in excess of 5432°F (3000°C). Because considerable heating occurs on the surface in the flame-cleaning operation, cleaning material less than $3/16$ in (4.762 mm) thick may be hazardous, as the metal may not maintain its dimensional stability and may warp. The deformation of flat metal sheets is particularly critical in flame cleaning. For this reason and because of safety factors, flame cleaning has lost favor and has been replaced by abrasive-blast cleaning and power-tool cleaning in most specifications.

Many persons believe that the main advantage of flame cleaning may be that moisture is removed from the surface and paint applied while the metal is still warm. Since heat has a tendency to make the paint less viscous, better penetration and adhesion result. The paint also sets up faster, resulting in a thicker coat and a faster drying time. Accordingly, a practical adaptation of the flame-cleaning process is the flame-dehydrating process, in which the primary purpose of the flame is to remove moisture from the surface and heat it slightly prior to painting rather than to remove surface contaminants.

Steam cleaning   Steam cleaning utilizes the heat and washing action of a jet of steam as it passes over the surface being cleaned. Steam or hot water is supplied to the nozzle of the steam gun at a temperature of approximately 300°F (149°C) and a pressure of 150 to 200 lb/in² (1,034,214 to 1,378,951 Pa). Water consumption is approximately 200 gal (757 l) per hour per nozzle. The steam or water solution can be fortified with a strong caustic for emulsifying and removing sound paint, dirt, grease, smudge, soot, and other oily residues or with a commercial detergent to remove less adherent contaminants.

Steam cleaning is generally used to remove heavy soil, dirt, and grease on structures that do not lend themselves to soak or spray cleaning and that because of their size, shape, or location would be difficult or impossible to solvent-wipe. Steam cleaning does not remove tightly adhering residues such as mill scale or rust scale. Unless it is used with a strong caustic additive, neither can old paint coatings be removed. In most cases, the results of steam cleaning are enhanced by wire-brushing or spot-sandblasting.

The advantages of steam cleaning are that the process is quite mobile and relatively inexpensive and that there is no dusting or abrasive contamination. Steam cleaning with a detergent or caustic additive is an excellent way of removing oil and grease residues. The initial cost of equipment is relatively low, and the equipment is quite portable and small. In addition, steam cleaning does not alter the metal surface, and surface dimensions and characteristics do not change. The disadvantages of steam cleaning are that it does not remove tightly adhering contaminants such as rust scale and mill scale and that in most cases it is difficult to remove adherent sound old paint. When a detergent is used, steam cleaning generally leaves an alkaline surface that is unsuitable for painting unless it is thoroughly rinsed.

Equipment for steam cleaning generally consists of a small flash-type boiler, a low-pressure atomizing burner, an engine-driven pump to supply water, and a gasoline-engine power supply. The entire unit can be mounted on wheels or a trailer to make it portable. The unit should be entirely self-contained in case municipal power and water are not available.

Various types of steam nozzles are available for the variety of cleaning problems encountered. For large, flat surfaces, a fan nozzle is most suitable. A round nozzle is best for smaller, inaccessible areas or for areas requiring more highly concentrated cleaning. If scouring is required to remove grime, a nozzle fitted with a fiber brush can be obtained. The steam gun itself is of two types, injection-feed and suction-lift.

Because a certain amount of liquid will remain on horizontal surfaces after cleaning, the final cleaning pass should be systematic from top to bottom. The last pass of the steam gun on any surface should be made to remove surplus solution. The operator should try to leave the area as dry as possible. It is important that concentrations of detergent or caustic do not become excessive or remain on the surface after cleaning.

In painting bridges over railroad tracks and other items subject to grease and grime, it often is best practice first to steam-clean and then to spot–blast-clean prior to painting.

Under no circumstances should paint be applied to a surface that may still be wet. Generally 24 h should elapse to allow moisture to evaporate from crevices in sheltered areas.

To determine whether a surface is clean, the operator must resort to feeling. An area that has not been cleaned should feel grimy and dusty. An area that has only been wet should feel slick, and a slight rubbing will show a smear. If an area has been properly cleaned, the surface will feel firm and somewhat tacky but should not be slick or grimy to the touch. A slight rubbing should not leave a dirty smear. Clean hands before testing.

Water-blast cleaning   Water-blast cleaning (Fig. 6.4) utilizes a high-pressure water stream to remove any surface contaminant that is not tightly adherent. The water-blast–cleaning unit will pump approximately 10 gal (37.85 l)/min at pressures up to 12,000 lb/in² (82,737,000 Pa), thus accommodating a wide range of pressures for various applications. Units are rather small and are highly mobile. Water-blast cleaning will remove substances such as loose and blistered paint, chalking, grease, and other cumulative residues. In most cases, oil and grease are removed sufficiently so that further solvent cleaning is not required. Water-blast cleaning is competitive with hand and power-tool cleaning and in general produces similar results. For heavy, well-adhered rust, scaling tools may be required before the water blaster is used.

The advantages of water-blast cleaning are that (1) it leaves no dust or other loose material on the cleaned surface and (2) no abrasives are

Fig. 6.4   Water-blast cleaning. [*KTA-Tator Associates, Inc., Coraopolis, Pennsylvania*]

necessary. Thus abrasive-contamination hazards as well as abrasive cleanup are avoided. Water blasting is most efficient when used on surfaces that are irregular in shape and inaccessible to hand or power tools. For example, it is excellent for use in cleaning expanded metal and open grating on floors and catwalks, which are hard to reach with pneumatic or hand tools. It can also be used readily on crevices, flanges, and back-to-back angles. In addition, it can be employed quite effectively on structural steel, floor plates, piping, and storage tanks as well as in cleaning concrete and masonry surfaces. The disadvantages of water-blast cleaning are that the entire work area becomes wet, the surface being cleaned must dry thoroughly prior to painting, and tightly adhering contaminants such as tight paint and mill scale are not removed. The high-pressure water lance should be held within approximately 3 in (76.2 mm) of the surface. This posture is quite tiring for the operator, who has a tendency to hold the lance farther from the surface, thus decreasing the effectiveness of the method.

Abrasive-blast cleaning   Blast cleaning results when an abrasive is propelled by air, water, or centrifugal force against the surface being cleaned. The abrasive impacts and abrades metal being cleaned, both removing contaminants from the surface and roughening it. The cleanliness obtained by blast cleaning is the greatest obtained by any mechanical method. The two major types of production blast cleaning are those using an air-propelled abrasive and those using a rotary-wheel–propelled abrasive.

Air-Propelled Abrasive-Blast Cleaning. Air-blast cleaning utilizes high-pressure compressed air, which expands through an orifice to propel the abrasive onto the surface being cleaned. Open air-propelled blast cleaning (commonly called sandblasting) may be the most thorough and economical means of cleaning scale and heavy contaminants from field structures. Conversely, in many cases blast cleaning in cabinets or rooms is the best method of cleaning or polishing small parts.

Blast cleaning can be divided into two basic systems, direct-pressure and gravity-blast, depending upon how the abrasive is introduced into the compressed-air stream. Vacuum blasting, wet blasting, and safety are other considerations.

- *Direct-pressure blast system (tank type).* The abrasive in a pressurized tank is fed directly into the compressed-air line, and the air-abrasive mixture is discharged through a nozzle. This system is the most widely used for industrial and marine field blast-cleaning operations because a great volume can be ejected at higher speeds and greater concentration than with the gravity method.

- *Gravity-blast system.* In a gravity-blast system, the abrasive drops from an elevated storage tank into the inner supply at the nozzle. Air enters the nozzle from a separate line, mixes with the abrasive, and propels it out of the nozzle. A wider spray pattern results, although the abrasive has less impinging force than in the direct-pressure type. Less abrasive is used, and because of the lower impingement speed there is a lower breakdown rate. As a result, gravity-blast systems have found extensive use in small installations and blast cabinets.

A variation of the gravity-blast system, the induction-suction type, moves the abrasive through a feed hose by suction created in an induction chamber behind the large nozzle by means of a jet of high-velocity compressed air that expands from a smaller nozzle. Such a unit is similar to the gravity-feed type but is much more versatile, in that it is readily portable, relatively inexpensive, and excellent for blast-cleaning small areas in the field at low cost. The advantage of this type over the gravity-feed type is that the compressed-air stream can lift and move the abrasive to the blast nozzle, whereas the gravity-feed type must have an elevated tank to introduce the abrasive.

Induction-suction equipment may also utilize gravity feed by raising the abrasive supply to a level above the nozzle. The gravity-feed type of blast equipment relies upon the flow of abrasive to the nozzle by gravity only. Suction created by the action of the jet is not relied upon to induce abrasive flow and, in practice, is wasted by the induction of secondary air into the nozzle around the entrance of the gravity-feed tube.

Adequate volume and pressure of compressed air are perhaps the most critical parts of the air-blast–cleaning operation. To be economically feasible, air-blast cleaning requires both high pressure (90 to 100 lb/in^2, or 6,205,000 to 6,895,000 Pa, at the nozzle) and high volumes of air (81 to 338 ft^3/min, or 0.038 to 0.160 m^3/s). The larger the air compressor, the larger the blast-cleaning nozzle that can be used. The larger the nozzle operating at the proper pressure, the more quickly the job can be done.

Compressed air is available either from a stationary engine-driven or electric compressor or from a portable gas or diesel engine compressor.

▪ *Vacuum-blast cleaning.* Vacuum-blasting systems clean steel in a manner similar to that described for direct-pressure blasting, but the abrasive is reclaimed from the immediate blast area by a vacuum. The blast nozzle is enclosed in a hollow cup with a rubber or brush seal around its perimeter. The blast nozzle through which the abrasive flows is in the center of the cup. The abrasive flows through the nozzle, impacts on the metal being cleaned, and is sucked up by the vacuum intake surrounding the nozzle. Abrasives with a low breakdown rate can be recycled, and abrasives with a high breakdown rate such as sand are vacuumed into an enclosed hopper or bin and discarded.

The obvious advantage of this sytem is that abrasive cleanup and dusting are almost completely eliminated. Major disadvantages are that the equipment is more expensive, the blast-cleaning production rate is quite slow, the system works best on flat pieces and plates (irregular shapes can be cleaned but only with difficulty), and the equipment is quite bulky and not nearly as portable as conventional open air blasting.

While this system should not be used for high-volume production work, it may often be the best method of surface preparation in areas where a high degree of surface cleaning is required but abrasive contamination must be eliminated, as in places where hydraulic equipment and electric motors are used.

▪ *Wet-blast cleaning.* Wet-blast cleaning also utilizes the principles of direct-pressure blast cleaning, with water added to the abrasive-air mixture to reduce or eliminate dust formation. Water, with or without a rust inhibitor, can be introduced into the abrasive-air mixture in either of two ways: (1) It is mixed with the blast sand directly in the pressure tank. The sand and water thus mixed should be used as quickly as possible, before the sand settles. (2) The water, with or without an inhibitor, is mixed with the blast sand either directly behind or directly ahead of the blast nozzle. Water can be introduced to a compressed-air–abrasive stream before it leaves the nozzle or be sprayed onto the air-abrasive mixture after it leaves the nozzle. The amount of water can be regulated at the nozzle, and as a result the abrasive can be thoroughly wet to eliminate almost all dusting or be only slightly wet to eliminate most dusting but to make cleanup easier.

In wet-blasting on a vertical surface, considerable water and sand run down the sides onto the surface below, piling up into a solid mass. These deposits of sand and water must be removed by shoveling or other means. A surface cleaned in this manner should be hosed down with either fresh or inhibited water immediately after blast cleaning. If the water used during wet-blast cleaning is inhibited, rinsing with fresh water will often suffice. If the water is not inhibited, rinsing with an inhibitor solution will ultimately be required to stop flash rusting.

Generally, 15 gal (56.8 l) of water per ton of sand is required to wet the sand in the pressure tanks for wet blasting, or 3 to 4 gal/min (0.19 to 0.25 l/s) will be sufficient to wet the abrasive suitably if the water is added after the abrasive has left the nozzle.

Effective rust inhibitors are 0.2% solutions by weight of chromic acid, sodium chromate, sodium dichromate, or potassium dichromate. In addition, a 2% solution of a mixture of 4 parts diammonium phosphate and 1 part sodium nitrite can be used. If experience shows that the concentration is not high enough to prevent rusting, the amount of inhibitor should be increased. It should be understood that the protection offered to the metal by any of these treatments is of very limited duration. Moreover, if any solutions containing chromic acid or dichromates are used, the operator as well as people in the vicinity of the operation should be protected from continuous exposure to the solutions or from breathing mists of them.

▪ *Personnel and safety accessories.* Protection of the blast-cleaner operator's eyes and respiratory system is a major safety consideration in any open blast-cleaning operation. Two types of helmets are commonly used in the blast-cleaning industry. One is a slipover protective device against ricocheting abrasive, usually made of canvas with a plastic face mask. Because this type has no provisions for eliminating dust particles from the air breathed by the operator, it should be used in conjunction with a respiratory filter fitting over the nose and mouth. The second type is an air-fed helmet, made of metal or plastic, into which a separate supply of air is fed. Because the hood is under positive pressure, the operator does not inhale any dust resulting from the blast-cleaning operation. A recent innovation, particularly in hot climates, has been the use of an air-conditioned hood or suit. Air to these units is not only filtered but cooled for greater comfort. Pay particular attention to standards for cleaning and handling respiratory air.

Helmet air purifiers take air from the compressor, regulate it to the lower pressures required for the operator's helmet, and remove dust, moisture, and oil fumes, thus providing dry breathing air to the blaster. However, the purifier does not remove carbon monoxide or other gaseous contaminants from the helmet air. In the past, it was sufficient to ensure that the air-intake manifold of the compressor be located away from the exhaust of any adjacent machinery. It is now mandatory that any automatic shutoff device controlled by the blast cleaner be installed on the blast machine. The device, commonly called a "deadman," will shut off the air supply to the machine if the spring-loaded control lever is released by the operator. This will prevent the dangerous whipping of an operating blast hose if an operator becomes disabled.

Rotary-Wheel Blast Cleaning. In centrifugal-wheel blast cleaning, abrasive particles are propelled against the metal being cleaned at high velocity by electrically driven wheels. A typical rotary-wheel machine has from 4 to 16 or more wheels, although in steel-fabricating shops 8-wheel machines are most common. A single wheel can discharge up to about 1600 lb/min (12.1 kg/s) of steel shot or grit, in contrast to air blasting, in which the nozzle discharge is approximately 100 lb/min (0.76 kg/s) of abrasives.

Because the impact energy of an abrasive particle varies with the square of its velocity and is directly proportional to its weight, the impact energy in a centrifugal-wheel blaster depends upon the weight of the abrasive only if all abrasive particles move at the same velocity. In air-blast cleaning the air stream accelerates larger particles more slowly, resulting in lower impact energy.

The advantage of centrifugal-wheel blast cleaning is that the rate of cleaning is much faster than with any other type of blast cleaning. All kinds of abrasives can be used, and all abrasives can be reclaimed and recycled. Dusting is eliminated, as is abrasive contamination of the work area. The degree of cleaning with centrifugal-wheel blasting conforms to requirements for white, near-white, commercial, and brush-off blast cleaning, depending upon the speed at which the object being cleaned is moved past the abrasive-throwing wheels.

The disadvantages of centrifugal-wheel blast cleaning are that the initial equipment cost is considerably higher than for any other cleaning method, although operating costs may be lower, and that the blast-cleaning units are not portable and work must be brought to them. Furthermore, objects exceeding a certain size may not fit into a particular machine. In many machines the interior of partially protected or enclosed surfaces such as box trusses cannot be adequately cleaned by centrifugal-wheel blasting. In these cases, surface contaminants must either be ground off or be removed by air-blast cleaning. Depending upon the size and capacity of the objects being cleaned, wheels can be added to make the blast-cleaning pattern more nearly complete.

Because of the nature of the operation, high wear is associated with the moving parts of the equipment, particularly when grit abrasives are used. Since the equipment is mechanically complex, it is more difficult to keep in operation than conventional blast equipment. In spite of these disadvantages, however, advantages far outweigh them in metal-fabricating and -finishing shops, in which fast, thorough surface preparation is required.

Blast Cleanliness. The degree of cleanliness is determined by the time during which the abrasive impinges on the metal surface being cleaned. Blast-cleaning standards have been established by the SSPC and the National Association of Corrosion Engineers (NACE). Both organizations describe the requirements for white blast cleaning, near-white blast cleaning, commercial blast cleaning, and brush-off blast cleaning. Excerpts from the SSPC *Surface Preparation Specifications* follow:

- *White blast metal cleaning (SP-5).* "A White Metal Blast Cleaned Surface Finish is defined as a surface with a gray-white, uniform metallic color, slightly roughened to form a suitable anchor pattern for coatings. The surface, when viewed without magnification, shall be free of all oil, grease, dirt, visible mill scale, rust, corrosion products, oxides, paint, or any other foreign matter. The color of the clean surface may be affected by the particular abrasive medium used."

- *Near-white blast cleaning (SP-10).* "A Near-White Blast Cleaned Surface Finish is defined as one from which all oil, grease, dirt, mill scale, rust, corrosion products, oxides, paint or other foreign matter have been completely removed from the surface except for very light shadows, very slight streaks, or slight discolorations caused by rust stain, mill scale oxides, or slight, tight residues of paint or coating that may remain. At least 95% of each square inch of surface area shall be free of all visible residues, and the remainder shall be limited to the light discoloration mentioned above."

■ *Commercial blast cleaning (SP-6).* "A Commercial Blast Cleaned Surface Finish is defined as one from which all oil, grease, dirt, rust scale, and foreign matter have been completely removed from the surface and all rust, mill scale, and old paint have been completely removed except slight shadows, streaks, or discolorations caused by rust stain, mill scale oxides or slight, tight residues of paint or coating that may remain; if the surface is pitted, slight residues of rust or paint may be found in the bottom of pits; at least two-thirds of each square inch of surface area shall be free of all visible residues and the remainder shall be limited to light discoloration, slight staining or tight residues mentioned above."

■ *Brush-off blast cleaning (SP-7).* "A Brush-Off Blast Cleaned Surface Finish is defined as one from which all oil, grease, dirt, rust scale, loose mill scale, loose rust and loose paint or coatings are removed completely, but tight mill scale and tightly-adhered rust, paint and coatings are permitted to remain provided that all mill scale and rust have been exposed to the abrasive blast pattern sufficiently to expose numerous flecks of the underlying metal fairly uniformly distributed over the entire surface."

NACE defines its blast-cleaning standards similarly to the SSPC definitions. In addition, it sells plastic-encapsulated steel panels that have been sand-, grit-, or shot-blasted to the specified standard. The SSPC sells photographic standards showing various degrees of surface cleaning (hand and power-tool cleaning and the four degrees of blast cleaning) over four rust grades of steel (adherent mill scale, rusting mill scale, rusted steel, and pitted and rusted steel). Other organizations have developed various photographic standards depicting degrees of blast cleaning and hand and power-tool cleaning and even the extent of weld grinding. Often job standards are prepared at a worksite to avoid arguments about interpretations of the degree of workmanship required. Such job standards have been prepared for blast cleaning, edge and weld grinding, and painted steel, showing the acceptable number of pinholes, degree of runs and sags, and other workmanship defects.

Surface Profile. The surface profile, or roughness, of the blast-cleaned surface depends upon the type of abrasive used and the force with which it impacts on the surface being cleaned. Although there is some question as to the best surface profile for a given coating system, it is generally recognized that too deep a surface-profile anchor pattern will result in peaks that may not be covered sufficiently by the coating system, causing corrosion-initiation sites. Similarly, too shallow an anchor pattern will

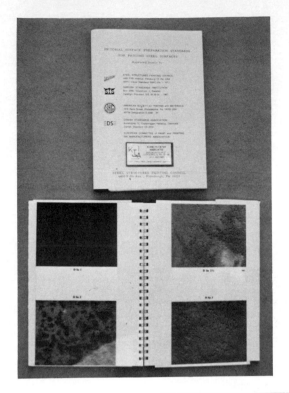

Fig. 6.5 NACE (left) and SSPC (right) blasting standards. [*KTA-Tator Associates, Inc., Coraopolis, Pennsylvania*]

result in a smooth metal surface, to which paint will not adhere satisfactorily. It is generally accepted that a surface profile of approximately 1½ to 3½ mils (0.0381 to 0.0889 mm), measured from the top of the highest peak to the bottom of the lowest valley, is suitable for most coating systems. Thick coating systems require a greater surface profile, and thin systems generally should have a lower surface profile.

Surface profile can be measured by a number of laboratory methods, using a high-powered vernier microscope, grinding and measuring with a depth micrometer, undertaking microscopic cross-sectioning, or employing electronic-stylus tracing equipment. Field determinations involve the use of surface-profile comparators, depth micrometers, and a replica tape.

Blast-Cleaning Abrasives. There are four general classifications of abrasives used in blasting operations (see Table 6.1). While the harder

TABLE 6.1   Commonly Used Abrasives

Metallic	Synthetic, nonmetallic, silica-free	Siliceous	Agricultural
Chilled cast iron	Silicon carbide	Garnet	Coconut shells
Cast steel	Aluminum oxide	Quartz	Black walnut
Malleable iron	Refractory slag	Silica	Pecan shells
Crushed steel	Rock wool	Decomposed rock	Peach-pit shells
Cut steel wire	by-products		Filbert shells
Aluminum shot			Cherry-pit shells
Brass shot			Almond shells
Copper shot			Apricot-pit shells
			Rice hulls
			Ground corncobs

SOURCE: *Steel Structures Painting Manual,* volume I: *Good Painting Practice,* Table IV, Steel Structures Painting Council, 4400 Fifth Avenue, Pittsburgh, Pa. 15213.

abrasives cut more deeply and more quickly than the softer abrasives, their brittleness gives them a higher breakdown rate. This rate will cause them to become embedded in the metal surface being cleaned. The embedded abrasives may contribute to coating failure when they ultimately loosen or, in the case of dissimilar metal particles used as abrasives, cause galvanic corrosion.

Blasting abrasives are classified by shape (shot or grit) or by material (steel, iron, aluminum oxide, slag, etc.).

■ *Shot.* These are particles which are spherical or nearly spherical in shape. Shot will hammer and peen the surface being cleaned and ricochet in enclosed areas. While most metals can be made into shot pellets, steel is the most commonly used. Because shot rolls more readily and therefore is easier to reclaim, it is widely used in rotary-wheel blast-cleaning machines. Normally, shot alone does not give an adequate surface profile.

▪ *Grit.* These are abrasives which have irregular sharp or semisharp surfaces. Metallic grit is almost always crushed cast iron or steel. When it is recycled, the numerous impacts will round the angular surfaces into a shot. Natural siliceous abrasives such as sand are semisharp abrasives that because of their cheapness and high breakdown rate usually are not recirculated.

Since metallic abrasives are generally more expensive, to be economically feasible they must be used more than once. Generally they can be recycled from 50 to 5000 times before they disintegrate and are no longer effective. Synthetic abrasives also are higher-priced than siliceous abrasives and usually should be recycled for maximum economy.

▪ *Sand (siliceous abrasive).* This material is inexpensive enough to use without recycling and for best results should be bought cleaned and bagged. Bulk sand, which is cheaper, should not be used unless it has been thoroughly cleaned of mud, dirt, and other contaminants. Agricultural abrasives, while inexpensive, are bulky, and their cost depends upon prevailing freight rates. Siliceous abrasives, particularly silica sand, are the most common abrasives for field air-blast cleaning because they are inexpensive and convenient and normally are not reclaimed. Siliceous abrasives, however, will break down upon impact into a fine dust, which when inhaled in sufficient quantity by a worker may cause silicosis, a lung ailment. Either workers should be protected from inhaling this dust, or another abrasive such as refractory slag, which is slightly more expensive but does not contain free silica, should be used. The U.S. Environmental Protection Agency (EPA) is preparing regulations to govern the control and use of sandblasting. Some local regulations already restrict its use.

In December 1980, the Occupational Safety and Health Administration (OSHA) plans to issue proposed sandblasting standards that could drastically affect surface-preparation procedures. Presently OSHA is preparing a regulatory analysis (costs, benefits, affected industries, workers at risk, feasible engineering controls, etc.) that will be reviewed by a standards advisory committee. Once the major issues have been resolved, details will appear in the *Federal Register.* The National Institute for Occupational Safety and Health (NIOSH), the research wing of the Department of Health, Education, and Welfare, is suggesting a sandblasting standard to be adopted by OSHA, which is part of the Department of Labor. Some suggest outright banning of sandblasting operations.

The health hazards attributed to sandblasting suggest that operators be required to use proper safety clothing and equipment. Recent epidemiological studies on refinery employees wearing proper equipment indicate a lower incidence of respiratory problems than among general plant workers. It is common practice in the paint industry for sandblasting operators to shed their equipment against the advice and instruction of job foremen. This practice should not be condoned.

▪ *Other abrasives.* Agricultural abrasives and plastic and glass beads are used primarily to remove flashings from aluminum and plastic castings and light scale such as carbon from intricate parts. These abrasives are employed when it is desirable to remove a minimum of material and to impart a bright finish to the parts being cleaned. The profile obtained by using these abrasives is minimal (less than 1 mil, or 0.0254 mm).

Recent Developments. Airco, Inc., a licensee of the Lockheed-California Company, is developing a new surface-blasting technique which utilizes solid carbon dioxide pellets instead of sand or other blasting media. The process provides a method of cleaning surfaces without producing potentially hazardous residues or requiring expensive cleanup operations. Carbon dioxide pellets have a relatively high degree of hardness and can be made at a relatively low cost from the naturally abundant raw material. Early tests proved the ability of 0.44-in- (11.18-mm) diameter pellets to prepare a painted surface in preparation for topcoating. However, these pellets were too large to be utilized with standard blasting equipment. Additional work is proceeding to develop commercially acceptable blasting hardware.

Three other recent developments are noteworthy. Williams Contracting Company, Chamblee, Georgia, has developed a portable rotary-wheel blast-cleaning facility that can be transported to a jobsite to clean and paint steel. Kue Engineering, Ltd., Yorkshire, England, has developed a blast-cleaning system that uses high-volume, low-pressure air mixed with abrasive and water in controlled amounts. The system is said to allow great blast-cleaning versatility with very little airborne contamination. R. T. Nelson Painting Service, Oklahoma City, Oklahoma, and Wheelabrator-Frye, Inc., Materials Cleaning System Division, Mishawaka, Indiana, have both developed portable self-propelled rotary-wheel blast-cleaning units designed for the field blast cleaning of tanks, ship hulls, and other large, flat surfaces.

6.2.2 Chemical cleaning methods   Chemical surface-preparation methods differ from mechanical methods in that the surface of the metal being cleaned is actually changed by chemical reaction with the cleaner. In solvent cleaning, the contaminants themselves are changed and removed from the surface, but the metal surface is not affected; in acid cleaning the contaminants also are dissolved, but so is a portion of the metal surface; and in alkali and emulsion cleaning the contaminants are usually saponified or emulsified and then removed. With the exception of solvent cleaning, all chemical cleaning methods change the surface of the metal by forming a surface complex with the residual base metal. In many cases this complex is beneficial; in others it is not and must be removed.

Solvent cleaning   Solvent cleaning is used to remove from the surface oils and greases, which would interfere with the adhesion of the paint film to the metal. Such oils and greases may be of two types: mineral oils and vegetable oils. Mineral oils are insoluble in acids and alkalies and are unsaponifiable. Their removal depends largely on a straight solution by the solvents or upon emulsification. Animal and vegetable oils are also insoluble in acids and alkalies, but they can be saponified by alkali cleaners. Their removal by alkali cleaning depends upon the extent of saponification and emulsification. However, both mineral and vegetable oils are solubilized by common solvents (mineral spirits, naphthas, ketones, etc.), and oils are often best removed by solvent cleaning.

Solving cleaning therefore is limited to the removal of common oils and greases from surfaces by dissolving them. The advantages of solvent cleaning are as follows:

- Solvents readily remove oil and grease.
- Solvents are easily applied, are inexpensive, and require little effort to clean small contaminated areas.

There are several disadvantages:

- A solvent, if reused, may become oil-contaminated and, instead of removing oil or grease, spread it over the surface being cleaned.
- Extensive solvent cleaning of large areas can be expensive and time-consuming.
- Most solvents present a potential fire or explosion hazard. The fumes of most solvents are toxic, and prolonged contact with the skin can result in irritation and a loss of skin-moisturizing oils.
- Only oils and greases can be removed. Rust, mill scale, and other nonoil contaminants are unaffected by solvent cleaning.

The SSPC *Surface Preparation Specification for Solvent Cleaning* states that cement spatter and other foreign matter should first be removed by brushing with a stiff fiber or wire brush, by scraping, or by cleaning with alkaline cleaners. If alkaline cleaners are used, they should be followed by a freshwater rinse. To remove oil and grease the solvent can be applied by wiping the surface with rags or brushes wet with solvent, by spraying the surface, by vapor degreasing, using stabilized chlorinated hydrocarbon solvents, or by immersing the surface in tanks of solvent. In all cases, the final solvent rinse shall be carried out with a clean, noncontaminated solvent; otherwise the oil or other contaminants will merely be spread over the surface being cleaned.

While the principle of solvent cleaning is quite simple, in many cases the required equipment is not.

Wiping. Wiping with clean cloths, sponges, brushes, etc., dipped into the solvent cleaner and scrubbed over the metal being cleaned is a common practice. The excess solvent from the cleaning and wiping operation must be removed either by draining it from the surface or by wiping it up with a rag or sponge. The last rinse, of course, must be made with clean solvent; otherwise a residual oil film may be left on the surface. The solvent used in the cleaning operation cannot be reclaimed when cleaning is done by this method.

Dipping. In dipping, the surface being cleaned is immersed in tanks of solvent, either at ambient temperature or heated. If only one tank is used, the solvent will become contaminated. Therefore, usually two or more tanks are used, the first tank containing the dirty solvent utilized for preliminary cleaning.

Spraying. The solvent is sprayed onto the metal being cleaned. It then drains back into a reservoir or sump and may be reused. Again, it is important that the final spraying be done with clean solvent.

Vapor Degreasing. In the simplest form of this method, solvent is held in a reservoir at the bottom of the vapor degreasing tank. The solvent is boiled to create vapors filling the entire inside of the tank. Work parts are laid in the hot vapor, which condenses on them and flows down, dissolving oil and grease in the process. The liquid solvent falls into the solvent reservoir, carrying the oil and grease with it. Because the heat is not great enough to vaporize the grease and oil, they stay in the reservoir of solvent. Since only clear vapor reaches the parts, no residue is left on them. When the work parts become as hot as the vapor, condensation stops and no further cleaning is accomplished. The work is then removed from the machine and allowed to cool.

This method is suitable only for parts that are lightly soiled and have enough mass to achieve cleanliness prior to reaching vapor temperature. Straight vapor degreasing is not suitable for light-gauge work that will reach vapor temperature before total cleaning is achieved. It is employed mainly for in-plant cleaning and not for field use.

The two major hazards in the use of solvents are fire and toxicity. The cleaner with the lowest flash point should be used when all other factors are equal. All solvents are toxic to some degree, the fluorinated hydrocarbons being less so than the others. Breathing the vapors will cause headaches, nausea, dizziness, and a general feeling of lassitude. Severe overexposure will cause vomiting, unconsciousness, and even death. Affected individuals should be brought into fresh air. If they are unconscious, they should have immediate medical attention.

SSPC SP-1 includes a list of recommended solvents (also see Table 6.2).

Alkali cleaning   This process involves the action of alkalies or synthetic detergents (surfactants), which lift the soil from the surface of the work

TABLE 6.2 Surface-Preparation Specifications

Specification	Subject	Photo SSPS Vis 1	Purpose
SSPC Vis 1-67T	Description of visual standard		Photographic standards used as optional supplement to SSPC surface-preparation Nos. 2, 3, 5, 6, 7, and 10
SSPC Vis 2-68T	Visual standard for degrees of rusting of painted steel		Linear numerical scale for evaluating degree of rusting of painted steel; illustrated by black-and-white dot diagrams and/or color photographs
SSPC SP-1	Solvent cleaning		Removal of oil, grease, dirt, soil, salts, and contaminants by cleaning with solvent, vapor, alkali, emulsion, or steam. (The specification itself gives detailed solvent recommendations)
SSPC SP-2	Hand-tool cleaning	B, C, D St 2	Removal of loose rust, loose mill scale, and loose paint to the degree specified by hand chipping, scraping, sanding, and wire-brushing
SSPC SP-3	Power-tool cleaning	B, C, D St 3	Removal of loose rust, loose mill scale, and loose paint to the degree specified by power-tool chipping, descaling, sanding, wire-brushing, and grinding
SSPC SP-4	Flame cleaning of new steel		Dehydrating and removal of rust, loose mill scale, and some tight mill scale by use of flame, followed by wire-brushing
SSPC SP-5	White-metal blast cleaning	A, B, C, D Sa 3	Removal of all visible rust, mill scale, paint, and foreign matter by blast cleaning by wheel or nozzle (dry or wet), using sand, grit, or shot (for very corrosive atmosphere when the high cost of cleaning is warranted)

Specification	Subject	Photo SSPS Vis 1	Purpose
SSPC SP-10	Near-white blast cleaning	B, C, D Sa 2½	Blast cleaning nearly to white-metal cleanliness until at least 95 percent of each element of the surface area is free of all visible residues (for high humidity, chemical atmosphere, marine or other corrosive environments)
SSPC SP-6	Commercial blast cleaning	B, C, D Sa 2	Blast cleaning until at least two-thirds of each element of the surface area is free of all visible residues (for rather severe conditions of exposure)
SSPC SP-7	Brush-off blast cleaning	B, C, D Sa 1	Blast cleaning of all except tightly adhering residues of mill scale, rust, and coatings, exposing numerous evenly distributed flecks of underlying metal
SSPC SP-8	Pickling		Complete removal of rust and mill scale by acid pickling, duplex pickling, or electrolytic pickling (may passify surface)

SOURCE: *Steel Structures Painting Manual,* vol. I: *Good Painting Practice,* p. 40, Steel Structures Painting Council, 4400 Fifth Avenue, Pittsburgh, Pa. 15213.

being cleaned and displace it so that it can be flushed away by either the cleaner or a water rinse. The speed and efficiency with which this action takes place depends upon heat and the level of agitation. The effect of agitation is far more significant when the contaminants contain finely divided particles or are viscous.

Synthetic Detergents. There are three different types of industrial detergents: alkaline detergents, acidic detergents, and emulsifiable-solvent detergents. Alkaline detergents are by far the most widely used and are the only type discussed here. These detergents are based on a sodium or potassium cation in combination with various anions. The important anions used are carbonates, phosphates, silicates, and hydroxides. These alkaline solutions are combined with surface-active or wetting agents to form the detergent. Depending upon the type and severity of cleaning for which they are designed, alkaline synthetic cleaners vary in alkalinity, heavy-duty types generally being more highly alkaline.

The detergent is generally placed in a tank and heated. The object being cleaned is then immersed in the tank and agitated, or the detergent solution itself is agitated. Alternatively, the detergent may be brushed or sprayed on the surface to be cleaned; best results are achieved when application of a detergent is accompanied by brisk rubbing with a stiff scrub brush. After the detergent cleaning, the cleaning solution should be thoroughly rinsed off with fresh water.

Alkalies are highly effective cleaners, readily removing oils, greases, and other soil as well as soluble rust stimulators. In addition, alkali cleaning is economical, and alkali cleaning solutions are free of toxic fumes.

Alkalies are less effective than solvents in removing heavy or carbonized oils, rust-inhibitive oils, etc. Used alone, they leave an alkaline surface which is unsuitable for subsequent painting. However, dilute-acid rinsing overcomes this defect.

Often detergents are mixed with water and an organic grease solvent such as kerosine or mineral spirits, which is then added to the alkaline paint stripper. Because of the solvent, the solution must be used at temperatures below the solvent's vaporization temperature. Generally, these solutions are used only in immersion, although occasionally they are employed in spray stripping.

Alkaline Paint Strippers. Some alkaline solutions such as caustic soda and sodium hydroxide are used to strip saponifiable paints such as oil-based paints, alkyds, or epoxy esters. The key process in paint stripping is saponification. The chemical reaction with strongly alkaline solutions such as caustic soda or sodium hydroxide converts animal or vegetable oil or grease into a water-soluble soap. Once the oil vehicle or binder has been converted into soap, the paint can be washed away. The three most common methods of alkaline paint stripping are as follows:

1. *Hot flow-on stripping.* The stripping solution is sprayed through a spray head such as a perforated pipe or rake positioned at the top of the surface to be stripped. The solution cascades down over the paint surface, and the used solution is then transferred to a central tank for heating and recirculation.

2. *Tank-immersion stripping.* This method consists of dipping parts into a hot or cold solution of the alkaline paint stripper. Agitation and heat will speed paint removal. Once the paint has been loosened, a high-pressure rinse completes paint removal.

3. *Steam-gun stripping.* This method consists of first wetting down a large area with a steam stripping solution. Then a short soak is applied for the stripper to react, and a smaller section of the wet area is reworked. The cycle should be repeated until the entire painted surface has been stripped. Steam is often used to remove loosened paint at

regular intervals. This method may be the only feasible one for removing paint from large equipment parts. Alkaline cleaners in the immersion process usually use concentrations of 4 to 10 oz/gal (29.9 to 74.9 kg/m³) of cleaner; temperatures range from 180 to 212°F (from 82 to 100°C).

Alkali concentrations used in power-spray washes are generally considerably lower than those used in other methods. The usual concentration is ¼ to 2 oz/gal (1.87 to 14.9 kg/m³), and the cleaner is changed somewhat more frequently than in the soak tank or dip cleaning to avoid excessive accumulations of dirt that would clog spray nozzles and soaps that would cause foaming. Recommended temperatures range from 160 to 200°F (from 71 to 93°C); cleaning time, from 15 to 75 s.

Electrolytic cleaning   Basically, electrolytic cleaning is tank-immersion cleaning with agitation provided by the movement of hydrogen and oxygen bubbles that are created by the electrolytic decomposition of water in the solution. By having the work serve as the anode, the normal activity (wetting, emulsification, etc.) of the detergent is aided by the fact that positively charged smuts and contaminants are repelled by the positively charged work. In addition, oxygen bubbles are liberated on the surface of the steel (as a result of the decomposition of water into hydrogen and oxygen gas), resulting in agitation or a scouring effect to remove soil. This technique is widely used prior to electroplating, in which the ultimate degree of cleanliness is essential for a chemically cleaned surface as a preparation for inorganic finishing. Electrolytic cleaning is not used to remove paint but only to remove oil and grease. Recommended concentrations are 6 to 14 oz/gal (44.9 to 104.8 kg/m³) of alkaline cleaning solution; temperature ranges from 180 to 210°F (from 82 to 99°C). At about 6 V, current density is of the order of 50 A/ft², the range being from 25 to 100 A/ft² of material being cleaned.

Acid cleaning   "Acid cleaning" is defined as the removal of oxides in the form of corrosion products from a metal surface by means of an acid solution. Acid-cleaning solutions are most often applied to prepare aluminum or galvanized surfaces for painting. They are usually based on a phosphoric acid or dilute hydrochloric acid solution and produce a clean, finely etched surface suitable for paint application. However, acid cleaners are also occasionally applied to steel surfaces to remove grease, oil, dirt, or other surface contaminants and to remove or neutralize rust stimulators and light rust.

The success of acid cleaning can be attributed to two chemical actions: (1) the cleaning and etching of the metal surface and (2) the surface formation of a phosphoric acid complex with the metal surface that produces a thin, insoluble, corrosion-resistant surface coating. This thin surface film offers a good base for subsequently applied paint coatings and retards spur corrosion beneath the paint.

Although acid cleaning will remove mill-scale and heavy-scale contaminants, hours or even days of immersion generally would be required for complete removal. Therefore, acid pickling is used for mill scale and heavy rust scale. This process is often followed by phosphoric acid cleaning to remove residually softened corrosion products and to impart a more passive surface to the cleaned metal. There are a large number of phosphoric acid metal cleaners and rust removers, each designed for a particular metal-cleaning job.

If heavy coatings of oil or grease are present on the metal being cleaned, it may be advisable first to remove them with suitable solvents to avoid premature fouling of the acid-cleaning solution. The acid-cleaning solution can be applied by a number of methods, usually either brushing it on the metal being cleaned, using a long-handled scrub brush or broom, spraying it, or immersing the object being cleaned in a tank of the solution. After the solution has had time to set and clean the metal, it is wiped or rinsed off with hot or cold water. The surface is then painted after drying.

The chief advantages of acid cleaning are that surface contaminants are relatively effectively removed and that the residues from phosphoric cleaning are generally beneficial. Costs are moderate, and application generally is not difficult. The disadvantages are that immersion tanks and spray equipment, if used, must be acid-resistant. Furthermore, good results depend directly upon the skill and care of the operator.

Phosphoric acid cleaners are not very corrosive to the skin, but constant contact with an acid cleaner will dissolve oil from the skin, causing dryness and cracking. Operators should be equipped with protective clothing, rubber gloves, and goggles.

Pickling   Pickling is carried out by immersing the surface being cleaned in a dilute acid. While various acids, such as sulfuric, nitric, hydrofluoric, and phosphoric, can be used in commercial pickling, sulfuric acid is used almost universally for the pickling of carbon steels because of its low cost, availability, boiling point, and suitability.

The acid, usually heated to around 200°F (93°C), readily attacks the ferrous metal and any metallic contaminants such as mill scale or rust. Besides dissolving the rust or scale, hydrogen bubbling from the acid attack provides surface agitation to enhance the rate of cleaning. Agitation, produced either by moving the piece within the tank or by adding steam, also increases the cleaning rate. As pickling continues, iron sulfate is formed in the pickling bath. As the iron sulfate concentration increases, the rate of pickling decreases. When the solution reaches approximately 3 lb/gal (359.5 kg/m^3), the pickling bath is usually discarded. (This concentration is equivalent to about 0.6 lb/gal, or 71.9 kg/m^3, of dissolved iron.)

Inhibitors are added to the bath to reduce the rate of attack on the base metal. While the inhibitors have little effect on the rate of descaling or rust removal, they function to reduce or slow acid attack to the clean steel. Without the use of sufficient quantities or the right types of inhibitors, overpickling may result. When this happens, the steel surface will become rough and pitted and there will be greater smut contamination on the piece being cleaned. (Smut is the contamination caused by carbon particles that are left loose on the surface when the iron that originally supported them is dissolved by the acid in the pickling bath.)

The advantage of pickling is that it is a relatively inexpensive and relatively fast and effective method of removing scale, rust, and other metallic contaminants. Most coating systems perform well over a pickled surface, although some, notably the inorganic zincs, should ideally have a blast-cleaned surface for greater adhesion. A disadvantage is that pickling must be done in a specialized facility; it is not a field cleaning technique. Second, hydrogen evolved during the acid cleaning may diminish the flexibility and ductility of the metal, causing what is known as hydrogen embrittlement. Furthermore, nascent hydrogen may be absorbed into the metal, where the individual atoms recombine to form hydrogen gas. Such a recombination results in blisters in the metal. Exactly how hydrogen penetrates the metal is unknown, but many demonstrations have proved that it will penetrate and pass quickly through sheet steel.

Conversion coatings   Pickled or chemically cleaned surfaces are often pretreated prior to painting. Such pretreatments are called "conversion coatings"; the most common are zinc phosphate conversion coatings and chromate conversion coatings. Phosphate conversion coatings are produced by treatment with solutions containing phosphoric acid that are saturated or supersaturated with acid zinc phosphate salts. Steel pickled in a sulfuric or hydrochloric acid bath often is further prepared by immersing it in a dilute phosphoric acid bath to form a conversion coating. Chromate conversion coatings can be produced on zinc, cadmium, aluminum, magnesium, copper, and brass. The chromate conversion film can be applied by immersion or, in some cases, by brushing or spraying. A number of suppliers manufacture proprietary chromate conversion coatings for commercial use.

Phosphate conversion coatings for steel or ferrous surfaces and chromate conversion coatings for zinc, aluminum, cadmium, and other metals provide an excellent base for subsequent painting. The adhesion of most organic coatings is improved when a conversion coating is present, and underfilm corrosion is substantially diminished. In fact, the presence of a chromate conversion coating on strip galvanized coil steel is almost mandatory for good paint adhesion and the reduction of the zinc corrosion called "white rust."

Paint strippers
Alkali Cleaners.   While all previously described cleaning methods remove contaminants from a metal, concrete, or wood surface or cleanse

such surfaces of grease and oil, it is often desirable to remove paint, lacquer, and other old coatings from the surface by chemical means. Alkaline cleaners used as dip or spray solutions have been described. Such cleaners can be considered adequate for paint removal when the caustic is strong enough and contact long enough to saponify and attack the old paint. Caustic removal of old paint is a lengthy process, and large pieces often require numerous immersions in hot caustic. While all paint coatings are attacked to some degree by hot caustic, this stripping method works best with oil-based coatings such as alkyds, epoxy esters, and some alkali-sensitive resins such as butadiene-styrene, polyvinyl acetate, and other resins containing the ester linkage.

Solvents.   Solvents can be used to remove paint, and most thermoplastic coatings can readily be removed by redissolving them in the appropriate thinner. Vinyls, chlorinated rubbers, and some acrylates can readily be removed by using appropriate solvents. Shellac, varnish, and similar furniture coatings can be removed by flooding the surface with denatured alcohol, allowing the alcohol to remain in contact with the surface for a short while, and then rubbing or scraping it with steel wool.

Other popular paint removers are based on methylene chloride, perhaps blended with trichloroethylene, and ethylene dichloride. Small amounts of paraffin wax, in combination with alcohol and some cellulose or cellulose acetate and/or thixotropes, may be used to give the mixture body and thus to keep it from running down vertical surfaces.

Paint-stripping operations, except for immersion methods, are generally quite slow and labor-intensive. For small areas, however, these methods can produce a very clean surface free of all old coating.

Recent developments: chelates and rust-converting coatings Some work is being done, in both Europe and the United States, to develop an organic coating or conditioner that is said to react directly with the metal surface to form an organometallic complex called a "chelate." The chelating agent reacts with various metallic oxides (generally iron oxides), tying them up and reducing their subsequent reactivity. The complexed iron oxides are said to be relatively inert and stable and to provide a good base for a conventional coating system. Chelate pretreatments are designed for use over rusting surfaces, although heavy rust scale and corrosion pit deposits must first be removed. The chelate solution may be applied by any conventional means (roller, brush, spray); after a suitable reaction time (generally 12 h) the excess solution is washed off, and the surface is ready for painting. The success of chelating solutions has not been suitably established, and this potentially promising method of stabilizing a rusty surface needs work.

6.3 Surface-Preparation Standards, Inspection, and Inspection Equipment

6.3.1 Cleanliness standards   Perhaps the best-known and most widely referenced surface-preparation cleaning standards are those published by the Steel Structures Painting Council (see Table 6.2). The National Association of Corrosion Engineers sells plastic-encapsulated blast-cleaned metal coupons cleaned to white, near-white, commercial, and brush-off degrees of cleanliness. Photographic standards depicting various steel starting conditions, followed by varying degrees of hand and power-tool cleaning and blast cleaning, are also sold by the SSPC (Fig. 6.5). These standards, originally developed by the Swedish Standards Institution in Stockholm, are commonly called the Swedish standards but officially are now covered under SSPC Vis 1.

The above-mentioned standards for surface preparation are the most widely used in the United States, but there are others that have limited use or appeal to specialized groups. The Maryland Department of Transportation has developed a photographic blast-cleaning standard depicting near-white and commercial blast cleaning that has been adopted by five other states. The Society of Naval and Marine Architects and Engineers has printed color photographs showing varying degrees of paint removal from a steel plate. There are European and Japanese photographic standards similar to the Swedish standards depicting degrees of hand, power-tool, and blast cleaning. Some of these standards also depict degrees of cleaning and/or grinding of welded areas.

Instead of using a nationally accepted photographic or standard definition, many specifiers stipulate that at the outset of a job a portion of the work be cleaned in the presence of the specifying engineer, inspector, and other concerned parties. A decision on the extent of cleaning required will be made in the field, and the blast-cleaning area will be preserved as well as possible, perhaps by spraying a clear coating over the surface or, in the case of a small panel, by encapsulating it in plastic. The job-prepared "standard" will then be the basis of acceptance for all future cleaning work.

As might be expected from the variety of cleaning standards, there is no clear-cut, incontrovertible method of specifying hand or power-tool cleaning or blast cleaning. Steel corrodes differently in different environments, and the amount and colors of rust, rust scale, mill scale, and corrosion pitting allow considerable departure from any defined pictorial standards.

6.3.2 Surface profile   The determination of surface profile also is somewhat arbitrary. The three most common field methods use a depth micrometer, a surface-profile comparator, and a replica tape.

Depth micrometer   The most common depth-micrometer technique is

Fig. 6.6 Depth micrometer.

to use a penetrating-needle dial micrometer with a flat base. The base of the instrument rests upon the tops of the profile peaks, while the needle penetrates below the base into the bottoms of the valleys. The depth of penetration below the plane of the base is recorded on the dial face of the instrument. Another, less used method is to grind a small area of the blast-cleaned surface until the bottoms of the valleys can no longer be seen. This can best be done by a milling machine, although it has been done successfully with a hand grinder. A depth micrometer can be used to measure the anchor pattern, or a standard micrometer can be used to take the difference between a reading of the profiled surface and the ground surface.

Surface-profile comparator   The surface-profile comparator consists of a blast-cleaned disc or series of coupons with a known profile depth. The profile standard is placed on the blast-cleaned surface, and a visual or tactile comparison is made. The Keane-Tator surface-profile comparator uses an electroformed disc that has been accurately measured with a microscopic technique by the SSPC. Three discs, for sand, shot, and grit abrasives, are available. Each disc has five segments, each of which has a different anchor-pattern height. A 5X illuminated flash magnifier can be used to compare the blast-cleaned surface more accurately with the disc standard. Also available are blast-cleaned stainless-steel coupons of different anchor-pattern depths. The coupons also are used for visual and tactile comparison.

Replica tape   The Testex Press-O-Film replica-tape method determines the anchor-pattern profile by compressing a foamlike tape onto

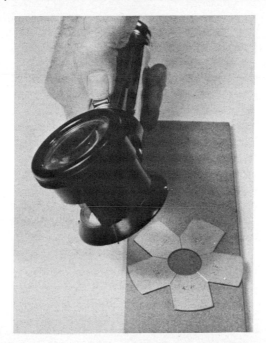

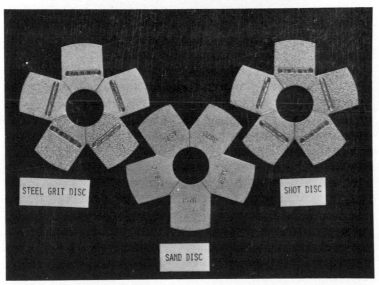

STEEL GRIT DISC

SHOT DISC

SAND DISC

Fig. 6.7 Keane-Tator surface-profile comparator: (above) the comparator in use; (below) the three discs.

the blast-cleaned surface. The tape crushes into the anchor pattern, and what is essentially a replica of the anchor pattern is produced. The profile height can then be measured with a specially modified spring micrometer. The advantage of this method is that a definite number, rather than a comparative estimate of the closest-appearing coupon or disc segment, is obtained.

As a general rule, thicker coatings require a greater surface-profile depth. It is felt that a surface profile ranging from 1.5 to 3.5 mils (0.0381 to 0.0889 mm), as measured by any of the above methods, is satisfactory for almost all coatings applied today primarily for corrosion protection.

6.3.3 Inspection   Surface-preparation and coating-application inspection are highly recommended to ensure that specification requirements are upheld. The old adage "A painter covers his mistakes" is often all too

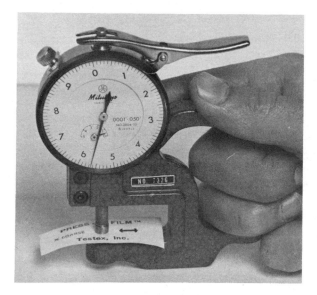

Fig. 6.8   Testex Press-O-Film ® replica tape.

true. After paint has been applied to a surface, it is all but impossible to determine whether or not the desired or specified cleanliness was obtained. Usually the paintwork will look good after application, but within a year or so premature rusting, pitting, and peeling of paint may occur if the surface was poorly prepared. Unannounced spot inspections are often sufficient to keep work quality high, although on important or large jobs full-time on-the-job inspection by a trained coatings inspector may be required.

INSPECTION CHECKLIST FOR SHOP-APPLIED COATINGS

1. The use of *cutting oils* shall be such that a minimum of oil is deposited on the steel. Any oil on the steel shall be carefully removed with solvents.

2. All *contact surfaces* connected by bolts shall be blasted and coated with a suitable primer before being shop-bolted. When carbon steel bolts (not cadmium-plated or galvanized) are used in the shop, the bolt heads and nuts shall be blasted and coated.

3. *Weld spatter* shall be removed before priming.

4. The *cleaning of shot or grit* before reuse shall be such that the blasted steel is free of smudge.

5. All *back-to-back angles* shall be blasted and primed before assembly.

6. *Slivers* shall be ground smooth before priming.

7. All *sharp, rough, or burred edges* shall be ground smooth.

8. All *abrasive* shall be *removed* before priming.

9. *Bucks* (*timbers*) shall be *cleaned* of foreign paints or coatings so as to avoid contamination of the coating system.

10. *Paint hoses* shall be *cleaned* of dried paint to avoid loose paint's falling into wet coatings.

11. Special care shall be used by painters to assure proper *thickness on flanges,* inside and out.

12. If *identification tags* are *welded* to the steel, the weld shall be continuous to avoid corrosion underneath the tags. *Metal marking devices* shall be selected so as *not to bleed* through plastic resin topcoats. Suitable examples are Nissen metal markers and Speedry Chemical Marker No. 2017.

13. *Dunnage* shall be used judiciously *to minimize damage* to coatings in loading, shipping, or unloading.

14. At the jobsite the steel shall be *unloaded* so as *to minimize coating damage.* Hooks at both ends should be used as often as possible. Chains wrapped around the steel should be minimized. Nylon slings are recommended. In stacking the steel at the jobsite, ample dunnage shall be used.

Courtesy Porter Coatings, Houston, Texas.

6.3.4   Guide specifications   Two surface-preparation specifications for the blast-cleaning surface preparation of the interior of a water tank are presented here. The first defines only the desired end result and allows the sandblasting contractor every latitude to achieve the specified degree of cleanliness. Such a specification is entirely adequate for most jobs.

SIMPLE SURFACE-PREPARATION SPECIFICATION

1.0 *Surface preparation.* The exterior surface preparation shall be as follows. The bottom three rings shall be fully sandblasted according to SSPC SP-10 near-white blast-cleaning specifications. Any rusted or abraded areas on the remaining areas of the tank shall be spot-sandblasted according to SSPC SP-10 near-white blast-cleaning specifications. Care shall be taken that the edges of the spot-blasted areas be sanded or feathered smooth so that a uniform substrate is available for painting. The remainder of the exterior surface shall receive an SSPC SP-7 blast. All debris from the blasting process shall be hauled to a suitable site arranged by the contractor.

The second specification not only defines the desired end result but also has specific requirements regarding ambient conditions, blast-cleaning air quality, air pressure, abrasive size and type, cleanliness after blast cleaning, etc. Such a specification may unduly restrict a contractor on many jobs, but used intelligently by a knowledgeable specification writer, it may eliminate dishonest contractors and improve work quality. Which is the best specification? The answer depends upon the knowledge of the specification writer and the nature of the job.

DETAILED SURFACE-PREPARATION SPECIFICATION

1.0 *Surface preparation*

    1.1 The subcontractor shall examine all surfaces to be coated to determine that such surfaces are suitable for specified surface preparation and coating. Any surface found to be unsuitable shall be reported immediately to the contractor in writing for corrective action. Subcontractor shall not start surface preparation until such surface has been corrected. Starting surface preparation shall preclude any subsequent claim by subcontractor that such surface was unsuitable for the specified surface preparation or coating.

    1.2 Blast-cleaning equipment including pots, hoses, and nozzles shall, as determined by the contractor, be clean and suitable for the work. Adequate clean compressed air in quantities and at pressures as specified by this specification shall be available for use. Effective oil and water separators shall be used in all main compressed-air lines and shall be placed as close as practicable to the equipment. Prior to using compressed air, quality of air downstream of the separator shall be tested at suitable outlets by blowing the air on a clean white blotter (minimum size, 8½ by 11 inches, or as approved) for 2 min to check for any contamination, oil, or moisture. Such test shall be performed at the beginning of each shift and at not less than 4-h intervals. Test shall also be made after any interruption of the air-compressor operation or as required by the contractor. Air shall be used only if the blotter test indicates no visible contamination, oil, or moisture on the blotter. In addition, separators shall be bled continuously or at intervals approved by the contractor. All lines shall be tested individually prior to use.

1.3   Prior to blast cleaning, contamination shall be removed from the steel surfaces. Oil and grease shall be removed by solvent cleaning in accordance with SSPC SP-1.

1.4   Steel surfaces shall be blast-cleaned as defined in SSPC SP-10 with compressed air and material as specified. Air pressure at nozzle shall be not less than 80 psig.

1.5   Material for blast cleaning shall be one of the following or approved equal:

    1.5.1   Abrasive for manual blasting shall be packaged dry Monterey sand, Ottawa sand, or Joplin flint of 30- to 60-mesh size.

    1.5.2   Abrasive for automatic centrifugal blasting shall be a composite mix of 1 part steel grit, SAE GL-40, and 1 part shot, SAE S-230 or S-330. Shot alone shall not be used.

1.6   Anchor profile shall be between 1½ and 3½ mils, as determined by Keane-Tator profile comparator.

1.7   Blast-cleaned surfaces shall be coated before any rust visible under a 5X magnification forms. Surface shall be spot-checked to ensure that rust has not formed prior to coating. Moreover, unless approved in writing, the total elapsed time from start of blast cleaning to application of coating shall not exceed the time tabulated below. (This table shall be used as a guide only.)

Maximum time elapsed (hours)	Relative-humidity range (percentage)
4	73.1–80
8	67.1–73
12	60.1–67
18	55.1–60
20	50.1–55
24	42.1–50

1.8   If contaminants become embedded in the steel surfaces, the steel shall be reblasted.

1.9   Immediately prior to coating, all surfaces shall be near-white-metal–blast-cleaned (SSPC SP-10), dry, free of dust, and loose particles removed by oil-free air blowoff followed by vacuum cleaning.

1.10   Surface-preparation operations shall be performed only under the following conditions:

1.10.1   No moisture shall be present on surface.

1.10.2   Relative humidity shall not exceed 80 percent.

1.10.3   Ambient and metal surface temperature shall be greater than 5°F above the dew point.

1.10.4   Blasting shall not be performed in the same area where coating or drying of coatings is in progress.

1.10.5   Equipment shall be in good operating condition, as approved by contractor.

1.11   Prefinished items and adjacent surfaces not to receive coating shall be masked and protected prior to surface preparation and during all operations.

1.12   Blasted surfaces shall be subject to approval by the contractor prior to the application of the prime coat.

BIBLIOGRAPHY

Bibliography on Chemical Cleaning of Metals, TPC Publication No. 6, National Association of Corrosion Engineers, Katy, Tex., 1976.

Effects of Surface Preparation on Service Life of Protective Coatings: An Interim Statistical Report, National Association of Corrosion Engineers, Katy, Tex., 1977.

NACE Surface Preparation Handbook, National Association of Corrosion Engineers, Katy, Tex., 1977.

Recommended Practices for Shop Cleaning and Priming, National Association of Corrosion Engineers, Katy, Tex., 1961.

Surface Preparation of Steel and Other Hard Materials by Water Blasting prior to Coating or Recoating, National Association of Corrosion Engineers, Katy, Tex., 1972.

Tator, Kenneth B.: *Inspection and Inspection Procedures,* Third International Symposium on Pulp and Paper Industry Corrosion, May 5–8, 1980.

Visual Standard for Surfaces of New Steel Airblast Cleaned with Sand Abrasives, National Association of Corrosion Engineers, Katy, Tex., 1970.

Visual Standard for Surfaces of New Steel Centrifugally Blast Cleaned with Steel Grit and Shot, National Association of Corrosion Engineers, Katy, Tex., 1975.

Chapter 7

Coatings for Steel

JON RODGERS
W. M. McMAHON
Reliance-Universal

GUY E. WEISMANTEL
Chemical Engineering

7.1   Elements of the Specification   To assure satisfactory performance the architect or corrosion engineer must specify the total system. This includes:

Surface preparation
Pretreatment, if needed
Application techniques, including timetables
Paint thickness (wet and dry)
Coating system (primers, intermediate coat, and finish coat)
Instructions for shop or field priming and painting
Inspection requirements
Procedure for dealing with problems
Color schedules
Other data as required (procurement, storage, etc.)

Design considerations can also affect coating performance. This is true of most ferrous structural shapes such as plates, beams, channels, pipes, bars, and thin-gauge sheets. For these surfaces, obviously, it is sometimes necessary to apply a protective coating before they are assembled as part of an overall structure. This is particularly true of a structure that has areas that become inaccessible as it is assembled. An example is the application of siding to the structural-steel framework of a building. Once the siding has been attached to the structure, it is impossible to reach the face of the steel in contact with the siding. The face must therefore be painted well before the siding is applied. Another problem arises because conventional practice often allows the use of skip welds instead of continuous welds for joining members; skip welds are chosen for their strength, but they create a crevice (the "skip") where the two surfaces being joined are not welded. The protective coating being applied over the surfaces must then bridge the skip and provide protection in the unpainted crevice. Many steel structures have failed in this area over a period of time because it has not been possible to stop the ingress of water and chemicals and subsequent crevice corrosion in the area where a coating could not be applied.

For industrial plants, surface-preparation and coating standards are in effect before the plant goes into operation. Structural steel and all steel vessels usually are shop-coated at the fabricating shop. Paint inspection must start at this point because the steel often is shipped with either a prime coat or prime and finish coats. Coatings are selected, among other reasons, for their resistance to damage in being handled by slings, wire ropes, etc. They must be capable of easy repair in the field with minimum surface preparation and with maximum longevity and protection.

In addition, many refinery and petrochemical plants face dangers of fire and explosion. No sparks of any kind are allowed, and the use of static-dissipating sandblasting hose and the electrical grounding of

equipment and hoses are therefore required. Because of this requirement, some operators elect to compromise on allowable surface-cleaning
techniques, and this in turn impacts on coating selection.

Welding is a problem before and after painting. If welding is required
after shop coats have been applied, the paint can interfere with the weld.
To avoid such problems, specify that the prime coat be held back a
minimum of 1 in (25.4 mm) on both sides from the edge prepared for
welding (Fig. 7.1). The problem is exacerbated when steel is prepainted
with a zinc-rich primer prior to fabrication. This prime coat can reduce
welding and gas-torch cutting speeds and, more important, reduce weld
strengths and corrosion protection at and near the seam.

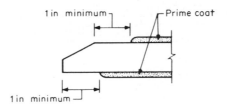

Fig. 7.1 Do not prime an area that requires welding.

Field welds used to assemble piping, equipment, and structural steel
which have been shop-primed in all cases require field touch-up of the
prime coat for a minimum distance of 1 in on each side of the weld. Such
painting should be done only *after* the removal of all weld splatter and
proper surface preparation. Painting over welding flux is a serious cause
of corrosion, lack of coating adhesion, and structural failure. *Remove all
weld splatter prior to sandblasting.*

The paint specification writer should coordinate these painting requirements with the work of the architect and the construction crews,
maintaining close contact via inspection to assure that other crafts do not

TABLE 7.1 Typical Surfaces That Are Not Painted

Nameplates, code inspection plates, and code stamps.
Machined surfaces and finished metal operating parts of machinery, valves, and other
  equipment such as valve stems, shafts, and flange faces. Use Cosmolene or other suitable
  grease-type rust inhibitors to give adequate protection from corrosion until such
  equipment is installed.
Galvanized or other plated surfaces not scheduled for painting.
Interior carbon steel surfaces of piping and equipment unless otherwise specified as
  equipment or piping spool drawings.
All carbon steel surfaces which are to be fireproofed.
Stainless-steel and nonferrous metal surfaces such as aluminum, copper, brass, and Monel
  Metal.
Glass, plastic, and tile or other ceramic surfaces.

create problem surfaces for painting (e.g., by weld spattering). An important function of a job specification is to designate which surfaces should *not* be painted (Table 7.1) and also to specify steel surfaces that require galvanizing (Table 7.2).

7.2 Mill Scale Generally, specifications referring to painting carbon steel surfaces concern steels which contain 9 percent chromium or less. If the coating is specifically chosen for a particular steel, it may be necessary, as with piping, to test the steel (by means of a wet chemistry patch test or a nondestructive testing instrument) to verify the alloy. Typically, 3 percent of the steel delivered from a mill is mislabeled.

TABLE 7.2 ASTM Galvanizing Specifications

Number	Description
ASTM A-123	Structural-steel members over ⅛ in (3.175 mm) thick
ASTM A-525	Coating class G-235; structural-steel members or steel sheets less than ⅛ in thick
ASTM A-153	Anchor bolts and all bolts, nuts, and washers used for general structural bolting
ASTM A-325-70a	Bolts, nuts, and washers used for high-strength structural bolting

Mill scale must be removed before painting. A combination of iron and oxygen, mill scale is frequently referred to as blue scale or magnetic oxide of iron. It is cathodic with respect to steel, and when mill scale, steel, and an electrolyte are coupled, a very active corrosion battery in which the steel acts as an anode is established. Mill scale appears on all hot-rolled plates or on structural shapes such as I beams or channels which are extruded from hot steel ingots. Mill scale is a very brittle material which is easily fractured. Further, it has a coefficient of expansion that is entirely different from that of steel and is broken as a result of the normal heating and cooling cycle of any piece of metal exposed to atmospheric conditions. Mill scale must always be removed before any other function, such as painting or coating, is performed.

If a paint or protective coating system is applied over a piece of hot-rolled steel on which mill scale is intact, it can be expected to fail in a short time. No coating film is a perfect moisture barrier, and as water permeates the film, corrosion cells will be established wherever the mill scale is broken. Further, as corrosion products accumulate, they will pry off the mill scale, which will carry with it any paint or coating film which has been applied over it. For a primer to develop adhesion, therefore, it must *contact or wet the surface.* Accordingly, the first requirement for good adhesion is a clean surface and a coating with proper viscosity for wetting.

The greater the surface to which the coating adheres, the greater the total adhesive strength compared with the tensile strength of the system. Accordingly, the second requirement for good adhesion is a great surface area without the high peaks and valleys that accompany a poor surface profile.

Experience confirms these basic considerations. Whatever interferes with wetting or reduces the effective surface area decreases adhesion. For these reasons, the use of sandblasted steel is recommended whenever possible. As bases for maintenance coatings, steel substrates may be ranked in the following order of desirability: (1) white sandblasted steel, (2) commercial sandblasted steel, (3) brush-off sandblasted steel, and (4) wire brushed weathered steel.

The Steel Structures Painting Council (SSPC) through its specifications recognizes nine methods of preparing steel for painting. Of these, four are abrasive-blasting procedures which are also recognized by the National Association of Corrosion Engineers (NACE; Table 7.3). Two widely used commercial standards are the surface-profile comparator (KTA-Tator Associates, Inc., Coraopolis, Pennsylvania; Fig. 7.2) and the Clemtex anchor-pattern standards (CAPS; Clemtex, Ltd., Houston, Texas; Fig. 7.3). These devices are designed for the quick field inspection of a blast-cleaned surface. CAPS are made from stainless steel with the same hardness as mild steel.

TABLE 7.3   Surface Preparation Specifications

SSPC	NACE	Description
SP 1-63	. . .	Solvent cleaning
SP 2-63	. . .	Hand-tool cleaning
SP 3-63	. . .	Power-tool cleaning
SP 4-63	. . .	Flame cleaning of new steel
SP 5-63	No. 1	White-metal blast cleaning
SP 6-63	No. 3	Commercial blast cleaning
SP 7-63	No. 4	Brush-off blast cleaning
SP 8-63	. . .	Acid pickling
SP 10-63	No. 2	Near-white blast cleaning

A uniform anchor pattern (Fig. 7.4) is vitally important to adhesion. It is obtained by using a clean, well-graded, and sized abrasive (Table 7.4). Sand or grit blasting is the ideal method of surface preparation since it removes not only all scale and corrosion products but chemical contaminants as well.

Fig. 7.2 Three surface-profile comparator discs used in conjunction with a lighted magnifier. [*KTA-Tator Associates, Inc., Coraopolis, Pennsylvania*]

TABLE 7.4   Abrasive Needed to Obtain a Specified Anchor Pattern

Anchor pattern (mils)*	Pressure blast or centrifugal wheel
½	80/120-mesh silica sand, 100-mesh garnet, 120 aluminum oxide, or G-200 iron or steel grit
1	30/60-mesh silica sand, 80-mesh garnet, 100 aluminum oxide, or G-80 iron or steel grit
1½	16/40-mesh silica sand, 36-mesh garnet, 50 grit aluminum oxide, or G-50 iron or steel grit
2	16/40-mesh silica sand, 36-mesh garnet, 36 grit aluminum oxide, or G-40 chilled iron or steel grit
2½	8/40-mesh silica sand, 36-mesh garnet, 24 grit aluminum oxide, or G-25 iron or steel grit
3	8/20-mesh silica sand, 16-mesh garnet, 16 grit aluminum oxide, or G-16 chilled iron or steel grit

Courtesy Clemtex, Ltd., Houston, Tex.
* 1 mil = 0.0254 mm.

Fig. 7.3 Six anchor patterns of different profile depths to be compared with an abrasive-blasted surface to determine the approximate profile of the surface. Clemtex Anchor Pattern Standards (CAPS). [*Clemtex, Ltd., Houston, Texas*]

General overview:
Proper surface preparation: dense uniform pattern (*a*)
Improper surface preparation: nonuniform pattern (*b*)

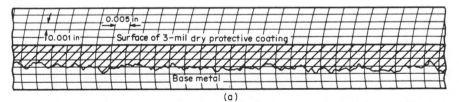

Fig. 7.4 Surface profile for mild steel, using various blasting abrasives. [*Clemtex, Ltd., Houston, Texas*]

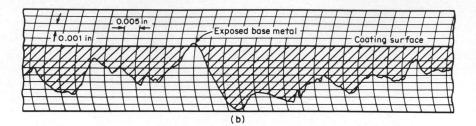

(b)

Silica abrasives

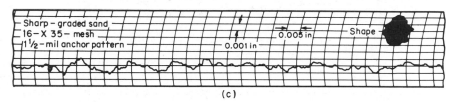

(c)

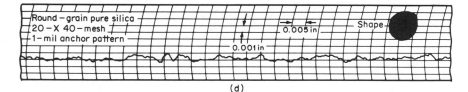

(d)

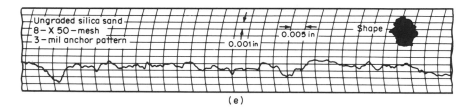

(e)

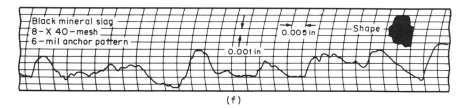

(f)

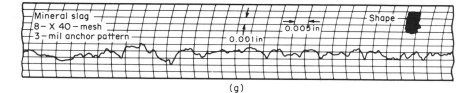

(g)

Other mineral abrasives

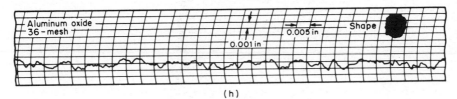

(h)

(i)

(j)

Metal abrasives

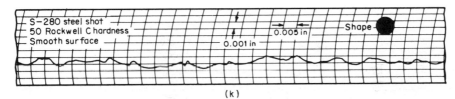

(k)

(l)

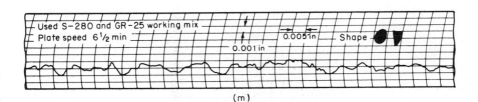

(m)

7.3 Surface Preparation

7.3.1 Sandblasting precautions   There is some confusion regarding blasting requirements. While requirements for SSPC No. 5 for a white-metal blast are well spelled out, this is not true of SSPC No. 6 for commercial blast cleaning. Here the specification calls for blasting 1 ft² (0.0929 m²) of surface every 3 min with a specific abrasive, a fixed volume and pressure of air, and a given nozzle. Obviously, the condition of the resulting surface will depend upon the degree of deterioration of the steel surface at the time that the blasting operation began. For most maintenance work we require or desire a surface which has been blasted free of all mill scale, old paint, and foreign material to yield a surface which is essentially gray in color, with no more than 10 to 20 percent of dark streaks which correspond to retained mill-scale binder. In all cases the specification should devote as much consideration to surface preparation and inspection as to choice of the coating itself.

SURFACE PREPARATION SPECIFICATION

A. When the preconstruction primer is mill-applied (after wheel abrasion) to a dry film thickness of 0.75 to 1.0 mil (0.01905 to 0.0254 mm).

   1. When steel is hot-rolled, retained heat after the rolling operation is adequate to burn off the oil transferred by the rolls, and no oil film is detectable, proceed as follows:

   a. Steel must be dry and free of ice crystals.

   b. If accidental oil or grease soiling occurs before blasting, the soil should be removed by solvent or steam cleaning.

   c. Grit employed in the blasting process must be clean. After each use of grit the collecting tower should be equipped with a clean-air stripping capacity adequate to remove all adhering mill scale, rust, dirt, etc., from the shot surface. If the grit is oil-contaminated, it should be discarded. Periodic grit inspection and sample collection will be conducted by the engineer's field representative. Any steel blasted and coated with an oily abrasive shall be reblasted.

   d. The quality of the cleanliness obtained by the blasting operation shall be judged by comparison with NACE comparison panels, which duplicate the standards specified and described by the Steel Structures Painting Council. The requirement is for SSPC Standard No. 10, near white.

   e. The blasting profile should be set to obtain an average of 2 to 3 mils (0.0508 to 0.0762 mm). The extreme range should not exceed 3 mils or be less than 1 mil (0.0254 mm) in depth.

Profile readings should be judged by comparison, employing the Keane-Tator profile-comparison gauge.

2. For sandblast operations, conditions required under Subparagraphs *A.*1.*a* and *A.*1.*b* remain the same.

   a. Since 30- to 60-mesh sand or grit is generally required to provide the desired profile, sand is discarded after an initial use. If the harder slag grits are to be used, the contractor must demonstrate the cleanliness and efficiency of the recovery system to deliver grit that is clean and oil-free.

   b. The requirements outlined in Subparagraphs *A.*1.*d* and *A.*1.*e* remain the same.

B. When steel is coated in oil, grease, or other material for protection against corrosion in transit. Galvanized surfaces such as conduit, pipe, and junction boxes shall be cleaned and degreased.

1. All such surface treatment when the solvent is recycled for continuous cleaning.

   a. By hot-solvent treatment in which the solvent is recycled for continuous cleaning.

   b. By steam cleaning with detergents of proven capability to remove surface contaminants, followed by a clear steam rinse.

   c. Cleanliness of the steel surface shall be determined by letting water flow over the surface. The water should completely wet the steel without beading, crawling, or breaking.

   d. If alkaline detergents are used in the cleaning process, distilled water should be applied to the surface, and either litmus or suitable pH paper should be brought into contact with the distilled water. If these papers indicate that the surface is alkaline, the steel must be rinsed further until the surface is completely neutral and the test in Subparagraph *B.*1.*c* repeated. For inorganic zinc, a phosphoric acid detergent followed by a clear steam rinse is adequate without pH tests.

2. If all the above tests are satisfactory, coating may proceed.

C. Coating with full film primers.

1. In shop applications all the required surface-preparation conditions are to be met with the exception of that outlined in Subparagraph *A.*1.*e.* If desired, the acceptable profile can vary from 2 to 4 mils (from 0.0508 to 0.1016 mm).

2. Since some of the blasting for full film application takes place in the field, the average of 3 to 4 mils (0.0762 to 0.1016 mm) is likely to prevail. This leeway is permissible.

D. Preparation for application of full film primer after cutting, welding, and assembly of the structure.

1. All damage and construction roughness must be treated as follows:
   a. All rough cuts and edges must be power-ground to eliminate all sharp ridges and points. The surface is then pencil-blasted to eliminate corrosion in pits or valleys. Work must be performed under dry weather conditions or under adequate roofing. These dry working conditions apply to all subsequent operations.
   b. All welds must be power-ground to eliminate any sharp protuberances. If any weld spatter still adheres to the surface, it must be removed. Corrosion at the edges and in the pits and valleys must be pencil-blasted or Vacu-Blasted down to an SSPC SP-6 commercial blast.
   c. The major preprimed surfaces must then be cleaned. Any areas where gouging or scraping has removed the pre-construction primer and corrosion has developed must also be pencil-blasted or Vacu-Blasted.
   d. The larger preprimed areas that remain must be cleaned to remove dirt and soil, grease or oil, crayon marks, etc. This is accomplished most satisfactorily and rapidly with a steam jet employing an acid phosphate detergent having a pH of 3.0 to 5.0 at the nozzle. The fan should be at about 12 in (304.8 mm) when it strikes the surfaces. This procedure is followed by a clear steam rinse. The degree of cleaning in this operation is readily visible, and any remaining soil is easily observed.
   e. Cleaning of soil and contaminants may also be accomplished with a sand sweep blast. This method is not as efficient as steam cleaning because if much oil or grease is present, a considerable amount of the coating film is removed. The film must be replaced by the second-coat application.
2. After approval of surface cleanliness, the second zinc-rich full film application may proceed.

E. Finish coat application.
   1. After allowing an adequate drying period for the inorganic zinc (usually 2 to 6 h, depending on weather conditions), the first coat of water-thinnable acrylic enamel may be applied, provided the inorganic zinc surface has been protected against further soiling or damage.
   2. If the inorganic zinc is allowed to stand several days without topcoating, the type of cleaning required to remove dust, salt deposits, or other contaminants can be determined in consulta-

tion with the owner's engineers. The first topcoat of acrylic enamel is usually applied in the constructor's shop.

3. The second coat of enamel is usually applied at the site after final erection. All construction damage that exposes bare metal must be treated to remove rust, dirt, or oil and spot-coated with a full coat, 2½ to 3 dry mils (0.0635 to 0.0762 mm), of inorganic zinc and then touched up with a coat of acrylic enamel.

4. The final coat of acrylic enamel may then be applied, provided the surface of the first acrylic-enamel coat is clean.

Brush-off blast requirements (SSPC No. 7) are also ambiguous. There is no definition of either the condition of the surface at the time that the blasting operation begins or the appearance of the surface at the time that the blasting operation is completed. A brush-off blast is recommended only for conditions in which a more or less conventional finish, such as an alkyd-based coating system, is applied to a steel structure in a low-corrosion area.

Because most protective coatings are thin films, it is important that grit size be controlled to limit etching of the steel substrate. For most maintenance coatings, the maximum desirable particle size of the blasting grit is 16- to 30-mesh, which generates an anchor pattern measuring approximately 1½ mils (0.0381 mm).

Steel shot used alone is *not* a recommended method of surface preparation. Smooth hills and valleys are produced, and coatings can peel off in sheets. The use of an 85:15 mixture of shot and grit or steel grit provides an excellent surface for protective coatings when the abrasive has the appropriate mesh size to achieve the desired anchor profile.

Centrifugal-wheel blast cleaning is in common use and is gaining in popularity throughout the world. It offers excellent surface preparation with a minimum of expense.

In determining the depth of the profile, or anchor pattern, on steel during blast cleaning, the size of the abrasive is the major factor. Too fine an abrasive can provide patterns which result in inferior bonding of the primer, while extremely coarse materials may produce peaks which will project through the coating. Reasonably close adherence to the coatings manufacturer's recommendation for an abrasive is a good policy. Determining the actual profile is difficult. Although some customers have called for profile-meter readings of a representative surface, such data confuse more than they enlighten. Profile meters are designed to measure surface roughness of machined metals, and the sensing stylus is more likely to go around a peak than to go over it. A visual device, the Keane-Tator surface comparator, is a more useful tool.

With a given abrasive, brush blasting at a rate of about 400 ft²/h (37.16 m²/h) gives a good profile. Most primers will adhere satisfactorily to such a surface. Unfortunately, the high preponderance of scale and rust makes this method a dangerous choice for any except the mildest service conditions.

Reducing the blasting rate to about 200 ft²/h (18.58 m²/h) should produce a commercial blast. Although this surface is safe for many less severe exposures, the user should keep in mind that mill scale can cause trouble. Furthermore, this rate is not satisfactory for previously painted steel if the new primer is an inorganic zinc. Inorganics do not adhere to organic surfaces. Also, in some cases placing inorganic zinc under insulation causes condensation and "sauna bath" conditions on piping and equipment. This leads to coating and/or insulation failure.

A near-white blast can usually be obtained at about 150 ft²/h (13.94 m²/h). It is reliable for most services.

A white blast usually is accomplished at a rate of 100 to 125 ft ²/h (9.29 to 11.61 m²/h), but with some steels even more time may be required.

7.3.2 Chemical cleaning   This method is widely used in the United States, particularly by original equipment manufacturers in the production of home appliances and automobiles. It can be either a completely satisfactory method of surface preparation or a trouble breeder, and there is much controversy on the subject. As commonly practiced, chemical cleaning consists of immersing a piece of hot-rolled steel in a 15% sulfuric acid bath maintained at a temperature of approximately 190°F (88°C). Iron enters the acid solution at those points where the mill scale is broken by the generation of hydrogen gas, which blows adjacent mill scale from the steel substrate. When chemical cleaning is properly done, the steel is free of mill scale, is a uniform gray in color, and contains a mild tooth, or etch. As commonly practiced in industry, however, it does not completely remove mill scale. The reason is very simple. The surface of virtually all steel members contains residual deposits of oil resulting from grease picked up during the milling or shipping operation or even from the hand prints of workers who have handled the metal. Since sulfuric acid has no ability to wet steel which is covered with oil, no scale is removed from beneath these oil deposits. To be completely effective, chemical cleaning of metal must be preceded by an effective degreasing operation.

One other aspect of this operation deserves comment. Most of the oil carried into a pickling or acid tank floats on the surface of the acid as the steel is immersed. As the member is withdrawn from the bath, the oil is redeposited on the surface, where it interferes with proper coating adhesion.

Since most protective coatings require a neutral or a slightly acid surface, it is common practice to rinse steel members in hot water or water containing 2% phosphoric acid following the acid-pickling operation. Many corrosion engineers confuse this type of rinsing operation with the common industrial metal-finishing process called phosphatizing. Furthermore, many corrosion engineers believe that if they wet down a sandblasted surface with a dilute phosphoric acid solution, they are in effect phosphatizing the surface. Nothing could be further from the truth. Phosphatizing can be accomplished only by immersing an article in a very carefully balanced chemical solution containing manganese dihydrogen phosphate; accordingly, it is generally restricted to shop operation.

For heavy-duty industrial coatings, a well-performed chemical cleaning operation is sufficient to receive coatings for maintenance work. In view of the hazards intrinsic to the operation, however, it cannot be recommended as a means of surface preparation for coatings which are to be in continuous immersion service.

One aspect of surface preparation should be mentioned in conclusion: the importance of proper sand and dust removal, particularly when coatings are applied to the interior of equipment which will contain water or corrosive chemicals. It is common practice for contractors to apply the prime, or first, coat by spraying. Such a practice is satisfactory if the surface has been vacuumed completely free of sand and residual dust particles. If this is not feasible, however, the first coat should be applied by brush whenever possible. The scouring action of brushing works dust and sand particles up into the body of the coating film, where they do little if any damage. On the other hand, if a coating material is sprayed over such contaminants locked in the anchor pattern, there obviously are many points where the coating material has no adhesion whatever.

Many experts consider the application of a wash primer as a part of the surface-preparation procedure, and we endorse this view. Primers of this type generate on the metal surface a thin layer of passive iron phosphate salts which minimize underfilm creep.

7.3.3 Wet sandblasting   Traditionally, dry abrasive and dry air are used in abrasive blasting to avoid rusting of the newly exposed steel surface. With the enforcement of more stringent environmental laws, wet sandblasting may become the most reasonable approach to cleaning steel surfaces. The injection of water into the sand flow at the nozzle or at the pot to reduce dusting is not a new procedure. Unfortunately, the addition of phosphoric acid to the water to inhibit rusting produced a loose, powdery surface contaminant which had to be removed prior to the coating application. Newer corrosion inhibitors with properties superior to those of phosphoric acid have become available in recent years. However, coatings manufacturers should be consulted whenever use of their products over these inhibitors is contemplated.

A newer technique of wet-abrasive blasting is provided by high-pressure water-blast machines, which are capable of ejecting a water stream at 10,000 psig. Such a stream can remove poorly adhering rust, scale, and coating from a surface, but for tight scale and for the efficient removal of bonded coating and rust sand is injected into the stream. Precautions are the same as for conventional wet sandblasting.

7.3.4 Solvent cleaning   Solvent cleaning is used to remove such contaminants as oil and grease from steel prior to a mechanical or acid-pickling treatment. While it does nothing to remove rust, scale, or old-coating residues, it is often a necessary first step in the preparation of steel since it is the only effective method of removing heavy deposits of lubricants.

7.4 Selecting the Coatings System   For most environments, standard architectural coatings provide a suitable performance on steel. While the University of Missouri at Rolla indicates that there is still no water-emulsion resin suitable for priming ferrous metals, ongoing research suggests that this situation may change. New carboxylated styrene-butadiene emulsions and new acrylic emulsions may prove themselves as primers for steel because they provide excellent moisture barriers. In the light of environmental constraints on solvent-based systems, advances can be expected in this area.

Industrial environments, with exposure to acids, alkalies, salts, solvents, and weathering, can limit the choice of coatings (Fig. 7.5). There can, of course, be combinations of these environments, and it does not necessarily follow that a coating or a lining material which is suitable for two environments individually would be serviceable under combined conditions or for the separate environments in sequence. For example, coatings based on an epoxy resin are excellent in alkaline and solvent environments, but they fail quickly when exposed to the two materials alternately. Coatings of this type endure some shrinkage under alkaline

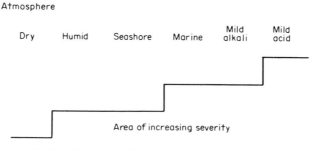

Fig. 7.5 Exposure effects on maintenance coatings.

conditions and some swelling when exposed to strong solvents. Repeated contraction and expansion of the coatings when exposed to alternate conditions can lead to cracking and subsequent failure. Prudence dictates the analysis of individual problems in great detail. The functional requirements of each component of the coatings system must also be considered (Table 7.5).

TABLE 7.5   Functional Summary of the Components of a Coating System

Coat	Main function	Specific requirement	Individual visual character	General requirement
Topcoat	Resistance to atmosphere	Atmosphere and/or environment resistance; bond to intermediate coat	Color	Thickness; strength; resistance; flexibility
Intermediate coat	Structure strength	Bond to topcoat; bond to primer	Color	Thickness; strength; resistance; flexibility
Primer	Adhesion	Bond to intermediate coat; adhesion to substrate	Color	Thickness; strength; resistance; flexibility
Substrate				

7.4.1   Primer coats   A primer should meet most, if not all, of the following requirements:

- Good adhesion to the surface to be protected when the surface has been cleaned or prepared according to specifications
- Appropriate flexibility-distensibility
- Satisfactory intercoat bonding surface for the next coat
- Ability to stifle or retard the spread of corrosion from discontinuities such as pinholes, holidays, and breaks in the coating film
- Enough chemical and weather resistance (inertness) by itself to protect the surface for a time period in excess of that anticipated before application of the next coat in the system
- In certain conditions, notably tank lining, chemical resistance equivalent to the rest of the system

Four types of primers commonly used over steel may be compared for their abilities to meet these requirements (Table 7.6).

TABLE 7.6 Comparison of Primers

Primer type and requirement	Alkyd or oil	Mixed resin	Resin identical to topcoat	Inorganic zinc
Bonding to surface	Usually have the ability to wet and bond to most surfaces and are somewhat tolerant of substandard surface preparation	Adhesive properties are the major consideration of formulation. Not quite as tolerant of substandard surface preparation as oil types	Adequate for proposed use when surface is properly prepared	Outstanding adhesion to properly cleaned and roughened surfaces
Adhesion of topcoats	Satisfactory for oil types; usually unsatisfactory for vinyls, epoxies, and other synthetic polymers. Are softened and lose integrity by attack from solvent systems of these topcoats	Formulated for a specific range of topcoats	Usually part of a specific generic system. Maximum permitted drying time before application of second coat must be observed	Fit into a wide range of systems. Tie coat may be required. Specific recommendation should be obtained for immersion systems
Corrosion suppression	Limited. Alkali produced at cathode and corrosion battery attacks film (saponification) and causes disbonding, resulting in the spread of underfilm corrosion	Usually formulated with good resistance to alkali undercutting; contain a chromate pigment for a degree of corrosion inhibition	Often contain chromate pigments for a degree of corrosion inhibition. Resistance to alkali undercutting variable	Outstanding in their ability to resist disbonding and underfilm corrosion. Anodic property of metallic zinc protects minor film discontinuities
Protection as single coat	Limited by severity of exposure	Limited by severity of exposure	Limited by severity of exposure	With very few exceptions will protect without a topcoat
Chemical resistance	Typical of alkyds	Resistance usually of a lower order than that of topcoat	Typical of system	Not resistant to strong acids and alkalies.

7.4.2   Intermediate coats   Intermediate coats may be required in a system to provide one or more of the following properties:

- Adequate film thickness of the system (body coat).
- Uniform bond between the primer and the topcoat (tie coat).
- Superior barrier with respect to aggressive chemicals in the environment. The coat may be too deficient in appearance or physical properties to be a satisfactory finish coat.

7.4.3   Finish coats   Finish coats are the initial barriers to the environment as well as the surfaces seen by management and the public. However, in some situations the barrier to the environment is primarily a function of the body or prime coat, while the finish coat serves to provide a pleasing appearance, a nonskid surface, a matrix for antifouling agents, or other specialized purposes. Obviously, the chemical resistance of the finish coat in such a situation must be sufficient to ensure its remaining intact in the environment.

7.5   Coating Materials   Coatings of a given generic type display common properties. Comparison enables coatings specifiers to choose the material best suited to their performance requirements.

7.5.1   Baked phenolic   Without exception, baked-phenolic coatings exhibit excellent resistance to acidic environments and to water. Most materials of this type also display excellent resistance to strong solvents. As a class the materials are very weak in alkaline environments. While their material cost is low, their applied cost is quite high since they demand the ultimate in surface-preparation application in multiple coats as well as high-temperature baking. High-baked phenolics are difficult to repair in the field.

7.5.2   Epoxy   Epoxy-resin coatings are available in three general types: oil-modified, catalyzed, and high-baked. The oil-modified varieties are commonly referred to as epoxy esters. They have properties intermediate between those of high-quality conventional enamels and truly chemical-resistant protective coatings. Since such products contain a drying oil, they are not suitable for exposure to strong alkalies. They display the high chalking rate characteristic of all epoxy coatings, and their use on exterior steel is limited. Their natural area of application is the interior surfaces of buildings exposed to fumes and mild alkaline cleaners. Catalyzed epoxies are offered in four common variations which differ in the catalyst employed. Amine-cured epoxies display the best solvent and acid resistance of coatings of this generic type. Polyamide-cured epoxies have somewhat less general chemical resistance but are superior to amine-cured epoxies in water resistance, weather resistance, and ability to adhere to difficult surfaces. Amine-adduct–cured materials are based on the prereaction of a portion of the catalyst with the epoxy resin. Such materials are less sensitive to climatic conditions than amine-cured epoxies and are considered the equal of either amine or polyamide variations in all environments. High-baked epoxies display the best chemical and solvent resistance of any type of epoxy but, as their name implies, require very high-temperature baking to achieve complete polymerization. Use of this type of epoxy is normally limited to the interior of tanks handling concentrated caustics or solvents. All epoxy-based materials are extremely weak in acid and water service, and they also show a marked tendency to discolor and chalk. The latter property alone has limited their employment in industrial plants in Southern latitudes. Inasmuch as catalyzed and baked epoxies are polymerized to form high–molecular-weight, very solvent-resistant films, they present overcoating problems. Most authorities agree that mechanical roughening of some type is imperative to secure good intercoat adhesion.

Epoxy coatings generally chalk upon exterior exposure, but new resins that reduce the chalking problem are becoming available. Actually, the chalk is a problem of appearance only, because once the surface chalks, the chalk itself provides a barrier to further degradation of the paint film.

Because epoxy systems often act as vapor barriers, they should not be applied to surfaces where transmitted vapor can condense under the coating and freeze. Moreover, since epoxy curing time is affected by temperature, the manufacturer's application limitations must be spelled out in the coatings specification.

Epoxy-polyamide systems have been threatened by environmental regulations that limit the amount of solvent in formulations. New formulations that increase the solids volume (Table 7.7) are under development. While it is true that certain epoxies have been formulated to be applied to wet steel or even to steel underwater, most manufacturers of these coatings qualify application at or below the waterline, noting that successful application must be determined in the field under actual conditions. Such underwater application is generally beyond the scope of this *Handbook*.

7.5.3 Vinyl   Vinyl-based protective coatings display the greatest versatility of all the commonly used resin types. These coatings have good resistance across the pH scale, excellent water resistance, and perhaps the lowest chalking rate of any organic product. The films are thermoplastic, thus ensuring ease of recoating. The main disadvantages of vinyl-resin coatings are their limited solvent and heat resistance (they are temperature-sensitive above 150 to 180°F, or 66 to 82°C). In addition, they display a marked tendency to be sensitive to intercoat contamination. The films have greater strength in tension than in adhesion, and if

TABLE 7.7 High-Solids Epoxy Coating Being Evaluated in the Field

Material	Pounds*	Gallons†
Base component		
TiO_2	425.7	12.77
Clay	106.7	4.96
Nuosperse 657	0.8	0.12
Thixatrol ST	10.5	1.30
Eponex DRH-151.3	249.0	27.39
Beetle 216-8	4.6	0.59
n-Butanol	221.9	32.87
Curing-agent component		
Versamid 1540	137.2	17.12
n-Butanol	19.4	2.88

P/B	1.3 : 1
Pigment-volume concentration	27.7 percent
Solids by weight	79.3 percent
Solids by volume	64.0 percent
Viscosity (No. 4 Ford cup)	61.1 percent‡

Courtesy Shell Chemical Co., Houston, Tex.
* 1 lb = 0.4536 kg.
† 1 gal = 3.785 l.
‡ After 1-h induction.

application is attempted over a contaminated surface, they tend to peel. In this respect they are substantially inferior to alkyd and epoxy coatings.

7.5.4 Inorganic zinc Inorganic zinc–silicate coatings display the best weather and solvent resistance of any type of protective coating. These products contain no organic matter and can be used in virtually any solvent, including chlorinated materials. The ultimate life expectancy of materials of this type in severe weathering service has not yet been established, but applications more than 20 years old still provide complete protection to the steel substrate. The main disadvantages of inorganic zinc silicates are their limited chemical resistance and critical application properties. Since the films contain zinc, they are suitable for neither strong-acid nor strong-alkali environments. Successful application requires good surface preparation and a minimum of special equipment.

Inorganic zinc–silicate coatings have three outstanding properties:

- They display the best resistance to petroleum products and chemical solvents of any known coating material.
- They display the best resistance to severe weathering environments of any known protective coating.
- They are not susceptible to underfilm corrosion even after exposures of a decade or longer.

This last property is of particular importance, accounting for the very widespread use of these coatings as primers in the petroleum, chemical, and petrochemical industries. In reporting on the behavior of organic paint and coating systems, it is commonly said that 90 percent of a failure results from underfilm corrosion and the remaining 10 percent from degradation of the surface of the film through oxidation. By contrast, inorganic zinc–silicate coatings completely debar underfilm corrosion and, within the limits of our ability to measure them, are also immune to surface oxidation.

The original inorganic coatings were packaged as three components. Still used to a very limited extent, they are commonly referred to as postcured products. The powder component is zinc dust in the 6- to 30-μm range. It may contain some red-lead or mineral dye to give the film a distinct color as an aid in application. The liquid component is made up primarily of sodium silicate (water glass), a completely inorganic mineral substance that has been used for many decades as an adhesive for cardboard boxes. In addition, the liquid commonly contains up to 10 percent sodium hydroxide, approximately 1 percent sodium bicarbonate or other carbon dioxide–releasing agents, and water. Photomicrographs indicate that when the powder and liquid components are mixed and subsequently are applied to a piece of sandblasted steel, the coating bonds both physically and chemically. When first applied and prior to curing, the coatings are water-soluble, but they can be rendered water-insoluble either by the application of heat (approximately 350°F, or 177°C, for 1 h) or by the application of a chemical curing solution.

Curing solutions are commonly mixtures of phosphoric acid and isopropyl alcohol. Effective solutions contain some inhibitor which will buffer the natural tendency of the acid in the curing solution to attack the zinc pigment in the coating. The principal difference in the quality of curing solutions being marketed in the United States pivots on the effectiveness of the buffering additive. When the acid-bearing curing solution is first applied to the hard yet water-soluble film of the inorganic zinc–silicate coating, the acid reacts with the sodium carbonate or sodium bicarbonate component to generate carbon dioxide gas, the reaction initiator. At the same time, the phosphoric acid in the curing solution reacts with the sodium silicate (water-glass) vehicle in accordance with the following equation:

$$3NA_2SiO_3 + 2H_3PO_4 \rightarrow 2NA_3PO_4 + 3H_2SiO_3$$

The last product in this equation is silicic acid, which has never been isolated but is known to exist. It is thought that this acid polymerizes immediately into a long-chain inorganic polymer, as indicated below:

$$3H_2SiO_3 \rightarrow 3SiO_2 + H_2O$$

Note that this inorganic polymer consists of SiO_2, which is sand coupled with water of hydration. The present thinking is that this polymer serves as the binder to hold the coating film together.

While the above reactions are proceeding, zinc and iron are reacting with phosphoric acid to form the respective phosphates. At the same time zinc and iron are also reacting with silicic acid and carbon dioxide to form insoluble agglomerates throughout the film in accordance with the following equations:

$$Fe + H_3PO_4 \rightarrow FePO_4 + 3H$$
$$3Zn + 2H_3PO_4 \rightarrow Zn_3(PO_4)_2 + 6H$$
$$Zn + 3CO_2 + H_2SiO_3 \rightarrow ZnSi(CO_3)_3 + 2H$$
$$Fe + 3CO_2 + H_2SiO_3 \rightarrow FeSi(CO_3)_3 + 2H$$

When heat, such as an open flame, is used in place of the acidic curing solution, the heat releases carbon dioxide gas, which permits the formation of silicic acid and the reactions mentioned above.

Since iron is one of the participants in the reactions leading to the production of the cured inorganic zinc–silicate film, it is obvious that the application cannot be made over any old paint or coating film or, for that matter, even over mill scale, which is an iron oxide. This explains the insistence of all manufacturers that application be attempted only over a white-metal sandblast.

Cured inorganic zinc–silicate films can quite properly be thought of as a cross between hot-dip galvanizing and a fused ceramic. From the standpoint of hardness and abrasion resistance they are quite comparable to 2-oz American Petroleum Institute (API)–quality hot-dip galvanizing. While they are somewhat inferior to galvanizing in flexibility, they have no trouble in following the normal contraction and expansion of steel surfaces. It is interesting to note that the hardness and abrasion resistance of inorganic films improve with aging, undoubtedly as a result of contact with carbon dioxide in the atmosphere. Carbon dioxide is the reaction initiator for this class of materials and tends to drive the reactions tabulated above to completion.

While the physical properties of the film closely approximate those of hot-dip galvanizing, there is quite a disparity between the two in chemical resistance and resistance to marine environments. This is not surprising when you consider that zinc is subject to normal atmospheric corrosion as well as to attack from both acid and alkaline fumes and spills. In the case of inorganic coatings, the zinc is partially encapsulated in a silicate matrix which has excellent chemical resistance. Long-term tests on offshore structures indicate that a 2½-mil (0.0635-mm) film of inorganic zinc silicate has at least 3 times the life expectancy of a 2-oz film of hot-dip galvanizing. While inorganic zincs cannot be recommended for services in which there is a spillage or contact with acid or alkalies, they are, relatively speaking, superior to hot-dip galvanizing in their ability to withstand these materials.

Properly formulated inorganic zinc films have been used in freshwater immersion for periods exceeding 8 years without displaying any signs of coating failure. Their performance in salt-brine immersion or seawater immersion, however, is not outstanding, and in most cases the films are destroyed in approximately 18 months. For such applications, it is common practice to apply a seal coat over the inorganic films. The rather short life expectancy of inorganic films in saltwater immersion is not difficult to understand when you consider the electrochemical sacrificial action of zinc toward steel. When zinc corrodes, the corrosion products are precipitated as a gelatinous deposit on the surface of the zinc coating. With water present to remove this corrosion film constantly, the zinc component of the coating will continue to expend itself at a fast rate.

Inorganic zinc–silicate coatings have an excellent history in hot, dry services such as on the exterior of mufflers and stacks. Used alone, they can be successfully employed at temperatures up to 800°F (427°C). When topcoated with a high-heat silicone coating, they can be used at temperatures up to 1000°F (538°C). In the latter case, it is imperative that the silicone be applied to the inorganic coating prior to the inorganic film's being heated to any temperature above 300°F (149°C).

There is a good deal of confusion about the generic types of paints and coating materials which can be applied successfully over inorganic zinc–silicate coatings. As a guidepost, it is well to remember that all zinc silicates are alkaline and can best be topcoated with materials which have a high order of alkali resistance. In this class are included converted epoxies and water-based latex materials. Conventional paints should not be used, as they contain oils which are saponifiable by the alkaline surface components on the inorganic film. The same stricture applies to long-oil epoxy esters which contain a high concentration of dry oil. From a technical point of view, the highest order of topcoat adhesion is secured when using an amine-catalyzed epoxy. This is not surprising when you consider that the amine-adduct catalysts used with these coatings are organic alkalines and, of course, favor alkaline surfaces.

The very long life in severe weathering conditions exhibited by inorganic coatings stems from the fact that they are dense films which do an effective job of debarring the transmission of water, oxygen, and ions. Even more important, they contain a very high loading of zinc, which is anodic to steel. Inorganic films provide a very favorable anode-to-cathode–area relationship when you consider the cathodes as the occasional pinholes which exist in these films. This cathodic-protection ability can be measured quantitatively in a laboratory by coupling coated panels to bare and sandblasted panels of the same dimensions, immersing the two panels in seawater, and measuring the cathode potential generated.

The effective inorganic coatings will exhibit a potential in excess of 0.8 V and maintain it for an extended period of time. Inferior products will display initial potentials in excess of 0.8 V but will drop in a short time below 0.75 V, at which point the cathode area begins to rust.

Inorganic coatings, when topcoated with chemically resistant organic films such as epoxies, chlorinated rubbers, and vinyls, represent the longest-lasting maintenance coating systems developed thus far for protecting steel in difficult chemical-plant environments. In such services the organic topcoats, in effect, prevent the aggressive chemical solutions from coming in contact with the zinc-filled primer. Water vapor does permeate the organic film, but no rusting of the substrate occurs, since the inorganic zinc–silicate film cathodically protects the steel in the presence of water. Combinations of this type have been used in very severe chemical-plant environments with no more than 0.5 percent failure after 10 years of service in coastal environments.

The principal large-scale applications for inorganic coatings in the United States are as follows:

1. Interior of petroleum-storage tanks
2. Interior of clean-oil tankers
3. Offshore platforms
4. Barges and small watercraft
5. Structural steel and piping for the petroleum, petrochemical, and chemical industries
6. Decks of floating roof tanks
7. Steel used in docks and wharves
8. Cooling-tower piping
9. Stacks and hot surfaces

Since the late 1960s many self-curing inorganic formulations have been developed. As a class, they are outstanding products. Accelerated tests and 10 years of experience indicate that the solvent-based, hydrolized ethyl silicate materials are uniquely superior to post-cured and self-cured heavy-metal silicates in several respects:

1. Better wetting properties; hence a higher order of adhesion to poorly prepared surfaces.
2. Improved temperature resistance.
3. Greater application latitude under adverse weather conditions.
4. Improved ability to protect a steel substrate under difficult corrosive conditions.
5. Improved overcoatability. The solvent-based formulations are essentially neutral in pH rather than strongly alkaline, as is the case with water-based products.

Many companies have developed exact performance standards for inorganic zinc coatings (see box). Table 7.8 is a guideline for choosing zinc-rich and other primers.

CARBIDE UNIFORM PAINT STANDARD: INORGANIC ZINC PAINT

1. SCOPE
    1.1   This standard describes the inorganic zinc paint to be used by Union Carbide for industrial painting. The paint may be used in a wide variety of conditions, some of which
        1.1.1   One coat paint for pipe, steel equipment and vessels in non-acidic and non-alkaline environments, or for high temperatures.
        1.1.2   Pre-erection, shop applied primer for subsequent finishes after erection.
        1.1.3   Inorganic zinc paint is not to be used under water, underground or under insulation.
    1.2   This standard does not describe:
        Thin-film preconstruction primers
        Zinc-rich weld-through primers
        Organic zinc-rich paints
        Post-cured zinc paints
        Specialized zinc-rich paints
    1.3   The standards described herein are intended to be minimum requirements. Prospective suppliers are encouraged to offer improved quality paints using proprietary techniques.
2. DESCRIPTION
    2.1   General—The paint shall be that commonly known as "inorganic zinc coating." The paint shall be a metal or alkyl silicate with high purity zinc powder. The paint may be water based or alcohol based.
        The water based vehicle shall be potassium silicate modified with collodial silicate.
        The alcohol based vehicle shall be partially hydrolyzed ethyl silicate in alcohol solvent, and may be acid hydrolyzed or amine promoted.
        Two-package and one-package paints are acceptable.
    2.2   Pigment—The zinc portion of the pigment shall be a finely divided zinc powder containing, by weight, a minimum of 96 percent metallic zinc and a minimum of 98 percent total zinc. The average particle size of the zinc powder shall not exceed 9.5 microns. Inert fillers such as barytes are not per-

mitted. Conductive fillers are not permitted under this standard. Low purity, filled, zinc materials may be tested by this standard; if acceptable, will be classified as "specialized zinc-rich paints."

The amount of zinc in the dried film shall be a minimum of 80% by weight and a maximum of 92% by weight.

2.3 Curing Agents—Post-cured inorganic zinc paints and "catalyzed" epoxy type zinc paints are not permitted under this standard; however, some zinc paints qualified under this standard may be optionally cured by heat or by a curing agent.

2.4 Modifiers—Some modification of vehicle is permitted:
   a. To prevent gassing
   b. To prevent settling
   c. To improve shelf or pot life
   d. To improve overcoatability.

2.5 Colors—Manufacturer's standard color is acceptable to provide contrast to sandblasted substrate; however, natural zinc powder color shall be available on request.

3. APPLICATION
   3.1 Carbon Steel—The paint shall be applied as one-coat to abrasive blasted steel to SSPC-SP-10, near-white condition. Dry film thickness: 2½ mils, minimum; 4 mils, maximum.

4. PERFORMANCE STANDARDS
   4.1 Panel Preparation—CUPE Manual, Lab Methods, Section 4.
   4.2 Tests and Minimum Scores:

	Components	
	Two	One
CTM-201, Salt Spray, 3000 hours	90	80
CTM-206, Adhesion	23	23
CTM-206, Heat Resistance	25	20

   4.3 Storage or Shelf Life—Shelf life at 77°F shall be a minimum of twelve months for two-package materials and six months for one package materials.
   4.4 OTHER MINIMUM TEST REQUIREMENTS
      4.4.1 General—Mixing and application characteristics will be determined by Union Carbide in accordance with Section 5, Paragraph 2, Carbide Uniform Paint Evaluation Manual (CUPE).
      4.4.2 Mixing and Preparation—Minimum test scores: one package—7, two-package—5. There shall be no gelation, gassing or settlement in the can containing the

> vehicle. Mixing will be according to manufacturer's instructions. Pot life of two package mixed material shall be eight hours.
>
> 4.4.3 Application minimum test score 5—Application should be accomplished without gun plugging, spitting or "fingers." Blisters, sags, runs and pinholes are not permitted. There shall not be excessive dry over-spray.
>
> 4.4.5 Drying time to rain shower resistance—One hour, maximum.
>
> 5. PRODUCT DATA—Essential product data and instructions are required.
>
> 5.1 Manufacturer's Product Data Sheet
>
> 5.2 UCC Paint Description Form, Appendix A, to include: Weight/gallon
> Weight solids
> Volume solids—Volatiles measurement method
> Percent zinc in dried film
> Percent lead in powder or pigment
> Zinc powder purity
>
> 5.3 Application Instructions
>
> 5.4 Material Safety Data Sheets
>
> 6. QUALITY ASSURANCE
>
> 6.1 After successful completion of all tests and acceptance of the paint, properties are required by Section 7, Quality Control, Monitoring of Paint (MOP), Table 7.1, may be determined and recorded. These properties are:
> IR Spectrophotometric scan of liquid components
> Weight per gallon of zinc powder or one-package paint
> Elemental analysis of powder or pigment
>
> Courtesy Union Carbide Corp.

7.5.5 Organic zinc   Some zinc-rich coatings are made by combining zinc dust with an organic resin or elastomer. Typically, these are epoxies, chlorinated rubbers, vinyl, acrylics, epoxy esters, phenolics, and organic silicones. There is no reaction between the zinc and the resin; typical loadings of zinc pigment reach 95 percent by weight or up to 70 percent by volume. Organic zinc-rich primers have been used successfully, but organic binders can encapsulate zinc particles, thus limiting the amount of galvanic protection to the amount of free zinc in the formulation.

As the organic resin degrades, loss of adhesion can occur. This is generally not a problem when the surface is topcoated with a porosity-free high-build topcoat. In this regard, acrylics show some benefit, for, unlike alkyds, they do not react with zinc in the primer. Because of this reaction alkyds can show signs of peeling after long exposure.

Zinc-rich primers have been used successfully to protect all kinds of exposed steel from corrosion according to the theory that metallic zinc dissolves anodically, offering cathodic protection to the steel substrate. What is still little known is that long-term protection is derived from a renewable barrier film of zinc reaction products. These zinc corrosion products are usually hydroxides, sulfates, carbonates, phosphates, oxides, and other nonconductive compounds. They are insulating materials that prevent the additional flow of zinc ions. (In effect, they protect the zinc from corroding, and no additional anodic loss of zinc takes place.)

Realizing that high zinc loadings can affect intercoat adhesion, weldability of primed steel, hardness of the coating, and, recently, total pigment cost, coatings chemists at Hooker Chemicals & Plastics Corp. in Grand Island, New York, are investigating a conductive pigment, diiron phosphide, that will reduce zinc loadings yet maintain corrosive resistance.

7.5.6 Furan   Furan coatings are among the most versatile and resistant organic films discovered to date, but because of application problems they have not found wide acceptance in industry. The resins are set with a strong-acid catalyst which makes it impossible to apply them directly to steel surfaces and limits their adhesion to any prime surface. Once cured, the films become extremely hard, and it is virtually impossible to maintain them.

7.5.7 Urethane   Urethane coatings are offered in a wide variety of formulations, varying from varnishes for wood surfaces to multiple-component materials for industrial service. They are finding increased acceptance with industry as cosmetic finishes and are used over epoxy protective coatings or primers because of their outstanding gloss and color retention. As corrosion barriers, they offer no demonstrable superiority to well-formulated activated epoxy coatings or high–molecular-weight vinyl coatings.

7.5.8 Coal-tar epoxy   As their name implies, coal-tar–epoxy coatings are mixtures of coal-tar pitch and low–molecular-weight epoxy resins. In the simplest terms, they are a combination of thermoplastic and thermosetting resins. Common formulations contain up to 35 percent of the epoxy resin. The resulting films are very resistant to acids and water, and they also have good resistance to solvents of moderate strength. While these coatings are never recommended for immersion in strong caustics, they do display good resistance to spills of mild caustics. Coatings of this

TABLE 7.8 Zinc-Rich and Other Primers

	Inorganic zinc, postcured	Inorganic zinc, self-cured, water-based	Inorganic zinc, self-cured, water-based, ammonium	Inorganic zinc, self-cured, solvent-based	Organic zinc, one-package	Organic zinc, two-package	Modified inorganic zinc primer	Special primer and tie coats
Cure type	Inorganic	Inorganic	Inorganic	Hydrolyzable organic silicate	Lacquer	Coreacting	Coreacting	Lacquer
Effect of sunlight	Unaffected	Unaffected	Unaffected	Unaffected	Surface chalking	Surface chalking	Very slow: chalk	Variable
Wet or humid environments	Outstanding	Outstanding	Outstanding	Outstanding	Very good	Very good	Very good	Usually good
Industrial-atmosphere contaminants:								
Acids	Requires topcoat	Requires topcoat	Requires topcoat	Requires topcoat	Requires topcoat	Requires topcoat	Requires topcoat	Requires topcoat
Alkalies	Requires topcoat	Requires topcoat	Requires topcoat	Requires topcoat	Requires topcoat	Requires topcoat	Requires topcoat	Requires topcoat
Oxidizing	Requires topcoat	Requires topcoat	Requires topcoat	Requires topcoat	Requires topcoat	Requires topcoat	Requires topcoat	Requires topcoat
Solvents	Outstanding	Outstanding	Outstanding	Outstanding	Limited	Very good	Very good	Limited
Spillage and splash of industrial compounds:								
Acids	Not recommended	Not recommended	Not recommended	Not recommended	Not recommended	Not recommended	Not recommended	See manufacturer's literature
Alkalies	Not recommended	Not recommended	Not recommended	Not recommended	Not recommended	Not recommended	Not recommended	
Oxidizing	Not recommended	Not recommended	Not recommended	Not recommended	Not recommended	Not recommended	Not recommended	
Solvents	Outstanding	Outstanding	Outstanding	Outstanding	Limited	Very good	Good	See manufacturer's literature
Water immersion	With suitable topcoat system on ship hulls and other marine structures				Not recommended	With epoxy topcoats on marine structures	Good	See manufacturer's literature

	Marine cargo and ballast tanks; fuel storage, including floating roof tanks			Not used	With suitable topcoat in marine cargo and ballast	See manufacturer's literature	See manufacturer's literature
Tank linings	Marine cargo and ballast tanks; fuel storage, including floating roof tanks			Not used	With suitable topcoat in marine cargo and ballast	See manufacturer's literature	See manufacturer's literature
Physical properties:							
Abrasion resistance	Outstanding	Outstanding	Outstanding	Good	Good	Excellent	Usually good
Heat stability	Outstanding	Outstanding	Outstanding	Good	Good	Good	Usually good
Hardness	Outstanding	Outstanding	Outstanding	Good	Good	Excellent	Usually good
Gloss	None	None	None	Flat	Flat	None	Flat
Colors	Gray or tints of gray	Gray or tints of gray	Gray or tints of gray	Gray or tints of gray	Gray or tints of gray	Gray or tints of gray	Usually dark oxide red or dark color
Additional notes	As primers for organic systems, provide greatly extended service life. Special application technique or use of tie coat may be required to avoid solvent bubbling in organic topcoat. Alkyds always require tie coat			Can be used to touch up inorganic primers compatible with topcoats	Choice of topcoat is critical	In some applications, satisfactory over mechanically cleaned surfaces. Excellent for touch-up of inorganic zincs	Used to ensure bond of system to substrate

type make excellent linings for water tanks and good coatings for equipment such as pipelines which will be buried in the soil. Their principal disadvantages are their color, which is always black, their high chalking rate when exposed to sunlight, and the difficulty of repairing or topcoating them. With age the films catalyze to such a point that refresher coats cannot be applied without mechanically roughening the film.

7.5.9 Chlorinated rubber   The principal chlorinated-rubber resin is manufactured by chlorinating natural rubber; it is almost always modified with alkyd resins in manufacturing protective coatings. Since commercial formulations generally contain a high percentage of the alkyd resin, they are much less sensitive than, say, vinyl coatings to intercoat contamination. Further, they are almost always offered in high-solids solutions and have easy application and recoating properties. Since formulations incorporate alkyd resins, they have limited resistance to strong caustics. Alkyd-modified chlorinated-rubber maintenance coatings are without exception inferior to well-formulated vinyl maintenance coatings. A limited number of manufacturers market unmodified chlorinated-rubber coatings. Such materials contain no alkyd resins and are outstanding protective coatings.

Liquid-applied chlorinated-rubber coatings should not be used in highly corrosive environments such as the interior of acid-storage tanks. The reason is that even though the coating is resistant to the chemical, rapid attack of the underlying surface will occur at any pinhole or discontinuity in the coating film. For extreme services, thick membrane linings such as rubber or polyvinyl chloride sheets are indicated. For severe abrasion or uses in which temperature is also a factor, the membrane should be overlaid with acidproof brick.

While the description of coatings presented above is not all-inclusive, it does cover the basic advantages and disadvantages in respect to their use on steel. Tables 7.9, 7.10, and 7.11 provide a more extensive guideline for a variety of coatings generally classified as lacquer-type, coreacting and condensation, and oil-type coatings.

7.6 Maintenance Finishes   Once a surface has been prepared for painting, priming should follow immediately. As a condition for performance, the specification should include a statement that carbon steel surfaces shall be primed within 6 h after completion of surface preparation, before rusting or contamination occurs.

The sharp edges of metal and the threads on pipes are difficult to cover adequately with primer. Frequently these are the first areas to rust after painting. Painting bolts and nuts with a zinc-rich primer followed by a Teflon topcoat is an economical maintenance aid because it prevents "freezing" the nut in place owing to corrosion. (The removal of corroded nuts on large equipment can cost more than $5 per bolt.)

Take particular care to apply a continuous coating on all metal surfaces. When sharp edges or protrusions can be power-ground flat and smooth, the extra work pays off in total paint performance. (Remember that welding leaves a hard-glaze deposit that should be removed by blasting or mechanical abrasion because the glaze does not support paint.) The specification itself should cover thinning instructions. Generally thinning is carried out only if necessary for the workability of the coating material and then only in accordance with the coating manufacturer's most recent printed application instructions. Do not substitute a solvent or thinner; use only that recommended by the manufacturer. Also, use the coverage rates recommended by the coatings manufacturer. Succeeding coats should be applied in a different color or shade from that of the preceding coat to help determine the uniformity and coverage of the coating. This practice also assists inspectors in detecting holidays (areas with too thin a film thickness). Both wet and dry film thicknesses of each coat and of the entire system should be specified. While detailed paint specifications should be in the hands of the contractor and inspector, coating-system information sheets (Figs. 7.6, 7.7, 7.8, and 7.9) should be prepared for use by the job supervisor and applicators.

Correct paint selection is important and a basic requirement for proper paint performance. Proper selection depends on knowing the types of corrosives, temperature conditions, and humidity conditions in the area. With many different kinds of paints on the market, selection is usually made by preliminary testing. Table 7.12 offers a guideline of various air-drying finishes used in maintenance paints.

While it is not practical for most architects to run a test-fence program, there are good reasons to monitor testing programs. One is that new paints are continually coming on the market, and there is always the possibility that one of them may lower painting costs significantly. If this paint is found, the cost of additional testing will have been well justified. Of course, investigations of new developments in paint application and maintenance procedures should also be continued. Here again, however, costs should be kept to a minimum.

A second reason for continuing testing is the desirability of determining alternative sources so that equivalent paints may be found.

A third reason for continuing paint testing is that testing reduces the amount of time spent with sales representatives in discussing the performance of their paints. For example, paint companies with poor-quality coatings will not try to make a sale if they know that specifications are based on performance and that this performance will be continually checked and monitored.

TABLE 7.9   General Properties of Lacquer-Types Coatings

	Petroleum hydrocarbon	Coal tar	Chlorinated rubber	Chlorinated rubber, oil- or alkyd-modified
Cure type	Lacquer	Lacquer	Lacquer	Drying oil
Effect of sunlight	Minor	Checking	Slow surface chalk	Slow surface chalk
Wet or humid environments	Excellent	Excellent	Excellent	Fair to good; will yellow
Industrial-atmosphere contaminants:				
Acids	Excellent	Excellent	Excellent	Fair to good
Alkalies	Excellent	Excellent	Excellent	Fair
Oxidizing	Good	Limited	Excellent	Fair
Solvents	Poor	Poor	Limited	Limited
Spillage and splash of industrial compounds:				
Acids	Good	Good	Very good	Not recommended
Alkalies	Good	Good	Very good	Not recommended
Oxidizing	Fair	Not recommended	Good	Not recommended
Solvents	Not recommended	Not recommended	Not recommended	Not recommended
Water immersion	Ship bottoms	Submerged and buried pipe	Ship hulls	Not recommended
Tank linings	Not used	Potable water	Swimming pool	Not used
Physical properties:				
Abrasion resistance	Poor	Poor	Good	Fair
Heat stability	Softens	Softens	Limited	Fair
Hardness	Soft	Poor	Good	Fair to good
Gloss	None	None	Semigloss to matte	Wide range
Colors	Black and aluminum	Black	Wide range	Full range
Additional notes	Water resistance and ease of repair result in wide use as ship-bottom systems		Traditionally low-solid and low-film build. New high-build available	

Vinyl-alkyd	Vinyl chloride copolymers	Polyvinyl chloride	Vinyl-acrylic	Acrylic
Lacquer	Lacquer	Lacquer	Lacquer	Lacquer
Slow surface chalk	Slow surface chalk	Slow surface chalk	Very slow surface chalk	Prolonged chalk resistance
Good; slight yellowing	Excellent	Excellent	Very good	Good
Good	Excellent	Outstanding	Very good	Good
Fair	Excellent	Excellent	Very good	Good
Fair	Excellent	Outstanding	Good	Good
Limited	Limited	Good	Limited	Limited
Not recommended	Very good	Excellent	Fair	Poor to fair
Not recommended	Very good	Excellent	Fair	Poor to fair
Not recommended	Good	Excellent	Fair	Poor to fair
Not recommended	Not recommended	Limited	Not recommended	Not recommended
Not recommended	Tank linings	Tank linings	Not recommended	Not recommended
Not used	Potable water and deionized water, glycol and alcohols, fuels	Caustic solution, fatty acids, alcohols, glycols, animals and vegetable oils	Not used	Not used
Fair	Good	Excellent	Good	Good
Fair	Limited	Limited	Limited	Limited
Fair	Good	Good	Good	Good
Limited range	Semigloss to matte	Gloss to matte	Semigloss	Gloss, semigloss
Full range	Full range	Most colors	Full range	Full range
Useful as primer or topcoat for more chemical-resistant vinyl copolymers	Available in high- and low-build formula. High-build more popular in maintenance applications	Low build. Economic limit to tank lining and other severe exposures	Wide use as finish coat with vinyl systems	Use as finish coat for epoxy systems for improved weathering

TABLE 7.10 General Properties of Coreacting and Condensation Coatings

	Coal-tar–epoxy or polyamide cure	Epoxy-amine cure	Epoxy-polyamide cure	Epoxy phenolic	Phenolic	Urethane-moisture cure	Urethane, two-package
Cure type	Coreacting	Coreacting	Coreacting	Condensation	Condensation	Coreacting	Coreacting
Effect of sunlight	Surface chalking	Yellowing and surface chalking	Yellow and surface chalking	Tank lining	Tank lining	See "Additional notes"	See "Additional notes"
Wet or humid environments	Excellent	Very good; may yellow	Very good; may yellow	Tank lining	Tank lining	Very good	Very good
Industrial-atmosphere contaminants:							
Acids	Excellent	Good	Good	Good		Good	Very good
Alkalies	Excellent	Excellent	Excellent	Tank linings	Tank linings	Good	Excellent
Oxidizing	Limited	Limited	Limited			Poor	Limited
Solvents	Limited	Excellent	Excellent			Excellent	Excellent
Spillage and splash of industrial compounds:				As tank linings:	As tank linings:		
Acids	Good	Fair	Poor to fair	Certain fatty acids	Limited minerals, acids	Fair	Good
Alkalies	Good	Excellent	Excellent	Very good	Not recommended	Fair	Good
Oxidizing	Not recommended	Not recommended	Not recommended	Not recommended	Not recommended	Not recommended	Not recommended
Solvents	Not recommended	Excellent	Very good	Very good	Outstanding	Good	Excellent
Water immersion	Marine piling; sewage basin	Tanks; sewage	Ship hulls	As tank linings	As tank linings	Not recommended	Ship hulls

	Marine ballast tanks	Marine cargo or ballast tanks; fuel storage tanks	Marine cargo or ballast tanks; fuel storage tanks	Wide range of solvents and foods; caustic cartage	Outstanding for solvents, aqueous solution and food products	Not used	Marine cargo or ballast tanks; fuel storage tanks
Tank linings							
Physical properties:							
Abrasion resistance	Limited	Good	Good		Good	Excellent	Outstanding
Heat stability	Excellent	Good	Good	Outstanding	Excellent	Good	Good
Hardness	Very hard	Very hard	Hard	Very hard	Excellent	Excellent	Excellent
Gloss	None	Wide range	Wide range	Excellent	Excellent	Range	Range
Colors	Black, red	Full range	Full range	Limited	Clear or dark shades	Limited range	Full range
Additional notes	Timing between applications critical in obtaining coat adhesion	Cure rate and pot life affected by temperature. All epoxies' appearance and weathering problems solved by use of acrylic finish coat		Usually high-baked coating. Low-baked type has decreased toughness and flexibility	High-baked coating. Lining should not be cleaned with alkali solution		Have ability to cure at low temperatures. Urethanes based on aromatic diisocyanates discolor and chalk rapidly in sunlight. Those based on aliphatic diisocyanates have prolonged resistance to yellowing and surface chalking. Care must be taken to avoid moisture contact before cure

TABLE 7.11   General Properties of Oil-Type Coatings

	Modified-oil phenolic	Epoxy ester	Alkyd	Uralkyd
Cure type	Drying oil	Drying oil	Drying oil	Oil type
Effect of sunlight	Yellowing; surface chalk	Rapid surface chalk	Slow surface chalk	Yellowing; surface chalk
Wet or humid environments	Fair to good; will yellow	Fair to good; will yellow	Poor to good; will yellow	Fair
Industrial-atmosphere contaminants:				
Acids	Poor to fair	Fair	Fair to good	Poor to fair
Alkalies	Poor	Fair	Poor	Fair
Oxidizing	Poor	Poor	Fair	Poor
Solvents	Limited	Limited	Limited	Good
Spillage and splash of industrial compounds:				
Acids	Not recommended	Not recommended	Not recommended	Not recommended
Alkalies	Not recommended	Not recommended	Not recommended	Not recommended
Oxidizing	Not recommended	Not recommended	Not recommended	Not recommended
Solvents	Not recommended	Not recommended	Not recommended	Not recommended
Water immersion	Not recommended	Not recommended	Not recommended	Not recommended
Tank linings	Not used	Not used	Not used	Not used
Physical properties:				
Abrasion resistance	Poor	Fair	Fair	Excellent
Heat stability	Fair	Fair	Fair	Good
Hardness	Fair	Good	Fair	Excellent
Gloss	Good	Wide range	Wide range	High gloss
Colors	Limited	Full range	Full range	Limited range

7.7 Shop Priming   The specification for shop priming must establish the *minimum* requirements for surface preparation and shop priming of equipment, structural steel, steel tanks, and piping. It should note that "shop priming" may not limit the supplier's responsibility for priming to the shop location. Field surface preparation and touch-up or painting of

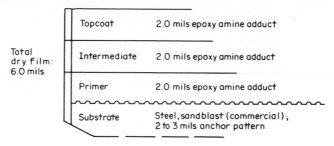

Fig. 7.6 Coating-system information: exposure, solvent spillage and alkaline dust; system type, adduct-cured epoxy.

Application

1. Clean the surface to a good commercial sandblast, using an abrasive size which will achieve the proper anchor pattern.
2. Apply the primer with an air or airless spray.
3. Allow sufficient time for cure before recoating.
4. Apply intermediate coat and topcoat to the specified thickness, using an air or airless spray. Allow sufficient time for cure between coats.

Inspection

1. Anchor pattern and cleanliness.
2. Appearance.
3. Thickness.
4. Cure.

Selection reasons

Primer. Adaptability to commercial surface; solvent resistance; compatibility with intermediate coat and topcoat; caustic resistance.

Intermediate. Compatibility with primer selected; solvent and caustic resistance.

Finish. Compatibility with intermediate coat; solvent and caustic resistance; color and appearance not important.

abraded or damaged areas of surfaces left unprimed can be part of the bid requirements.

A shop-priming painting schedule must identify the extent of shop priming of both the insulated and the noninsulated portions of equipment. It will also identify surfaces that are to be galvanized (generally structural steel that is not fireproofed is galvanized). Also included are surface-preparation procedures as well as the complete identification of all classes and types of material requiring painting. This is coupled with a complete shop-priming schedule (Fig. 7.10).

7.7.1 Paint application   Primers spread on the surfaces which they contact. If these surfaces are dirt, dust, scale, rust, oil, or moisture, the bond of the protective system to the structure can be only as good as the bond of the contaminant to the real surface. Furthermore, primers are formulated to stick to metals, concrete, wood, and masonry rather than to surface contaminants. The need for scrupulous surface cleaning prior to shop priming has become more important since spraying has replaced brushing as the common method of application. Brushing does permit a painter to work the primer into the surface, displacing the less adherent

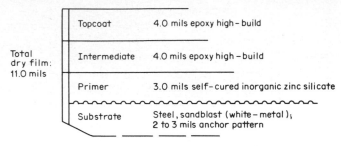

Fig. 7.7 Coating-system information: exposure, severe marine; system type, inorganic zinc–epoxy.

Application

1. Clean the surface to a white sandblast, using an abrasive size which will achieve the proper anchor pattern.

2. Apply the primer with air-spray equipment.

3. Apply intermediate coat and topcoat to the specified thickness, using an airless spray.

Inspection

1. Anchor pattern and cleanliness.

2. Appearance.

3. Thickness.

4. Cure.

Selection reasons

Exterior exposure. Marine conditions.

Primer. High degree of abrasion and marine resistance; compatibility with intermediate coat and topcoat; resistance to undercreep.

Intermediate. Compatibility with primer selected; high abrasion and saltwater resistance.

Finish. High abrasion and saltwater resistance; compatibility with intermediate coat; ability to build thickness; appearance not important.

contaminants. During application environmental conditions are critical. It is necessary to select a material that will dry or cure unaffected by any fluctuations in weather. Include the following items in the paint application portion of the specification:

▪ Painting must be performed in a dry environment free from blowing dust.

▪ Surfaces to be primed shall be dry and free from oil, grease, weld slag, weld spatter, and residues left by brush cleaning when paint is applied.

▪ Application shall be in accordance with the paint manufacturer's recommendations, including storage, mixing, handling, environmental conditions, surface temperature, and additives for accelerated drying if necessary.

▪ Paint shall be applied at the recommended spreading rates, but thickness shall not be less than the minimum dry film thickness specified. Should the spreading rate fail to produce the required thickness in one coat, additional paint shall be applied until the minimum requirements are met even if this work must be done in the field.

▪ When additional prime coats are required, the drying time between

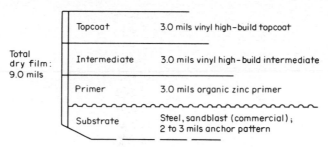

Fig. 7.8 Coating-system information: exposure, mild industrial; system type, inorganic zinc–vinyl intermediate and finish.

Application
1. Clean the surface to a good commercial sandblast, using an abrasive size which will achieve the proper anchor pattern.
2. Apply the primer with an air or airless spray.
3. Allow sufficient time for solvent evaporation before recoating.
4. Apply intermediate coat and topcoat to the specified thickness, using an airless spray.
Inspection
1. Anchor pattern and cleanliness.
2. Appearance.
3. Thickness.
Selection reasons
Exterior exposure. Mild industrial atmosphere; long-term appearance.
Primer. Resistance to undercreep; compatibility with intermediate coat and topcoat; applicable to commercial surface.
Intermediate. Compatibility with primer selected; water and chemical resistance.
Finish. Compatibility with intermediate coat; water and chemical resistance; color and gloss retention; resistance to mold; recoatability.

coats shall be held strictly within the limits recommended by the paint manufacturer.

- The prime coat shall be applied not more than 6 h after surface preparation. Should it become necessary to leave cleaned surfaces unpainted overnight, such surfaces shall be recleaned before painting.
- The prime coat on steel where subsequent welding is required shall be held back from the edge not less than 4 in (101.6 mm). Under no condition shall paint be applied to edges prepared for welding.
- Paint only on dry surfaces. Do not paint exterior surfaces in damp weather or when the temperature is less than 50°F (10°C) or below the dew point.
- All paint materials shall be delivered to the applicator in the manufacturer's original containers, unopened and with the label bearing manufacturer's name, product identification, and application instructions.

7.7.2 Inspection and troubleshooting Contractors and fabricators should be informed that shop priming is subject at any time to inspection, including the following:

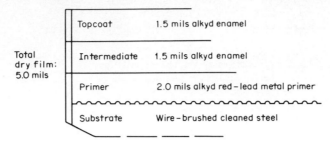

Fig. 7.9 Coating-system information: exposure, mild inland atmosphere; system type, alkyd.

Application
1. Prepare the surface with a hand or mechanical wire brush.
2. Apply an alkyd red-lead primer in full wet passes, using an air or airless spray, to the proper thickness.
3. Allow 18 to 24 h before recoating.
4. Apply intermediate coat and topcoat to the proper thickness, using an air or airless spray.
5. Allow 18 to 24 h before coating.

Inspection
1. Surface.
2. Appearance.
3. Thickness.

Selection reasons
Exposure. Mild inland atmosphere; dry; no corrosive atmosphere.
Primer. Poorly prepared surface; noncorrosive environment; compatibility with intermediate coat.
Intermediate coat and topcoat. Compatibility with primer selected; noncorrosive atmosphere; gloss and color retention; low-cost system; recoatability.

- Inspection of surface-preparation materials, equipment, and application
- Examination of prepared surfaces prior to paint application
- Examination of applied paint for discontinuities, lack of bond, and dry film thickness

In addition, the specification should identify the instruments that will be used to verify that the surface-preparation and priming requirements of this specification are met; for example:

Swedish pictorial standards, SSPC Visual No. 1, to check the degree of surface cleanliness

Clemtex profile comparators to check the surface profile of the blast-cleaned shape

Wet-film-thickness gauge as furnished by the contractor to check the wet film thickness

Mikrotest gauge to check the dry film thickness of the prime-coat material

Also, assure that all defects shall be corrected at no additional cost to the buyer, using the following procedures as applicable:

- At discontinuities, exposed metal shall be cleaned by wire-brushing and paint applied to cover completely and bond well into adjacent coating.
- Unbonded areas, as evidenced by blistering, cracking, flaking, or other forms of lift-off, shall be completely removed, and surfaces prepared by the method specified for the paint system and recoated.
- If the dry film thickness is less than that specified for the paint system, additional coating shall be applied over sound original paint. Unacceptable areas shall be repaired before overcoating.

SHOP PRIMING SCHEDULE (Continued)

Group I - Uninsulated Carbon Steel (Continued)

CARBON STEEL SURFACES	OPERATING TEMPERATURE RANGE ($^{\circ}$F) NORMAL ENVIRONMENT Not In Acid Areas		
	250 And Below	250 to 700	700 To 1000
VI. FURNACES AND EXCHANGERS			
a. Casing Steel	P.S. I (3)	P.S. 2 (4)	P.S. 3 (5)
b. Buckstays	P.S. I	P.S. 2	P.S. 3
c. Nozzles	P.S. I	P.S. 2	P.S. 3
d. X-Over Piping	P.S. I	P.S. 2	P.S. 3
e. Breeching	P.S. I	P.S. 2	P.S. 3
f. Stacks	P.S. I	P.S. 2	P.S. 3
g. Structural Steel	P.S. I	P.S. 2	P.S. 3
VII. STRUCTURAL STEEL			
a. All Structural Steel - Except as Otherwise Specified	(1)	---	---
b. Grating and Checker Plate	(1)	---	---
c. Stairway Hardware	(1)	---	---
d. Ladders and Cages	(1)	---	---
e. Toe Plates and Stringers	(1)	---	---
f. Handrails	(1)	---	---
g. Bolts, Nuts & Washers	(1)	---	---
h. Anchor Bolts	(1)	---	---
VIII. PIPING & FITTINGS			
a. Pipe	P.S. I	P.S. 2	P.S. 3
b. Fittings	P.S. I	P.S. 2	P.S. 3
c. Flanges	P.S. I	P.S. 2	P.S. 3
d. Valves	(2)	(2)	(2)
e. Spring Hangers	(2)	---	---
IX. PUMPS & MOTORS			
a. Casing	(2)	(2)	(2)
b. Nozzles	(2)	(2)	(2)
c. Baseplates	P.S. I	P.S. I	P.S. 2
d. Motors	(2)	(2)	(2)

NOTES:

(1) These surfaces shall be galvanized.
(2) Manufacturers Standard Finish is acceptable.
(3) P.S. I - Paint Standard No. 1 as included in Appendix.
(4) P.S. 2 - Paint Standard No. 2 as included in Appendix.
(5) P.S. 3 - Paint Standard No. 3 as included in Appendix.

Fig. 7.10 Typical page from a shop priming schedule. [*Courtesy Ralph M. Parsons Co., Pasadena, Calif.*]

TABLE 7.12 Characteristics of Air-Drying Maintenance Finishes*

General type	Generic type	Solvent	Type	Drying speed	Resistance				Temperature resistance	Application
					Alkalies	Water	Acids	Weather		
Air-dry; solvent evaporation	Polyvinyl chloride solution	K/AR	T,I	F	G	E	G	E	AM	High-durability topcoats
	Polyvinyl acetate emulsion	W	T	F	Q	A	A	E	AM	High-durability topcoats
	Chlorinated-rubber solution	AR	T,I	F	E	E	E	G	AM	High-durability topcoats and intermediate coats
	Polyacrylates	ES/AR	T	F	Q	A	G	E	AM	High-durability topcoats and intermediate coats
	Cellulosic lacquers	ES/AR	T	F	Q	A	A	G	AM	Topcoats
	Silicone solutions	AR	O	F	A	E	E	E	VH	Heat-resistant finishes
Dry; oxidative	Alkyds	M/AR	GP	D	Q	A	A	G	MEL	General utility; mild conditions
	Epoxy esters	M/AR	GP	D	A	G	G	A	MEL	General utility; reasonably severe conditions
	Silicone alkyds	M/AR	T	D	A	G	E	E	MEL	High gloss and color retention
	Oleoresinous	M	GP	D	A	Q	A	A	MEL	Mild conditions only
Air-dry; catalytic	Catalytic urethane	ES/AR	T,I	D	G	G	A	A	MEL	Abrasion-resistant finishes
	Moisture-cured urethane	ES/AR	T,I	D	G	G	A	A	MEL	Abrasion-resistant finishes

Epoxy amine	K/AR	P,I,T	D	E	A	G	A	MEL	General chemical resistance; high adhesion
Inorganic zinc silicate	W	P	F	Q	E	Q	E	H	Single-coat primer; salt-spray resistance
Polyvinyl butyral	AL	P	F	E	E	E	A	AM	Primer; high adhesion
Polyesters	S	T	D	G	G	E	G	MEL	Gloss topcoats

* A = acceptable; AL = alcohols; AM = ambient temperatures (to 170°F, or 77°C); AR = aromatic; D = dry in one day; E = excellent; ES = ester; F = fast; G = good; GP = general-purpose; H = high; I = intermediate coat; K = ketone; M = mineral spirits; MEL = medium elevated temperature (to about 300°F, or 149°C); O = one coat only; P = primer; Q = questionable; S = styrene; T = topcoat; VH = very high; W = water; / = OK.

7.8 Field Painting   The specification for field painting follows that for shop painting, but the inspection requirement to assure quality control becomes more important. Use appendixes to cover details of paint systems and coating schedules.

For each paint system include such information as the substrate, service conditions, operating-temperature range, minimum surface preparation and profile, manufacturer, and prime coat and/or final coat or coats as necessary. For each coating schedule show a complete listing of all surfaces to which paint is to be applied, together with equipment or item number and designation, surface temperature, paint system to be used, and manufacturer's code for the final color indicated.

In the field, carbon steel surfaces which have been primed with organic or inorganic zinc–rich materials shall be hosed down with clean salt-free water to remove surface salts and other contaminants and allowed to dry prior to topcoating. Sufficient drop cloths, shields, and other protective equipment shall be used to prevent surface-preparation abrasives, paint overspray, or paint droppings from fouling adjacent surfaces.

There are conditions that limit paint application in the field. Among them are:

- Paint shall not be applied when the ambient temperature is below 40°F (4°C) or above 110°F (51°C).
- Paint shall not be applied when the ambient temperature is expected to drop to 32°F (0°C) before the paint has had time to dry.
- Paint shall not be applied in rain, snow, fog, or mist.
- Paint shall not be applied in extremely windy or sandy conditions.
- Paint shall not be applied when the steel temperature is at or near the dew point.
- Paint shall not be applied in the vicinity of field sandblasting or grit-blasting operations.
- Paint shall not be applied when the steel surface temperature exceeds limits specified by the paint manufacturer for the specific coating materials being used.

Typical coating systems and coating recommendations are found in Table 7.13.

7.9 Inspection and Testing   In field measurements for required film thickness, some companies use the Elcometer exclusively (this device contains a preaged permanent magnet in a U shape plus a pointer that responds to magnetic flux). A word of caution is needed to explain the seeming discrepancy when the recommended profile for automatic blasting exceeds the specified film thickness for inorganic zinc. When an Elcometer is set for a given piece of steel, the shim rests essentially on top of the profile; thus the Elcometer will read the thickness of paint film above the profile tips in subsequent tests. Mikrotest units calibrated on smooth steel may read the paint film thickness from as much as 1 mil (0.0254 mm) below the profile tips on steel blasted to a profile of about 4 mils (0.1016 mm).

After the complete coating system has been applied, it should be checked for film thickness and film continuity. The first of these tests is normally made with a Mikrotest or Elcometer dry-film-thickness gauge. Both of these instruments are magnetic gauges, and the Elcometer must be standardized on sandblasted steel before taking film-thickness readings. It is important to remember that the accepted limit of accuracy of these instruments is 10 percent of the indicated thickness, plus or minus 1 mil. The accepted procedure for determining the average film thickness of a paint system is to take 10 readings per 100 ft² (9.29 m²) selected at random and to calculate a mathematical average. For immersion surfaces, the inspector should insist that all the coating film be above a certain minimum standard and in no case be less than 5 mils (0.127 mm).

Normally calibrated, a Mikrotest unit will read at least 0.5 mil (0.0127 mm) below the blast profile tips on steel that has been blasted to a 3-mil (0.0762-mm) profile. Therefore, about half of the 3-mil profile is filled with steel, and half is void. The voids (amounting to 1.5 mils, or 0.0381 mm) over the surface of the steel must be filled with primer.

Since a Mikrotest unit reads at least 0.5 mil below the profile tip, this fact is included in the minimum specification. The unspecified and unmeasured primer required to fill the profile is about 1 mil, and this must be added to the specified thickness in calculating material requirements and costs.

Unless a compressor of sufficient size is used, there will not be enough pressure at the nozzle to perform the work efficiently. This causes slower production and increases the cost per square foot for sandblasting. In addition, the anchor profile is reduced, resulting in a shorter coating life.

Inspection must be continuous, or problems will appear. In one gas-turbine project, the steel plate delivered to the construction site was clean, with all mill scale removed, and was properly shop-primed. During construction the steel plate was abused to the point that its protective prime coating was abraded. No attempt was made at the time of construction to reprime the areas which had been scratched and gouged. The seam welds should have been reprimed immediately after welding.

Competent inspection of surface preparation and coating application is a very specialized skill which is acquired by study, experience, and willingness to pay particular attention to details. In a great many cases, the life expectancy of a coating film will vary in direct proportion to the competence of the inspector. This is particularly true when work is completed by outside contractors operating under hard-money contracts. In

TABLE 7.13 Typical Coating Systems for an Industrial Plant

Area of use	System classification	Generic type	Dry film thickness (mils)*	Surface preparation	Theoretical coverage (square feet/gallon)†
Structural steel	Alkyd	Red oxide zinc chromate alkyd Long-oil alkyd	2½ 1½ ―― 4	SSPC SP-6-63 (Commercial blast)	295 480
	Vinyl	Modified vinyl primer Vinyl	2 3 ― 5	SSPC SP-6-63 (Commercial blast)	200 162
	Rubber	Chlorinated rubber Chlorinated rubber	2 3 ― 5	SSPC SP-6-63 (Commercial blast)	225 150
	Zinc (best protection)	Inorganic zinc Epoxy-polyamide	3 4 ― 7	SSPC SP-6-63 (Commercial blast)	333 233
Nonsubmerged steel near water	Vinyl	Modified vinyl primer Vinyl	2 3 ― 5	SSPC SP-6-63 (Commercial blast)	200 162
	Rubber	Chlorinated rubber Chlorinated rubber	2 3 ― 5	SSPC SP-6-63 (Commercial blast)	225 150

Service	Type	Coating	Thickness, mils	Surface preparation	
	Shop-primed	Inorganic zinc	½	SSPC SP-6-63 (Commercial blast)	308
		Vinyl	4		135
			5½		
	General	Epoxy–coal tar	8	SSPC SP-10-63 (Near-white-metal blast)	150
		Epoxy–coal tar	8		150
			16		
	Shop-primed	Inorganic zinc	1½	SSPC SP-5-63 (White-metal blast)	308
		Epoxy–coal tar	8		150
		Epoxy–coal tar	8		150
			17½		
Steel piping; interior (solvent or hydrocarbon service)		Inorganic zinc	3	SSPC SP-5-63 (White-metal blast)	333
Steel piping; exterior; above ground	Vinyl	Modified vinyl primer	2	SSPC SP-6-63 (Commercial blast)	200
		Vinyl	3		162
			5		
	Rubber	Chlorinated rubber	2	SSPC SP-6-63 (Commercial blast)	225
		Chlorinated rubber	3		150
			5		
	Zinc (best protection)	Inorganic zinc	3	SSPC SP-6-63 (Commercial blast)	333
		Epoxy–coal tar	8		150
			11		
Steel piping; exterior below ground	Traditional	Epoxy–coal tar	8	SSPC SP-10-63 (Near-white-metal blast)	150
		Epoxy–coal tar	8		150
			16		

TABLE 7.13 Typical Coating Systems for an Industrial Plant (Continued)

Area of use	System classification	Generic type	Dry film thickness (mils)*	Surface preparation	Theoretical coverage (square feet/gallon)†
	Shop-primed	Weldable inorganic zinc	1½	SSPC SP-5-63	308
		Epoxy–coal tar	8	(White-metal blast)	150
		Epoxy–coal tar	8		150
			17½		
Galvanized steel; exterior		Self-priming vinyl copolymer	4	SSPC SP-1-63 (Solvent cleaning)	129
Galvanized drain pipes, etc.		Self-priming vinyl copolymer	4	SSPC SP-1-63 (Solvent cleaning)	129
Galvanized gratings	Epoxy-polyamide	Epoxy-polyamide	4	SSPC SP-7-63	190
		Epoxy-polyamide	4	(Brush-off blast)	190
			8		
Structural steel	Alkyd	Red oxide–zinc chromate Alkyd	2½	SSPC SP-6-63	295
		Long-oil alkyd	1½	(Commercial blast)	450
			4		
	Vinyl	Modified vinyl primer	2	SSPC SP-6-63	200
		Vinyl	3	(Commercial blast)	162
			5		

Service	Coating	Surface preparation	Thickness, mils	Coverage, sq ft/gal
Rubber	Chlorinated rubber Chlorinated rubber	SSPC SP-6-63 (Commercial blast)	2 $\frac{3}{5}$	225 150
Zinc	Inorganic zinc Epoxy-polyamide	SSPC SP-6-63 (Commercial blast)	3 $\frac{4}{7}$	333 233
General-service steel chemical tanks; exterior Vinyl	Modified vinyl primer Vinyl	SSPC SP-6-63 (Commercial blast)	2 $\frac{3}{5}$	200 162
Rubber	Chlorinated rubber Chlorinated rubber	SSPC SP-6-63 (Commercial blast)	2 $\frac{3}{5}$	225 150
Shop-primed	Inorganic zinc Vinyl	SSPC SP-6-63 (Commercial blast)	1½ $\frac{4}{5½}$	308 135
Steel tanks for special chemical service	Polyester containing fiber glass Polyester containing fiber glass	Steel: SSPC SP-5-63	20 $\frac{20}{40}$	80 80
Water tanks (potable water); interior	American Water Works Association type	Iron oxide/vinyl Modified vinyl copolymer Modified vinyl copolymer Modified vinyl copolymer	SSPC SP-5-63 (White-metal blast)	2 1½ 1½ $\frac{1½}{5½}$	160 215 215 215

TABLE 7.13 Typical Coating Systems for an Industrial Plant (Continued)

Area of use	System classification	Generic type	Dry film thickness (mils)*	Surface preparation	Theoretical coverage (square feet/gallon)†
	Field-primed	Epoxy-polyamide	5	SSPC SP-5-63 (White-metal blast)	334
		Epoxy-polyamide	5		334
		Epoxy-polyamide	5		334
			15		
	Shop-primed	Inorganic zinc	1½	SSPC SP-5-63 (White-metal blast)	308
		Epoxy-polyamide	5		334
		Epoxy-polyamide	5		334
			11½		
Exterior	Alkyd	Red-lead/alkyd	2	SSPC SP-6-63 (Commercial blast)	295
		Long-oil alkyd	1½		480
		Long-oil alkyd	1½		480
			5		
	Where overspray is a problem	Red-lead–alkyd primer	2	SSPC SP-6-63 (Commercial blast)	295
		Modified alkyd	3		214
		Modified alkyd	3		214
			8		
	Rubber	Chlorinated rubber	2	SSPC SP-6-63 (Commercial blast)	225
		Chlorinated rubber	3		150
			5		

Category	Service	Coating system	Coats (mils)	Surface preparation	No.
	Zinc (maximum protection)	Inorganic zinc Epoxy–polyamide Modified acrylic	1½ 4 3 8½	SSPC SP-6-63 (Commercial blast)	308 233 185
Process equipment	Shop-primed	Inorganic zinc	3	SSPC SP-6-63 (Commercial blast)	333
	Vinyl	Vinyl	4	⋯	135
	To 750°F (399°C)	Modified silicone (aluminum)	2	⋯	300
	To 1000°F (538°C)	Modified silicone	2	⋯	240
Incinerators	To 750°F (399°C)	Inorganic zinc Modified silicone	3 2 5	SSPC SP-6-63 (Commercial blast)	333 300
Insulated flues and ducts		Modified acrylic latex Modified acrylic latex	2 2 4	SSPC SP-6-63 (Commercial blast)	305 305
Steel manhole covers	Traditional	Epoxy–coal tar Epoxy–coal tar	8 8 16	SSPC SP-10-63 (Near-white-metal blast)	150 150
	Shop-primed	Weldable inorganic zinc Epoxy–coal tar Epoxy–coal tar	1½ 8 8 17½	SSPC SP-5-63 (White-metal blast)	308 150 150

TABLE 7.13 Typical Coating Systems for an Industrial Plant (Continued)

Area of use	System classification	Generic type	Dry film thickness (mils)*	Surface preparation	Theoretical coverage (square feet/gallon)†
Structural steel in cooling towers	Traditional	Epoxy–coal tar Epoxy–coal tar	8 8 ― 16	SSPC SP-10-63 (Near-white-metal blast)	150 150
	Shop-primed	Inorganic zinc Epoxy–coal tar Epoxy–coal tar	1½ 8 8 ― 17½	SSPC SP-5-63 (White-metal blast)	308 150 150
General-purpose steel and steel accessories	Alkyd	Red oxide/zinc Chromate alkyd Long-oil alkyd	2½ 1½ ― 4	SSPC SP-6-63 (Commercial blast)	295 480
	Vinyl	Modified vinyl primer Vinyl	2 3 ― 5	SSPC SP-6-63 (Commercial blast)	200 162
	Rubber	Chlorinated rubber Chlorinated rubber	2 3 ― 5	SSPC SP-6-63 (Commercial blast)	225 150
	Shop-primed	Inorganic zinc Vinyl	1½ 4 ― 5½	SSPC SP-6-63 (Commercial blast)	308 135

NOTE: Pickling may be substituted for sandblasting for shop-primed steel. When this is done, the final acid rinse must be omitted.
* 1 mil = 0.0254 mm.

many cases, jobs are put out for bid and contracts awarded to the low bidder, who frequently lacks not only the efficient equipment and competent personnel to complete the work properly and expeditiously but in many cases even the financing to bring all the material together to the jobsite before initiating work. In cases of this type, the buyer's protection obviously rests in the hands of a capable inspector.

In other cases, work is awarded to financially responsible, capable contractors who are debarred from bringing a competent labor force onto the jobsite to complete the work by restrictive union requirements that they staff their crews with local personnel. Such rules obviously make the responsible contractor little more than a labor broker at the mercy of the hiring halls in seeking competent personnel. In this case, too, the consumer's protection rests in the hands of a capable inspector.

If an architect maintains his or her own inspectors, the need for competent inspection is reduced, but it nevertheless exists. The following inspection tips and comments on inspection tools should prove useful in meeting the paint specification.

If surface preparation is to be completed by sandblasting, the inspector obviously should inspect the sandblasting equipment before initiating any work. If possible, the inspector should require the use of rotary rather than reciprocal compressors for sandblasting, since they produce a continuous volume of low-temperature air. Reciprocating compressors not only produce a pulsating flow of air but also generate air at a very high temperature. When this heated airstream expands rapidly on leaving the sandblasting nozzle, rapid cooling results, which in periods of high humidity can induce moisture condensation on the steel surface. Furthermore, the inspector should confirm that the right size of sandblasting grit has been provided for the work and that the grit has been washed with fresh water and is free of clay, caliche, salt, and other contaminants. The inspector should make certain that sandblasting sand is used only once.

After sandblasting has been completed, the inspector should make certain that all dust is removed from the surface by brushing, vacuuming, or blowing the surface with compressed air.

The priming operation should be completed as soon as possible after preparing the surface, and the sandblasted areas should never be allowed to remain uncoated overnight. It is a wise precaution to leave a 4-in (101.6-mm) margin of cleaned steel around the perimeter of the primed areas so that there is no danger of priming on the following day areas which have not been cleaned adequately. For organic coatings which will be immersed, it is always a wise precaution to brush the prime coat. The scrubbing action of the paintbrush bristles tends to work the coating down into the anchor pattern and, at the same time, to lift any residual particles of sand or dust up into the body of the coating film, where they will do little harm. If priming is completed by brush, an inspection should be made the following day to ensure removal of all paintbrush bristles embedded in the paint film. The inspector should take wet-film-thickness readings during the priming operation to ensure proper coverage and confirm these readings the following day by taking dry-film-thickness readings.

After the prime coat has dried, the inspector should confirm that it is of the proper thickness and free of holidays. Then the inspector should determine whether or not the painted surface is contaminated before permitting the application of the next coat of the paint or coating system. This is a particularly important point when painting steel in modern chemical plants, which are almost constantly exposed to fumes from adjacent operating units or other industrial plants in the area. A check for surface contamination by acid or alkaline materials can be made by adding a few drops of an appropriate indicator to the surface of the primer. Commonly used indicators are phenolphthalein and potassium ferrocyanide. An alternative procedure consists of wetting a small area of primed steel with distilled water and placing in this water a piece of calibrated litmus paper. If contamination is found, it can normally be removed by flushing the surface with fresh water and proceeding with the application of the second coat immediately after the surface has dried. In an area with very heavy fumes in which surfaces are contaminated almost immediately, the preferred procedure is to clean the primed surface with saturated steam. The hot water or steam flushes from the surface very quickly and permits immediate topcoating. With the possible exception of catalyzed epoxies, all paint and coating materials prefer—indeed, demand—a neutral surface to achieve a high order of adhesion.

7.10 Recoating Previously Painted Surfaces   For ferrous or zinc-coated metals, if the existing finish is sound and shows only normal chalking, sand lightly and wipe surfaces dust-free. Whenever surfaces are badly checked and paint is cracked, alligatored, peeling, or in a generally poor condition, the best way to eliminate rust is to remove as much as possible by scraping, wire-brushing, or sandblasting if necessary. When surfaces have been wiped clean of dust, apply a coating totally compatible with that on the surface: this will usually be the same paint as that on the surface.

For galvanized-iron pipe rails, doors, gutters, flashings, and downspouts, if the existing finish is sound and shows only normal chalking, sand lightly and remove all dirt, oil, and grease by wiping surfaces with a cloth well saturated with mineral spirits. Spot-prime with one coat of an acrylic or other primer suitable for galvanized surfaces. Allow to dry overnight before applying the finish coat. Whenever surfaces are badly checked, cracked, alligatored, or peeling, hand-sand, wire-brush, or revert to sandblasting if necessary. Wipe the surface with mineral spirits to remove dust. Proceed with priming, allow the primer to dry, and then apply the finish coat.

BIBLIOGRAPHY

Alkyd Coatings for Prevention of Atmospheric Corrosion, National Association of Corrosion Engineers, Katy, Tex., 1965.

Amine Cured Epoxy Resin Coatings for Resistance to Atmospheric Corrosion, National Association of Corrosion Engineers, Katy, Tex., 1970.

Chemically Cured Coal Tar Coatings for Atmospheric Exposures, National Association of Corrosion Engineers, Katy, Tex., 1963.

Comparative Properties and Performance Chart for Coil Coating Finishes, Technical Bulletin IV, National Coil Coaters Association, Philadelphia, 1974.

Frye, S. C.: "Corrosion and Methods of Protecting Structural Steel," presented at Structural Engineering Conference, Miami, 1966, American Society of Civil Engineers, New York.

A Guide to the Shop Painting of Structural Steel, Steel Structures Painting Council, Pittsburgh, Pa., and American Institute of Steel Construction, New York, 1972.

1980 Sweet's Catalog File: Products for General Building, Vol. 8, Secs. 9.9 and 9.10, McGraw-Hill, Inc., New York, 1980.

Organic and Inorganic Zinc Filled Coatings for Atmospheric Service, National Association of Corrosion Engineers, Katy, Tex., 1973.

Performance of Alternate Coatings in the Environment, Steel Structures Painting Council, Pittsburgh, Pa., 1979.

Straight and Modified Phenolic Coatings for Atmospheric Service, National Association of Corrosion Engineers, Katy, Tex., 1967.

Urethane Protective Coatings for Atmospheric Exposures, National Association of Corrosion Engineers, Katy, Tex., 1973.

Vinyl Coatings for Prevention of Atmospheric Corrosion, National Association of Corrosion Engineers, Katy, Tex., 1963.

"Waterborne Coatings for Metals," *American Paint Journal,* June 11, 1979, p. 58 ff.

Zinc Filled Inorganic Coatings, National Association of Corrosion Engineers, Katy, Tex., 1961.

Chapter 8

Coatings for Metals Other Than Steel

DONALD R. PESHEK
PPG Industries

8.1   Introduction   Most metals are found in nature in a combined form with other elements such as oxygen, sulfur, phosphorus, etc. By adding energy during the processing of these metallic ores, the metal is raised from the natural state to the more metallic state. In this condition, the metal can be usefully employed in a vast number of applications. The metal, however, prefers to return to the more stable form in which it was found in nature, such as an oxide, a sulfide, etc. This process is called corrosion. Some metals form a tight oxide film on the surface which further limits corrosion and are said to be passivated. Among them are aluminum, cadmium, chromium, lead, and nickel. Zinc also forms an oxide layer, but corrosion continues at a reduced rate. Control of corrosion is based upon the prevention of chemical reactions that lead to destruction of the metallic state. The character and extent of the prevention measures depend upon the nature of the metal and the environment to which it is exposed. There are many ways to protect, decorate, or otherwise modify the surface of metals, one of which is the use of protective coatings. These coatings systems may be inorganic or organic in nature. The environmental service life and the degree to which the surface can be prepared are key requirements in the selection process.

Before choosing the exact coatings system for a particular metal surface, the particular alloy must be identified. In many cases the manufacturer of a particular metal will suggest coatings systems that perform well under given conditions.

Most metals other than iron or steel degrade by chemical reaction when in contact with other elements. This effect is also called "corrosion" or "tarnishing." Oxides or salts of the metal form on the surface. Depending on the environment and contact with other metals, the corrosion products may act to stop further reaction from occurring, contribute to an increased rate of corrosion, or have no effect one way or the other. In some cases the corrosion products alone provide sufficient protection and have a good enough appearance so that no other protective measures are needed. In most cases coatings are used for corrosion protection.

Corrosion is an electrolytic process. Anodic and cathodic cells form either on the metal surface or between two dissimilar materials in the presence of an electrolyte, usually water. An electric current passes through the cell and corrosion proceeds. The electrode potential varies with the temperature (Table 8.1). The metals at the top of the table are generally more reactive because their ions more readily go into a solution such as water. The lower end of the table indicates less dissolution of the metal in the liquid phase; gold, the least reactive of the metals listed, is at the bottom.

When choosing a particular coating system to be applied to one of these metals, there are sometimes size, shape, or weight limitations plus

TABLE 8.1   Electromotive-Force (EMF) Series

Electrode	Electrode potential (volts at 25°C)
Magnesium	−2.4
Aluminum	−1.7
Zinc	−0.76
Iron	−0.44
Cadmium	−0.40
Nickel	−0.23
Tin	−0.13
Lead	−0.12
Hydrogen	0.00
Copper	+0.34
Silver	+0.80
Gold	+1.5

considerations related to the method of application. Aesthetic values in regard to color, gloss, and metallic appearance should also be considered. In most respects a factory application can be much more thorough than a field-applied coating, and in many cases heat, ultraviolet light, or electron-beam curable thermosetting systems offer advantages in long-term durability and chemical, stain, and corrosion resistance over other coatings such as thermoplastic, oxidative, or chemically cured films. All these factors must be considered when selecting the best coating for a particular job. The architect and recommended coatings supplier should be consulted to obtain optimum performance and appropriate surface preparation.

8.2   Exterior versus Interior Exposures   The kind of exposure is usually the determining factor for the generic type of polymer used in the coating. Because exterior exposures are generally more severe than most interior exposures, a coating that is good enough to be used outdoors usually is satisfactory indoors. Heat, ultraviolet light, oxygen, rain, dust, abrasion, wind, salts, industrial gases, cold, ice, expansion and contraction, grease, and solvents are some of the conditions that a coating must endure in exterior applications. However, many indoor applications have their own very severe exposure conditions such as steam, cutting oils, temperature cycling, and corrosive chemicals. The accompanying form can be used as a good guide to obtain needed information on coating conditions. The information can then be used in discussions with engineers or coatings suppliers.

8.3   Pigmented or Clear Coating   The corrosion rate of metals can be reduced by both the pigment and the binder portions of a paint film. If,

Project No. _____
Date _____

PAINT SYSTEM REQUIREMENTS

Site _____

Specific location _____

Environmental conditions

  1. Interior _____ Exterior _____

    a. Gases, smoke, dust, etc.

          Types   Concentrations

    _____   _____

    _____   _____

    _____   _____

    _____   _____

    b. Liquids and aqueous solutions

          Types   Concentrations

    _____   _____

    _____   _____

    _____   _____

    c. Humidity conditions _____

    d. Variation or cycling exposure (approximate periods) _____

  2. Surface-temperature conditions

    a. Average _____°F   Maximum _____°F   Minimum _____°F

    b. Temperature cycling (approximate periods) _____

  3. Abrasive or erosive conditions _____

Application conditions

  1. Object to be coated _____

  2. Type of surface (e.g., aluminum, zinc) _____

  3. Condition of surface _____

  4. Permissible cleaning methods _____

  5. Permissible application methods

    _____ Brush   _____ Roller _____ Spray _____ Other

  6. Maximum drying time _____

  7. Permissible number of coats _____

Appearance requirements

  1. Color(s) _____

  2. Reflectance _____   3. Gloss _____

Minimum performance requirements

  1. Film integrity, to repaint _____ years

  2. Film appearance, to repaint _____ years

  3. Other _____

Applicable specifications _____

Remarks and recommendations _____

_____

_____

Courtesy *Plant Engineering Directory and Specification Catalog.*

however, you do not wish to destroy the aesthetic effect of the base metal itself, a clear coating can be used. When clear coatings are employed, multiple application and higher film builds are preferred to ensure film protection properties. The specification must assure that the clear finish will perform under the exposure conditions.

By far the greatest volume of coatings for metals other than steel is of the pigmented type. A pigmented system usually has the ability to reflect a large portion of the ultraviolet radiation, which is a damaging ingredient in exterior exposures. Other properties that pigments contribute to a coating are aesthetic appeal in the form of broad color selection, abrasion and antislip resistance, hardness, reinforcement, hiding, corrosion resistance, and viscosity or rheology control to ensure an adequate film build, etc., when the coating is applied by brush, roller, or spray equipment.

8.4 New Coating versus Recoating   Different parameters govern the type of coating to be used on new and recoating applications. Most recoating applications do not specify coatings that cure by heat or radiation. However, it is important to know what type of coating is already present in order to gain proper adhesion to the topcoat. Maintenance records should be consulted before writing a specification, or patch tests should be applied to assure that the solvent or thinner from one paint does not lift or otherwise damage the film under it. Inspection during surface preparation and application pays dividends in the long run.

With new applications of paint to bare metal, the coating will perform only as well as the degree of surface preparation exercised. A systems approach is recommended.

8.5 Durability   For any job, the life expectancy of the coated surface must be considered. If a piece of equipment or a surface is expected to have a serviceable life of only 3 years, there is no need to apply a multiple-coat system with a life expectancy of 10 or more years. In the same regard, you should not underdesign a coating system. The labor involved in surface preparation and recoating is more expensive than using a better coating in the first place.

In writing a specification for a coating for metal, it is essential that thorough surface preparation be included and that coating thickness be documented. Also include assurances that proper weather conditions prevail at the time of application. For critical performance areas, a periodic inspection program should be set up before, during, and after application.

8.6 Surface Preparation   It is generally accepted that a coating is only as good as its associated surface preparation. This important point cannot be overstressed. Often a coating is used to cover up soil, grease, or a previous failure or deficiency. For ultimate performance of a coating, the surface should be in the optimum condition dictated by the coating to be applied. Chapters 5 and 6 thoroughly cover surface preparation of metals; consult them for guidance.

8.7 Coatings   Thicker coatings in general provide an added measure of corrosion resistance. High-build coatings that give greater film thicknesses with fewer coats and reduced labor costs are now available. Painting on a surface that is in direct sunlight or applying too thick a film in one application can cause blistering. Blistering results when the surface of the film dries rapidly, thereby entrapping solvents that continue to vaporize; blisters or pinholes form as the solvent escapes from the film. For some applications, however, film thicknesses must be strictly limited, especially when high heat or thermal cycling is encountered. Most efficient coatings systems are based on two or more compatible coatings, a primer and a topcoat (sometimes an intermediate coat is used to increase total film build or ensure adhesion), and this practice greatly extends the range of possible coating choices.

8.7.1 Primers   Several kinds of primers are suitable for metal surfaces.

Wash primers   In many cases it is necessary to protect a metal surface temporarily from contamination which could lead to unwanted corrosion products. One method of providing this protection, while also enjoying the benefit of modifying the metal surface to improve its acceptance of other coating systems, is to use a wash primer. Such a primer modifies the metallic surface to produce a tightly adherent film composed of chromates, phosphates, or oxides.

Wash primers contain two separate components that are mixed just prior to application and have a workable application time or a pot life of about 8 h. Polyvinyl butyral is the usual organic portion in combination with a pigmentation system based on zinc chromate. A phosphoric acid catalyst solution is slowly added to the pigmented base while the mixture is being agitated to promote film cure. Extra thinner is sometimes necessary to obtain better spray application properties under conditions of high temperature and high humidity. Coatings of this type are particularly suitable for steel, aluminum, zinc, cadmium, chromium, tinplate, and terneplate, and they provide a smooth, tough priming base for a uniform-appearing, high-luster finish. Since 0.3 to 0.5 mil (0.0762 to 0.127 mm) is the normal film thickness, a suitable corrosion-resistant primer or intermediate coat must be applied over it within 2 days to ensure adequate protection. Wash primers are not recommended for use on surfaces subject to continuous in-service surface temperatures above 150°F (66°C). For best results, they must be applied only on bare, grease-free, clean metal.

Wash primers have been topcoated successfully with vinyls, phenolics, alkyds, nitrocellulose and oil-type products, and others.

Conversion coatings   Inorganic protective coatings for zinc, iron, aluminum, and magnesium often require that the surface be treated with phosphate, chromate, or other oxidizing solutions called "conversion coatings." On the surface of the metal is formed a reaction product which becomes an integral part of an object so treated. Most conversion-coating processes are proprietary and usually are applicable to only one metal. The coatings form a nonreactive barrier of low solubility between a metal and its environment. Adhesion to subsequent paint coats and ability to absorb protective oils and waxes, in addition to their own property of high corrosion resistance, are the benefits obtained. Conversion coatings are formed as soft gels, but when dried they become hard, tough, and flexible. However, overheating upon drying must be avoided because the coatings may deteriorate. Generally speaking, conversion coatings are factory-applied instead of being applied by the contractor on the site.

Zinc-rich coatings   Some primers are zinc-dust types. They are available in one-, two-, and three-component systems, with water or solvent as the diluent. The dried film usually contains between 82 and 92 percent metallic zinc by weight and provides the coated metal with chemical resistance similar to that of galvanized steel. These primers will resist weathering for several months before topcoat application, and some types will withstand continuous dry heat up to 500°F (260°C). Saltwater resistance is good, and epoxy, vinyl, chlorinated-rubber, or urethane coatings are recommended as topcoats when corrosive atmospheres are to be encountered. Organic primer types include epoxy or chlorinated-rubber vehicle systems, although silicate types usually are preferred in a water or solvent base. Zinc-rich primers are normally used to protect steel, but they may be applied on other metals below zinc in the electromotive series (see Table 8.1). Depending on the binder used, these primers can be classified as either organic or inorganic. Silicate types are inorganic, whereas epoxy and similar types are organic.

Zinc chromate primers   Based mainly on the pigment zinc potassium chromate, these primers are used on aluminum and magnesium as well as on ferrous metals. Their inhibitive qualities are due to the soluble chromate ions available. Being slightly basic, these ions can neutralize acids formed as corrosion by-products. Zinc chromate primers are not suited to highly acidic atmospheres or as primers for baking systems.

Primer binder systems can be based on alkyd, linseed-oil–alkyd, or phenolic resins. Most primers of this type are one-component in nature, although two-package catalyzed epoxies are available.

8.7.2 Finishing coats

Over the years coatings have become more specific for particular applications. Now environmental restrictions on solvent-based coatings are placing an added burden on the specification writer to choose a coating system that not only meets exposure requirements but conforms to environmental regulations. For this reason, you should work with your coatings suppliers to make the best possible recommendations.

Alkyds   As a class alkyds comprise a large portion of the organic coatings being used today, but water-based systems are replacing them. Still, alkyds are employed in a wide variety of industrial and consumer finishes.

Alkyd drying is accomplished by both solvent evaporation and autoxidation (the adsorption of atmospheric oxygen that results in a cross-linking reaction of the alkyd resin) and condensation (a chemical reaction which gives water as a by-product). By the use of plasticizers and control of the molecular makeup, coatings that range from hard and brittle to those that are softer and flexible can be prepared. They can be formulated to serve as exterior or interior coatings cured by forced heat or room temperature. High-gloss, adhesive, durable coatings are obtained, but in general alkali resistance is somewhat lacking. If a baking finish can be used, it is usually advantageous to do so since properties are improved over an air-dry finish.

Vinyls   Vinyl coatings vary considerably according to type. They are available as solutions, dispersions, or emulsions. By the use of different monomers and molecular weights, it is possible to obtain a wide range of properties. Common names of vinyl vehicles are polyvinyl chloride, polyvinyl acetate, polyvinyl butyral, etc. Most of these vehicles are copolymers or a mixture of two or more monomers. They are available as both air-drying and baking coatings when blended with suitable alkyd or plasticizing, flexibilizing resin modifiers.

Vinyl chloride coatings have excellent chemical resistance and flexibility, but since formulations are low in solids, multiple coats are needed to get desired film thicknesses. These materials are used mostly as intermediate coats and topcoats, although primers are also available. They are employed extensively in marine atmospheres.

Color retention on exterior exposure is very good for the acetate type, but prolonged immersion in water is not recommended. Many interior latex paints are based on this polymer because performance is adequate and cost per gallon is relatively low.

Cellulosic coatings   As the name implies, these coatings systems have a derivative of cellulose as their binder. They are known chemically as nitrocellulose, ethyl cellulose, cellulose acetate butyrate (CAB), etc. Cellulosic coatings belong to the family of coatings called lacquers since they dry quickly by solvent evaporation. They have good film-forming properties and dry times and exhibit good hardness and mar resistance when properly formulated.

When employed as industrial coatings today, they are usually modified with other polymers to improve adhesion, weathering, moisture resistance, chemical resistance, etc. Since they are thermoplastic, they remain soluble in solvents and are softened by heat. They are used most successfully as industrial coatings, for cabinetry, and as automobile topcoats.

Acrylics   Both thermoplastic and thermosetting coatings based on acrylic polymers are available; they vary widely in physical properties. Although moderate in cost, they are highly resistant to ultraviolet-light degradation and therefore are used extensively in exterior as well as interior applications. Large volumes are employed in the factory coating of aluminum siding, aluminum extrusions, air conditioners, and machinery parts. Acrylics also perform remarkably well when used as clear coatings to protect bright metals against tarnishing and dirt pickup and other miscellaneous objects against deterioration. Pigmented coatings retain their true fidelity over a wide range of weather conditions.

Epoxies   Although widely used in primer formulations, epoxy-based coatings are also used as finishing coatings for metals, especially for interior applications. They are most widely noted for their adhesion, chemical resistance, hardness, and overall toughness. Their biggest shortcoming is a liability to color fading, chalking, and checking in exterior exposures. Advantages include abrasion and alkali resistance.

Urethanes   Modified and unmodified urethane polymers (polyurethanes) are widely used when chemical and abrasion resistance is of prime importance. They retain a high gloss and maintain great flexibility at high hardness levels. Their main deficiency is that some types yellow severely when exposed to sunlight. Urethanes are packaged either as a two-component system or as one package which cures with atmospheric moisture or oxygen. Depending on the application, they provide an excellent coating system for metals. Costs per gallon are very high, but they are reduced on a life-expectancy basis in aggressive environments. As a class, these coatings are the fastest-growing type in the marketplace today.

Silicones   Silicone coatings are noted for their exceptional heat resistance and color and gloss retention on exterior exposure. Many specification-type products are based on the use of silicone vehicles even though the cost is quite high. These coatings can be pure silicone types or

TABLE 8.2 Relative Paint Film Properties

Coating type	Adhesion	Hardness	Flexibility	Fade resistance	Mar resistance	Relative cost per square foot	Solvent resistance	Salt-spray resistance	Stain resistance	Humidity resistance	Chemical acid and alkali resistance	External durability
Alkyd amine	Good	Good	Fair	Fair	Good	1.3	Fair	Fair	Fair	Good	Fair	Good
Epoxy polyamide	Excellent	Excellent	Fair	Poor	Excellent	2.0	Excellent	Excellent	Excellent	Excellent	Excellent	Poor
Epoxy ester	Excellent	Good	Fair	Poor	Good	1.7	Fair	Excellent	Good	Excellent	Good	Poor
Acrylic	Excellent	Good	Good	Good	Excellent	3.0	Good	Excellent	Good	Excellent	Good	Good
Silicone alkyd	Good	Good	Fair	Excellent	Good	5.0	Good	Good	Good	Good	Good	Excellent
Polyvinyl fluoride	Good	Good	Excellent	Excellent	Fair	9.0	Excellent	Excellent	Excellent	Excellent	Excellent	Excellent
Polyester (oil-free)	Good	Excellent	Good	Good	Excellent	1.4	Good	Excellent	Good	Good	Fair	Good
Alkyd	Good	Fair	Fair	Good	Fair	1.0*	Fair	Fair	Poor	Good	Poor	Good
Polyvinyl chloride	Good	Excellent	Excellent	Good	Good	1.5	Good	Excellent	Good	Excellent	Good	Good
Chlorinated rubber	Good	Excellent	Good	Good	Good	1.5	Good	Excellent	Good	Excellent	Good	Good

* Alkyd = 1.0 as a basis.

be modified with alkyds, polyester, acrylics, phenolics, etc., although the modified types do not perform as well as the pure types. Pure silicone-based coatings require baking at temperatures up to 500°F (260°C) metal temperature to develop optimum performance capabilities. The properties of modified types such as the silicone-alkyds are superior to those of straight-alkyd types at a more economical cost (there are some compromises in resistances to heat; that is, 300 to 450°F, or 149 to 232°C, in continuous or intermittent service).

Other coatings   There are many other coating types, among them phenolics, polyesters, ureas, melamines, fluorocarbons, styrene-butadienes, chlorinated-rubber coatings, oils (tung, linseed, etc.), and polysulfides. Each has found its own area of performance in industry. Depending on the environment, these alternative coatings often perform well.

Table 8.2 lists the coating properties and relative costs for some commercial formulations of coatings. The selection of paint is a compromise, and you should judiciously choose the weight of each particular factor in determining the proper coatings system.

8.8 Painting Metals Other Than Steel   Besides knowing the qualities of the coatings themselves, a specifier must understand the qualities and painting requirements of metals.

8.8.1 Galvanized steel   When the inherent-strength–cost factor of steel favors it over other metals, a good means of protection is galvanizing. Hot-dip galvanizing is the process by which steel or iron is dipped in a coating of molten zinc. Other processes are electrogalvanizing and cold galvanizing. To the strength and low cost of steel is added the advantageous corrosion resistance of zinc. The zinc coating serves as a barrier between the steel and the atmosphere, which in certain locations can be quite corrosive. Also, the zinc protects even exposed areas of steel via galvanic action should corrosion develop or occur at an edge because of cutting.

Some steel companies produce galvanized sheets with a special surface treatment (either by heat or with a phosphate solution) which they claim makes the sheets accept paint more readily. Otherwise, most companies recommend that galvanized steel be allowed to weather for 6 months or more to oxidize temporary inhibitive coatings that have been applied to give protection against staining by moisture during shipping or storage. Even after 6 months painting without attention to proper surface preparation can be a problem.

Why paint galvanized steel   Since galvanized steel is already protected, the question is often asked: why is painting needed? Answers include matching background colors, general appearance, a remote location that makes repairs difficult, heat reflection, end-use application such as a sign, and many others, but probably the most important is added corrosion protection, which gives the coated object a much longer life. The protection offered by painting over galvanized surfaces exceeds the sum of the separate paint and zinc coatings. Of course, for best results the surface to be painted must be free of dirt, grease, wax, or other contaminants. *ASTM Test Method D-2092* covers the preparation of zinc-coated steel surfaces for painting.

Primers for galvanized steel   When painting galvanized surfaces, the correct type of primer (or topcoat, if a one-coat application is specified) must be chosen to prevent peeling. Alkyd or oil-type house paints may react with a zinc surface to form brittle zinc soaps (this condition is especially severe if moisture or high humidity is present), which result in poor adhesion. One solution is to specify an oil-type paint containing a pigmentation system of approximately 80 percent metallic zinc dust and 20 percent zinc oxide or portland cement.

In some countries, paints with calcium plumbate (a lead pigment) are used to good advantage, but they are not recommended for total-immersion conditions in tap water or seawater. This is the only colored leaded pigment suitable for galvanized surfaces. Rust-inhibitive iron or steel primers such as red lead in oil or red oxide should not be used to coat galvanized surfaces because they exhibit poor adhesion and fail rapidly. For welded or damaged areas of galvanized surfaces, a zinc-rich paint (with a high loading of zinc dust) is recommended (see section "Zinc-Rich Coatings"). Zinc-rich coatings also serve as excellent primers for galvanized surfaces.

The practice of using latex paints as primers on galvanized metal is growing. Latex paints develop good adhesion to galvanizing that is clean and free of grease or oil. On interior surfaces with no unusual atmospheric conditions, very good results may be obtained. Good results may also be obtained on primed galvanized gutters and downspouts on residences, provided the topcoat is an alkyd or oil-based house paint or a semigloss or gloss latex topcoat. Flat paints are not preferred. Latex paints based on polyvinyl acetate and acrylic-type binders slightly etch the zinc surface and give satisfactory adhesion properties. In chemical areas and in areas of high humidity and extreme temperature changes, the practice of using latex paint as the primer is discouraged.

White rust (zinc oxide), which is the final corrosion product of a zinc surface in air, must be removed before painting weathered surfaces.

Topcoats over galvanized metal   The topcoat must be compatible with the primer so that it will not lift the primer. Depending on the severity of atmospheric conditions and the desired life expectancy, any exterior coating of good quality can be used. Two coats of paint over a properly primed galvanized surface should give excellent anticorrosion protection. Asphalt coatings are not ordinarily recommended for painting galvanized steel. However, when in contact with high-acid soils, poultry or barnyard manure, and similarly corrosive substances, all exposed areas should be covered with a heavy coating of asphalt. Galvanized steel is not affected by mild alkaline conditions and need not be protected against mild ammonia fumes or concrete.

8.8.2 Aluminum   Aluminum is an excellent substrate for organic coatings if the surface is properly cleaned and prepared. On exposure to air, the surface of bare aluminum changes visibly during the slow formation of its protective oxide film. The oxide film, although slowing further corrosion, may be rough or irregular in appearance and so cause deposits of dirt and grease or other encrustations, giving the surface a poor appearance if it is not cleaned regularly. For some applications, such as indoor decorative items, a coating may be applied directly to a clean aluminum surface. However, a suitable prime coat such as a wash primer or a zinc chromate primer usually improves the performance of the finish coat. Properly cleaned and treated to convert its surface to oxides, aluminum serves as an excellent substrate for organic finishes. Since bare aluminum has excellent resistance to corrosion, edges produced in the manufacture of coated items by sawing, drilling, etc., usually are of no major concern if they are not visible. Coatings designed for aluminum show excellent adhesion and can be formulated to be formed and drawn to the capability of the metal. Aluminum is not a difficult metal to paint and never carries mill scale (as steel does) to cause paint-adhesion problems. Any oxide on the surface of aluminum adheres tightly and holds paint well.

For applications involving outdoor exposure or for indoor applications in which a part is exposed to impact or abrasive forces, a surface treatment or modification such as anodizing or chemical conversion coating is required prior to the application of a primer and a finish coat. Anodizing in sulfuric or chromic acid electrolytes provides an excellent surface for organic coatings. Usually only thin anodic coatings are required as a prepaint treatment. Decorative parts for home appliances generally are anodized before painting to assure good paint adhesion over an extended period. A sulfuric acid anodic coating is used when only part of the surface is painted for decorative effect; it protects the unpainted portions of the surface. Conversion coatings usually are less expensive than anodic coatings, provide a good base for paint, and improve the life of the paint by retarding corrosion of the aluminum substrate material. Adequate coverage of the entire surface by the conversion coating is important for good paint bonding.

One of the simplest chemical treatments is a cold aqueous solution of phosphoric acid mixed with emulsifying agents and solvents. Preparations of this type are sold as proprietary products; usually they are diluted with water before applying them for 5 to 10 min. The surface is then rinsed and dried before painting.

One of the earliest metals to be coil-coated, in the 1940s, was aluminum in slats for venetian blinds. Today there is the active National Coil Coaters Association (NCCA), whose staff and members help promote the market for coil-coated aluminum. In coil coating, there are basically three main operations. The first is pretreatment, the second is coil coating, and the third is baking. The process is fully automated, and continuous control monitoring ensures the manufacture of a suitable product.

Pretreatment consists of five stages in which solutions and rinses are sprayed on both upper and lower surfaces. In one system a warm (150°F, or 66°C) alkali precleaning solution is applied, then a rinse, followed by a conversion (chromate) coating process at 105°F (41°C) and by cold and hot rinses (including a neutralizer). Other pretreatments not used on coil are a chromate-phosphate (MIL-C-5541) conversion coating and a wash-coat primer polyvinyl butyral–zinc chromate (MIL-C-8514), as well as inclusion of an acid wash composed of 10% nitric acid and 1% hydrofluoric acid. Suitable topcoats are silicone acrylics, polyesters, siliconized polyesters, fluoropolymers, acrylic latex, plastisols, and epoxy-urea. Patterned vinyl (6 to 14 mils, or 0.1524 to 0.3556 mm) is sometimes laminated to aluminum coil for use in many commercial products.

Although different grades of purity are available for aluminum, paintability characteristics are very similar, and the same types of surface treatments can be used. Phosphoric acid treatments, either cold or at 150°F (66°C), are used in many applications to prepare the surface for painting. In areas where aluminum products are subject to corrosive industrial fumes or salt-laden air, an inhibitive primer is required. A rapid-drying zinc chromate primer meeting military specification MIL-P-6889 is sometimes used for this purpose. For marine applications, alkyd-based zinc chromate primers meeting government specification JAN-P-735 give very good results. Primers based on lead pigments should not be used for aluminum because pitting occurs in wet environments. If a bare aluminum part is to be in contact with a dissimilar metal such as steel or copper, galvanic action could occur and lead to failure. It is important to coat both metals with two coats of a suitable primer and then to seal the joints with a sealer such as one based on silicone.

Coatings for aluminum siding are factory-applied and baked to form relatively hard and very durable finishes. Under normal conditions, these coatings may be expected to protect and decorate the surface and continue to be functionally serviceable for from 10 to 25 years, depending upon the quality of the product. After some years of exposure, the factory finish may be expected to develop a slow rate of chalking and change slightly in color as the surface erodes. Failure will ultimately occur by erosion to a very thin film rather than by gross defects such as blistering, cracking, or corrosion.

The weathered surface of most factory finishes for residential siding usually is an adequate base for subsequent repainting after appropriate cleaning to remove any surface film of chalk, dirt, and mildew. The less durable qualities of coating will tend to chalk at a fast rate, especially on south-facing exposures in the northern hemisphere. If the loose film of pigment from the eroded surface is not removed, it will interfere with adhesion upon repainting. Removal of the loose chalk face is readily accomplished by washing the surface with water, followed by brushing with a bristle brush. A long-handled car-washing brush is ideal for this purpose. The same washing and brushing treatment will usually remove the dirt film that tends to collect on the more weather-resistant finishes or in protected areas such as those under eaves. In heavy industrial areas, this dirt film may have a greasy nature that will interfere with adhesion. In this case, the surface should be cleaned with a detergent solution and then thoroughly rinsed.

Mildew is a serious problem in certain areas and must be removed to prevent infection of the repainted surface. In its active form, mildew is clearly distinguished by colonies that look like dark spots. Under low-power magnification, these spots will be seen to produce a network of threadlike filaments, or hyphae, over the surface. Dormant mildew will often be difficult to distinguish from dirt. The hyphae disappear, and even under magnification the residue has the appearance of a dirt particle. A simple test for the presence of mildew is to apply a drop of bleach to the soiled surface. The organic mildew growth will be oxidized and lose its dark color: the spot will bleach out. Inorganic soil will be largely unaffected. Chalk, dirt, and mildew may be effectively removed by using a light blast with fine sand or the following cleaning solution:

3 qt (2.84 l) warm water
⅓ cup (0.079 l) detergent (Tide)
⅔ cup (0.158 l) trisodium phosphate (Soilax)
1 qt (0.946 l) sodium hypochlorite, 5% solution (Clorox)

Use rubber gloves, goggles, and appropriate safety precautions when handling this solution. Stronger concentrations of cleaners may prove harmful to the coating surface. Cleaning must be followed by thorough rinsing with water to remove all traces of the cleaning solution in order to prevent blistering and loss of adhesion when repainting.

Properly cleaned siding may be repainted with acrylic latex or alkyd-based house paint. The repainted surface will not weather at the same rate or remain the same color as the original factory finish. For this reason, avoid repainting a section of a larger surface. Even if the refinishing paint is perfectly color-matched, it will soon weather to a glaringly different appearance.

The repainting of commercial industrial siding on steel or aluminum represents a different set of problems. It is not always feasible to repaint an entire structure because of size and expense. Also, the factory finishes employed are usually premium silicone polyester and fluorocarbon coatings that are intended to serve for the maximum possible time and do not lend themselves as readily to repaint adhesion. In these cases, the siding must be repainted with materials specifically formulated to achieve adhesion and weather at the same rate as the original finish. The pigments must be the same as in the original material to minimize color difference on weathering. Acrylic lacquer, siliconized alkyd, acrylic-urethane, and air-drying fluorocarbon coatings have been developed to complement the original baked enamels in these premium qualities for repair of damage that sometimes occurs during installation. Acrylic latex touch-up materials are also available for repair of installation damage on residential construction. These materials are available only from the suppliers of the original factory finishes and are intended to be suitable for use only with their specific product lines.

8.8.3 Magnesium   Magnesium is used for many applications (especially aircraft) for which its light weight is of greatest importance. This metal is very reactive and generally requires additional finishing to meet service requirements.

A choice of several surface treatments is available for magnesium received from a mill. It is important to identify the particular treatment because the surface treatment to be given later will be governed by the processing treatment applied at the mill. *ASTM Test Method D-1732* covers the preparation of magnesium-alloy surfaces for painting.

Like aluminum, magnesium alloys react in the presence of air to form a self-protective film. This film, which is composed of magnesium carbonate and oxides, is fairly alkaline and, unlike the aluminum film, continues to increase in depth when exposed to dampness or high-humidity atmospheres. Normal painting over such an alkaline surface would usually fail. Therefore, it is imperative that magnesium surfaces which will be subjected to corrosive and moist conditions be cleaned thoroughly. If conditions are dry and only a shelf-life extension or mild exposure is to be encountered, a clear or colored lacquer coating should be sufficient.

If an organic-coated magnesium article is to be subjected to severe environmental conditions, not only must it be thoroughly cleaned, but a chemical or electrochemical treatment should be given to the surface. The different types of treatments are spelled out in detail in government specifications; some require a license for use. Since the surface of magnesium presents an alkaline environment, any surface treatment employed should be alkali-resistant and, preferably, slightly acidic. This aids adhesion between the surface and the treatment. It is desirable to avoid, as much as possible, using magnesium in direct contact with other metals.

Likewise, primers must be carefully chosen for their alkali resistance. Red-lead and other leaded pigments (except lead chromate) that perform excellently for steel may actually promote the corrosion of magnesium. Satisfactory primer vehicles have been based on phenolic–modified alkyd resins, phenol–tung-oil varnishes, epoxy baking resins, chlorinated rubber, and vinyl copolymer resins. Baked primers are preferable to air-drying types. In some cases organic topcoats are painted directly over surface treatments without the use of a primer.

When applying a topcoat or topcoats, it is important that they be compatible with the primer if one is used. Topcoats with the best results to date are those known to be highly resistant to alkali and to have low permeability to water vapor. Those having vehicles based on polyurethanes, acrylics, epoxies, polyvinylidene chloride, and fluorides are preferred. Alkyds can be used in mild service conditions, provided a good surface treatment and a suitable primer are applied. For most requirements, dry film builds of 2.0 to 4.0 mils (0.0508 to 0.1016 mm) are satisfactory. When temperatures of 500 to 600°F (260 to 316°C) are to be encountered, a silicone-based topcoat is preferable to an anodic surface treatment. If applicable, baking usually will improve the performance of a topcoating by increasing its density or by causing some cross-linking.

8.8.4 Lead   Although competitive materials are more widely used, lead is sometimes employed for roofing, sound attenuation, downspouts, fasciae, and other purposes. As lead ages, its surface becomes dull because of a formation of an oxide or sulfate. This thin layer is beneficial because it prevents further corrosion.

In many industrial applications, lead is preferred because of its good chemical resistance to sulfuric and hydrochloric acids, brines, etc. An added advantage is that the protective film repairs itself by forming new oxides if it is damaged. While lead does not usually need additional protection, it is sometimes painted to improve its appearance.

Before painting, the surface should be thoroughly cleaned and degreased. If an industrial atmosphere is to be encountered, a chromate pretreatment should be applied. Finishing systems that have performed well on lead surfaces are as follows:

1. Zinc dust–zinc oxide primer with alkyd topcoat
2. Vinyl primer with chlorinated-rubber topcoat

3. Red-lead epoxy primer with silicone-alkyd topcoat
4. Red-lead acrylic latex primer with acrylic latex topcoat

Straight-alkyd or drying-oil systems should not be used directly over lead surfaces because their drying oxidation mechanism is catalyzed by lead. On aging, the systems become brittle, which causes peeling and flaking.

8.8.5 Silver   Attractive in appearance, silver has been used for decorative purposes since ancient times. In most instances, silver is not painted or coated with organic finishes because a coating is not needed or desired. However, silver does tarnish upon exposure to sulfur-bearing industrial atmospheres or to materials containing organic sulfur compounds such as eggs. Oxide films produced on silver appear to be converted to sulfide by subsequent exposure to air containing sulfurous gases. To protect silver from tarnishing use a suitable chromate treatment or a protective clear lacquer, or both. As with all other metals, the surface must first be thoroughly cleaned of all contaminants such as tarnish, oil, or grease before coating.

8.8.6 Titanium   This versatile light metal, which is resistant to attack by most common acids and bases, is used in the aircraft, missile, food, and chemical industries. Its relatively high cost has kept it out of many applications in the past, but recent price reductions have expanded its use into the tank-truck field, among others. In numerous instances, titanium and its alloys are used without a coating because of their resistance to many chemicals. For certain applications, however, a coating may be needed to stop contamination or be desired for its aesthetic effect.

British patent 1,100,912 covers a process which provides a surface coating for titanium or a titanium alloy essentially as nonreactive as titanium itself by a stepwise electrolytic treatment in aqueous sulfuric acid to build up successive oxide layers. Paints and lacquers may subsequently be applied. If titanium is to be painted without such an electrolytic surface treatment, the surface should first be sandblasted. A wash primer is recommended as the next step. Finally, depending on the use and environmental conditions of the object, a suitable topcoat or topcoats should be applied.

8.8.7 Copper   The use of exposed copper metal is widespread, encompassing doors and door frames, roofs, windows, handrails, architectural trim and fittings, and many other decorative items. However, copper's relatively high cost and the difficulty of maintaining the original attractive appearance have restricted its use from being greater. Metal alloys which contain copper are brass and bronze, and most of the recommended coatings for copper can also be used on them.

Like many other nonferrous metals, copper does not always need painting for protective purposes, but it can be coated for aesthetic reasons or to prevent tarnishing of high-luster surfaces. Clear lacquers are usually necessary for the adequate life and service of chemical films used as decorative finishes outdoors. Finishes for exposure indoors can be the same as those used outdoors but in most cases need not be as durable or as thick. Typical outdoor exposures of clear lacquers over copper should be approximately 2 mils (0.0508 mm) in thickness when dry, while an 0.6-mil (0.0152-mm) dry film is adequate for indoor performance. As mentioned previously, the performance of thermosetting lacquers is superior to that of air-drying types. Of greatest importance, the surfaces of copper objects must be cleaned of all contaminants. If an opaque finish is to be used, sanding, acid-etching, or other surface roughening should be carried out, or a wash primer can be applied after thorough cleaning.

Uncoated copper subject to atmospheric conditions will form layers of corrosion products, which serve to protect or to reduce the rate of further surface deterioration. A good example is the green patina that is evident on the Statue of Liberty in New York Harbor.

When choosing clear coatings for metals, it is important to select binders that do not dry by oxidation as they darken and form acids on aging. Also, the solvents used should be free of sulfur or other chemicals that can lead to tarnishing. High baking temperatures may also lead to darkening of the metal.

In the 1960s the International Copper Research Association (INCRA) sponsored laboratory work to overcome the darkening of copper underneath clear films. It was found that about a 1½ percent benzotriazole level in the vehicle solids of an acrylic lacquer using a xylene-toluene solvent system could effectively prevent discoloration. This coating now has the association's trademark INCRALAC A™; it lasts up to 5 years. It is suggested for use on interior or exterior architectural copper and brass and is available from manufacturers licensed by INCRA.

If hardness and scratch resistance are important, as on decorative objects, furniture, and architectural trim and fittings, polyurethane vehicles have been successfully used.

Another development sponsored by INCRA is the use of a polyvinyl fluoride film laminated to copper, which can subsequently withstand forming operations; the film is expected to last from 20 to 30 years. In addition to outstanding weathering, it has excellent abrasion and impact resistance. Epoxy-acrylic and epoxy-butyral thermosetting adhesives are used as bonding agents.

8.8.8 Zinc   The painting of zinc surfaces has often led to premature failure through flaking or peeling of the coating. This condition is usually caused by a reaction of the paint's binder with zinc metal to form zinc soaps at the metal-coating interfaces. In the cycle of expansion and contraction, adhesion is lost, and the coating flakes off. For best results, therefore, it is important to clean the zinc surface thoroughly and then to pretreat it with a conversion coating before finishing. Anodizing zinc is also a means of improving corrosion resistance. The coatings recommended for zinc are usually the same as those recommended for galvanized steel.

Because sheet-metal roof and flashing zinc has a tendency to creep (stretch) with time, new zinc alloys have been developed. One of these, a zinc-copper-titanium alloy, has better creep resistance and has had some success, especially in Europe.

Zinc metal can be polished to a high gloss and then protected with a clear lacquer film based on an acrylic or polyurethane binder with a zinc complexing agent (rubeanic acid, or dithio-oxamide) and an ultraviolet absorber. The protection of zinc is especially important in conditions of chemical contamination and in marine environments.

8.9 Testing Methods   While there are many performance tests which could be performed on a particular coating or coating system, usually only certain specific tests need be performed for an intended application. If the environmental conditions for a particular coating can be determined, you should be able to specify which tests and limits are applicable for a coated metal. Some of the more widely used tests are covered in Chapter 2. For metals other information is as follows.

8.9.1 Salt spray   To accelerate the corrosive conditions experienced by many coated metals, the salt-spray test has become accepted as one of the more reliable testing methods. A 5% solution of sodium chloride (common table salt) in purified water is vaporized into a fog in a closed chamber. The temperature is maintained at 95°F (35°C) for the duration of the test. Test specimens (coated panels) are then exposed to the salt-fog vapor for a certain time (results usually are expressed as hours of exposure). After definite intervals of time the panels are examined for corrosion-resistance properties. Once a coating has passed a definite length of time in the chamber without failure, the test is usually discontinued (after 1000 h for coated aluminum). ASTM B-117 or ASTM B-201 should be consulted for specific testing information.

8.9.2 Adhesion   Many different methods are used to measure the adhesion of coatings, and some of the simplest ones, such as using a knife or scratching with a suitable instrument, can give very good comparative results. However, more sophisticated methods are available, and a widely accepted one is the cross-hatch test. In this test, 100 squares, $^{1}/_{32}$ in (0.794 mm) square, are cut into a coating. A piece of cellophane tape is applied over the squares and then rapidly pulled off. By counting the number of squares of paint removed, a relative degree of adhesion is obtained. *Federal Test Method Standard No. 141b: Method 6301* and *ASTM D-2197: Methods A and B* cover methods of testing the adhesion of coatings. Another method used to determine adhesion under more physical conditions is the reverse impact test in combination with taped adhesion.

8.9.3 Film thickness   From the standpoint of both the contractor and the customer, it is important that the specified thickness of a coating be properly applied and that a suitable means of measuring it be available in case of a guarantee question or a coating failure. Usually the job specification requires that a minimum dry paint film thickness be applied and mentions the instrument to be used to measure it. There are instruments that can measure the wet thickness and others that can measure the dry thickness. ASTM test methods that cover measurement of film thickness are D-1212, D-1005, D-1186, and D-1400.

8.9.4 Hardness   This property of a coating measures resistance to deformation, whether it be scratching, indentation, marring, wearing, or the like. A crude method of checking hardness uses a fingernail. Other methods use calibrated pencils, indenters, scratching instruments, a pendulum rocker, and knife-type testers. ASTM D-1474 covers the indentation hardness of organic coatings.

8.9.5 Durability   The best test for a coating that is intended for exterior application is to conduct an evaluation under the conditions that will be encountered. How well the coating or coating system holds up in regard to changes in color and gloss, erosion, cracking, chalking, peeling, etc., is a measure of its durability. Since this type of testing is not always possible, paint companies have set up panel-testing sites under various climatic conditions, and they also utilize instruments which can somewhat simulate exterior conditions. These instruments, which accelerate weather testing, are very useful for separating obviously poor-performing coatings. The most common instruments of these types are known as Fade-O-Meters, Weatherometers, Dew Cycle Weatherometers, the QUV cabinet, and the Cleveland Condensing Humidity Cabinet. ASTM G-23 and D-822 cover accelerated-weathering tests.

8.9.6 Drying   The most common test for drying is to feel the surface with the finger. Continued practice will enable an inspector to judge drying qualitatively. For quantitative measurements, drying instruments are available. While testing for dryness in the finished film is important, especially if additional coats are specified, it is also necessary to assure that original surfaces are dry before painting. Drying can be a problem with cast iron, which is porous and will absorb moisture or other liquids that it touches. To drive out absorbed material from the pores, cast iron should be heated before blasting and painting. This is done by placing it in an oven for 8 to 12 h at 300°F (149°C) or by heating it with torches until this temperature is reached.

BIBLIOGRAPHY

Acrylic Latex Coatings for Resistance to Atmospheric Corrosion to Metal Surfaces, National Association of Corrosion Engineers, Katy, Tex., 1976.

The Analysis and Composition of Aluminum Corrosion Products, National Association of Corrosion Engineers, Katy, Tex., 1960.

Burns, Robert M.: *Protective Coatings for Metals,* Reinhold Publishing Corporation, New York, 1955.

Here's How to Paint Galvanized Steel, Zinc Institute, New York, 1974.

Identification of Corrosion Products on Copper and Copper Alloys, National Association of Corrosion Engineers, Katy, Tex., 1959.

Paintability of Galvanized Steel, American Iron and Steel Institute, New York, 1970.

Painting Guidelines No. 1: Paint Systems for Galvanized Steel, International Lead-Zinc Research Organization, New York, 1974.

Precautionary Procedures in Chemical Cleaning, National Association of Corrosion Engineers, Katy, Tex., 1959.

Prevention of Corrosion of Metals, SAE Report HSJ477a, Society of Automotive Engineers, Warrendale, Pa., 1977.

Spring, Samuel: *Metal Cleaning,* Reinhold Publishing Corporation, New York, 1963.

Standish, J. V., and F. J. Boerio: "Anodic Electrodeposition of Paint for Coil Coating Galvanized Steel," *Journal of Coatings Technology,* April 1980, p. 29 ff.

Chapter **9**

Exterior Coatings for Wood

BENSON G. BRAND

GUY E. WEISMANTEL
Chemical Engineering

9.1 Introduction Organic coatings are applied to wood surfaces for three major purposes: (1) to improve appearance, (2) to protect the wood surface from deterioration, and (3) to cover up defects in the surface. Wood is a variable and nonuniform natural material. Its coating therefore presents special problems that depend on the kind of wood being coated, the way in which the surface is prepared for coating, and the type of coating being put on the surface. These problems are complicated by consideration of the expected results to be obtained by coating the wood.

The type of coating (varnish, filler, sealer, stain, paint) should be selected on the basis of the function of the coating and the special problems presented by the particular kind of wood being coated. The kind of wood is not only the type (i.e., softwood or hardwood) but the degree of finishing, the shape, and the past history.

Wood is an extremely popular material for construction. Since many kinds are used, selection of the optimum coating is a complex problem. Wood composition varies widely not only with respect to the type of wood but within the wood structure itself. This variability creates unique problems that are not usually present in the coating of metal, plastics, or other materials with homogeneous surfaces. This chapter will discuss selection of paint types as a function of the type of wood surface to be coated, as well as the end-use requirements imposed by specific conditions.

9.2 Factors in Durability An organic coating for wood is exposed to a wide range of forces and stresses that subject it to deterioration. Factors causing deterioration are wood structure, moisture, ultraviolet and infrared light, mechanical and chemical forces, mildew, and shock.

9.2.1 Wood structure Although wood is one of the most widely used construction materials, it is also one of the most variable. Since it is a natural product, it varies with conditions of growth as well as with the time of year. A close examination of wood, its structure, and the way in which it is formed will give a better understanding of its properties and some of the basic reasons underlying the selection of materials used in coating wood.

Wood classification   Wood is divided into two broad categories, termed "hardwood" and "softwood." These names are somewhat misleading, since many hardwoods are softer than some softwoods, and vice versa. Hardwoods are cut from deciduous trees, those that shed their leaves during the colder seasons. These trees include oak, maple, walnut, birch, cherry, ash, hickory, and mahogany. Many varieties of wood are found in this category. Basswood (the material used for model-airplane construction), for example, is much softer and more porous than most softwoods. Softwood is a product of evergreen, or coniferous (cone-bearing), trees. Certain species of softwoods such as white and yellow pine, cedar, and redwood are important because they are used very widely for construction purposes.

Both categories of wood are subdivided into open-grained and close-grained woods. This classification is based on the degree of porosity of properly cured wood. Porosity is a major factor in the proper selection of coating materials. For example, oak and mahogany are very porous, and their pores must usually be filled to produce a smooth finish. On the other hand, birch and maple have only a few, usually very fine pores, so that in most cases no filler is needed to obtain a smooth finish.

How wood is formed   A tree grows in girth, or circumference, by the formation of a new layer of wood each year. This layer is formed just beneath the inner bark. In spring, when growth conditions are best, the moisture content of the earth is high and sap flows freely. The cells formed in this layer are large. Growth is rapid, and the layer being formed is soft, open-celled, and porous. This layer is called early wood. During the later part of the growth period, the growth rate decreases, the cells become smaller, and the resulting wood is denser, harder, and frequently of a different color (it is called late wood). Layers thus formed are known as annual rings, which show up as the "grain" when the wood is cut into lumber. Wood grain contributes both to the appearance and to the durability of wood. While the distinction between springwood and summerwood is not pronounced in such woods as birch, gum, and maple, trees such as oak, fir, and most pines have very definite, wide annual rings, giving rise to a very pronounced grain.

The annual rings, because of their varying porosity, permit the development of a very pleasing appearance by the application of stains or dyes to the wood before coating. The stains penetrate the open, porous springwood more deeply than they do the dense summerwood, thus increasing the contrast of the grain and developing beautiful patterns. On the other hand, this difference in density causes a difference in the rate of expansion and contraction due to moisture changes, which creates strains on any coating film applied over the surface of highly grained wood.

Structure of wood   Examining a piece of wood under a microscope shows that it is made up of numerous small cells with rigid walls and hollow centers. These cells, which are usually about 10 times as long as they are wide, are aligned with their long axes vertical in the growing tree. The cell walls are composed largely of cellulose. The cells are cemented into bundles, or fibers, by a complex material composed mainly of lignin, tars, resins, and coloring matter.

When a log is cut into lumber, it is cut so that the surface to be coated is parallel to the fibers and to the long dimension of the cells.

9.2.2   Moisture   Many paint chemists insist that moisture is the most destructive force in the deterioration of paint films. This is especially true in the case of organic coatings that are applied over wood and exposed to an outdoor environment. Wood, because of its porous nature, allows moisture to penetrate the voids and travel over quite long distances through the normal moisture passages (capillaries). In addition, exposure to the combined action of moisture and ultraviolet light from the sun generates a synergistic effect, in which deterioration is much greater than the sum of the two factors would appear to warrant.

It is common practice to formulate paints for exterior application to wood so as to allow moisture to pass freely back and forth across the film. If this is not done, moisture accumulates behind the film, developing hydraulic pressure and forming a blister, which then causes the paint film to fail.

If we consider that moisture is also a destructive chemical that acts on the resinous components of a paint film, we can readily see why its destructive force is as great as it is. Moisture tends to react with oils, alkyds, and other resins that contain ester groupings, active hydroxyl or acid groups, or other similarly reactive chemical structures. If the resin is built by esterification, as is the case with alkyds, oils, polyesters, etc., this reaction with moisture (hydrolysis) tends to decrease molecular size and molecular weight, and the resulting fragments of this decomposition are frequently water-soluble. When these water-soluble fragments are washed out of the film, they leave a void, which can be the first step in the general physical deterioration of a paint film. The film gradually loses its integrity, strength, and homogeneity and becomes useless as a protective agent for the wood substrate.

It is imperative that resinous or binder components of organic coating films be selected to achieve maximum water resistance. Consequently, water-resistant materials such as epoxies and polyurethanes deserve attention even though their cost is much higher than that of materials previously used for exterior-paint binders. However, most epoxy coatings do have chalking problems.

Moisture can influence the behavior of the wood substrate so that deterioration of the paint film is increased. Springwood and summerwood are of different densities and porosities, and moisture enters springwood more readily and to a greater extent than it enters summerwood. This difference gives rise to a differential swelling, called "grain raising" or "grain swelling," that puts undue localized mechanical strains on coatings that are applied to the wood. Another factor to consider is the moisture content of the wood. Experimental work shows that the most nearly perfect condition for painting wood occurs when the moisture content is about 12 percent. When the moisture content falls below about 8 or 9 percent at the time of painting, early failure by cracking can be expected because the wood will swell by absorbing moisture from the atmosphere on exposure to highly humid conditions. On the other hand, if the moisture content of the wood rises above 15 or 20 percent, coatings fail by blistering and peeling because the wood is so damp that good adhesion of the paint film does not develop and proper penetration of the paint liquid into the pores is not obtained.

Redwood and cedar contain up to 6 percent of water-soluble substances known as tannins. Unless special care is taken, these can stain.

9.2.3 Ultraviolet light   The energy distribution of sunlight is divided into three major components (Fig. 9.1). The high-frequency energy por-

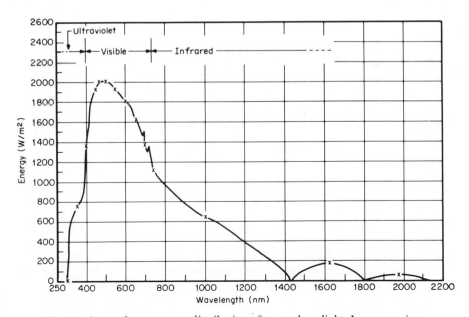

Fig. 9.1 Approximate energy distribution of natural sunlight, June noontime.

tion is ultraviolet radiation with a wavelength shorter than 400 mm; it constitutes about 15 percent of total radiation.

Being of high frequency or short wavelength, the ultraviolet radiation excites organic molecules, causing increased vibration of electrons and interatomic bonds. When the excitation becomes great enough, the bond can rupture, giving rise to the formation of free radicals and initiating chemical reactions that are normally expected. This leads to additional polymerization or depolymerization and other modes of molecular disintegration or rearrangement.

The combined effects of ultraviolet light and moisture cause much greater deterioration than does the action of either component alone. In addition, other components of paint films, for example, titanium dioxide, can undergo oxidation and reduction under the influence of ultraviolet light and moisture. This oxidation-reduction cycle increases the rate of decomposition of the vehicle that binds the titanium dioxide particles together into a paint film. The resultant failure mode is chalking.

Moisture and ultraviolet light in combination and in the presence of zinc oxide give rise to the formation of hydrogen peroxide, a very strong oxidation agent that causes decomposition of organic polymers. Similar reactions are known to occur with other pigmenting materials. For example, mercury compounds such as mercuric sulfide can undergo a decomposition that leads to fading.

This synergistic action of ultraviolet light and moisture has been used in the design of the dew-point Weatherometer. This instrument maintains a paint film at the dew point (100 percent relative humidity) and at the same time subjects it to the action of ultraviolet light. The dew-point Weatherometer is an excellent tool for determining relative rates of deterioration by chalking. This is a factor of vital importance in the evaluation of certain types of protective coatings that are used for exterior durability.

Coatings can contain ultraviolet absorbers or ultraviolet inhibitors. (The latter name is not correct but is frequently used.) These compounds do not inhibit the radiation of ultraviolet light but do inhibit its decomposing action. They can also perform in a somewhat different manner. For example, polyvinyl chloride decomposes under the influence of ultraviolet light by dehydrochlorination, or by splitting off hydrochloric acid, thus forming an unsaturated organic compound. The hydrochloric acid generated by this reaction is a catalyst for the further decomposition of vinyl chloride. The process is known as autocatalysis. If a compound that immediately absorbs any hydrochloric acid or other decomposition product as soon as it is formed is added to the polyvinyl chloride, the decomposition process is markedly hindered. Such compounds used in vinyl chloride formulations are known as vinyl stabilizers. However, decomposition initiation is a function of ultraviolet energy. If vinyl chloride films, for example, are shielded from ultraviolet radiation, dehydrochlorination does not occur, or if it does, it occurs at a very greatly reduced rate. Consequently, ultraviolet absorbers are frequently used in conjunction with stabilizers. In view of the complicated nature of ultraviolet failure, specification writers should visit test fences that compare exterior products under actual use conditions. If time permits, there is substantial value in doing one's own field testing.

Sometimes an ultraviolet absorber is used to benefit a paint film. Many of these compounds reradiate the absorbed energy from ultraviolet light in the form of visible light. These materials are known as fluorescent compounds. The color of the fluorescence depends on the compound. If it is white, the compounds are often known as optical brighteners; if they fluoresce in some specific color, they are classed as fluorescent pigments. Fluorescent pigments are used quite frequently to obtain special effects and more brilliant colors. While they tend to decrease the sensitivity of the film to ultraviolet-light deterioration, they easily lose fluorescence upon exterior exposure.

9.2.4 Infrared radiation (heat)   Referring again to Fig. 9.1, we see that a sizable part of the radiant energy from the sun lies in the long-wavelength, infrared end of the spectrum above 740 mm, beyond the visible portion. Practically all organic materials absorb infrared radiation as a particular function of their chemical composition. Materials having a large number of chemical groups with strong infrared-absorption bands will absorb a large portion of this thermal energy, resulting in a temperature rise that will lead to film decomposition.

Heat also causes chemical reactions that might not otherwise occur. Therefore, energy in the infrared spectrum has a very marked effect on the rate of deterioration of organic coatings (paints) exposed to the external atmosphere, especially to direct sunlight.

Sunlight falling on a painted surface can create temperatures hot enough to be uncomfortable to the hand. This is especially true if the paint is a dark color. Black films, of course, absorb much more energy than do white films, in which a large portion of the energy is reflected or reradiated. Aluminum flakes can be extremely effective in reflecting solar energy. The ability of a coating to absorb or reflect heat (emissivity) is becoming an important factor in design for both energy and environmental reasons.

Another factor is involved in the action of infrared radiation on painted wood. When the temperature of the wood increases, the vapor pressure of the moisture in it increases at an exponential rate. This, in turn, increases the pressure exerted against the paint film from the wood side and can result in rupturing the adhesion of the film to the wood.

Such a rupture creates a pocket in which additional moisture can accumulate, giving rise to liquid-filled blisters. This phenomenon is particularly noticeable when improper construction has allowed an undue amount of moisture to penetrate the wood fibers, where it can follow the wood grain by capillary action. Although this action of temperature causes the paint film to fail, failure is indirect, since loss of adhesion, rather than the chemical action of moisture on the paint film itself, is the primary cause of failure.

To avoid this problem, select a primer that is low in viscosity and hence will have a high degree of penetration into the wood and so will decrease the possible tendency to blister. For the same reason, it is good practice to select coatings for wood that "breathe." Paint should not be applied in direct sunlight; otherwise it may dry rapidly before all the solvent escapes. This leads to blistering.

9.2.5 Mechanical forces   Organic coatings, when applied to wood surfaces, are subject to quite high mechanical stress. The coefficient of linear expansion of organic coatings is as much as a hundredfold different from that of wood. When a piece of coated wood is subjected to temperature changes, as is the case under normal outdoor-exposure conditions, a terrific strain is placed on the paint film, quite frequently resulting in grain cracking and rupture.

9.2.6 Chemical factors

Oxygen   Some of the chemical factors involved in the deterioration of paint film exposed to the outdoor atmosphere have been mentioned in connection with moisture, ultraviolet and infrared radiation, etc. In addition, the action of oxygen contributes to film deterioration. Paint films that use "drying oils" such as linseed oil cure or harden by an oxidation process. The oxygen combines chemically at the points of unsaturation in the fatty-acid component of the linseed oil to form peroxides, which decompose to give cross-linking. This process causes the linseed oil to "dry." However, the oxidation process does not stop when the paint film is dry. It continues, and the long-chain fatty acids are decomposed to smaller chain lengths called fragments. These fragments are frequently water-soluble and are leached out of the film and removed by a rainstorm, heavy dew formation, periods of high humidity, etc., resulting in the decomposition of the film. Moreover, continued oxidation and cross-linking increase the brittleness and hardness of the paint film. The paint becomes fragile, and failure is caused by cracking or chipping or by mechanical damage.

Fumes and pollution   Recent studies in connection with air pollution investigations disclose that the deterioration of paint films in an outdoor atmosphere is increased by the presence of air pollutants. The type of deterioration is a function of the type of pollutant present in the air. We are all familiar with the deterioration or discoloration of paint films formerly experienced in soft-coal–burning areas where sulfur compounds and sulfides could attack paint films, leading to early discoloration and deterioration because white lead pigments turned·into blue lead sulfide. This circumstance led to the development of fumeproof paint systems. Where air pollution is serious, architects may have to choose specialized functional coatings. The problem of the effect of air pollutants on paint films is being attacked from two directions, by the improvement of paints to resist deterioration and by the removal of pollutants from the air.

9.2.7 Microbiological factors Microbiological factors affect paint durability with mildew stains, loss of adhesion, and failure due to flaking and chipping. Failure is often a function of bacterial decomposition of the wood layer immediately under the paint film rather than a loss of adhesion of the paint film itself. The life processes of bacteria involve manufacturing enzymes which can attack natural materials and decompose them into small fragments that are used as nutrients by the organisms. A favorite source of nutrients for some bacteria is wood cellulose. Moisture conditions in the wood immediately beneath a paint film frequently are almost ideal for the reproduction and prolific growth of bacteria.

Microbiological organisms also attack cellulosic and other organic materials or agents that are used for viscosity control, particularly in emulsion paints. Bacteria attack these materials, generating gaseous decomposition products. Not only does the paint deteriorate by loss of viscosity, but a gas which leads to explosions and rupture of the paint can is generated. This susceptibility was particularly true of older formulations based on casein and of some improperly formulated aluminum paints. Water reacts slowly with aluminum powder to produce aluminum oxide and hydrogen gas that exerts a pressure of up to 10 psi (68,948 Pa) on the container lid. Some new aluminum pastes containing inhibitors are now available for water-based systems. Some companies suggest that these systems will meet the new, stringent environmental guidelines of the Environmental Protection Agency and some states. Even today specification writers may include pH and fungicide requirements as part of the paint specification. This portion of the specification should be written with the help of a consultant familiar with formulation.

A third area of concern is the unsightly growth of mildew on paint surfaces. Mildew is often particularly noticeable on the north side of a house, behind bushes or shrubbery, or in shaded areas, where moisture is liable to be retained on the paint surface for a relatively long period of time.

Many materials have been used to control the growth of microbiological organisms. For many years paint formulators have employed such compounds as phenylmercuric oleate and phenylmercuric acetate to control both bacteria and fungi. However, recent ecological considerations have made it desirable to consider the complete removal of mercury and mercury compounds from paints and paint films. A major consideration in selecting a paint system is the identity and performance of the agents used to control microbiological attack.

9.2.8 Shock   The effects of shock are twofold. One aspect is the alternate freezing and thawing of the liquid paint, which affects paint quality for both oil-based and water-thinned systems. The freezing and thawing of the liquid paint while still in the can can cause coagulation of pigments, precipitation of resins, coagulation of colloids, and breaking of emulsions. All these factors result in improper performance of the paint on application and in poor quality of the paint film after it has been applied. Many formulation methods are used to control the harmful effects of freezing and thawing, but the paint-system specification should include temperature requirements for transportation, storage, and application as a sure way to avoid this problem.

Thermal shock is also a problem. For example, when a wood surface which has been exposed to sunlight is suddenly drenched with a cooling rain, a thermal shock which introduces severe stresses is generated in the paint. The pores of the paint film have been opened because of expansion due to heat, and the sudden application of cool water can cause a rapid ingress of moisture into the paint film. This ingress brings about a swelling effect, while at the same time the sudden cooling produces a shrinkage effect. These effects create high internal stresses in the paint film which gradually weaken film integrity and can cause early failure of the film.

Of course, paints also are exposed to physical shocks such as bangs, dents, scrapes, and abrasions. Although these may not have any visible effect on the surface of the film, its internal structure is frequently disturbed, inducing early failure. A preventive maintenance program will normally take care of these problems.

9.3 Modes of Failure   To prevent normal modes of failure in paint films applied to wood, architects must consider exposure conditions. Knowing these conditions, they should anticipate the mode that is most likely to occur (Table 9.1).

It is not the purpose of this section to discuss the causes and effects of paint film failure in detail. Suffice it to say that items 1 and 2 are usually the result of mechanical stresses and strains. Item 3 is usually a combination of mechanical stress and chemical changes, and items 4, 5, 6, and 7 more or less involve chemical changes occurring within

TABLE 9.1   Common Failure Modes

1. Cracking
2. Checking
3. Loss of adhesion
    a. Blistering
    b. Peeling
    c. Flaking
4. Fading or loss of color
5. Staining or discoloration
6. Chalking
7. Erosion or loss of thickness

NOTE: All paints fail. The secret in good specification writing is to attain failure by erosion without cracking. The first sign of failure is loss of gloss; chalking comes next, and finally erosion. It is the intermediate failures (e.g., blistering) that create repaint problems.

the paint film. The selection of a paint system to be applied to a wood substrate and subsequently to be exposed to the outdoor environment appears to involve careful selection of materials that are both mechanically and chemically stable under the conditions to which the paint film is expected to be exposed. In special cases this procedure may include careful selection not only of each raw material involved but also of the processing procedures to be used to incorporate these raw materials into a liquid paint. Architects may have to deal with these conditions if commercial products to fit the conditions are not readily available.

9.4   Classification According to Use   Paints may be classified in many ways, among them (1) use, (2) type of resin or binder, (3) degree of gloss or smoothness, and (4) substrate to which they are to be applied. This chapter classifies protective coatings for wood on the basis of their expected exterior use (see Table 9.2).

9.4.1   Outside house paints   Probably the largest volume of coatings for wood is used to paint the exterior of frame or wood houses. The preponderant color is white, which is available in both oil-based and emulsion- or latex-type paints. Test exposures indicate that these two types of paints are about equally durable under similar conditions of exposure, provided there is proper surface preparation. This prepara-

TABLE 9.2   Recommended Painting Systems for Various Surfaces

Surface	Sheen	Binder type	Finishing schedule
Wood siding[a]	Gloss	Alkyd	1 coat exterior wood primer 1 or 2 coats gloss house and trim paint
	Satin	Acrylic latex[b]	1 coat exterior wood primer 1 or 2 coats satin latex house paint
	Semigloss	Acrylic latex	1 coat exterior wood primer[c] 1 or 2 coats semigloss latex house and trim paint
	Flat	Alkyd or oil	1 coat exterior wood primer 1 or 2 coats long-oil alkyd or alkyd-oil house paint
	Flat	Latex[d]	1 coat exterior wood primer 1 or 2 coats copolymer latex, flat
	Flat	Acrylic latex	1 coat exterior wood primer 1 or 2 coats acrylic latex, flat
	Flat stain	Alkyd or latex	1 or 2 coats semitransparent alkyd or latex stain[e]
	Clear gloss	Varnish	2 or 3 coats marine spar varnish
	Clear satin flat	Latex	2 or 3 coats clear latex wood finish
Door and trim	Gloss	Alkyd	1 coat exterior wood primer 1 or 2 coats gloss house and trim paint
	Semigloss	Alkyd	1 coat exterior wood primer 2 coats alkyd semigloss enamel
	Semigloss	Acrylic latex	1 coat exterior wood primer 1 or 2 coats latex semigloss enamel
	Clear gloss	Varnish	2 or 3 coats marine spar varnish
	Clear gloss	Polyurethane[f]	2 or 3 coats gloss urethane varnish
	Clear satin	Polyurethane	2 or 3 coats satin urethane varnish

Surface	Sheen	Binder type	Finishing schedule
Masonite	Gloss	Alkyd	1 coat exterior wood primer 1 or 2 coats gloss house and trim paint
	Semigloss	Alkyd	1 coat exterior wood primer 1 or 2 coats alkyd semigloss enamel
	Semigloss	Acrylic latex	1 coat exterior wood primer 1 or 2 coats latex semigloss enamel
	Satin	Acrylic latex	1 coat exterior wood primer 1 or 2 coats satin latex house paint
Rough wood or shakes	Flat	Alkyd or oil	1 or 2 coats alkyd or oil stain or shake
	Flat	Latex	1 or 2 coats latex stain or shake
Flooring and decking	Gloss	Phenolic alkyd	1 coat porch and deck enamel (thinned 10 percent) 1 coat porch and deck enamel (without thinning)
	Satin flat	Acrylic latex	1 coat exterior wood primer 1 or 2 coats latex floor enamel

[a] Researchers are recommending low-PVC acrylic emulsion primers for southern pine. This is horrible wood to paint, especially tangential surfaces that expose early- and late-growth bands. One cannot expect a suitable paint job on southern pine, and architects should choose Douglas fir if the surface requires a coating.

[b] The word "latex" has evolved to mean all types of water-based coatings, not just the original butadiene-styrene resins.

[c] As of October 1979, the U.S. Department of Agriculture, Forest Products Laboratory, Madison, Wis., did not recommend water-based primers for wood, although test fences at the Department and at the American Plywood Association, Tacoma, Wash., show that some commercially available water-based primers give good performance on wood. For background information on their work, refer to the *Journal of Coatings Technology*, July 1979, p. 53 ff. Use latex primers only in accordance with manufacturers' recommendations.

[d] For some services vinyl acetate or vinyl acetate-acrylic systems give suitable service. One usually expects premium-quality 100 percent acrylic systems to give better weather protection and longer life.

[e] Commercially available latex stains modified with emulsified oils give excellent service life (10-year durability) in tests by PVO International, Richmond, Calif.

[f] Traditional polyurethane varnishes, even with ultraviolet absorbers, do not hold up well under exterior exposure. Depending on exposure conditions, revarnishing is often needed in 6 months. If failure occurs, recoating can be a problem because failure is usually dandruff-type flaking and not erosion.

tion includes wire-brushing, chipping, sanding, and other means to obtain a uniform, durable surface. Excessive amounts of chalk on old paint films should be removed. The specification must include the proper primer for use on the surface to be coated, and the primer must be applied in the prescribed manner. Finally, the topcoat is applied to the primed surface after a suitable drying period.

There are three general finish coats for exterior wood: alkyd-based, oil-based, and water-based. Water-based finishes are generally flat, while alkyd and oil paints are eggshell or glossy. The water-based finishes use different emulsions for the binder; the most common are acrylic, polyvinyl acetate–acrylic copolymers, and emulsified oil. Many new acrylic formulations are self-priming except over staining woods such as redwood and cedar.

Outside house paint should have a moderate degree of chalking to "clean" itself as it is washed by rain. Chalking-type paints should not be used where the washings could run down over masonry or paint films of another color, which would cause serious streaking and staining problems that would be very difficult to remedy. For this application specify trim and trellis paints or paints that are formulated so that little if any chalking occurs and no washing or streaking takes place.

Oil-based outside house paints are usually pigmented with titanium dioxide, zinc oxide, calcium carbonate, talc, barites, or other white materials or combinations thereof. The binder portion of the paint is usually a variety of refined linseed oils. Driers are added to the formula so that the paint will dry in a reasonable time.

Recent regulations restrict the use of lead compounds in paint formulations. Many manufacturers of outside house paints formerly included lead in some form in their paints. The lead was usually in the form of leaded zinc oxide, basic lead silicate, or basic lead sulfate and lead driers. These lead compounds are no longer permitted in paints that may come in contact with the general public. Do not specify lead-containing paints in these areas. Obtain details on lead regulations from the U.S. Environmental Protection Agency (EPA), the Department of Health, Education, and Welfare, and the Occupational Safety and Health Administration (OSHA).

Organic solvents or thinners are usually added to oil-based paints to adjust the body or viscosity to a point that gives good application properties. Be sure to specify the exact amount and kind of thinner or solvent that may be added to effect minor changes in brushability. Avoid the addition of large volumes of thinner, which would result in the application of a film that would be too thin for good durability and usually would result in a paint that would apply very sloppily and hide poorly. The need for large amounts of solvent to yield a paint suitable for brushing should be reason to question the quality of a paint. On very hot days a slow-evaporating solvent (even kerosine) is sometimes added to oil paint to slow drying and prevent lap marks. The slow solvent gives the paint a "wet edge" so that the painter can move laterally and brush back into a just-painted section without a heavy paint buildup at the lap. Otherwise unsightly brush marks will show up in the dry finish coat.

"Emulsion paints" and "latex paints" are synonymous names that are applied to a relatively new class of coatings. This class is based on a latex binder, a suspension of very finely divided particles of resins or oil in water. When the water evaporates, the resin particles unite to form a continuous film. Pigments and extenders in the suspension are thus held together by a solid resin matrix in the dried film, producing a coating with the protective qualities of an oil-type paint film.

The advantages of emulsion paints are many. They may be thinned with water, and application tools may be cleaned with water. Application is easier since there is less drag on the brush. Paint films tend to be more porous, permitting more breathing and thus decreasing the tendency of moisture to accumulate behind the films, causing blistering. The absence of organic solvents or thinners results in a low odor, decreased toxicity, and practically no fire hazards.

The architect should carefully specify repaint jobs when using emulsion paints. The surface may require a special primer to seal a chalky surface of old paints; otherwise undue peeling and flaking can result. This primer is usually either a specially formulated composition containing sealing ingredients or an oil-based primer.

Most of the problems encountered with oil-based outside house paints are also present with emulsion paints. Each requires viscosity-control agents to prevent excessive flow when the paint is applied to a vertical surface and fungicides to prevent attack by mildew. Discoloration can also create problems in job acceptance. Carefully cover the requirements for color durability, especially if paints are exposed to industrial fumes. Keep swatches of the actual paint, but be aware that even swatches change color when they are filed and kept in the dark.

With pollution regulations aimed at controlling solvent emissions, many manufacturers now make water-thinnable exterior coatings for wood. Typically, a straight acrylic-based house paint has a more flexible film than other copolymers. This more flexible film has less tendency to crack or peel when applied to wood surfaces that may expand or contract as the moisture content varies. Although acrylic resins are more resistant to yellowing and discoloration, they are somewhat more sensitive to degradation by sunlight. In severe exposures the architect can consider calling for inclusion of ultraviolet absorbers in the formulation. Recent practice in emulsion-paint formulation includes the use of alkyd modifiers to improve adhesion and durability. Many clear polyurethanes are ultraviolet-sensitive and require ultraviolet absorbers. Pigmentation normally eliminates the need for these absorbers. Some newer types of urethane resins are not so sensitive to ultraviolet degradation.

All liquid emulsion paints tend to deteriorate by alternate freezing and thawing. Specify the necessary protection, and avoid application at low temperatures, in which drying does not proceed satisfactorily. This problem is more severe now that the EPA is recommending lower limits on freeze-thaw agents such as ethylene glycol. These lower limits are suggested to reduce the air pollution caused by the evaporation of organics. Generally, the minimum temperature for satisfactory application is about 45 to 50°F (7 to 10°C). Requiring an application temperature is very important in two-component systems (e.g., epoxy-polyamide) because the curing reaction is temperature-dependent and low temperatures slow curing substantially.

Because emulsion-paint films are more porous than films from oil-type paints, they are frequently more subject to staining. Stains may result from external sources, such as corroding screens, rusting metals such as nails, decomposing leaves, and drippage from trimmed areas. They may also be due to internal sources, among them knots in the lumber, the bleaching out of water-soluble ingredients from such staining woods as cedar and redwood, and condensation of moisture on the unpainted back of the wood. Keep this characteristic in mind before specifying an emulsion-based paint.

The application of exterior latex paints necessitates certain precautions. Good performance usually requires application over a suitably primed and prepared surface. Both water-thinned and oil-based primers are suitable, but for best results use the primer expressly designed for the topcoat system. Oil-based primers are usually recommended over fresh unpainted wood to prevent grain raising. They generally require a minimum of 48 h of drying time before application of the topcoat. Latex primers are frequently topcoated after drying overnight. A typical analysis of an exterior white emulsion-based house paint is given in Sec. 9.7.

9.4.2 Trim and trellis paints   Formulas for trim and trellis paints are very similar to corresponding formulas for exterior house paints. The major difference is that trim and trellis paints are designed to have the least possible amount of chalking in order to prevent any streaking over lower surfaces due to the washing down of the chalk particles. They are available in either oil-based systems or water-thinned or emulsion systems. Trim and trellis paints often come in dark or dark pastel colors. Their major use is for trimming window frames, door frames, etc., in a color that either harmonizes or contrasts with the major area of the house.

In oil-based systems, the vehicle for trim and trellis paints is usually a long-oil alkyd. The major pigment component is usually a colored pigment rather than white. A trim and trellis white or pastel would not contain a chalking pigment. Specification writers should not allow tinting exterior white paints that contain chalking-type titanium dioxide.

9.4.3 Shake and shingle paints and stains   Two classes of materials, stains and paints, are normally used for coating shakes and shingles. It is well to remember that stained shingles can be painted, but painted shingles should never be stained.

Shake and shingle stains are blends of oil, resins, and driers, plus a wood preservative (usually creosote) and coloring material. Their function is to penetrate the wood (usually cedar) in order to prevent its deterioration and, at the same time, to impart a general color to the wood surface. Since stains often contain creosote or pentachlorophenol and water repellent, application of paint over a stained surface can require the use of a good sealing-type primer to prevent bleeding of the creosote through the paint to the film surface, carrying with it some of the coloring matter, usually dye, from the original stain. It is therefore usually recommended that stained shingles be refinished by restaining. However, painting can be done by using aluminum paint or similar sealing-type coatings before the paint is applied. The application of stains over paints is not recommended, since their hiding power is low and coverage cannot be obtained. Also, their viscosity is quite low so that maximum penetration of the wood can be obtained. The application of stains over tight, nonporous paint films is seldom satisfactory.

Shingle paints are very similar to normal house paints. In fact, normal house paints are frequently used to paint shingles. Special shingle paints are designed to give a flat, somewhat porous film that facilitates breathing and permits the escape of moisture from the thin wood of the shingles. They usually have a heavier body than normal house paints, permitting the application of a somewhat thicker film.

Shake paint is a full-bodied, oil-modified, alkyd-based flat finish for use on shakes, shingles, hand-split cedar, rough-surfaced vertical or horizontal siding, wood fencing, and similar rough-surfaced wooden materials like board-and-batten. The spreading rate depends on the quality of the product. Some shake paints attain 450 ft²/gal (11.04 m²/l) on smooth surfaces. The rate drops to as low as 150 ft²/gal (3.68 m²/l) on rough surfaces. Under normal conditions these coatings dry to touch in 8 h and may be recoated in 48 h. Normally, rough new wood is primed with shake paint thinned with 1 pt (0.473 l) of boiled linseed oil per gallon (3.785 l). For best performance, however, prime bare wood with an all-weather exterior wood primer.

Shakes and shingles are often stained. Some shake paints are suitable for this purpose. For staining shingles, thin shake paint with ½ gal (1.893 l) of boiled linseed oil and ½ gal of paint thinner per gallon (3.785 l) of shake paint. Allow 48 h before recoating.

Another method of finishing shakes, shingles, and rough siding makes use of exterior stains. This painting system, sometimes referred to as shingle stain, produces a flat sheen but provides greater penetration into the substrate. Stains are available as solid hide stains and as semitransparent stains. Solid hide stains are used in refinishing previously stained exterior wood surfaces. Semitransparent stains, which let the wood grain show through, are used on new exterior wood surfaces or on surfaces that have previously been coated with a semitransparent stain in the same color range. Apply stains at full strength. The drying time is approximately 24 h.

9.4.4 Barn paints   Barn paints are normally formulated for low cost and good protection. Since appearance is secondary to protection, the use of low-cost pigments is common practice. Low-quality iron oxides which have been ground to the required fineness (in some cases, even brick dust) are used as the coloring agent. The resinous or binder material is usually refined linseed oil. Among other ingredients are driers and necessary thinners to obtain application viscosity. Since the composition is quite simple, the goal of good protection with low cost is readily attained. The color is usually restricted to red, the natural color of most low-cost pigments.

Barn paints may be applied by brush or roller and, with additional solvent, by spray gun. Application by spray gun tends to reduce costs, since the largest portion of the cost of painting a structure is the labor required for application. Spray-gun application is many times faster than brush or roller application, and the overall cost of the job thus is greatly decreased.

Because barns and other farm structures are functional, the main concern is protection rather than appearance. Farm structures are also large and contain extensive unbroken areas; hence, the interest in keeping costs low.

The wood of old barns is frequently very dry and absorbent. Quite often the application of two coats of barn paint will lengthen service life by as much as 5 years. In this case a common practice is to specify 1 pt (0.473 l) of boiled linseed oil and ½ pt (0.237 l) of paint thinner to each gallon (3.785 l) of paint, followed by a full-strength (or slightly thinned) finish coat.

9.4.5 Lawn-furniture paints   Paint films that are applied to lawn furniture present a special problem. Since the coating is usually applied to relatively thin sections of wood, it must have optimum resistance to moisture penetration and extremely good durability. Small amounts of moisture can cause greater warping and cracking of thin sections of wood than of thicker sections Consequently, formulations used for furniture normally are based on high-quality enamels designed to have minimum moisture permeability.

The resinous or binder portion of these paints is selected for its durability. Normally, medium and medium-long oil-extended alkyd resins are used for this application. These resins contribute excellent durability but are somewhat higher in cost than the normal linseed-oil type of vehicle that is used for painting large areas such as houses or barns.

Exterior enamels become hard and brittle. Some wood is hard, and unless the specification calls for penetration into the wood, the paint will chip and peel. For high-quality work one or more "wash coats" of paint highly thinned will be specified. The wood will act like a sponge, soaking in the wash coat. When the topcoat is applied, it melts into the primed surface rather than sitting on top of the wood. This technique is suitable for many hard, semipermeable surfaces.

9.4.6 Porch and deck finishes   As their name implies, these coatings are designed to be used on wooden porch floors. They are characterized by good resistance to marring and abrasion, good durability under outdoor conditions, and a relatively high gloss. Synthetic resins are usually employed for the binder. To obtain the higher gloss normal for this type of formulation, the resin content is usually increased to about 50 percent of the paint's total solids content. As in lawn-furniture paints, this causes the formulation to operate near its optimum resistance to moisture penetration. While this capability is necessary to protect the wood substrate, it can create a penetration problem during application if the wood itself is very hard. In this case wash coats may be a necessary part of the specification.

To obtain quality appearance, the specification writer should spend an equal amount of time in writing the specification for the wood itself. Select kiln-dried planks (e.g., fir) free of knots, sap, and other defects. Prepare the surface properly to assure paint penetration, and backprime and paint the edges in addition to priming the exposed surface. For new work the architect can expect about equal exterior durability and resistance to wear from both water-based and oil-based floor and deck finishes, with a slight edge in colorfastness in favor of oil-based finishes. Repaint work requires knowledge of the finish already on the surface, especially if a water-based system is under consideration. Water-based paint may not penetrate the old surface and may merely sit on the top of it, eventually peeling off in sheets.

Additional slip resistance is attainable by adding pumice or clear sand to the paint. The manufacturer's recommendations should be followed, and the finish stirred often during application. In some cases, depending on the quality and durability desired, simply scattering sand onto the wet paint gives adequate slip resistance.

Porch and deck enamels are available in a wide range of colors, since like trim and trellis paints they are generally used to furnish contrast to the color of the body of the house. The most popular colors are light blues, slate grays, browns, tans, greens, and reds.

9.4.7 Marine finishes   There is a whole class of coatings specifically designed for marine use. Formulations are selected on the basis of whether a coating will be exposed primarily to salt water or to fresh water (Chapter 14). Failure of marine paints is usually related to the passage of moisture back and forth through the film because of the osmotic pressure of the water. Very large hydraulic pressures can be built up under paint films, causing loss of adhesion by blistering. It is apparent that the osmotic pressure experienced in fresh water is much higher than that found on immersion in seawater, in which the salt content tends to decrease the pressure. Therefore, special precautions should be taken in specifying finishes that are designed for small wood vessels used primarily in fresh-water areas. Especially thick films are helpful in decreasing moisture transfer.

9.4.8 Clear finishes   The recent trend to rustic-type home construction calls for increased use of "natural" wood finishes for exterior surfaces. Redwood is quite often specified, and to give it some protection from the elements a clear finish is usually applied to the properly prepared surface. Redwood is a very porous wood, with a high content of water-soluble dyes and other chemicals. It expands and contracts with changes in moisture content and with temperature. Therefore, it is extremely difficult to obtain good durability from any of the normal clear finishes when applied over redwood. No really satisfactory material is available for use on redwood exposed to exterior environments. The failure of these clear finishes usually occurs with the development of very tiny cracks in the coating, giving the entire surface a milky-white appearance. When this situation occurs, the old coating must be removed down to bare wood and the surface completely refinished. This is a very expensive process, and because of the poor durability of modern finishes for this purpose it must be repeated more often than is desirable.

Pigmented coatings for application over redwood are based largely on soybean drying oils, with a recent trend toward flexible synthetic materials including copolymers of styrene and butadiene. Acrylic formulations show some promise of durability.

The durability of a clear redwood finish cannot approach that of properly pigmented materials. When a clear coating is selected for its aesthetic value, some sacrifice in durability must be tolerated.

9.4.9 Wood preservatives   Wood preservatives are products which make wood uninhabitable or unpalatable to wood-destroying organisms. They are normally added to wood to deter attack by fungi, marine organisms, insects such as termites, bacteria, or even blue mold. Wood preservatives normally are not considered to be coating materials. They are usually applied by an impregnation process involving both vacuum and temperature. However, two phases of wood preservation make it necessary to consider these products in a discussion of exterior coatings for wood.

Quite frequently it is necessary or desirable to paint over preserved wood, but many products that are used for wood preservation interfere with the proper application of paints over their surface. It is necessary to use special sealers or primers to seal in the preservative in order to have a properly performing paint film applied over the surface. "Knot sealers" are often found to be effective for this purpose. They include certain aluminum-pigmented paints for sealing in wood preservatives that would otherwise bleed through the paint film and mar its appearance as well as interfere with its proper performance.

It is sometimes necessary to preserve wood after it has been installed in place in construction. In this case, the wood preservative is applied by brush from a very thin (dilute) solution to obtain deep penetration into the wood structure. Unfortunately, preservatives work well only when they are well dispersed throughout the wood. Application of a thin coating of preservative on the surface slows down but does not completely stop attack by deteriorating influences.

Wood preservatives contain toxic materials such as pentachlorophenol, soaps of heavy metals (copper naphthenate or zinc naphthenate), and coal-tar derivatives (creosotes, wood oils, and Carbolineum). Among other preservatives are water-soluble salts such as copper chromate and chromated zinc chloride. Arsenic compounds were used in the past but now are not recommended. From their chemical identity it is apparent that these materials easily contribute to the poor performance of any organic coating that might be applied over the surface of wood which contains them and that their toxic nature requires special specifications and handling.

Present termite-control practice involves the treatment of wood installed in place in wooden structures, particularly older structures. It frequently entails drilling holes into lumber and injecting into the holes under pressure solutions of materials obnoxious to termites. This practice can often cause failure of paint films applied to the wood so treated. Unfortunately, it also frequently interferes with proper repainting of such failed areas by the use of normal techniques. It is quite often necessary to use special sealers over these areas in order to get good paint adhesion and good paint film performance.

Coating systems that are to be applied over preserved wood should be selected with extreme care, even to the extent of pretesting potential systems for their behavior over the candidate surface.

Architects or specification engineers who require specific details on preservation techniques should get in touch with the Society of American Wood Preservers, Inc., 1401 Wilson Boulevard, Arlington, Virginia 22209.

For pole treatment, get in touch with the Electric Power Research Institute, 3412 Hillview Avenue, Palo Alto, California 94303. Its research funding at Oregon State University's Forest Research Laboratory, Covallis, Oregon 97331, concentrates on the use of Vapam (sodium N-methyldithiocarbamate) and Vorlex (20% methylisothiocyanate in dichloropropenes) to extend the life of Douglas fir and red cedar by 10 to 15 years. On the basis of a replacement cost of $1500 per pole, the Bonneville Power Administration estimates an annual saving of $2,250,000.

9.4.10   Knot sealers   When paint is applied over wood knots, particularly in high-resin woods such as pine, the resins from the knots bleed through the paint film and present unsightly spots on the paint surface. It is necessary to precoat any knots in wood siding, etc., with a coating composition that seals the resinous content of the knots into the wood and prevents its bleeding through the paint film. These compositions are known as knot sealers; they form a barrier preventing the discoloration from migrating to the surface. Some sealers are applied locally over the knots before the primer system is applied on new wood. Others can be used over the entire surface of such aesthetic materials as knotty pine which are to receive a clear, unpigmented topcoat.

Knot sealers are normally based on one of three types of resinous materials. The oldest and probably the most widely used is a shellac composition. Shellac by itself does not have good outdoor durability, but when it is coated with another resin or a paint, it is extremely durable. The knot sealer is usually a 4-lb cut of shellac. That is, it consists of 4 lb (1.8 kg) of orange or bleached shellac dissolved in 1 gal (3.785 l) of denatured ethyl alcohol. Shellac must be freshly prepared before application; shellac solutions that are more than a few months old will not dry properly, the surface remaining tacky. This condition is caused by a reaction which takes place very slowly between the shellac and the alcohol solvent.

A second composition that is used to a considerable extent is based on polyvinyl butyral. This resin is water-white, has extreme flexibility, and is also alcohol-soluble. Like the shellac, it is applied from an alcohol solution but usually is made up in a lower concentration. Normal procedures call for about 2 to 3 lb (0.9 to 1.36 kg) of resin per gallon (3.785 l) of alcohol.

One of the most promising materials for knot sealing is Formula WP-578, developed by the Western Pine Association. This sealer is based largely on an alcohol-soluble phenolic resin and a small amount of polyvinyl butyral. The ratio is about 10 parts phenolic resin to 1 part vinyl butyral. The resins are dissolved in alcohol, with about 1 lb (0.45 kg) of resin to 2 lb (0.9 kg) of alcohol or about 3 lb (1.36 kg) of resin per gallon of alcohol. This formulation is also recommended quite highly as a sealer for shingle stains or for sealing asphaltic varnishes to prevent their bleeding or to solve similar bleeding problems that are encountered with tars, wood preservatives, etc.

Do not specify the widespread use of knot sealers. These materials are extremely good sealing agents; they present a good barrier to the passage of moisture, especially from the wood outward to the atmosphere. However, the use of wood sealers over an entire surface will decrease the rate of moisture expiration, leading to a buildup of moisture behind the paint film. Under some circumstances, this could lead to failure of the paint film by blistering. This is particularly true in houses in which inadequate moisture barriers are present (see box "Priming and Backpriming").

PRIMING AND BACKPRIMING

All wood trim and cases to be installed before plastering shall be primed before installation. Wood frames, trim, and other woodwork to be installed against masonry, concrete, or plaster shall be backprimed. Natural-finish woodwork shall receive the first seal coat before being installed. The first coat of paint, stain, or finish shall be applied as soon as possible after the woodwork has been fitted and erected. Shop coats of paint shall be touched up prior to the application of priming called for under this contract. Highest-quality specifications call for backpriming wood trim and for assurances that sash, doors, and trim be stored under cover until used and that all sides and edges be primed before installation. Backpriming of wood siding is desirable as a safety factor to reduce the amount of moisture that can migrate through the board to the paint film interface. However, this is seldom done except on construction built to the very highest standards.

It is of the greatest importance to prime the edges, ends, and both sides of porch flooring and steps before they are laid. Priming will keep the boards from cupping and prevent the peeling of paint caused by ground moisture.

9.4.11   Railroad-trestle and wood-bridge coatings   Railroad trestles and wood bridges present a special painting problem. They are usually constructed of heavy wood timbers, around which tumbleweeds and windblown dry vegetation tend to collect, making the structures prone to fire. Once a fire starts, it can smolder for many days in heavy wood timbers, and the structure may collapse under a heavy load.

When this problem exists, the specification writer should consider intumescent coatings (Chapter 13), which are formulated so that heat causes the dried paint film to expand to form a cellular structure. This structure then chars, forming a layer of thermal insulation on the surface of the wood that prevents the wood underneath from igniting.

The active ingredients in intumescent coatings are of several types. One type is formulated from isano oil, hydroxylated oil whose source is a nut native to South Africa. Upon heating, this oil decomposes to form water, which blows "steam" bubbles in the coating to form a cellular insulating structure. The dried isano oil is distensible and flexible enough to permit these bubbles to form in a continuous fashion rather than blowing holes in the coating film. One advantage of isano-oil intumescent coatings is durability, especially when exposed to outdoor conditions.

Other materials that have been used with some degree of success in intumescent coatings include inorganic salts with a high water-crystallization content, which lose water on heating, and organic compounds, which decompose under the influence of heat to form gases. The disadvantage of these materials in exterior finishes is that they tend to be water-soluble and to leach out of the film, leading to a loss in intumescence as the film is exposed.

9.5 Classification According to Composition   Paint systems available for application to wood substrates can be classified according to the composition of the resinous or binder portion of the paint. The formulations are divided largely into two broad categories, solvent- or oil-based systems and water-thinnable systems. The resinous components are the major point of variation in these formulations. For the most part, pigmentation is quite similar from formulation to formulation, depending upon the hiding power (opacity) desired, the critical pigment-volume concentration, and the color.

9.5.1 Solvent-thinned systems   Systems that are based on oils or organic solvent-soluble resins can be thinned only with the proper organic solvent.

Drying-oil vehicles   Drying oils are vegetable oils that react with oxygen to polymerize and harden into a tough, durable film. Many oils undergo this particular chemical reaction, but most of them have serious deficiencies that restrict their use as the sole resin component of a paint system.

Drying-oil–bodied-oil vehicles   Bodied oils improve drying time slightly, decrease the set-to-touch time, improve the gloss of paint, promote good leveling qualities so that brush marks are less prominent, and tend to make the finished film more flexible, thus decreasing the tendency of the paint film to grain-crack when applied over wood.

Drying-oil–alkyd vehicles   Long-oil alkyd resins (containing as much as 60 to 65 percent oil) are useful as the sole vehicle component of high-quality nonchalking paints of extreme durability. Found largely in trim and trellis paints, high-quality porch and deck enamels, etc., these resins have also been promoted as a replacement for all or more of the bodied linseed oil in the standard bodied-raw–linseed-oil combination. These paints have good durability.

Alkyd vehicles   As mentioned in the preceding section, long-oil and extremely long-oil alkyds containing from 40 to 60 percent oil are used in fairly widespread amounts as the resin component of high-quality finishes for wood to obtain extreme durability. These finishes include trim and trellis paints, lawn-furniture paints, and other paints for areas where extreme resistance to weathering is desired.

Epoxy vehicles   The use of epoxy resins in wood finishes, particularly for outdoor applications, has not been overly successful. Many epoxy-based paints chalk upon exterior exposure. Once chalk is present, further chalking stops unless the surface is washed clean by rain or mechanical means. Some newer resins exhibit improved outdoor-exposure characteristics. Epoxies are well suited for deck coatings in small craft, for some applications in which extreme chemical resistance is required, for food-processing plants, and for some components of wooden cooling towers. Acrylic-modified epoxies deserve some attention as outside house-paint resins.

Polyurethane vehicles   First-generation polyurethane resins deteriorated under the influence of ultraviolet light. They were based on aromatic materials. Newer urethanes based on aliphatics provide tough, durable finishes with excellent gloss and color retention as well as good chemical resistance for both interior and exterior use. Before specifying urethanes, check the resin type, and, if possible, examine test-fence data.

Phenolic vehicles   The use of phenolic vehicles is restricted largely to alcohol-soluble phenolic resins as a component of knot sealers. As a rule, phenolic resins are not flexible enough to permit satisfactory use over wood substrates.

Vinyl resins   Vinyl resins are seldom if ever used in solvent-thinned coatings intended for application to wood. Their use is restricted largely to metal finishes. The only exception is polyvinyl butyral, which is used as a component of knot sealers.

Miscellaneous other resins   Several other resins have, of course, been used for application over wood to obtain special properties. While baking systems are practically never used over wood substrates, baking-type resins are employed in factory-applied primer systems to wood siding.

The lumber mill applies a primer coat to properly cured lumber at the optimum moisture content. The coated lumber is then baked at about 160°F (71°C) to develop a hard, tough topcoat after the lumber has been installed in the house. Preprimed lumber offers construction savings, but topcoats must be compatible with the primer to avoid problems of intercoat adhesion. Finish coats should be applied before the primer begins to chalk or deteriorate.

9.5.2 Water-thinned systems   Water-thinned systems fall into two major types: (1) the emulsion, or latex, type, which is based on the suspension of very finely divided particles of resins or oils in water; and (2) water-soluble coatings, which are based on resins which are soluble in water but which, on drying, loose their solubility toward water and become insoluble. Most people consider emulsion paints and latex paints to be synonymous, but the word "latex" strictly refers to an emulsion of rubber particles. The broader term "emulsion paints" is preferred.

With emulsion paints the film is formed when water evaporates from the surface or is absorbed by the substrate. The resin and pigment particles are "welded" into a solid-resin matrix that encloses the pigment particles and forms a continuous, durable film. Emulsion paints have several advantages. They may be thinned with water, which is low in cost compared with organic solvents, and application tools may be cleaned with water. Application is easier since there is less drag on the brush. The resulting films are more porous, permitting breathing, or freer passage of water vapor back and forth across the film, and decreasing the tendency of the film to blister. The absence of organic solvents or thinners results in a low odor, decreased toxicity, and practically no fire or pollution problems. The major disadvantage is that some emulsion paints require a special primer or sealer system to seal chalky surfaces and thus prevent undue peeling and flaking of the newly applied film. This primer is usually a specially formulated composition designed to work with the specific topcoat being applied. Usually it contains special sealing ingredients. It may or may not be a water-emulsion system.

Some emulsion-paint films tend to be more porous than films from oil-type paints; they are frequently subject to staining. Stains may result from external sources such as corroding screens, rusting metals (nails, gutters), decomposing leaves, ornamental iron, and drippage from trim areas. They may also arise from internal sources such as knots in the lumber, the leaching out of water-soluble ingredients from such staining woods as cedar or redwood, and the condensation of moisture on the unpainted back of the wood.

On the other hand, newer acrylic-based emulsions are especially well qualified for use on metals and staining woods, making them suitable as all-surface primers.[1] Do not become confused and specify traditional acrylic paints for special problems (e.g., stains). The advanced formulations are based on new acrylic latex vehicles (such as MV-23, manufactured by Rohm and Haas Inc.) and require appropriate pigmentation and new thickening and dispersing agents.

Emulsion paints usually include a minimum temperature of about 45 to 50°F (7 to 10°C) for satisfactory application; otherwise the film will not coalesce properly. Since most of the resinous binders in emulsion paints are synthetic, they are not usually nutrients for microbiological organisms. However, the porous nature of these films permits seepage of nutrients from the wood substrate to the surface or, in case of mildew, penetration of the film by mycelial growth of the fungi to the nutrients in the wood. In addition, certain additive materials used to obtain desirable application properties are nutrients; examples include cellulosic compounds used to control viscosity. Hence, mildew growth is a major problem with emulsion paints. Fungistatic agents such as zinc oxide and calcium carbonate tend to improve the efficiency of most mildewcides by controlling the alkalinity of the paint film surface. The growth of mildew, particularly *Aureobasidium pullulans,* is discouraged at a pH above 7.8.

Acrylic-based paints Acrylic resins are polymers of esters of acrylic acid, methacrylic acid, or other homologues. These resins have exceptional color and gloss retention. They are characterized by a water-white color and complete transparency. In the solid state, they form the basis for Plexiglass and Lucite plastics. Paints formulated from emulsions of these resins usually have very clean colors, good flexibility, and all-around durability. However, they sometimes exhibit a tendency toward poor color retention in dark tints.

Quite frequently, acrylic resins are modified by blending them with other resins. Some paints combine alkyd-resin emulsions with the acrylic emulsion to obtain better wetting and adhesion over chalky surfaces and improved intercoat adhesion between topcoat and primer. The addition of an alkyd resin, with its vegetable-oil content, introduces additional problems with mildew growth.

Polyvinyl acetate paints Polyvinyl acetate emulsion paints are used to some extent in exterior water-thinned house paints. They are lower in cost than their acrylic counterparts and have excellent color retention. Since pure polyvinyl acetate as a sole resin does not exhibit outstanding durability, it is customary to use a copolymer polyvinyl acetate or to

[1] Andrew Mercuris and Roy Flynn, "Latex-Based All-Surface Primers," *Journal of Coatings Technology,* July 1979, pp. 45–51.

blend the vinyl acetate emulsion with other emulsions such as acrylics, linseed oil, or alkyd-resin emulsions. When resin blends are properly formulated, satisfactory outdoor durability can be obtained.

Styrene-butadiene paints   Styrene-butadiene copolymer resin emulsions are not normally used for exterior paints for wood. They generally tend to become yellow as they age, and grain cracking and other flexibility failures occur when they are applied over wood. Today very few manufacturers produce these products.

Other resin-emulsion systems   Linseed-oil emulsion is becoming popular for use in exterior emulsion systems. It has several obvious advantages. It is relatively easy to make, the raw-material cost is low, and since it is easy to handle, the paint production cost is low, resulting in a low-cost paint system. The durability of linseed oil has been well established, and its use as an emulsion system tends to improve paint durability.

9.6   Surface-Preparation Systems and Primers

9.6.1   Surface preparation   Exterior wood, both new wood and wood to be repainted, requires careful surface preparation. The specification should call for inspection prior to painting to ensure that all problems have been corrected. Such correction includes filling all gouges, deep scratches, and knotholes, as well as large cracks that appear in the surface, with putty, caulking compound, or knot filler. (Knot filler is a commercial preparation designed to fill large gaps in wood surfaces.) All rough areas should be sanded smooth and dusted clean. All spatters and splashes of paint and plaster from previous operations should be carefully removed. Pitch pockets or sappy portions need to be shellacked or sealed with knot sealer. Nail holes, cracks, or defects should be carefully puttied after the first coat with putty matching the color of the stain or paint.

GENERAL INSTRUCTIONS FOR PAINTING EXTERIOR WOOD SURFACES[1]

A. If the surface is unpainted new wood, proceed to step *C.*

B. If the surface is being repainted:
1. Thoroughly prepare the surface.
    a. Remove loose paint, chalk, dirt, and residual surface gloss by brushing, scraping, or using a blowtorch.
    b. If rotted areas exist, either replace the affected areas completely with new wood or
        i. Cut or chip out affected small areas.

      ii. Treat with a biocide such as pentachlorophenol or copper napthenate solution.

  2. Fill voids, chips, or holes with an appropriate durable exterior wood filler, and sand smooth.

  3. Caulk openings in seams, around windows, etc.

  4. If excessive blistering has occurred, consider adding additional soffit ventilation and/or ventilators at the wall base to reduce moisture levels within wall cavities.

C. Apply sealer to new surfaces such as plywood to reduce excessive penetration of the primer. This procedure will also reduce grain raising with the subsequent application of water-based primers.

D. Apply an appropriate knot sealer to knots, pitch pockets, and other highly resinous areas to reduce bleed-through in subsequent topcoats.

E. Apply primer to the surface.[2,3]

F. Make one or more applications of topcoats, depending on the desired durability and the manufacturer's instructions. Always allow adequate drying time between topcoats.

SOURCE: IFT Inc., Berkeley, Calif.

[1] All applications of sealers, fillers, caulking compounds, primers, or topcoats must be made when the surface is dry and free of surface moisture.

[2] Always use a primer-topcoat combination specified by the manufacturer.

[3] In applying oil-based paints, try to do so when surface temperatures are at their daily maximum to reduce heat blistering. To reduce moisture effects on topcoats, avoid working when dew is heavy (early or late in the day) or in foggy conditions.

The wood surface should be dry before a prime coat is applied. This is especially important for oil-based solvent-thinned systems. The presence of undue moisture prevents desired penetration of the paint into the pores of the wood, causing inadequate adhesion and early failure of the paint system by flaking or blistering. There should not be more than 15 to 18 percent moisture in the wood at the time of painting. Readings above 20 percent are too high: at this point the wood is definitely too damp to paint. Moisture-meter testing in January, February, and March gives a more accurate picture of moisture conditions than testing in hot summer months. Pay careful attention to areas where water can penetrate, such as around window flashings and doorjambs. If necessary, seal these areas with a caulking compound or similar composition that will prevent moisture from penetrating the back surface of the wood.

CAUTION: In some cases, especially if chalking has occurred, factory-primed hardboard should be reprimed before a finish coat is applied.

EFFECT OF MOISTURE ON PAINT FILM

Blistering and peeling are usually the result of moisture in back of the paint film. Common external sources of moisture entry behind the paint film are leaking roofs, leaking eaves, cracks around doors and windows, split siding, and downspouts and gutters stopped up with leaves, snow, or ice.

Moisture accumulates inside a house from a damp basement, a wet crawl space, humidifiers, gas heating or cooking without proper venting, and steam from cooking, laundering, and bathing. Moisture vapor always seeks an area of lesser pressure. Consequently, moisture-laden air inside a house fights its way, by one path or another, to the outside atmosphere. In a tightly closed house that is constructed without a vapor barrier, moisture will migrate through the interior walls to the back of the siding and through the siding to the exterior.

When moisture vapor from inside the house migrates through the interior walls and reaches the colder surface of the back of the siding, it condenses and migrates through the siding to the exterior paint interface. This action destroys the adhesion between the wood surface and the paint film, causing blistering and peeling.

Houses built today are more tightly sealed and better insulated than those built some years ago. (There is also more moisture from washing and drying and from greater use of baths.) Consequently, moisture vapor from internal sources, in seeking the outside atmosphere, must migrate through the siding. As a rule, small houses with less cubic feet of air show the greatest damage. In climates with long and cold winters, the problem is more severe. Blistering can occur on any side of a house, but it usually shows up first on the south and west sides if they are exposed to sunlight.

Good design techniques will prevent paint problems from arising. Unfortunately, good construction practices are often eliminated or overlooked to reduce costs. Installation of a vapor barrier to prevent moisture migration is one practice that is often omitted, yet it is very helpful in reducing paint problems. A vapor barrier is most effective when applied to the back of the internal wall, where it prevents excessive moisture vapor from entering the interwall space.

Once a problem occurs, it is possible to release moisture vapor in the interwall space by "wedging." Drive small wedges under the siding at each stud on two or more siding boards just above floor level and again on two or more boards just below the ceiling. Lift the siding $\frac{1}{8}$ to $\frac{1}{4}$ in (3.175 to 6.35 mm) by using a wide-blade tool, such as a painter's spatula or a cement chisel, placing it at the nail and gently tapping it under the siding up to the nail. After forcing the siding out, place the wedge under the siding beside the nail, and then tap down the nail. Before repainting the surface, allow approximately 2 months for drying out. Certain types of siding cannot be wedged. In this case, use tube breathers or louvers, although experience indicates that they are less effective than wedges.

To ventilate attic space adequately, there should be a minimum of 1 ft² (0.093 m²) of louver opening, divided between two or more gables, for every 300 ft² (27.87 m²) of floor space in the attic.

To reduce moisture in a damp crawl space, specify a vapor barrier at the time of construction. If this has not been done, the best practice, if possible, is to cover the ground in the crawl space with roll roofing paper or polyethylene film. The cover should be complete, with at least a 4-in (101.6-mm) overlapping of the sheets at the edges. Openings in the foundation should not be completely sealed.

The repainting of old, previously painted surfaces requires considerably more attention than the painting of new unpainted wood surfaces. Any paint left on a wood surface should be tightly adhering and continuous. If the old paint has flaked badly, exposing large areas of bare wood, the exposed wood surface should be sanded smooth and reprimed with a primer recommended for use with the intended topcoat system. Any nailheads that have worked their way up through the paint film should be resunk, and the heads coated with a knot sealer or another sealing type of paint system.

If the work is quite old and the paint film has been built up to a very thick layer, it may be desirable to remove all the old paint film with a blowtorch. This is a very hazardous procedure for persons unskilled in the art of handling blowtorches. If it is not feasible, the use of nonflammable paint and varnish removers is strongly urged. This procedure involves considerably more work and is much more expensive than the use of blowtorches, but it may save the building.

If specific boards or localized areas fail by flaking or blistering, the problem is almost always caused by moisture entering the wood from the back, as around window flashings or from leaks in roofs. The first step is to attempt to remove the source of moisture by changing the method of construction, repairing damaged items such as roofing and flashing, or sealing off the area with a caulking compound, etc.

In repaint work, inadequate surface preparation is the usual cause of loss of adhesion of the new paint to the previous paint. If the previous surface is still glossy, the new paint may not adhere. This happens most frequently on protected, unweathered areas such as the surfaces under overhangs and on ceilings of porches and carports. Wash and sand these glossy surfaces before painting. Since excessive chalking of the previous paint may interfere with adhesion of the new paint, remove the chalky substance before repainting. If mildew on the old paint is not removed, it may cause the new paint to peel.

9.6.2 Primer systems   The purpose of a primer system is to promote proper adhesion to the wood substrate and intercoat adhesion between the primer and the topcoat. Consequently, all paint manufacturers have available specific primer systems designed for use with their topcoats. It is common sense to employ the primer that was designed for use (either over bare wood or over precoated wood) with the topcoat system.

By and large, solvent-thinned primer systems are used for solvent-thinned topcoats, and water-thinned primers for water-thinned topcoats. Even today, however, many paint manufacturers still recommend that optimum adhesion, particularly over chalky surfaces, is obtained by using a particular oil-based, solvent-thinned primer.

Proper design of the primer system involves several major considerations, probably foremost of which is the control of moisture permeability. The common practice of adding extra linseed oil to a primer alters the ratio of pigment to oil and upsets moisture-penetration characteristics, resulting in a less porous film with less pigment and hence in less transfer of moisture across the film. This technique actually increases paint film failure due to blistering.

The purpose of a properly designed paint system (primer and topcoats) fundamentally is to protect a vulnerable wood surface from deterioration and to improve its appearance. Since the function of the primer is to improve adhesion by control of surfaces, it must be specially designed as the intermediate between the wood and the topcoat. Therefore, the architect should specify compatible primer and topcoats, usually from the same manufacturer.

9.7 Typical Formulations   Typical formulations of paint systems are given below as a guide to a potential purchaser and are not intended to serve as working formulas to be used as starting points by formulators. Only a general type of "label" analysis is given. Working formulas intended for paint manufacturers are much more complex, including not only weight but volume relationships.

9.7.1 Outside white oil-based, lead-free

	Percent
Pigment, 60.0 percent	
Titanium dioxide	20.2
Magnesium silicate	62.4
Mica, water-ground	17.4
	100.0

Vehicle, 40.0 percent
Refined linseed oil   35.5
Medium-bodied linseed oil   29.8
Lead-free driers   2.5
Mineral spirits   32.2
                                      ─────
                                      100.0

9.7.2 Outside white acrylic water-thinned

Percent

Pigment, 39.2 percent
Rutile titanium dioxide   41.0
Anatase titanium dioxide   2.2
Talc   21.6
Calcium carbonate   35.2
                                     ─────
                                     100.0

Vehicle, 60.8 percent
Acrylic resin   28.2
Antifreeze   2.8
Surfactants   2.3
Ammonium hydroxide   4.2
Water   62.5
                                     ─────
                                     100.0

9.7.3 Outside primer: acrylic for staining woods

Percent

Pigment, 36.8 percent
Rutile titanium dioxide   45.5
Calcium carbonate   26.1
Basic silicate white lead   28.4
                                     ─────
                                     100.0

Vehicle, 63.2 percent
Acrylic resin   31.2
Surfactants   1.1
Antifreeze   3.3
Bodying agent   4.0
Pine oil   0.5
Water   59.9
                                     ─────
                                     100.0

9.7.4 Shake and shingle paint, gray

Percent

Pigment, 49.2 percent
Titanium dioxide   27.0
Talc   58.3
Mica   13.4
Carbon black   0.4
Aluminum stearate   0.9
                                     ─────
                                     100.0

Vehicle, 50.8 percent
Modified linseed oil   63.5

Soybean lecithin	1.1
Mineral spirits	33.0
Driers	2.4
	100.0

9.7.5 Trim and trellis paint, green

Percent

Pigment, 28.8 percent	
Chrome green	36.4
Calcium carbonate	63.4
Clay	0.2
	100.0
Vehicle, 71.2 percent	
Long-oil alkyd (linseed)	45.0
Soybean lecithin	0.6
Mineral spirits	51.7
Driers	2.7
	100.0

9.7.6 Barn paint, red

Percent

Pigment, 60.8 percent	
Venetian red	23.5
Zinc oxide	5.9
Calcium carbonate	70.6
	100.0
Vehicle, 39.2 percent	
Raw linseed oil	40.5
Bodied linseed oil	28.9
Soybean lecithin	1.5
Mineral spirits	27.4
Driers	1.7
	100.0

9.7.7 Porch and deck enamel, red

Percent

Pigment, 18 percent	
Red iron oxide	100.0
Vehicle, 82 percent	
Medium-oil alkyd (42 percent soybean oil)	50.5
Mineral spirits	48.5
Driers	1.0
	100.0

9.7.8 Clear redwood finish

Percent

Chemically modified soybean oil	58.5
Driers	2.4

Mildewcide	0.5
Mineral spirits	38.6
	100.0

9.7.9 Knot sealer

	Percent
Alcohol-soluble phenolic resin	30.0
Polyvinyl butyral	3.0
Denatured alcohol	66.0
Catalyst	1.0
	100.0

9.8 Economics Differences in price usually are reflected in differences in the quality of paint performance. Economists calculate that the cost of paint represents about 15 percent of the price of a paint job. On this basis it is false economy to save 20 percent of the purchase price of the paint and lose 20 percent of the paint's durability. Early failure of any paint film necessitates more frequent repainting. Since 85 percent of the cost of painting is labor, the less frequent the repainting, the greater the savings.

Only in special cases should fewer than two coats of paint be applied to exterior wood surfaces, particularly new wood surfaces. These coats normally consist of one coat of primer and one coat of topcoat. Frequently three coats are recommended over previously unpainted wood (a primer and two applications of topcoats). This procedure will usually result in film thicknesses of 3 to 5 mils (0.0762 to 0.127 mm), depending on the application technique. A good three-coat system over sound wood, if moisture from the reverse side does not contribute to early failure, should last 5 years before repainting. Earlier failure or longer life may be expected from any given paint film, depending upon the condition of the substrate, the application, and the conditions to which the film is exposed.

9.9 Troublesome Conditions

9.9.1 Difficult surfaces Wood varies according to type; for example, redwood behaves differently from yellow pine when both are painted with the same paint. Species of wood vary in paint-holding properties. Redwood and cedar, when properly cut, hold paint best. Southern pine (or yellow pine), Douglas fir, and western larch are poor paint-holding woods.

Depending on the circumstances prevailing during the growth of the tree, southern pine varies greatly in physical characteristics. The wood has soft white sections (spring growth) and hard brown resinous sections (summer growth); the hard sections can comprise almost half of the board area. Paint adheres well to the soft sections but may scale from the hard resinous sections, especially as it weathers and becomes less flexible.

Southern pine expands abnormally when it absorbs water and shrinks greatly when it dries out, thus accelerating breakdown of the paint. As a result of the moisture loss and shrinkage of fiber, it develops wood breaks or slits (wood checks) as it ages. Paints will slit when the wood itself develops checks. Water from rains can get into the wood when the wood checks, causing further paint scaling.

Redwood   Redwood, like cedar, is an extremely porous wood with a very high water-soluble content. The water-soluble materials in redwood are largely tannin dyes. These dyes are also oil-soluble and will penetrate paint film, causing reddish or reddish brown stains to appear on the surface of the film. Special primers which react with or seal off the dyes to prevent their migration to the film surface are usually designed for application to redwood. In addition, the high porosity causes redwood to expand and contract significantly with changes in water content and even with the humidity of the air. This characteristic puts undue mechanical strains on paint films and quite frequently leads to premature failures by cracking. This cracking, which follows the lines of the wood grain parallel to the fiber structure, is usually called "grain cracking."

Clear coatings are a particular problem on exterior redwood. When failure occurs, especially with urethanes, the coating flakes off as skin does on a sunburned person. The only remedy is sanding until the surface is intact and then reapplying the clear topcoat. The California Redwood Association (CRA) in San Francisco does not recommend glossy varnishes of any type for exterior use. It has drawn up redwood coatings specifications including methods to prevent bleeding through pigmented finishes. Redwood usually is used with clear finishes to expose the color and texture of the wood. For best service and wood protection use a penetrating stain chosen from a brand list supplied by the CRA.

Yellow pine   Yellow pine is another wood that frequently presents surface problems in the application of paint. Although it is a dense wood, it is very susceptible to moisture uptake. With changes in moisture content yellow pine also exhibits significant swelling and shrinking differences according to the grain. At the same time, its density hinders the complete penetration of liquid paint to obtain good mechanical adhesion. The wood also has a high resin or pine-tar content and consequently often presents considerable problems with staining. Staining is especially likely over pine knots; it is very difficult to seal in completely the resins of pine knots. Yellow pine is also quite subject to moisture deterioration and therefore must be completely protected from the elements.

Coatings on yellow pine usually fail by cracking, flaking, peeling, and blistering. In most cases the failure is related to dimensional changes caused by moisture. These movements create wood stress, particularly at the junctions of summerwood and springwood. Coatings on yellow pine should be able to breathe, that is, to allow controlled rates of water passage in and out of the wood.

Some companies turn to factory priming to improve coating performance on yellow pine. Such priming does not reduce the need for a tight specification on the coating to be used or eliminate the requirement for inspection at the point of paint application. Specifications on factory priming should always include the required mil thickness on the total board, including the edges and ends as well as the face. Slash grain siding may require special priming to assure adhesion to both hard and soft lignin.

According to the Southern Pine Association, three-coat systems using an acrylic emulsion vehicle work well on yellow pine. Clear systems fail. Recent advances in wood preservatives and finishes show promise for some waterborne materials.[2]

Plywood, Masonite, and composition board   Plywood presents a very special problem. Most plywood is surfaced with fir, a very soft, porous wood with an extremely pronounced grain. It is almost impossible to apply water-thinned finishes directly over fir plywood because of the intense grain raising, or localized swelling of certain portions of the grain, which results in a rough and unsightly appearance. Consequently, special plywood sealers have been developed for application over fir plywood to minimize grain raising. These sealers are also extremely useful in preventing undue absorption or penetration of the resin content of prime coats. If an oil-based primer is applied directly to fir plywood, practically all the resin content of the primer will penetrate the fir, leaving on the surface a binder-poor coating that has very poor film integrity or film strength. The application of topcoats over prime coats of this nature often results in very early failure. If the surface is properly sealed with fir sealers, this penetration does not occur and satisfactory paint films are obtained.

In the early 1960s the American Plywood Association (APA) began a test program that led to certifying certain coatings for plywood. Many of these coatings are approved for exterior use, and a complete listing of certified coatings is available from the APA. Most of the approved formulas are rubber-based high-build (up to 20 mils, or 0.508 mm) coatings, although at least one 8-mil (0.2032-mm) urethane-vinyl copolymer system is available for bare plywood including southern pine. Architects and engineers should be familiar with APA specifications to ensure that correct coatings are applied to plywood. It is common practice to use exterior plywood on many surfaces such as garage doors and siding and

---

[2] For detailed information on this work, get in touch with the Mississippi Forest Products Laboratory, Mississippi State, Miss.

to coat the surface with standard exterior paints, sometimes without priming and almost always without painting the edges of the plywood. This practice often leads to checking and cracking that are not the fault of the paint. The problem is also common with flat-grain wood. Since the contributing condition (checking and grain slitting) is in the wood itself, these surfaces may be expected to require periodic scraping, sanding, repriming, and recoating.

Masonite is a hard variety of preprimed composition board that is less susceptible to swelling and checking than the softer types are. However, sawtooth cuts, sharp protrusions on the trailing edge of these boards, are inadequately sealed by the factory primer. Unless they are sealed, dew and rain droplets hang on the edges and wick into the board, causing cracking, flaking, and peeling of the paint. This condition usually is most noticeable from 2 to 3 in (from 50.8 to 76.2 mm) above the trailing edge of the board. Belt-sanding the affected areas to remove damaged paint down to the bare board is recommended, followed by priming the exposed surface with a marine spar varnish reduced by 50 percent with turpentine (particular attention should be paid to sealing the trailing edge). Sand lightly to reduce the gloss after drying. Then apply a house-paint pigmented primer before the finish coating. (Some imported Masonite cannot be painted.)

There are many varieties of pressed composition board, but all are similar in that they consist of pressed fibrous materials that readily swell in the presence of moisture, eventually checking or cracking the face of the boards themselves. Once this has occurred, repainting will bridge the cracks and checks temporarily, but they will reopen with weathering. Water-soluble dyes in the composition materials leach to the surface, and the painted siding begins to take on an overall rough, checked, and shadowy-stained appearance. Unfortunately, little can be done to correct this condition. The best solution is preventive maintenance: repriming the siding immediately following installation, making sure that all edges have been adequately coated, and keeping sufficient paint on the surface at all times to seal out moisture.

It is well to remember that there are many grades of plywood, and that plywood intended for use under outdoor conditions *must* be an outdoor grade or a marine grade; otherwise delamination will occur. Although painting tends to protect the wood, it does not prevent delamination of regular interior-grade plywood exposed to outdoor conditions. In addition, specifications for painting exterior plywood should always call for sealing the edges with a mil thickness adequate to prevent the entrance of moisture.

9.9.2 Difficult situations

Design problems   Paint performs differently according to its exposure to the atmosphere. The area directly under eaves, which is protected from rainfall, usually tends to collect a large amount of dirt, since moisture is not present to cause chalking or to wash off the chalked material and take the dirt with it. Also, since sunlight does not hit this area directly, it is subject to mildew accumulation. The dew formation does not dry off early in the morning, and the extra moisture tends to encourage the growth of mildew. This is especially true on the north side of houses, where sunlight does not hit the wood to dry off the dew formation properly.

Moisture Moisture problems have been discussed previously in this chapter. They are due largely to improper construction techniques or to the failure of some component that is supposed to keep moisture away from the back of the wood siding.

Staining problems The use of bronze screening presents a major staining problem. This screening tends to oxidize, yielding water-soluble or water-leachable materials. These materials, usually copper salts, wash down over the paint film, where they are absorbed because of the porous nature of the film or where they react chemically with some ingredient in the film, causing an unsightly green or greenish brown stain. One of the major subjects of paint formulation research is the development of ways to combat staining from bronze screen wire. This staining appears to be independent of the type of paint. Solvent-thinned oil-based paints appear to stain the same color and in the same manner that water-thinned emulsion systems do. One technique to combat this staining is, of course, to prevent the bronze wire from oxidizing. This is frequently done by coating the screen with a lacquer-type clear varnish; or a plastic screen may be chosen. Plastic screens, however, are subject to insect attack that causes degradation.

Insect problems A major problem with the application of paint in summer is the tendency of paint to attract insects. A freshly painted surface, because of its pure-white nature, frequently attracts large swarms of insects, which light on the paint film, where they become embedded and die. This not only causes an unsightly appearance but gives rise to weak spots which can cause early failure of the paint film, since the paint is quite a bit thinner in these areas. Water-based emulsion paints appear to have some superiority over solvent-thinned oil-based paints in this respect, because they dry much faster and the number of insects embedded in the surface appears to be considerably smaller. Some work has been done with insect repellents such as citronella in the paint formulation. However, it seems that the advantages are balanced by a slight loss in the durability of paint films containing citronella. This is a field that still needs research attention.

Shrubbery problems The presence of shrubbery around a frame building presents special problems. Shrubbery encourages the growth of mildew, bacteria, and other microorganisms, which are transferred from the foilage to the paint film, where because of their enzymatic exudation they deteriorate the film. In the case of mildew, there is also the problem of defacement or discoloration. Shrubbery also tends to prolong the presence of moisture on the surface, since it shades the side of the building from the sun, and dew and rain do not evaporate quickly. The transpiration or exudation of moisture from the leaves of the plants also increases humidity in the immediate area, resulting in a slower rate of evaporation and a longer period of wetness. It is good practice to ensure free space between shrubbery and painted surfaces.

Mildew growth   Dirt and mildew look much alike on the painted surface. Mildew, however, appears in colonies as dark, brownish black splotches. Under magnification, mildew is easily distinguishable as a fibrous growth. The most prevalent organism defacing paint films appears to be the fungus *Aureobasidium pullulans:* research workers have found that approximately 95 percent of mildew defacement is due to this organism.

Mildew feeds on the oil in a paint. Other conditions being equal, it grows more rapidly on soft paints with a high oil content than on harder paints. In areas where mildew is a problem, specify paints containing mildewcidal pigments or additives.[3]

Before repainting always remove mildew with a dilute sodium hypochlorite solution. Even if specifications call for repainting with a "mildew-resistant" paint, existing mildew must be removed so that it will not contaminate the fresh paint film. Under certain conditions, mildew may continue to grow under and through the new paint film.

The surface should be washed with a solution consisting of approximately 2 oz (0.057 kg) of household chlorine bleach added to 1 gal (3.785 l) of water. After the surface has been washed with this solution, it should be rinsed thoroughly with clean water, preferably by the use of a hose. Another removal technique is to wash the surface with a solution of sodium hypochlorite in order to kill the mildew before paint is applied. Scrubbing with trisodium phosphate can also help remove old mildew. The application of paint over living mildew usually results in the failure of intercoat adhesion and in poor durability of the paint system. Intercoat adhesion problems such as peeling also occur if too long an interval of time elapses between application of the undercoat and the finish coat and the undercoat then becomes too hard to form a bond with the finish coat or if there is moisture (rain, dew, frost) on the undercoated surface when the finish coat is applied.

After the surface has been washed and rinsed, it should be permitted to dry thoroughly (approximately 48 h) before being painted. However,

---

[3] To test for mildew, apply a 6% Clorox solution to the suspect surface. Mildew will bleach out, but dirt will not.

it should be painted as soon as it is sufficiently dry in order to prevent the growth of new mildew on the surface.

9.9.3 Environmental concerns   Control technique guidelines for architectural coatings are now available from the EPA; and some states, for example, California, have already adopted regulations to control volatile organic materials (VOM) from these coatings. California requires that such field-applied products contain less than 250 g/l of VOM (D/L Laboratories, in New York, is investigating the status of current paint technology to meet these regulations.)

The California regulation, Rule 1113, which went into effect on September 2, 1979, applies to architectural coatings; however, there are a number of exemptions that will continue until 1982, when they will be reviewed. The rule prohibits the sale of coatings that do not conform to the VOM limitation. This prohibition means that architects and others involved with coatings that fall under the rule face specifying new classes of materials that reduce the amount of organics and/or have higher solids. Consequently, there are appearing on the market new types of coatings, some untested for life expectancy and many that require special application and surface-preparation techniques.

As a result, it will behoove specification writers to include detailed inspection procedures on jobs requiring these paints and to keep abreast of coating developments for new paints that meet the rule. This advice applies to states other than California because many states will follow EPA guidelines in this area in order to publish acceptable state implementation plans to meet *National Ambient Air Quality Standards.*

Rule 1113 highlights   The rule is not applicable to coatings supplied in containers of 1 l or less or to traffic coatings used on public streets (it *does* apply to paints used on curbs, driveways, and parking lots). It does apply to architectural coatings (no more than 250 g of VOM per liter of coating minus water is permitted) and to coatings for interior use, which are allowed to contain 350 g of VOM per liter of coating (minus water) until September 2, 1980. In addition, there is a small-business exemption. Other coatings exempt until September 2, 1980, include varnish, lacquer, or shellac; semitransparent stains; opaque stains for use on bare redwood, cedar, mahogany, or Douglas fir; primer-sealers and undercoats; wood preservatives; fire-retardant coatings; tilelike glaze coatings; waterproofing materials; industrial maintenance finishes; metallic-pigmented coatings; swimming-pool coatings; graphic-arts coatings; and other miscellaneous groups.

Calculating VOM   While architects generally are not involved with formulation details, they should be familiar with the solvent restriction problem in anticipation of specification problems. Some standard products may no longer be available after the adoption of the strict environmental regulations. In addition, architects may want manufacturers to complete paint test-formula calculations (Table 9.3) in order to assure that the coatings specified are in compliance.

TABLE 9.3 Premium-Quality Gloss White Enamel

Material	Formula		Total solids		Gallons of water	Pounds, VOM
	Pounds	Gallons	Pounds	Gallons		
Tamol 731—25%	10.8	1.31	2.70	0.34	0.97	
Nopco NDW	2.0	0.27	1.00	0.14		1.00
Propylene glycol	59.8	6.94	· · ··	· · ·		59.80
Ti-Pure R-900	269.6	7.86	269.60	7.86		
Disperse at high speed to N.S. 7. Then add, at low speed:						
Propylene glycol	57.8	6.71	· · ·	· · ·		57.80
Rhoplex Ac-490 (46.5%)	555.9	62.71	258.50	27.01	35.70	
Super-ad-it ⎫	1.0	0.12				
Water ⎪	15.2	1.82	· · ·	· · ·	1.82	
Texanol ⎬ Premix	15.7	1.98	· · ·	· · ·		15.70
Triton G-R-7 ⎪	2.0	0.23	2.00	0.23		
Nopco NDW ⎭	2.9	0.40	1.50	0.20		1.50
2.5% solution Q.P. 4400	80.4	9.65	2.00	0.24	9.41	
Total	1073.1	100.00	537.30	36.02	47.90	135.80

Calculations

Total gallons of water:	$100.00 - 47.90 = 52.1$
Pounds VOM per gallon (minus water):	$135.80 \div 52.1 = 2.61$
Grams VOM per gallon (minus water):	$2.61 \times 454 = 1184.94$
Grams VOM per liter (minus water):	$1184.94 \div 3.785 = 313.06$
Carb rating:	O.K. for *interiors* until Sept. 2, 1980

SOURCE: John A. Gordon, Jr., University of Missouri–Rolla, Extension Division. Refer to Rohm and Haas Inc., IG-90-2.

BIBLIOGRAPHY

Advisory Recommendation: Maintenance of Structural Timber Framing, American Institute of Timber Construction, Englewood, Colo., 1971.

APA Qualified Coatings Institute, American Plywood Association, Tacoma, Wash., 1979.

Banov, Abel: "Unsuitable Coatings Can Cause Early Failure of Hardboard," *American Paint Journal,* May 5, 1980, p. 26 ff.

Barnes, H. M.: "Coating Durability on Organolead-Treated Southern Pine," *Journal of Coatings Technology,* vol. 51, no. 651, 1979, pp. 43–45.

Barquest, G., and J. Black: *Exterior Finishes for Wood: A Slide Presentation,* Forest Service, Forest Products Laboratory, Madison, Wisc., 1977.

Browning, Bertie L.: *The Chemistry of Wood,* Interscience Publishers, New York, 1963.

Finishing Shingles and Shakes, Red Cedar Shingle and Hardsplit Shake Bureau, Seattle, Wash., 1976.

1980 Sweet's Catalog File: Products for General Building, McGraw-Hill, Inc., New York, 1980.

"Porch and Deck Paints," *Consumer Reports,* July 1976, pp. 417–420.

Solvent Emission Control Laws and the Coatings and Solvents Industry, Environmental Science Services, New York, 1976.

Technical Guide: Painting over Knots, Western Wood Products Association, Portland, Oreg., 1972.

Thompson, Warren S., and H. M. Barnes: "Advances in Wood Finishing and Wood Preservation Technology," *Proceedings—Complete Tree Utilization,* Forest Products Research Society, Madison, Wisc., 1978.

Zicherman, J. B.: "Painting Southern Pine," *Forest Products Journal,* vol. 19, no. 4, 1969, pp. 44–51.

———: "Painting Yellow Pine," *Western Paint Review,* May 1969, p. 18*A* ff.

Chapter **10**

Interior Architectural Coatings

HOWARD W. GOETZ
Consultant

10.1 Introduction Proper specifications for the interior construction of commercial buildings and houses must include details for surface preparation and application instructions. Moreover, each substrate, whether previously painted or unpainted, requires an acceptable painting system. Unless a specific primer or coating is absolutely required for a particular substrate, the choice of the paint system is often left to the specification writer. A detailed explanation of the advantages and disadvantages of the different types of paints or paint systems for each substrate is presented later in this chapter.

Specification writers should have some understanding of the effect of formulation techniques on the performance of coatings. They should also realize that the thickness of paint films has a tremendous bearing on performance. Wet and dry film thicknesses can be important parts of a specification.

10.2 Factors Determining Selection of Coatings   Architects, maintenance engineers, and those involved in writing paint specifications can draw up more satisfactory specifications for paint systems if they recognize the important performance requirements. The selection of one paint or paint system for all conditions is not practical, for many factors determine proper paint selection (Table 10.1).

In many situations one coat of paint will not produce a satisfactory finish. The painting procedure will then involve multiple coats of paint, making it necessary to consider each coat as part of a system.

10.3 Wall Coatings   Before applying a coat of paint over painted surfaces, determine the condition of the existing coating. Only a thorough examination of the paint film and the substrate will provide the informa-

TABLE 10.1   Factors That Affect Paint Selection

For new work
  Type of substrate
For repainting
  Condition of the surface
  Type of coating already on the surface
  Solubility characteristics of the pigments in the coating to be painted
  Gloss of the surface being painted
Method of application
Durability desired
  Resistance to staining
  Scrubbability
  Washability
  Resistance to burnishing
  Resistance to chemicals
  Resistance to abrasion
  Resistance to oils and solvents
Price of the coating being applied
Cost of maintaining the coating in an acceptable condition

tion needed to write a proper repainting specification. Use the following guidelines.

10.3.1 Previously painted walls in good condition   A surface painted with a flat paint that has good adhesion to the substrate should be washed to remove dirt, grease, or any other foreign matter with a solution of 3 oz (0.085 kg) of trisodium phosphate per gallon (3.785 l) of water. If mildew is present, add 1 qt (0.946 l) of a bleaching agent such as Clorox to 1 gal of the cleaning solution. The presence of mildew can be detected on most surfaces by subjecting it to a few drops of Clorox. The mildewed area is bleached toward a light tan or off-white color, while the color of any dirt present will not be changed. After washing, thoroughly rinse the surface and allow it to dry. Surfaces that are wet or damp from washing and rinsing should not be painted.

The preparation of painted surfaces that have been coated with a semigloss or high-gloss enamel should follow the same procedure. After washing the painted surface, sand to dull the existing glossy surface. Remove the dust caused by the sanding operation with a tack rag. Previously painted surfaces conditioned for recoating as outlined can then be painted with alkyd or oleoresinous paints. Previously painted surfaces to be coated with latex should have the dust resulting from the sanding operation removed with a rag dampened with water. Some tack rags may be impregnated with a resinous material that could be deposited on the substrate during the removal of the dust generated by sanding. This resinous material could have an adverse effect on the adhesion of latex paints applied over it.

10.3.2 Previously painted walls in poor condition   Plaster, concrete, concrete block, or drywall which has been painted with a flat paint that has lost adhesion to the substrate or to the previous coating must be wire-brushed and sanded to remove all loose and scaling paint. Widen cracks in plaster to form an inverted V. Before filling in the cracks with a water-mix or latex-type patching plaster, thoroughly dampen the cracks and surrounding edges. This wetting of the area to be patched will prevent the absorption of water from the patching compound into the dry plaster. Such a loss of water would leave the compound crumbly, with poor adhesion to the plaster. After the patches have hardened, normally in 16 to 24 h, sand them smooth.

Cracks or openings in concrete building blocks or concrete walls should be repaired before repainting. Allow a minimum curing time of 24 h before painting. Before recoating painted drywall, countersink nails and fill all holes with tape-joint cement. Sand to a smooth finish. All loose and scaling paint must be removed, preferably by sanding, but take care to avoid scuffing or breaking the top layer of paper of the drywall. Use the procedure outlined in Sec. 10.3.1 to remove dirt, grease, or other foreign matter present on the surface of either the paint or the substrate.

Prepare surfaces coated with a semigloss or high-gloss enamel for repainting by following the directions for surface preparation outlined above. After patching has been completed and any mildew eliminated, sand the surface to remove the gloss.

Spot-prime patched areas and those from which loose or scaling paint has been removed, using a primer that is compatible with the topcoat specified. Prime the entire wall if excessive failure of previous coats of paint has resulted in a large number of patched areas. In either situation, tint the primer to match the topcoat color and thus assure adequate hiding over the primer. Topcoats formulated to provide holdout and color uniformity over substrates of uneven porosity can be used as self-primers, thereby eliminating the need for a tinted primer.

10.3.3 Unpainted plaster   Plastered walls should be cured for a 30-day period at a minimum temperature of 60°F (16°C). (Heat should be provided if prevailing temperatures fall below 60°F.) This suggested curing time depends upon the prevailing temperature, humidity, and ventilation. Thorough curing will reduce the alkalinity and the moisture content to a safe level for painting. The drying oils in primers that cure by oxidation will be saponified by the alkali present in inadequately cured plaster, resulting in a loss of adhesion and poor holdout of succeeding coats of paint.

Plaster that has had the white coat wet-troweled to a hard finish and has had a minimum curing time of 30 days can be primed with a latex, alkyd, or oil-based primer-sealer. Tinting the primer-sealer to match the color of the topcoat is recommended to ensure hiding and color uniformity in a two-coat system. Topcoats that can be applied over primer-sealers include water-based–emulsion, alkyd, and oleoresinous types. They range in gloss from a flat though a semigloss to a high-gloss finish.

Plaster that has not had adequate wet troweling of the white coat will not have a hard finish. Instead, a chalky, porous residue with a weak bond to sound plaster will be present on the surface. This condition will also exist if the plaster was subjected to freezing while the white coat was being cured or hardened.

Often, water-based primers will not adhere satisfactorily to plaster on which a chalky deposit is present. This type of primer-sealer does not wet the chalky material that acts as a barrier between it and the surface. Poor wetting causes loss of adhesion. An alkyd or oil-based primer-sealer will wet a crumbly powder of this type and provide better adhesion. However, it is good practice to remove as much of the chalky residue as possible before applying the primer to the plaster.

Also, some keen-finished plasters are smooth and hard. An additive is used to impart this hardness. Conventional latices will not grab onto such surfaces, and you must resort to alkyd wall primer-sealers.

10.3.4 Unpainted concrete building blocks above grade

The rough texture of these blocks can be eliminated by filling voids with a latex-type block filler. Tinted to the topcoat color, this filler can be applied with a roller or a brush. Either of these methods is preferred to application with a spray gun. Spraying block filler on concrete blocks does not provide maximum filling unless it is followed by rolling. Remove any excess filler remaining on the surface with a squeegee if a smooth finish is desired. After a drying time of 24 h, topcoat the block filler with a water-emulsion, alkyd, oleoresinous, or catalyzed-epoxy coating. For maximum resistance to the abrasion of scrubbing and washing to which corridor walls are subjected, a catalyzed-epoxy coating is recommended.

The texture of the building blocks can be retained by eliminating the latex (water-based–emulsion) block filler. Replace it with a latex wall primer-sealer tinted to the topcoat color. Apply a latex, alkyd, or oleoresinous topcoat after the primer-sealer has dried for 24 h.

The texture of the blocks can be retained in a catalyzed-epoxy system by replacing the latex wall primer with a thinned catalyzed-epoxy coating. Reduce the initial coat in a ratio of 1 part catalyzed epoxy to 1 part epoxy solvent. After the initial coat has been cured for a minimum of 24 h at temperatures above 60°F (16°C), apply the unreduced epoxy coating as the second coat.

10.3.5 Unpainted concrete building blocks below grade

For a smooth finish, fill concrete building blocks with a catalyzed-epoxy block filler, using the application procedure suggested for a latex-type block filler in Sec. 10.3.4. This type of filler is specified in preference to the latex type because it provides maximum water resistance. Maximum performance is realized by topcoating with a catalyzed-epoxy coating blocks filled with a catalyzed-epoxy block filler.

Substituting a catalyzed-epoxy coating reduced in the ratio of 1 part epoxy solvent to 1 part catalyzed epoxy for the filler will maintain the rough texture of the blocks. Following this primer (after a drying time of 24 h) with the same unreduced catalyzed-epoxy coating is suggested to maintain the water, chemical, and abrasion resistance of an epoxy system.

Either of the systems outlined above is recommended for shower stalls or other areas where moisture is a problem. Use proper ventilation.

10.3.6 Unpainted concrete walls

Forms used for concrete walls usually are coated with compounds to permit easy release of the forms from the concrete after it has hardened. These compounds, which adhere to concrete walls, must be removed to assure a satisfactory bond of the paint to the substrate and to prevent retarding the drying rate of the coating.

Oil, which frequently is used as a form release, can be removed from a concrete wall by scrubbing the wall with a strong solution of hot water and trisodium phosphate. The washing should be followed by thorough rinsing. Other types of form release may require different methods to remove them. Manufacturers should be consulted for the preferred procedure to accomplish complete removal of their form releases.

After conditioning the concrete wall, apply a latex-type primer-sealer. Then topcoat with a latex, alkyd, or oleoresinous paint, which may have a flat, semigloss, or high-gloss finish.

The substitution of a catalyzed-epoxy system for those suggested above will provide a coating that is more resistant to alkalies, efflorescence, and the abrasion of scrubbing and washing. The catalyzed-epoxy topcoat thinned in a ratio of 1 part topcoat to 1 part epoxy thinner should be used as the primer. Then apply the unreduced catalyzed-epoxy topcoat over the cured primer.

10.3.7 Unpainted drywall   Countersink nails, and fill in nail holes with tape-joint cement. Sand-tape joints and filled nail holes to a smooth finish. Do not break or scuff the top paper layer of the drywall because any broken or scuffed areas will have a rough texture when painted. Paints applied over such areas can dry to a lower gloss and a different color if they are sensitive to surfaces of uneven porosity. Remove dust resulting from the sanding.

Prime with a latex-type primer-sealer. Since alkyd or oil-type primer-sealers will raise the nap of the dry, they should not be used for this type of surface. Recoat after drying with a latex, alkyd, or oleoresinous type of topcoat, which may have a flat, semigloss, or high-gloss finish.

10.3.8 Ceilings   To minimize the effect of paint on acoustical tile or plaster, latex-based paints are preferred to the alkyd type. High-PVC latex paints are recommended for both acoustical tile and plaster. These low-binder coatings are more porous, thereby exerting less effect on the acoustical properties of the substrates. If adequate hiding is obtained in one coat, do not specify an additional coat; a single coat will minimize the effect of paint on acoustical properties.

Ceiling texture coatings are suitable for coarse texture on concrete, stucco, or wallboard ceilings. While some of these finishes are flaky and cannot be touched or cleaned without damaging the surfaces, some hard, durable ceiling texture coatings based on acrylic latex emulsions are available. These films breathe, conceal minor defects, and can be broomed, vacuumed, washed, or repainted. They simulate an acoustic surface. New surfaces do not require a prime coat. Follow normal precautions with previously painted ceilings.

10.4 Wood Surfaces   Interior wood surfaces require little treatment other than sanding them smooth and clean. A pigmented finish such as a flat, semigloss, or gloss enamel should be preceded by a primer (e.g., an enamel undercoat) that will fill the wood and can be sanded. This type of coating provides a smooth base for application of the enamel topcoat.

Wood surfaces to be stained or finished with only a clear coating also require sanding. Fill porous woods to prevent excessive absorption of the stain or the clear coating. Filling porous woods before staining prevents uneven absorption of the stain, and a lighter, more uniform appearance will be achieved. Follow the stain with a clear coating of the gloss preferred.

Oil- or resin-based stains can react or peel (lift) when lacquer-based topcoats are applied. New water-based stains alleviate that problem and also avoid solvent emissions. While varnishes and stains are exempt from new California restrictions until 1982, some companies are now marketing waterborne systems based on acrylics and PVA.

10.5 Metal Surfaces   Iron and steel surfaces must be free of rust, mill scale, oil, and grease before they are primed. Primers should be applied immediately after the cleaning operation because clean steel can begin to rust within a few hours. Primers formulated with a rust-inhibiting pigment will prevent continued oxidation of the metal. This type of pigment includes red lead, zinc chromate, basic lead silicate, and dibasic lead phosphate, to name a few. Selection of the vehicles in these primers depends upon drying and exposure requirements. For more corrosive environments specify a rust-inhibitive primer formulated with the vehicle that will be the most resistant to them. Apply two coats of primer for the more severe exposures. Topcoat the primed metal with a coating that is compatible with the primer and will also exhibit the same resistance to the conditions which governed selection of the primer.

Much of the metal used in construction arrives at the building site preprimed. Primers formulated to provide satisfactory protection against corrosion of the metal can be topcoated after dirt or grease is removed and chipped areas are spot-primed.

Metal primers that serve only as decorative coatings and provide no rust-inhibitive properties require a barrier coat if latex topcoats are to be applied over them. Many water-based paints used over primers formulated without rust-inhibiting pigments will cause the metal to rust if the water in the latex paints penetrates the primer. Application of solvent-type topcoats over this type of primer will provide a paint system that offers marginal protection to the metal if it is not subjected to corrosive conditions. Specifications for primers for metals should include an adequate level of rust-inhibitive pigments that will provide protection against oxidation of the metals.

Galvanized-metal surfaces are protected against rusting, but exposure to corrosive conditions and weathering will eventually destroy galvanic protection. While painting galvanized metal offers additional protection to the metal and enhances its appearance, conventional paints will not adhere to new galvanizing. Before painting, therefore, galvanized metal must be chemically treated or etched. This treatment can be eliminated (after removing the oil present) by using primers that do not depend upon the etching treatment for adhesion. These primers include latex metal primers designed specifically for this type of surface and oil-, alkyd-, or varnish-based primers pigmented with zinc dust or portland cement. Such primers provide a satisfactory base for the usual air-drying type of topcoats.

10.6   Floor Paints   Clear finishes are applied to hardwood floors to preserve their appearance. This type of finish prevents soiling the surface, simplifies cleaning the floors, and brings out the beauty of the wood. There are two types of clear finishes: surface coatings and penetrating sealers. If a wood-floor finish must be decorative as well as protective, an opaque floor paint can be used.

Opaque finishes for concrete floors are usually applied to hide the color of the concrete and prevent dusting caused by traffic. Penetrating-type concrete-floor sealers are used to reduce dusting if the color of the floor is not part of a color scheme.

Conditions governing the selection of floor finishes for both repainting and new work are discussed below.

10.6.1   Previously painted floors in good condition   Refinishing a floor that has a coating requires only removing any dirt, grease stains, or wax before repainting if the floor is in good condition. Remove all dirt, grease, or stains by scrubbing the floor with a water solution of 3 oz (0.085 kg) of trisodium phosphate. Rinse the surface thoroughly with clean water, and allow it to dry.

Remove all wax. This can be accomplished by dissolving the wax in mineral spirits. After the wax has softened, the solvent and dissolved wax must be wiped up immediately; otherwise the wax will be redeposited on the surface. Rags used to wipe up the solvent and wax should be changed frequently. If they become saturated with wax, removal will be incomplete. Any wax remaining on the surface will have an adverse effect on the drying rate and the adhesion of floor paints applied on it.

After cleaning the floor, sand it to dull the gloss of the finish and eliminate rough spots. Pick up all dust with a vacuum sweeper, and clean the floor thoroughly with a tack rag. (If latex floor paint is used, substitute a rag dampened with water for the tack rag.) Spot-prime small areas that have been worn to the substrate before the final coat is applied.

10.6.2   Previously painted floors in poor condition   Remove completely floor finishes applied on wood that have failed extensively through loss of adhesion and erosion caused by traffic. Normally, sanding is the fastest and least expensive method of preparing a floor for refinishing. Initial sanding can be done with a coarse grade of sandpaper. Follow this operation with progressively finer grades, using a fine sandpaper for the final sanding. After all dust has been removed, the floor can be refinished with an opaque coating, a stain, or a clear or penetrating type of sealer. Follow the procedures outlined for finishing new wood.

The removal of badly worn concrete-floor finishes requires mechanical abrasion. The complete removal of dust should be followed by the patching of chipped areas. After the patches have cured, apply the concrete-floor coating suggested for painting new concrete.

10.6.3   New wood floors   Before applying the initial coat, sand the floor to remove dirt and to smooth out any imperfections caused by installation. Remove dust from the sanding operation with a vacuum cleaner. After the floor has been prepared for the application of the initial coating, it must be protected from dirt, dust, and moisture. Therefore, it is good practice to apply the first coat immediately after the sanding and cleaning operations have been completed.

Stains or dye coats   These are applied on wood floors to change their color. Allow them to dry overnight before recoating. If the color of the wood is acceptable, this step is omitted.

The more porous hardwoods such as oak may require filling if a smooth surface is preferred. Fill the pores in the wood with a paste or liquid filler. Instructions on the filler label for thinning, application, and drying should be followed.

Either type of filler is applied with a brush, first by brushing it across the grain of the wood and then by immediately brushing it with the grain to fill the pores. The drying of the filler must be observed closely so that the wiping operation will remove from the surface the excess but not the filler that has been worked into the pores. Wipe the filler with a burlap bag, wiping across the grain after the filler has lost its gloss. The filler should dry 24 h before recoating.

Clear finishes   If neither staining nor filling of a wood floor is desired, the initial coat of clear finish, thinned according to the manufacturer's instructions, should be applied immediately after sanding and cleaning have been completed. The following coats of finish are normally applied without thinning. Sand lightly between coats, and allow each coat to dry 24 h before recoating.

Floor varnishes that range from a flat to a high gloss are available. The gloss selected is a matter of preference, but higher-gloss coatings are more resistant to wear scuffing and require less maintenance.

Opaque floor paints   Although hardwood floors are normally finished with clear coatings to bring out the effect of the grain, opaque floor paints also are used. Surface preparation, as previously described, should be part of the specification. Thinning the initial coat according to the manufacturer's recommendations will provide a system which has better adhesion to the substrate because the coat penetrates the wood. The second coat normally is applied without thinning.

Penetrating sealers   A fourth method of finishing hardwood floors involves penetrating floor sealers. As their name implies, these sealers are absorbed by the wood and are not surface coatings such as clear finishes or opaque floor paints. Penetrating sealers, which remain beneath the surface of hardwood floors, impart wear resistance and prevent excessive soiling of the wood.

This type of finish also offers the advantage of easier touch-up. Traffic will cause the sealer and the wood to wear together, giving the impression that breakdown of the finish is minimal. In contrast, surface coatings show the effects of wear in heavy traffic areas. If only these areas are refinished, the partial refinishing usually is readily apparent, for it is difficult to blend the freshly applied surface coating into the old finish.

10.6.4 New concrete floors   The deterioration of concrete-floor paints is caused by mechanical abrasion or loss of adhesion to the substrate. Excessive alkalinity and the presence of water are contributary causes to failure. The passage of moisture through a concrete floor by capillary action or hydrostatic pressure is prevented by installing adequate drainage and by pouring the floor on a moisture barrier.

The finish given a concrete floor during construction has an effect on the durability of the paint system applied. Acid-etching is mandatory for a steel-troweled, hard, smooth finish and is also recommended for a porous finish to achieve a good bond between the paint and the concrete.

Before treating the concrete floor with an acid etch, remove all dirt, grease, and oil by scrubbing the floor with a strong solution of hot water and trisodium phosphate. Repeated scrubbing may be necessary for stubborn areas. Rinse the floor thoroughly with clean water. Floors that have absorbed large amounts of grease or oil may require a second scrubbing if grease or oil comes to the surface after the floor has dried for several days.

Concrete hardeners prevent acid-etching the surface. Since there are many types of hardeners, consult the manufacturer for the most satisfactory method of removal. If chemical removal is not possible, mechanical abrasion may be necessary.

After cleaning the concrete floor, etch with a 10% solution of muriatic acid (a commercial grade of hydrochloric acid). Allow this acid solution to remain on the surface for 10 to 15 min or until bubbling stops. This treatment neutralizes surface alkalinity, removes the layer of fine cement powder known as laitance, and opens pores in the concrete floor to improve the bond between the paint and the substrate. The removal of laitance by acid-etching is essential, as this layer of powder has a very weak bond to sound concrete. After the reaction between the acid and the floor has ceased, thoroughly rinse the floor to remove all traces of acid. Allow the floor to dry before painting.

Alkyd, latex, oleoresinous, or chlorinated-rubber types of floor paints, thinned as instructed by the manufacturer, can be used as the primer. Thinning the first coat assures penetration of the paint into the concrete and a better bond of the paint system to the substrate. The second coat is normally applied without reduction.

Catalyzed-epoxy floor paints require greater reduction than latex or solvent-type floor paints do. The eventual hardness of catalyzed-epoxy films can cause delamination of the concrete floor if the initial coat is a surface coating rather than a penetrating type. This form of failure can be prevented by appreciably thinning the catalyzed epoxy. A minimum ratio of 1 part epoxy solvent to 1 part catalyzed epoxy is suggested; ratios up to $1\frac{3}{4}:1$ are recommended for more nearly impervious concrete surfaces. After the initial coat has cured for 24 h at a temperature of 60°F (16°C), recoat with the same unreduced catalyzed-epoxy coating.

Clear finishes for concrete floors require the same surface preparation described for opaque coatings. The more commonly used finishes penetrate the concrete, serving as a sealer rather than a surface coating. Their primary function is to keep dusting to a minimum. Areas protected with a penetrating sealer will not show the effects of traffic as readily as those painted with a surface type of coating do.

10.7 Types of Paint   Broadly classified, paints fall into flat, semigloss, and high-gloss categories. To some degree, choice of gloss is a personal decision, but it can have a bearing on the performance of paints formulated with comparable vehicles and pigments. Within this limitation, higher-gloss finishes will clean up better and have greater abrasion resistance. However, they will accent substrate imperfections. Low-gloss paints, particularly those with low angular sheen, are suggested to obscure defects in walls and ceilings. Table 10.2 relates types of paints (systems) to the various surfaces previously discussed. In choosing one of the various systems, select a sealer that will not raise the nap of the surface being painted. For all practical purposes, alkyd wall primer-sealers have all but disappeared except on keen-finished plasters and as sandwich coats when there is a bleeding problem.

Gloss is controlled by the ratio of the volume of pigments to the volume of vehicle solids or binders in a paint. The term used to define this ratio is "pigment-volume concentration (PVC)." As the PVC increases,

TABLE 10.2 Recommended Specifications for Interior Architectural Surfaces

Surface	Finish	Topcoat	Finishing schedule
Drywall, Sheetrock, cellulose board, and skip texture[a,g]	Flat	Alkyd	1 coat, PVA sealer or 1 coat, alkyd wall primer-sealer 1 coat, alkyd flat enamel
		Latex	2 coats, latex wall finish[b]
		Acrylic deep-tone latex	2 coats, acrylic deep-tone latex flat[b]
	Stipple gloss	Alkyd	1 coat, PVA sealer or 1 coat, alkyd wall primer-sealer 1 coat, alkyd gloss stipple[a]
	Stipple semigloss	Alkyd	1 coat, PVA sealer or 1 coat, alkyd wall primer-sealer 1 coat, alkyd semigloss stipple[a]
	Stipple eggshell	Alkyd	1 coat, PVA sealer or 1 coat, alkyd wall primer-sealer 1 coat, alkyd eggshell stipple[a]
	Stipple flat	Alkyd	1 coat, PVA sealer or 1 coat, alkyd wall primer-sealer 1 coat, alkyd flat stipple[a]
		Latex	1 coat, latex flat stipple 1 or 2 coats, latex wall finish
		Acrylic deep-tone latex	1 coat, latex flat stipple 1 or 2 coats, flat acrylic deep-tone latex
Ceiling trusses and beams (rough surfaces)	Flat	Polymerized oil	1 or 2 coats, mill white (spray application)
Acoustic tile (Perforated tile may be rolled; fissured gypsum should be sprayed)	Flat	Alkyd	1 or 2 coats, alkyd flat enamel
		Latex	1 coat, high-PVC latex
		Acrylic deep-tone latex	1 or 2 coats, flat acrylic deep-tone latex
Acoustical plaster	Flat	Alkyd	1 coat, alkyd flat enamel (spray application)
		Latex	1 coat, high-PVC latex

Surface	Finish	Topcoat	Finishing schedule
		Acrylic deep-tone latex	1 or 2 coats, flat acrylic deep-tone latex
Woodwork[c]	Gloss	Alkyd	1 coat, enamel undercoat 1 coat, mixture of enamel undercoat and alkyd gloss enamel (optional) 1 coat, alkyd gloss enamel
	Semigloss	Alkyd	1 coat, enamel undercoat 1 coat, mixture of enamel undercoat and alkyd semigloss enamel (optional) 1 coat, alkyd semigloss enamel
			1 coat, enamel undercoat 1 or 2 coats, alkyd semigloss deep-tone enamel
		Acrylic latex	1 coat, enamel undercoat 1 or 2 coats, semigloss latex enamel
			1 coat, enamel undercoat 1 or 2 coats, deep-tone latex semigloss enamel
	Eggshell	Alkyd	1 coat, enamel undercoat 1 coat, mixture of enamel undercoat and alkyd eggshell enamel (optional 1 coat, alkyd eggshell enamel
	Flat	Alkyd	1 coat, enamel undercoat 1 or 2 coats, alkyd flat enamel
Wood floors, steps, and porches[e]	Gloss	Phenolic alkyd	1 coat, floor enamel (thinned 10 percent) 2 coats, floor enamel
	Satin flat	Acrylic latex	1 coat, enamel undercoat 2 coats, latex floor enamel
Masonry, concrete, plaster, brick, and concrete block[d]	Gloss	Alkyd	1 coat, alkyd or PVA wall primer-sealer or 1 coat, enamel undercoat[e] 1 coat, alkyd gloss enamel
	Semigloss	Alkyd	1 coat, alkyd or PVA wall primer-sealer or 1 coat, enamel undercoat[e] 1 coat, alkyd semigloss enamel

TABLE 10.2   Recommended Specifications for Interior Architectural Surfaces   (*Continued*)

Surface	Finish	Topcoat	Finishing schedule
	Semigloss (*continued*)	Alkyd deep-tone	1 coat, alkyd or PVA wall primer-sealer 1 or 2 coats, alkyd semigloss deep-tone enamel
		Acrylic latex	1 coat, alkyd or PVA wall primer-sealer 1 or 2 coats, latex semigloss enamel
		Acrylic deep-tone latex	1 coat, alkyd or PVA wall primer-sealer 1 or 2 coats, latex semigloss deep-tone enamel
	Eggshell	Alkyd	1 coat, alkyd or PVA wall primer-sealer, or 1 coat, enamel undercoat[e] 1 coat, alkyd eggshell enamel
	Flat	Alkyd	1 coat, alkyd or PVA wall primer-sealer 1 or 2 coats, alkyd flat wall enamel
		Latex	1 coat, alkyd or PVA wall primer-sealer 1 or 2 coats, latex wall finish
		Acrylic deep-tone latex	1 coat, alkyd or PVA wall primer-sealer 1 or 2 coats, flat acrylic deep-tone latex
	Stipple gloss	Alkyd	1 or 2 coats, alkyd or PVA wall primer-sealer 1 coat, alkyd gloss stipple
	Stipple semigloss	Alkyd	1 or 2 coats, alkyd or PVA wall primer-sealer 1 coat, alkyd semigloss stipple
		Alkyd deep-tone	1 coat, alkyd or PVA wall primer-sealer 1 coat, alkyd semigloss stipple 1 or 2 coats, alkyd semigloss deep-tone enamel
		Acrylic latex	1 coat, alkyd or PVA wall primer-sealer 1 coat, alkyd semigloss stipple 1 or 2 coats, latex semigloss enamel

Surface	Finish	Topcoat	Finishing schedule
	Stipple eggshell	Alkyd	1 or 2 coats, alkyd or PVA wall primer-sealer 1 coat, alkyd eggshell stipple
	Stipple flat	Alkyd	1 coat, alkyd or PVA wall primer-sealer 1 coat, alkyd flat stipple
		Latex	1 coat, alkyd or PVA wall primer-sealer 1 coat, latex flat stipple
		Acrylic deep-tone latex	1 coat, alkyd or PVA wall primer-sealer 1 coat, latex flat stipple 1 or 2 coats, flat acrylic deep-tone latex
Concrete floors	Gloss	Phenolic alkyd	1 coat, floor enamel (thinned 10 percent) 1 or 2 coats, floor enamel
		Two-component epoxy-polyamide	1 coat, epoxy-polyamide (thinned 1 : 1) 1 or more coats, epoxy-polyamide
	Satin flat	Acrylic latex	1 or more coats, latex floor enamel
Metal	Gloss	Alkyd	1 coat, rust-inhibiting primer[f] 1 coat, enamel undercoat 1 coat, alkyd gloss enamel
	Semigloss	Alkyd	1 coat, rust-inhibiting primer[f] 1 coat, enamel undercoat 1 coat, alkyd semigloss enamel
		Alkyd deep-tone	1 coat, rust-inhibiting primer[f] 2 coats, alkyd semigloss deep-tone enamel
		Acrylic latex	1 coat, rust-inhibiting primer[f] 1 coat, enamel undercoat 1 coat, semigloss latex enamel
		Acrylic deep-tone latex	1 coat, rust-inhibiting primer[f] 2 coats, latex semigloss deep-tone enamel
	Eggshell	Alkyd	1 coat, rust-inhibiting primer[f] 1 coat, enamel undercoat 1 coat, alkyd eggshell enamel

TABLE 10.2 Recommended Specifications for Interior Architectural Surfaces (*Continued*)

Surface	Finish	Topcoat	Finishing schedule
	Flat	Alkyd	1 coat, rust-inhibiting primer[f] 1 or 2 coats, alkyd flat enamel
Metal flooring, steps, catwalks, etc.[f]	Gloss	Phenolic alkyd	1 coat, alkyd rust-inhibiting primer 2 coats, floor enamel
	Satin flat	Latex	1 coat, alkyd rust-inhibiting primer 2 coats, latex floor enamel
	Gloss	Phenolic alkyd	1 coat, latex rust-inhibiting primer 2 coats, floor enamel
	Satin flat	Latex	1 coat, latex rust-inhibiting primer 2 coats, latex floor enamel

[a] For the finish coating of skip-textured surfaces, the stipple finish indicated is to be replaced by using a suitable alkyd or latex wall coating of the sheen desired. Some heavy-bodied PVA coatings can act as both sealer and topcoat if the pigment-volume concentration is low.

[b] For first coat on new untextured Sheetrock, use PVA sealer.

[c] For specifications for clear coatings on wood paneling, trim, doors, cabinets, and floors, see Table 16.1 in Chapter 16.

[d] If a smooth surface is desired, substitute a latex block filler for the first coat of alkyd wall primer-sealer. If 2 coats of alkyd wall primer-sealer are specified, substitute 1 coat of latex block filler followed by 1 coat of enamel undercoat.

[e] For deep colors, eliminate the undercoat and apply 2 coats of enamel.

[f] Prime ferrous metal with an alkyd metal primer. Prime galvanized metal and aluminum with a PVA sealer, latex metal primer, or other primer specified for these metals.

[g] For enamels, follow a procedure similar to that with woodwork, but substitute a wall primer-sealer for an enamel undercoat.

the percentage of free binder decreases. The reduction of free binder results in lower gloss in the dried film. This type of paint has a low angular sheen, which minimizes imperfections in walls or ceilings.

Flat paints can be formulated with a PVC beyond the critical PVC to gain the advantage of dry hiding and reduced cost. These types of paint, which contain insufficient vehicle solids for the pigment portion, will stain more readily and clean up poorly. Moreover, the film is more readily broken down when subjected to scrubbing.

In addition to classifying paints as flat, semigloss, or gloss types, they can be defined by describing the vehicle. (For typical formulations see Table 10.3.) Among the vehicles used in interior finishes are oleoresinous, latex (water-based), alkyd, epoxy ester, catalyzed-epoxy, and polyurethane. Many products formulated with these vehicles can be used interchangeably for the protection of interior substrates. The similarities of paints made with these vehicles are discussed below.

10.7.1 Oleoresinous paints   The use of paints formulated with oleoresinous vehicles is declining. Films of enamels made with these vehicles (broadly defined as "varnishes") have a tendency to yellow on aging. In a solvent type of paint, on the other hand, the alkyds offer superior nonyellowing properties.

10.7.2 Latex emulsions   Latex emulsions are the vehicles of most water-thinned or latex paints. Latex-type primers are recommended for drywall construction because they do not raise the nap present on this type of substrate. Primers such as solvent types that raise the nap of drywall require sanding before application of the second coat of paint.

Flat latex wall paints offer the advantage of good color uniformity over surfaces of uneven porosity. Two coats can be applied in 1 day, the second after the water has left the initial coat (when the gloss has disappeared). If latex paint has been properly formulated, the dried film can later be touched up after being marred or scuffed. Ease of application and easy cleanup with water of tools and spatters on adjacent surfaces are attractive features. Furthermore, water-based paints are not normally associated with environmental problems related to control of hydrocarbons.

Initial coats of flat latex wall paints can be cleaned up or washed as well as flat alkyd wall paints can, but their washability gradually becomes poorer as additional coats are applied over the years. The buildup of the water-soluble wetting and dispersing agents used in the manufacture of this type of paint probably contributes to this failing. As new coats of latex are applied, these wetting agents, leached out of previous coats, can become more highly concentrated in succeeding coats. This situation can be corrected by the application of a barrier coat of flat alkyd paint to seal off the previous coats of latex paint. When the surface is repainted at a later time, flat latex paints can be successfully used. Multiple coats of semigloss and gloss latex enamels do not lose their excellent washability and scrubbability.

Latex paints are sensitive to the condition of the surface upon which they are applied. This is true of flat latex paints and to a much higher degree of semigloss and gloss latex enamels. Gloss, dirt, soap, and grease adversely effect the adhesion of these paints. This characteristic must be considered if they are to be applied on kitchen or bathroom surfaces. If the surface has not been adequately prepared, crawling, alligatoring, or sagging of latex paints will result regardless of gloss.

Primarily to conform to environmental regulations, water-based paints using new emulsions (vehicles) are being introduced by many manufacturers. The specification writer should not arbitrarily begin to use these products without comparative field testing.

Many semigloss and gloss latex enamels are formulated with glycols

TABLE 10.3   Typical Formulations for Interior Architectural Coatings*

Alkyd wall primer

	Percent
Pigment, 59.5 percent	
Titanium dioxide, Type II	21.57
Calcium carbonate	78.43
	100.00
Vehicle, 40.5 percent	
Soybean alkyd resin	18.37
Heat-treated linseed oil	14.19
Ester gum	5.42
VM&P naphtha	6.06
Mineral spirits and driers	55.96
	100.00

Vinyl acetate-acrylic latex wall paint

	Percent
Pigment, 35.66 percent	
Titanium dioxide, Type III	70.59
Calcium carbonate	8.24
Silicates	21.17
	100.00
Vehicle, 64.34 percent	
Vinyl acetate acrylic resin	23.26
Methyl cellulose	.65
Glycols	4.85
Water	71.24
	100.00

Alkyd flat wall paint

	Percent
Pigment, 50.90 percent	
Titanium dioxide, Type II	39.47
Calcium carbonate	15.79
Silicates	39.47
Silica	5.27
	100.00
Vehicle, 49.10 percent	
Soybean alkyd resin	22.21
Heat-treated linseed oil	1.29
Soybean oil	2.40
Odorless mineral spirits and driers	74.10
	100.00

Alkyd satin enamel

	Percent
Pigment, 46.59 percent	
Titanium dioxide, Type II	57.14
China clay	19.05
Calcium carbonate	11.90
Silicates	11.91
	100.00

Vehicle, 53.41 percent
 Soybean alkyd resin   38.12
 Processed soybean oil   3.41
 Mineral spirits and driers   58.47
               100.00

Alkyd enamel undercoat

               Percent

Pigment, 61.35 percent
 Titanium dioxide, Type II   25.93
 Calcium carbonate   74.07
               100.00

Vehicle, 38.65 percent
 Soybean alkyd resin   37.30
 Mineral spirits and driers   62.70
               100.00

Alkyd gloss enamel

               Percent

Pigment, 31.88 percent
 Titanium dioxide, Type II   98.49
 Zinc molybdate   1.51
               100.00

Vehicle, 68.12 percent
 Soybean alkyd resin   50.87
 Mineral spirits and driers   49.13
               100.00

---

* New pollution regulations govern the amount of volatile organics that can be used in certain architectural coatings. In states where these regulations apply, the specification writer should request a formulation analysis similar to that given in Table 9.3 of Chapter 9.

and water as a part of their solvent system. These glycols are water soluble. In the absence of adequate ventilation or in the presence of high humidity, they release the water too slowly, causing the enamels to run or sag. In extreme cases, this phenomenon can take place in a flat latex film during its drying stage, although the amount of glycol present is appreciably lower than in an enamel. Sufficient ventilation, using fans, is required to prevent sagging if humidity is high or if evaporation of the water from the latex film raises the humidity to a point at which continued loss of water from the film is prevented or retarded. New pollution control laws set limits on the amount of glycols that can be included in a formulation.

10.7.3 Alkyd paints Alkyd flat wall paints have good application properties and, if properly formulated, dry to a uniform color and gloss when applied on surfaces of uneven porosity. In some instances, they do not offer the touch-up characteristics of latex paints or provide cleanup with water. However, multiple coats of alkyd flat paints do not lose their initial ease of cleanup.

Alkyd semigloss and gloss enamels are not as nonyellowing as acrylic latex semigloss emulsion enamels. Although alkyd enamels, like alkyd flats, are not highly sensitive to the surface upon which they are applied, adequate surface preparation must be made. Humidity does not affect their application to the degree experienced with latex semigloss enamels.

10.7.4 Epoxy ester paints   Epoxy ester paints generally have approximately the same properties as alkyd paints. Like alkyd paints, they cure or dry by oxidation. They should not be confused with catalyzed-epoxy coatings, which have appreciably greater chemical resistance.

10.7.5 Catalyzed-epoxy coatings   These coatings cure by the reaction that takes place between the epoxy resin and the catalyst. They are mixed prior to application to initiate the reaction necessary for adequate cure of the film. Premixing improves the chemical resistance of the cured film and its adhesion to the substrate.

Since temperature has an effect on the speed of the reaction, a minimum temperature of 60°F (16°C) for both air and substrate is recommended. Below this temperature, the cure of the coating is retarded. Catalyzed-epoxy coatings provide chemical, water, and abrasion resistance superior to that of the conventional oxidizing type of enamels and latex paints, but they have a greater tendency to yellow. The strong solvents needed to dissolve the resins in catalyzed-epoxy systems make it necessary to provide ventilation during their application. (Water-soluble catalyzed-epoxy coatings equaling the performance of the present solvent type will eliminate the odor problem.)

Solvents for catalyzed-epoxy coatings can soften or lift previous coats of paints, necessitating tests for this type of failure. Should this problem occur, the specifier may have to turn to an enamel formulated with solvents that will not soften the previous coat of paint. Softening of previous coats can cause loss of adhesion to the substrate.

NOTE: Patch tests to determine the solvent resistance of previous coatings do not always indicate that an adhesion problem may exist because the hard-drying catalyzed-epoxy coating may cause the initial coat of paint to lose its adhesion to the substrate at a later date. Although a catalyzed-epoxy system offers the advantages of excellent chemical, water, and abrasion resistance, its application over other types of paint can lead to adhesion problems. Therefore, the optimum performance of this type of coating is realized when it is applied on an unpainted surface.

10.7.6 Polyurethane coatings   Polyurethane coatings have found wide application as surface coatings for wood floors and as penetrating sealers to lay dust on concrete floors. Applied as a surface coating on wood, a polyurethane coating offers good mar and abrasion resistance, but it has some limitations. It cannot be successfully applied over shellac, some types of finishes containing natural resins, or stearate-bearing coatings. Application over these types of finishes can result in the paper peeling of the polyurethane from the previous finish. Because of the difficulty of determining the composition of the old coating, polyurethane finishes can present problems when applied over old finishes. Complete removal of previous coats of paint is recommended for best results.

10.8 Computing Coverage and Performance   The performance of a paint system depends upon the amount of paint applied to the surface that it is designed to protect. The film thickness of a primer affects its ability to seal the substrate and eliminate areas of uneven porosity. An excessive spreading rate per gallon of topcoat results in a thin film of paint that normally does not hide the surface or the previous coat of paint satisfactorily. A thin coat lessens the scrubbability of wall paints; a thin film shows the effect of frequent washings earlier than thicker films of a wall paint do. A heavier coat of floor paint appears to have greater resistance to abrasion because the thicker film does not wear through as rapidly to expose the substrate. Adequate film thickness of rust-inhibiting paint systems is important to realize maximum protection to the metal substrate.

It is just as important to assure that a paint system arrives at the desired dry mil thickness without resorting to massive applications of a single coat of paint. Several thin coats, with adequate drying between them, is the preferred practice. This method eliminates the blistering or cheesy films that accompany too heavy an application.

Specifications for a successful paint system therefore must refer to the film thickness of each coat of paint. The amount of coating to be applied can be expressed in a spreading rate, by specifying the square feet of surface to be covered with a gallon of paint, or in terms of wet or dry film thickness.

Because the dry film, which consists of the solids in the vehicle and the pigment, is that part of the liquid paint remaining on the surface after the film has dried, its thickness should be of primary importance. The solvents of the thinners that leave the paint film during the drying stage reduce the viscosity of the solids to make it possible to apply them and thus are not included in the calculations to determine the dry film thickness of a paint film. Although solvents are not a part of a dried paint film, however, their composition is important, for they affect the application properties of a paint—the flow and leveling which contribute to the final appearance of the coating.

After the dry film thickness of a coating required to protect the substrate and provide the physical characteristics expected of the paint film has been determined, the amount of paint needed in square feet per gallon can be calculated from the following, letting x = the wet film thickness:

$$\text{Mils wet} \times \text{percent volume of solids}[1] = \text{mils dry}$$

Example: Final dry film thickness must be 2.2 mils (0.56 mm).

$$\text{Volume of solids in paint} = 55 \text{ percent}$$
$$x \times 55 \text{ percent} = 2.2$$
$$x \times 0.55 = 2.2$$
$$0.55x = 2.2$$
$$x = \frac{2.2}{5.5}$$
$$x = 4 \text{ mils (0.102 mm) wet}$$

The wet film thickness can then be converted to the square-feet-per-gallon application rate, using the following formula. Let a = spreading rate per gallon (3.785 l), and convert the 4 mils to 0.004 in:

$$\frac{\text{Cubic inches per gallon of paint}}{\text{Spreading rate} \times \text{square inches per foot}} = \text{wet film thickness}$$
$$\frac{231}{a \times 144} = 0.004$$
$$a \times 144 \times 0.004 = 231$$
$$0.576a = 231$$
$$a = \frac{231}{0.576}$$
$$a = 401 \text{ ft}^2/\text{gal (9.84 m}^2/\text{l)}$$

This formula can also be used to determine the dry film thickness of a paint film if the spreading rate per gallon and the volume of solids are known. As an example, the wet film thickness of a paint film applied at a spreading rate of 401 ft²/gal can be determined by using the formula for calculating wet film thickness. Let x = the wet film thickness:

$$x = \frac{231}{401 \times 144}$$
$$x = \frac{231}{57,744}$$
$$x = 0.004 \text{ in (4 mils)}$$

The dry film thickness can then be determined by inserting the known value of 4 mils into the formula for calculating dry film thickness. Let z = the dry film thickness and the volume of solids in the paint = 55 percent:

---

[1] Percent volume of solids in paint supplied by the manufacturer.

$$4 \times 55 \text{ percent} = z$$
$$4 \times 0.55 = z$$
$$z = 2.2 \text{ mils dry}$$

Table 10.4 relates wet film thickness to spreading rate per gallon, eliminating the need to calculate these data. The formula for calculating either of these values is given to provide an understanding of the derivation of this table and also to make it possible to calculate the data if desired.

The theoretical dry film thickness of a coating can be determined during application by measuring the wet film thickness with gauges available to the industry. The formula

Mils wet $\times$ percent volume of solids $=$ mils dry

can be used to calculate the dry film thickness.

TABLE 10.4   Spreading Rate and Wet-Film Thickness

Spreading rate (square feet per gallon)	Film thickness (mils)	Spreading rate (square feet per gallon)	Film thickness (mils)
1604	1.0	321	5.0
1458	1.1	308	5.2
1337	1.2	297	5.4
1234	1.3	286	5.6
1146	1.4	277	5.8
1069	1.5	267	6.0
1000	1.6	259	6.2
945	1.7	251	6.4
891	1.8	243	6.6
802	2.0	236	6.8
729	2.2	229	7.0
700	2.3	223	7.2
668	2.4	217	7.4
617	2.6	211	7.6
573	2.8	206	7.8
535	3.0	201	8.0
501	3.2	196	8.2
472	3.4	191	8.4
446	3.6	182	8.8
442	3.8	178	9.0
401	4.0	174	9.2
382	4.2	171	9.4
365	4.4	167	9.6
349	4.6	164	9.8
344	4.8	160	10.0

BIBLIOGRAPHY

American Standard Specifications for the Application and Finishing of Gypsum Wallboard, GA-202-72, Gypsum Association, Evanston, Ill.

"Can an Interior Paint Conserve Energy?" *Consumer Reports,* February 1980, p. 45 ff.

Flick, Ernest W: *Water Based Paint Formulations,* Noyes Data Corp., Park Ridge, N.J., 1975.

"Priming Standards," *Finish Carpentry, Bulletin No. 6,* Wood Moulding and Millwork Producers, Portland, Oreg., 1973, p. 24.

Recommendations for Finishing and Maintaining Maple, Beech and Birch Flooring, Maple Flooring Manufacturers Association, Glenview, Ill., 1974.

The Selection of Paint, Bulletin 796, National Paint and Coatings Association, Washington, 1976.

Specifications for Construction, Care and Maintenance for Wood Flooring, Wood and Synthetic Flooring Institute, Glenview, Ill., 1975.

Chapter **11**

Exterior Masonry Surfaces

HARRY M. HERR
Paint Engineering

11.1 Introduction Coatings are applied to exterior masonry to protect and decorate both new and old construction. The architect and the specification writer have the responsibility of furnishing complete proprietary specifications. Proprietary specifications are specifications bid by the contractor which will not be subject to change-order requests for either products or procedures at any time between the start and the completion of the project. Architectural specifications should be written around systems rather than around products. This is particularly true of exterior masonry surfaces because these surfaces are among the most widely used. Improper specifications can easily lead to problems. Problems often arise when products for the primer, intermediate coat, and finish coat are obtained from different manufacturers.

In addition to furnishing a coating system, the architect should furnish systems for masonry, roofing, waterproofing, and sealants. While each of the systems employed may be complete in itself, it is essential that it be chemically compatible with all the other systems used on the project, especially if the materials come in contact with each other.

It is unrealistic to expect coatings to adhere if they are applied over certain release agents, curing compounds, and other soils or contaminants which may be present on the surface or within the substrate. In the same vein, if water from any source is permitted to migrate behind the coatings, hydrostatic pressure will eventually delaminate the coating.

The term "masonry" originally meant a building component constructed by a mason from individual units such as brick, stone, tile, or concrete blocks. Today, cast concrete, asbestos and cement board, stucco, exterior plaster, and plasterboard are also included under the general term. In discussing painting of the masonry surface, "coating" and not "finish" should be used as the term to describe the treatment.

The three most important properties of a masonry wall that affect coatings are (1) the chemical property, (2) the surface property of smoothness, and (3) the porosity of the substrate and the moisture content. When a surface is heavily alkaline or acidic or is too smooth or too rough, too wet or too dry, resulting problems affect the coating material and the length of service. The architect is the major party responsible for the successful conclusion of a building project. For this reason a job captain working out of the architect's office must see that specifications are followed exactly.

Before specifying a painting system for exterior masonry it is necessary to understand the types of surfaces (Table 11.1).

11.2 Stucco   There are several types of stucco (Table 11.2), and many coating systems are suitable for painting. Regardless of the paint used, the coatings specification should include a notation on stucco curing time before paint application. A prime problem is the alkalinity of the surface

TABLE 11.1 Exterior Masonry

Type	Definition	Comments
Concrete	A mixture of cement, sand, aggregate, and water made into a semifluid and placed in a mold. It will dry or set into a hard, stonelike mass	This product may be cast as pour-in-place, precast, or prestressed concrete. Paint problems can occur when an improper mixture is used

Type	Definition	Comments
Masonry or concrete block	A mixture of cement, sand, aggregate, and water of lean and dry consistency molded into block form. This mixture may have a soap solution added as a mold-release agent	Block is used in all kinds of construction of rigid, fireproof, and permanent buildings. Walls of concrete block can have an advantage over poured concrete, in that the blocks are thoroughly seasoned before they are used in a building
Mortar	A bonding mixture made up of cement, lime, and sand. Water is used to bind brick, block, or stone into a building	A formula and physical characteristics to include flexural strength and shrinkage should be included in the specifications. Mortar has many possible ingredients. The kind of mortar, the composition, the strength, the use, and the cost should fit the purpose
Stucco	A plaster consisting of a mixture of cement, lime, sand, and water proportioned to make a workable consistency so that it can be applied over old or new buildings as a finish	This product is trowel- or spray-applied in one to three coats over a wire-reinforced substrate. Stucco does not make a building fireproof, but it adds materially to fire protection. It is a better insulator than wood
Brick	A fired-clay block	Specifications should include compressive-strength requirements
Decking or flooring	Concrete, brick, or tile used as the ground surface or horizontal divider in a building	Specifications are written to cover the use or purpose intended
Clay tile	A clay and water mixture which may be air-dried or baked	Specifications are written to cover the use or purpose intended
Adobe	An air-dried clay block consisting of clay, water, and reinforcing or binding fibers	Use is limited to areas in the southwestern part of the United States
Aggregate	River rock cast in concrete and exposed for decorative purposes	This is a process or system which can be used on walking surfaces, precast wall panels, or pour-in-place walls
Asbestos siding (glazed)	An extruded panel consisting of cement, asbestos, and resins	As this product is often hard-glazed, repainting requires special specifications. Generally, it is not wise to paint glazed-asbestos siding
Asbestos siding (unglazed)	An extruded panel consisting of cement, asbestos, and resins	A number of painting specifications are available for coating this product. New findings on asbestos's toxicity may make asbestos products less viable as construction materials

TABLE 11.2 Types of Stucco

Name	Definition	Comment
Integrated color	Iron oxides (browns, red, yellows) mixed in the stucco prior to application	Pastel tones are generally employed. Deep tones are seldom specified
Full stucco	One scratch coat: first coat on wire or paper One brown coat: second coat One finish coat: skip-troweled on dash coat	It should be ¾ to 1 in (19.05 to 25.4 mm) thick
Skip coat	Skip-troweled by hand	Various appearances such as swirled or Spanish-style rough are produced
Dash coat	Blown on over brown coat	It is applied by machine to give a textured appearance
Emulated stucco	Texture paint applied by airless-spray system	This is generally a good coating. The majority of paint manufacturers will supply a warranty or a guarantee with each paint job. Specify only reputable products

if the coating is applied too soon after the stucco has been placed. If surface wash water shows a pH of 8.0 or higher, the surface should be washed down with a neutralizing agent such as zinc or sodium sulfate (1 lb/gal, or 119.8 kg/m³, of water). Any hairline or structural cracks should be filled or patched prior to a coating application.

Some finish stucco coats contain integral color which has been tallow-carried, or tallow or other animal oils have been applied to the surface to prevent water penetration. These surfaces are very hard to repaint, and the oils must be removed or stabilized before painting. To remove this tallow or oil scrub with a hot-water solution of trisodium phosphate followed by thorough flushing and rinsing with clean water. The cleaning agent must be thoroughly removed, or the final condition of the surface will be as bad as the oily condition. Peeling will result. If the oil or grease is very heavy, the only solution to the problem may be sandblasting.

When a stucco wall is to be painted, it will usually require a two-coat system. If a blind stucco coat is used, it is a nonstructural finish coat for decorative purposes only. A full stucco finish should be between ¾ and 1 in (between 19.05 and 25.4 mm) in thickness. It should be applied first as a "scratch" coat from ⅜ to ½ in (from 9.525 to 12.7 mm) thick and allowed to set for 24 h. This first coat is heavily cross-hatched to provide a strong mechanical key. The second "brown" coat should also be from ⅜ to ½ in thick and should be brought to a true and even surface. The specification should assure that proper curing moisture is maintained for at least 3 days, the surface being wet as needed.

In today's stucco application by machine, the base coat is usually pumped and sprayed on as a dry mixture that might be more closely related to gunite. If the mixture is too dry, it will be loosely packed and porous. If it is too wet, it can have plastic shrinking cracks. The quality of a good stucco surface will depend largely on the skill of the plastic worker, and it therefore is wise to have inspectors on the site during application.

Specifications for stucco should include a basic formula.[1] The specification writer should give special attention to the total thickness of stucco required and the method of application. If the surface requires a tie coat, intermediate coat, and finish coat by trowel application, this requirement should be specified. Even greater care should be given to spray-applied single or two-coat systems. Paints create stress as the film dries, and if the stucco itself adheres poorly to the surface, the result is cracking, chipping, and flaking, not of the paint to the stucco but of the stucco to itself on the original surface. This is *not* a paint problem, and it is often possible to distinguish between paint peeling and stucco peeling by examining the back of the paint flakes. If the stucco is breaking loose, it usually adheres to the unexposed side of the paint chip.

11.2.1 Inspecting stucco prior to painting   In all cases, plaster or stucco surfaces should be properly cured, dry, clean, and free of all contaminants before they are painted. Remove dirt, dust, and loose plaster. Neutralize surfaces that have high alkalinity or efflorescence. Rake out and fill cracks, scratches, holes, and other imperfections with a suitable patching compound to match the textural profile of the surface. Test the surface (Table 11.3) to assure that it is ready for painting. Inspect the surface prior to painting to ensure proper surface preparation.

Many specifications establish general requirements for preparing each surface but avoid detailed methods of preparation. For example, the statement "Clean free of dust, dirt, oil, grease, and wax" lists certain general requirements but does not describe in detail the methods to be used to clean the surface. Using this approach enables the painting contractor to prepare the surface in the most practical and efficient manner consistent with good painting practices. General specifications, however, open the door to faulty workmanship and liability arguments if problems occur because directions are not specific.

---

[1] Stucco information is available from the Stucco Manufacturers Association, 14006 Ventura Boulevard, Sherman Oaks, Calif. 91423.

TABLE 11.3 Testing Exterior Masonry Prior to Painting

Test name	When to use or specify test	Test description	Comments
Test to assure proper curing	When surface looks wet or when in doubt	Moisture meter; any instrument designed to test water content of masonry substrate	If surface is too wet, sealer or paint will penetrate for good adhesion
Test for alkalinity	When masonry is fresh	Use pH-indicating papers*	Wet surface and paper before reading
Water-column test	To determine if water is penetrating the substrate and by how much	Use Dickey Meter Masonry Research, 2550 Beverly Boulevard, Los Angeles California 90057	Take a mean average of several readings
Moisture pickup when surface is wet	Porous masonry or concrete block	Preweigh sample, immerse in water for 24 h, and measure water pickup in weight	Use this test to determine the amount of waterproofing needed to seal the substrate
Test for tallow or oil on the surface	Stucco and other masonry surfaces to be painted	Remove some surface coating and test for vegetable oil. If no oil is present, suspect silicone	General contractor is responsible for wall condition
Test for surface smoothness	Tiltup or hard-troweled masonry	Paint test area and pull off tape to determine adhesion	Surface may require acid-etching prior to coating application

* pH papers: pHydron Papers, Micro Essential Laboratory, Inc., Brooklyn, N.Y. 11210.

11.2.2 Painting systems for stucco   Oil- and alkyd-based paints are available for masonry, but with the imminence of more stringent pollution regulations that will limit the amount of solvent which can be used in oil and alkyd formulations, emulsion primers and finish coats will become even more widely specified. While many of these emulsion-based paints are suitable for stucco, the stucco surface must be thoroughly cured.

A minimum of 30 days is recommended to reduce both water and alkali content to a safe level. Painting too soon will trap moisture. Since alkali is activated by moisture, there is an immediate danger of "hot spots" that will burn the binder in the paint coating. While latex-emulsion primers and finish coats are not as sensitive to alkali as is solvent-thinned paint, painting too soon traps moisture and interferes with the curing process through which stucco derives its strength. New stucco should be wiped clean of any fine powder that is left on the surface during troweling as well as of any dirt and dust that have been deposited. While forced heat and ventilation may be provided to aid the curing of interior surfaces, weather conditions dictate the additional length of time that may be required to achieve a paintable outdoor surface. The surface should be dry and clean. Removal of dirt, powder, and loose mortar can be achieved by a combination of wire-brushing and scraping and, in the case of building block, by rubbing the surface with a brick or a piece of block. Cracks or breaks should be repaired with a water-mix grout or with an epoxy patch. The edges around the area being patched with water-mix grout should be dampened so that water will not be absorbed from the mixture and produce a powdery, crumbling patch that will not support paint.

Once masonry surfaces have been prepared for painting, the performance of all succeeding coats will depend heavily on the ability of the prime coat to adhere and to retard the damage that can occur when alkali is activated. The specification writer should assure compatibility of each topcoat with the previous coat as well as with any sealants or patching compounds. For most coatings this is best accomplished by utilizing paint from the manufacturer of the other coating materials.

In light of environmental regulations that limit the amount of organic compounds that can be used in paints (such as mineral spirits in solvent-based paints, ethylene glycol in water-based paints, and ketones and esters in lacquers and epoxies), "hot" surfaces (those that are extremely alkaline) pose a special specification problem in some parts of the United States. It behooves the architect to emphasize the stucco specification. The stucco surface is just as important as, if not more important than, the coatings specification. When formulations are limited by environmental restrictions, only limited guarantees can be expected if efflorescence or severe alkaline conditions exist.

EFFLORESCENCE

Surfaces composed of mortar, brick, building block, and concrete frequently develop efflorescence. Soluble salts such as calcium and magnesium sulfates, when dissolved by water and carried to the surface, are deposited there as the water evaporates, leaving a crystallized wet or dry and powdery substance called efflorescence. Remove efflorescence from the face of bricks, building block, and mortar lines by wire-brushing. For large areas of brick surfaces, removal by washing with phosphoric acid and thorough water flushing are recommended. Persistent efflorescence is often associated with a water problem. Water leaches out salts from the masonry, constantly depositing them on the surface. This condition cannot be cured by using paint.

A stucco painting system includes a sealer, a primer, and a topcoat for hot surfaces or a primer and a topcoat for normal surfaces. The sealer is formulated to seal the surface of masonry against water penetration and hydrostatic pressure acting against the rear of topcoating films, to stabilize a possible chalking condition, and to tie the entire coating system to the masonry substrate. It must penetrate the masonry substrate rather than bridge the capillary system.

For highly alkaline surfaces, for which the topcoat will be alkyd- or oil-based, the most reliable formulas for sealers are based on chlorinated-rubber binders that have good penetration qualities and high flexural strength. As environmental regulations push these binders out of the market, an alternative for hot surfaces is traffic paint based on chlorinated rubber if the pigment-volume concentration (PVC) is in the same range as that of the sealer.

Some sealer formulas are based on a long-oil alkyd which has qualities that include penetration and high flexural strength. The pigment volume for such products is low and is included in the formula to serve as an application guide rather than to act as a filler or as a hiding coat. These products are thinned to spray or brushing consistency with mineral spirits or paint solvents. They should never be applied over a wet or damp surface because they may seal residual water within the substrate to a point at which escaping water may delaminate interior wall paints or coatings. This type of product must never be applied over loose dirt or delaminating paint.

In many cases paint suppliers indicate that sealers are not guaranteed to prevent alkali burn if a severe alkali condition exists. The best approach to overall coating specifications employing an alkyd- or oil–based sealer is to inform the paint manufacturer of the conditions expected and then follow the manufacturer's recommendations. Some oil or alkyd sealers saponify on exposure to alkalies when moisture is present in the concrete, creating adhesion problems. It is apparent that in the future it will become more important to specify the masonry formula closely in order to avoid alkaline conditions from the outset.

Most stucco surfaces do not present a painting problem, and standard painting systems (Table 11.4) can be employed.

11.2.3 Painting previously painted stucco   If the surface is tight and nonchalking, grease, oil, chalk, and other foreign matter should be removed. Regardless of any other surface preparation, cracks should be filled. For a sound but lightly chalked surface, after removing grease and as much loose chalky material as possible by scraping with a wire brush, wash or hose down the surface. Once the cracks have been filled, the surface is ready for painting. Heavy chalking will interfere with adhesion. Remove heavy chalk by washing with a mild detergent solution

TABLE 11.4 Masonry Painting Systems

Surface	Finish	Type	Product	Film (mils)	Comments
Stucco and emulated stucco	Flat	Acrylic	Prime: 1 coat concrete and masonry primer-sealer	0.5	Acrylics have good color and gloss retention in comparison with oil-based systems. Dark or bright clear colors may require additional coats
			Finish: 1 or 2 coats acrylic latex	1.5	
	Semigloss	Acrylic	Acrylic semigloss house paint	1.5	NOTE: Modified acrylics and copolymer polyvinyl acetate latex paints can give adequate protection, depending on exposure conditions, but these are generally considered inferior to acrylics
Poured concrete; tilt-up concrete cement	Flat	Acrylic	Prime: 1 coat concrete and masonry primer sealer	0.5	See note above
			Finish: 1 or 2 coats acrylic latex	1.5	Concrete form-release agents prevent a good bond between concrete and paint. The most frequently used agent is oil, which must be removed by scrubbing with a strong solution of hot water and detergent followed by thorough flushing with clean water. If other types of form release have been used, removal by other methods may be required
Concrete and cinder block	Flat	Acrylic	Prime: 1 coat PVA block filler or 1 coat sand-modified primer	4.0	For total bridging of holes and cracks, use PVA block filler containing aggregate or fibrous filler
			Finish: 1 or 2 coats acrylic latex	1.5	

TABLE 11.4 Masonry Painting Systems (*Continued*)

Surface	Finish	Type	Product	Film (mils)	Comments
Concrete floors	Low sheen	Acrylic epoxy	Finish: 2 coats 100 percent acrylic latex floor paint or 1 or 2 coats of acrylic-epoxy floor paint	1.5	No primer is required. This breathing-type film is not recommended for use on floors where vehicles are parked
	Gloss	Phenolic alkyd	Finish: 2 coats floor paint	1.5	
	High gloss	Two-component epoxy polyamide	Prime: 1 coat mixture (epoxy and polyamide) thinned 100 percent	0.5	
			Finish: 1 coat mixture at full strength	1.5	
Asbestos cement siding and shingles	Flat	Acrylic	2 coats acrylic latex	1.5	Only weathered surfaces without glaze are paintable. The presence of glaze or silicone treatment on asbestos cement can be detected by pouring water over the surface. Absorption will be noted if the surface is paintable
Concrete clear	Clear	Silicone	Finish: first and second coats of 5 percent silicone water repellent	. . .	For use on above-grade small-pore or dense masonry. It will not waterproof concrete cinder block and can interfere with repainting

Surface	Finish	Paint	Coverage	System	Remarks
Concrete waterproof	Clear	Stearated oil	. . .	Finish: 1 or 2 coats water retardant	Clear, penetrating water-repellent compound for all masonry surfaces
Concrete asphalt	Traffic paint	Chlorinated-rubber acrylic or acrylic epoxy	7.5 / 1.5	Finish: 1 or 2 coats traffic paint or 1 or 2 coats acrylic or acrylic-epoxy floor paint	No primer is required / Resists bleed-through of asphalt sealers
Previously painted masonry	Flat	Acrylic	1.5 / 1.5	Prime: 1 coat surface conditioner, with or without aggregate / Finish: 1 or 2 coats acrylic latex	Use surface conditioners on light, chalky surfaces only / See note under "Stucco and Emulated Stucco"
Textured coatings	Flat	Acrylic texture coating	0.5 / 20.0 / 10.0	Prime: 1 coat concrete and masonry primer-sealer / Finish: 1 coat acrylic rough or coarse-textured coating or / 1 coat acrylic sand-textured coating	Use primer-sealer to form a moisture barrier resistant to alkalies / NOTE: Fill concrete block with 4-mil film of PVA block filler

followed by extensive rinsing. Painting should follow within 24 h to assure that new chalk does not begin to develop upon weathering.

Masonry surfaces such as concrete, stucco, block, brick, asbestos siding, and shingle generally use the same painting system. However, repainting cement-based water paints poses problems if the surface has dried out or shows excessive chalking. It is wise first to remove as much of the heavy chalk accumulation as possible by wire-brushing. Repair cracks, and for the first coat of paint use a water-based acrylic latex thinned with 1 pt (0.473 l) of 55 percent nonvolatile acrylic emulsion per gallon (3.785 l) of paint. This helps promote adhesion by assuring that the dry surface does not soak up the binder. If polyvinyl acctate (PVA) paints are used, thin with 1 pt of PVA per gallon.

CEMENT-BASED PAINTS

If improperly applied or cured, cement paints dust excessively and can develop shrinkage cracks. They require thorough moist curing, and they cannot be applied at or below freezing temperatures. These paints have low resistance to abrasion and should not be used on floors. They are not suitable for concrete surfaces which must be cleaned frequently.

Cement paints are available in pastel or intermediate shades, but do not specify deep, solid colors. Cement-based paints require protection from direct contact with water containing sulfates. Commercially mixed cement-based paints are likely to provide better color uniformity than job-mixed paints. They contain additives which improve their water resistance and workability. These paints generally have good exposure characteristics, including durability. They mask surface imperfections and can be applied to damp concrete surfaces and to surfaces which will be damp frequently. They also have good alkaline resistance and good freeze-thaw characteristics.

A typical cement paint consists of 1 part hydrated lime and 5 parts white portland cement, water being added until the mixture has the consistency of condensed milk. Iron oxide coloring is often added to obtain light tinting, and 2 parts fine sand are added to fill the pores of rough cinder block completely.

To apply, dampen the surface and then spray or brush with a short, stiff-bristled brush. The paint should dry slowly for proper curing. After it becomes firm, it must remain damp with sprayed water for about 48 h. Surfaces painted with this paint require a sealer before they can be repainted with other types of paint.

When repairing exterior stucco, a water mix of latex-type patching plaster should be used under latex paint, since the oils present in patching compounds will bleed into and discolor the latex film. When using a water-mix patch, the surrounding edges of cracks and damaged spots should be thoroughly dampened so that moisture is not absorbed from the patch mixture, leaving a crumbly and unsound surface to support paint.

11.3 Concrete   Concrete placed by pouring into forms, prestressed concrete, and precast concrete may or may not have form oils, curing compounds, and other chemical products which may have to be removed prior to coating, particularly if the products present are not chemically compatible with the coating system (Table 11.5). The porosity of the surface must be considered, especially if a stain is to be applied. Test absorption by spraying with water. A large degree of surface porosity indicates that the surface will absorb the paint binder, opening the way to chalking or adhesion problems.

Concrete surfaces are generally of three types: *poured* against forms, such as concrete walls; *hand-troweled,* such as floors or sidewalks; and *gunited,* produced by spraying.

Surface preparation for poured concrete is difficult because of rough spots and pitting. Rough spots must be smoothed off by scraping; then holes and pits must be filled by sacking. Use a knife or a sharp, pointed object in pinholes to ensure that large voids do not exist behind the pinholes. Finally, acid-etch the entire surface.

SURFACE-PREPARATION CHECKLIST FOR POURED CONCRETE

- Scrape off sharp projections with a trowel, scraper, or putty knife.
- Rinse the surface and allow excess water to drain off.
- Prepare a mortar, using 1 part cement and 2 parts sand, with enough water to form a creamy paste.
- Rub mortar into pores, pits, and holes with a coarse cloth, cement sack, or wood float.
- Wipe off excess mortar so that a smooth, dense surface is obtained.
- After sacking, keep the surface damp for several days to cure the mortar.
- Etch the entire surface with a mixture of 1 part hydrochloric (muriatic) acid and 2 parts water. Apply acid with a brush or a garden spray. Wet the surface thoroughly. The solution will bubble vigorously for $\frac{1}{4}$ to $\frac{1}{2}$ min as the concrete is etched. When the bubbling ceases, the acid has been neutralized. CAUTION: Workers should wear rubber gloves when applying acid. If acid is spilled on the skin, it should be washed off immediately with water.

TABLE 11.5   Release Agents, Bond Breakers, and Curing Compounds

Item	Definition	Types	Where and why each type is used
Release agent	Coating applied to wood or metal forms for pour-in-place concrete	Waxes	When painting is not specified
		Silicones	When painting is not specified
		Urethane coatings	When a smooth surface is desired
		Stearated oil or resin	When painting is specified
Bond breaker	Material used to release concrete poured against concrete slabs in tilt-up construction	Soaps and detergents	To be removed before painting
		Silicones	When painting is not specified
		Stearated oil or resin	When painting is specified
Curing compound	Material used to hold water of hydration within masonry to complete curing	Pigmented waxes	For concrete roadways and walkways; not to be painted
		Short-life– resin cures	Resin destructs within several weeks, and surface can then be painted

- Rinse the surface with water. Brush during washing to remove salts.
- A properly etched surface is slightly granular and free from glaze. If any glazed areas remain, repeat the etching procedure.
- Allow the surface to dry completely before paint application.

Troweled concrete normally does not have air pockets or pinholes. In this circumstance, etch the surface thoroughly to clean and remove any glaze. Treat it with a mixture of 1 part hydrochloric (muriatic) acid and 2 parts water, washed with clean water and dried, as described for poured surfaces. A properly etched surface will have a perceptible roughness when rubbed with the fingers. Repeat the etching process if the surface is still smooth.

Observe caution if a concrete floor has been treated with a hardener to produce a more dense and smooth surface. Paint application over hardened concrete is not reliable and should be avoided unless test patches prove that there is satisfactory adhesion. Sometimes it is necessary to etch a hardened floor with concentrated acid to remove the glaze. This is a hazardous operation and is not normally recommended. (See box.)

A gunite surface is prepared by spraying mortar from a nozzle. When the proper thickness has been attained, the surface is usually smoothed off with a wooden float or trowel. When the finish is dense, as on hand-troweled concrete, it can be prepared by acid-etching. If it has not been troweled, it becomes necessary to plaster the surface with a cement mortar similar to that used in sacking poured concrete.

Surface preparation is largely dependent on environmental restrictions. If sandblasting is not permitted, chemical etching is an alternative, provided wash-off of the treatment does not contaminate local water supplies. If a water-blasting system is used to prepare the surface, allow time for complete drying before application of the coating system. Chemical cleaners should be based on mild acids such as phosphoric or acetic acid rather than hydrochloric acid. Hydrochloric acid should never be used except in etching areas such as greasy garage floors.

Some synthetic tires contain plasticizers that will migrate into, stain, and/or remove both solvent-thinned and latex floor paints. Allowing as much as 5 days' additional drying time before permitting traffic on the paint is helpful, but it is not possible to predict results. Even epoxy coatings may be affected by plasticizers. Do not apply paint in tire lanes if the possibility of paint failure is objectionable. When preparing a concrete surface for painting, the specification writer should assure that all materials of the painting system are compatible. These include hardeners, release agents, grout aids, bond releases, patching compounds, and curing compounds.

ETCHING CONCRETE FLOORS

Concrete must first be clean and free of hardening agents. To remove dirt, grease, and oil scrub the floor with hot water, using a strong detergent solution. Rinse by flushing thoroughly with clean water. For stubborn areas repeat the procedure until they are clean. Leave the surface unpainted for several days after scrubbing, since deeply penetrated oil and grease may continue to rise to the surface and can then be scrubbed away.

Once the floor is clean, mop on a 10% solution of muriatic acid.

Allow the solution to remain on the surface for 10 to 15 min, until all bubbling ceases. This action neutralizes the alkalinity of the surface and at the same time gets rid of laitance and opens pores in the concrete to permit a good bond between paint and substrate. The profile of a properly etched concrete surface should resemble the texture of fine sandpaper. Repeat etching if necessary. Flush the surface thoroughly with clean water and allow it to dry completely before painting.

WARNING: During the etching process workers must wear protective goggles for the eyes and rubber gloves and boots. If the solution is splashed onto the skin or in the eyes, flush immediately with cold water and call a physician.

Architects or engineers writing specifications should understand that hydrostatic water pressure behind below-grade installations cannot be corrected with paint; instead, drainage must be corrected to eliminate the pressure. Also, floors laid on grade without a moisture barrier offer only a marginal surface for paint adhesion, since water is drawn into the concrete by capillary action.

Concrete is finished in two ways after pouring. One is a broom-float finish which leaves a slightly porous surface. The other is a steel-troweled finish which is hard and smooth. Both should be acid-etched prior to painting, but acid-etching is mandatory for a steel-troweled floor if there is to be a good bond between concrete and paint.

Laitance   A thin layer of fine cement powder called laitance is present on the surface of all poured concrete floors. Laitance has a poor bond to the sound concrete beneath and is a weak surface. Unless removed by acid-etching before painting, this structurally weak layer will crumble and scale, and along with it the paint. Acid-etching is necessary to correct this condition.

Concrete hardeners   These materials slow the rate of water evaporation to obtain structurally strong concrete. Some hardeners are mixed in the concrete; others are applied over the surface as a membrane. Hardeners prevent etching of the concrete to prepare it for painting and will in themselves prevent a good bond between paint and concrete. Removal by mechanical abrasion may be necessary, or in some cases a solvent wash can be used prior to etching or painting. Consult the manufacturer of the concrete hardener to ascertain the best removal method.

Freeze-preventive compounds   Compounds such as calcium chloride are added to the concrete mix to prevent freezing. These draw moisture. Hollow metal door bucks filled with concrete for stability in emplacement have been known to rust out as the calcium chloride draws moisture from surrounding blocks and air. When such compounds are used in a mortar mix, the continued presence of water promotes efflorescence and keeps the alkali active. In such a situation unmodified latex coatings hold up better than oil- or alkyd-modified latex or solvent-thinned coatings. An alkali-resistant prime coat based on chlorinated rubber is required. In extreme cases, it may be necessary to apply an alkali-resistant all-epoxy system.

New concrete floors should be allowed the equivalent of 3 months of dry summer weather for aging before being painted. The floor should be bone-dry before paint is applied. In a basement, drying may take as long as 1 or 2 weeks, depending on heat and ventilation. If the architect desires only a clear product to prevent surface wear and dusting, use a sealer based on a modified-epoxy–resin vehicle that penetrates the pores and hardens the surface.

The specification therefore should cover surface-preparation details. These include:

- Instructions on how to cure, dry, clean, and free the surface of all contaminants
- Tests to assure that bond breakers and curing agents are compatible with paint coatings
- Instructions for form-oil removal
- Details to neutralize and clean surfaces of high alkalinity, lime salts, efflorescence, or mortar stains
- Instructions for removal of grease, oil, mildew, stains caused by corroded metals, scale, excess mortar, and loose particles and any powder residue from sacked areas
- Repair procedure to fill cracks, scratches, holes, and other imperfections with suitable masonry patching compounds to match the textural profile of the surface

As a further guide to selecting coatings for concrete, refer to *Guide for the Protection of Concrete against Chemical Attack by Means of Coatings and Other Corrosion-Resistant Materials.* [2]

11.4 Masonry Block   Many manufacturers use lightweight aggregate and a soap solution in their mix water for blocks. As a result, the blocks become quite hydroscopic. Water penetration into these blocks is greater than into denser blocks. For some jobs there may be an economic incentive to specify all additives, wetting agents, mold-release agents, integral colorants, and other surface contaminants which may be present. It is

[2] American Concrete Institute Committee 515, Title 63-59, Detroit, Mich. 48219.

also advisable to specify the exact amount of water absorption which will be tolerated. All these factors affect the coating-system specification.

There are two kinds of masonry block: concrete and fired clay. The concrete block should be handled in the same manner as reinforced concrete. Keep in mind, however, that many materials are used as aggregate and that any change in the mix changes the absorption characteristics of the block. If water penetration is extreme, specifications should call for application of a block filler to the exposed surface prior to application of either a water retardant or a coating system. Specifications should also call for the complete removal of any industrial soils such as roofing tar or mortar which may have dripped onto the wall surface during construction. For exterior work a good block filler will usually stop the entry of water from wind-driven rain even on lightweight cinder block. However, prevention is dependent on a good filling operation. In a hard-driving rain, even a hole the size of lead in a lead pencil will permit the entry of water.

For below-grade interior block where chronic moisture is a problem, a water-resistant and alkali-resistant epoxy filler and epoxy finish coats should be used rather than latex, alkyd, or oil types. For above-grade interior block in areas where moisture may be a problem, as in shower-room walls, an all-epoxy system should also be used. Block fillers should not be confused with the special materials that are sold as waterproofing-sealing compounds to stop hydrostatic water pressure below grade.

STEARATED WATER RETARDANTS: OIL-PHASE AND WATER-PHASE

Use   These products are used to seal completely the masonry substrate against water penetration and to act as a seal coat for topcoating application. In addition to acting as sealants, they prevent soil accumulation if no coating is applied. This quality is an advantage when natural masonry is desired as a visual finish.

Qualities   These products react chemically with free salts contained in the masonry substrate to form plugs in the capillary system which are insoluble to attraction by water. They are recommended for their antispalling characteristics. The products are best applied by spray at low pressures to allow a 6- to 8-in (152.4- or 203.2-mm) rundown over the entire surface treated.

NOTES OF CAUTION:

• Oil-phase materials must be applied to dry surfaces. Water-phase materials must be applied to damp surfaces.

• Oil-phase topcoatings can be applied over any dry, previously water-retardant–treated surface. Water-phase topcoatings can be applied over damp or dry, previously treated water-phase water retardants.

■ Oil-phase finish coats can be applied immediately after application of an oil-phase water retardant.

■ Water-phase finish coats can be applied 4 days or less after application of an oil-phase water retardant.

■ Water-phase top or finish coats can be applied immediately after application of a water-phase water retardant.

Typical Formulas for Stearated Water Retardants

Oil-phase		Water-phase	
Nonoxidizing resins	50 lb	Nonoxidizing resins	50 lb
Metallic stearates	75 lb	Organic stearates	75 gal
Alcohol	100 gal	Alcohol	25 gal
Mineral spirits	900 gal	Water	900 gal

Both these products may either be used as clear coatings or be tinted if masonry stain and waterproofing are required.

Comments   The formulas given are not complete but are offered as aids to architects to further their understanding of coating products. When masonry substrates are properly sealed against water penetration, the extended life of the top finish coatings will be assured and the cost of maintenance lowered.

All the problems discussed under the heading "Concrete" generally apply to concrete masonry block. The additional features that must be kept in mind are workmanship and the fact that concrete masonry blocks require mortar. Here two separate masonry materials are combined, presenting two possible sets of painting problems.

11.5 Mortar   New or fresh mortar is usually alkaline. Mortar generally can be a good guide to potential problems. For example, when a brick wall is to be painted, the mortar joints are usually flush-cut. Surface smoothness differs between brick and mortar, as do porosity and texture. Furthermore, brick probably has neutral alkalinity while mortar has maximum alkalinity. In many cases moisture content will be at extremes, the brick being completely dry and the mortar of maximum wetness. It is little wonder that masonry surfaces of block or brick and mortar present difficult painting problems.

To avoid painting problems the specification writer should call for field inspection of mortar joints prior to application of water retardant or coatings. All joints must be clean and free of bee holes or other imperfections. If repointing or patching is required, this work should be completed before further application of other materials. There is a definite advantage in specifying the desired formula for mortar as well as in calling for the compressive and flexural strength of the cured mortar. While mortar is generally a balanced mixture of cement, lime, and sand (1 part cementing material and about 3 parts aggregate), there are many formula variations. Any number of additions or admixtures can be put into mortar to increase its strength, bonding ability, waterproofness, or durability or even to change its color.

In some cases it may be necessary to specify different painting practices for mortar and block on the same wall. Regardless of whether joints are raked or flush, joint areas may be the weakest points of any wall and should therefore be given careful consideration in specifications.

11.6 Brick   According to most experts, a brick masonry wall should never be painted. Brick in itself is a finished material, and any painting of brick is often due to a problem. Correcting a problem with a coating material introduces the possibility of additional problems. (Some brick homes, however, are painted white to reflect heat, with estimates of energy savings reaching 10 percent.)

Before a painting system is specified for brick, several questions should be considered. First, the brick will normally be neutral, but the mortar will be extremely alkaline. The surface smoothness of the brick can and will vary, while the surface of the mortar can be expected to be rough. The porosity of the brick and usually that of the mortar can be very low, so the bond between the two surfaces may be questionable.

It is also important to know the exact type and kind of brick wall that is being considered. Are the bricks made of clay, adobe, concrete, plastic, sand, or lime? Is the kind of mortar used standard or standard plus one or more admixtures? Is the wall veneer, load-bearing, load-bearing and shear-resistant, retaining, or a less common type? Has the wall ever been treated before, and if so, with what? Answers to these questions determine the surface preparation and the painting system.

While brick is seldom painted, brick walls often receive a coat of water retardant and sometimes a uniforming glaze coat. Glazed brick, like glazed asbestos cement, will not hold conventional oil, alkyd, and latex paint and enamel. The paint manufacturer's specifications should be consulted. The presence of a glaze or silicone treatment on asbestos cement can be detected by pouring water over the surface; if the water is absorbed, the surface is paintable.

Concrete, building block, and brick, like plaster, should be given a minimum of 30 days' curing or drying time to reduce water and alkali to a safe level. After the mortar has dried, the wall surface should be free of dust and other soil prior to the application of any surface treatment. On block and brick surfaces, damage frequently follows mortar lines since mortar mixes are usually more alkaline than block or brick. An alkali-resistant primer helps retard damage caused by efflorescence, but when the efflorescent condition is extreme, some damage may be expected until the surface is neutralized with age.

Once a brick surface has been prepared for painting, the performance of all succeeding coats will depend on the ability of the prime coat to adhere and to retard the damage that can occur when alkali is activated. Painting practice follows the same guidelines as masonry block (Table 11.4).

Many imported bricks, especially those imported from Mexico, have not been fired properly. They chip easily and erode. General painting practice for these materials is similar to that for painting adobe, but good paint performance cannot be expected if bricks continue to degrade. A crumbling brick will simply take the paint off as it breaks away from the surface.

11.7 Concrete Flooring and Decking   Specifications to cover the preparation of masonry decking prior to application of any type of coating must be explicit in every detail. This is important because a failure in decking application disrupts traffic and is costly to correct. A completely successful initial paint application is therefore advantageous. Concrete ground slabs and concrete hanging decks are often finished with a steel trowel. In some cases these areas have also been treated with a hardening compound known as Pozzolan. Among other chemical products which may be present and must be removed before applying deck coating are curing compounds, wax or resin, and other materials that often are applied to fresh concrete to protect the surface from marring during construction. Hard, smooth surfaces must be opened by sand or water blasting, power sanding, or chemical etching. Surface contaminants must be removed. Waxes and oils can be removed with alkali in a water solution. Heavy oils may have to be lifted from the substrate by an initial treatment of diesel oil (the specification should elaborate when and how this is to be done), which is then washed off with a warm-water solution of alkali. After selecting a deck coating, the architect should work closely with the coatings manufacturer. Inspectors should supervise surface preparation and application.

Exterior decking may be stained, painted, or covered with a high-build traffic-resistant coating. In any event, specifications for deck finishing must be exact in every detail as to the preparation of the surface and substrate and proper procedures for applying the coating system. Correcting a problem caused by improper placing of a deck finish system can be disruptive and expensive.

11.8　Miscellaneous Masonry Surfaces　Among other masonry surfaces are clay tile, adobe, aggregate and natural stone, and asbestos siding.

11.8.1　Clay tile　Whether specifications call for a flooring tile or a roofing tile, preparation prior to coating is the same. Tiles which have been fire-glazed or coating-glazed should have their glaze removed or softened prior to coating. Since clay tiles are generally considered to be a finish in themselves, the mere fact that a coating has been proposed should lead to careful consideration of the reasons. If the coating is specified to change the color or finish, extra consideration probably will be necessary. However, moisture or water problems, excessive wear, efflorescence, oil treatments, or waterproof coatings can all cause coating problems. Waterproof coatings cannot correct leaking or weeping tile surfaces. If the attempt to correct a failure or change a problem is recognized in advance, proper steps can be taken.

A clay tile is defined as a molded-clay body, either solid or hollow, that has been thoroughly fired. The unit is generally considered to be a thin baked clay tile used as a decorative covering for floors, walls, roofs, and other purposes. Tiles may be hard and vitreous or soft and absorbent. The texture may be glazed or unglazed, enameled, sanded, ceramic-glazed, or treated by a combination of two or more of these. Recognition may be difficult for those unfamiliar with the processes. Much care should be employed before coatings are used. When clay tiles must be painted, personal inspection is necessary to determine why a coating is desired or required. As a general rule, older clay-tile surfaces can be repainted by using the surface-preparation and painting system for poured concrete. If the surface is very porous, prime coats may require an additional vehicle such as approximately 1 pt (0.473 l) of acrylic emulsion per gallon (3.785 l) of paint. The specification should call for caution in respect to hostile environments and areas of heavy traffic.

11.8.2　Adobe　Adobe is a sun-dried clay brick in which the clay consists of exceedingly fine or small particles. It will absorb large amounts of water and, in so doing, dissolve, expand, and melt. Any really good or long-lasting adobe-brick unit will have an asphaltic bituminous stabilizer. While the stabilizer will not keep out all the water, it will keep out the greater part. Coating asphalt must be part of the information on coating adobe. In looking at adobe with a view to coating, it is essential to remember its physical makeup: high absorbency, low strength, and extremely fine particle size. The asphaltic bituminous stabilizer can work against many applications. Adobe surfaces may be scaly and yet not be easily brushed. Although adobe may have picked up any amount of dust, dirt, smoke, smog, soot, etc., cleaning could change or destroy the surface. The fact that the asphalt stabilizer is a potential coating hazard must also be considered.

Adobe normally is not painted unless it must withstand the effects of weathering. In this case the surface may be loose and flaky or otherwise unstable (colloidal clays tend to swell when wet and shrink when dry). So the condition of the wall calls for careful inspection. The coating must be carefully selected to adhere completely without overwetting the surface and producing a waterproof cover. Furthermore, an asphalt bituminous stabilizer has probably been added to the manufacturing mix, and this can cause bleeding and adhesion problems.

Adobe, like brick, is seldom painted. Heat-reflective coatings are usually whitewash. Most adobe is treated with a water retardant and, in some cases, with a low-solids glaze coat. In any case, preparation consists of washing down the surface with a mild detergent in warm water and of washing off residual soap by spraying with clean cold water. Surface and substrate must be dry prior to application of the coating system.

WHITEWASHING

Whitewashing is a relatively simple and inexpensive way to brighten adobe and the interior of livestock and other service buildings. Apply whitewash with a brush or a spray gun.

Surface preparation   Remove all dirt, scale, and loose material by scraping or brushing with a wire brush. Many whitewashing jobs have proved quite satisfactory without further surface preparation. For the best job, wash off as much of the old coat of whitewash as possible with hot water and vinegar or a weak acid solution.

Dampen the walls before applying whitewash. Unlike most paints, whitewash improves in application and adherence when the surface is slightly damp.

Mixing   Lime paste is the basis of whitewash. Protect eyes and skin during mixing. The material may be prepared by either of the following methods:

▪ Soak 50 lb (22.7 kg) of hydrated lime in 6 gal (22.7 l) of water. Refined limes such as chemical hydrate, agricultural-spray hydrate, finishing lime, and pressure-hydrated lime have fewer lumps and make a smoother paste.

▪ Slake 25 lb (11.3 kg) of quicklime in 10 gal (37.85 l) of boiling water. Cover and allow to slake for at least 4 days.

Each of these preparations makes about 8 gal (30.3 l) of paste.

Different whitewash mixes are suggested for different surfaces. Smaller batches may be prepared by reducing the ingredients in equal proportion in the formulas given below.

General Woodwork.   Dissolve 15 lb (6.8 kg) of salt in 5 gal (18.9 l) of water. Add this solution to the 8 gal (30.3 l) of paste, stirring constantly. Thin the preparation to the desired consistency with fresh water. To reduce chalking, use 5 lb (2.27 kg) of dry calcium chloride instead of salt.

Brick, Concrete, or Stone. Add 25 lb (11.3 kg) of white portland cement and 25 lb of hydrated lime to 8 gal (30.3 l) of water. Mix thoroughly to a thick slurry, and thin to the consistency of thick cream. Mix only enough material for a few hours' use. To reduce chalking, add 1 to 2 lb (0.45 to 0.91 kg) of dry calcium chloride dissolved in a small amount of water to the mix just before using.

Plaster Walls. One of three formulas is recommended:

- Soak 5 lb (2.27 kg) of casein in 2 gal (7.6 l) of water until it is thoroughly softened, in about 2 h. Dissolve 3 lb (1.36 kg) of trisodium phosphate in 1 gal (3.785 l) of water, add this solution to the lime, and allow the mixture to dissolve. When the lime paste and the casein are thoroughly cool, slowly add the casein solution to the lime, stirring constantly. Just before use, dissolve 3 pt (1.42 l) of formaldehyde in 3 gal (11.37 l) of water, and add this solution to the whitewash batch, stirring constantly and vigorously. Do not add the formaldehyde too rapidly. If the solution is added too quickly, the casein may form a jellylike mass, thus spoiling the batch.
- Dissolve 3 lb (1.36 kg) of animal glue in 2 gal (7.6 l) of water. Add this solution to the lime paste, stirring constantly. Thin the mixture to the desired consistency.
- The first formula, given above for use on plaster walls, is a time-tested, long-life mix that is also suitable for general use. The following formula is also suitable: Dissolve 6 lb (2.72 kg) of salt in 3 gal (11.37 l) of boiling water. Allow the solution to cool, and then add it to the lime paste. Stir 3 lb (1.36 kg) of white portland cement into the mix.

Coloring Pigments may be added to whitewash to provide color. The following have proved satisfactory:

- *Black.* Magnetic black oxide of iron
- *Blue.* Pure precipitated brown oxide of iron or mixtures of black oxide of iron with turkey or Indian red
- *Green.* Chromium oxide, opaque, or chromium oxide, hydrated
- *Red.* Indian red made from pure ferric oxide
- *Violet.* Cobalt violet and mixtures of reds, white, and blues
- *White.* Lime itself
- *Yellow.* Precipitated hydrated iron oxides

Application Some surfaces may require two coats of whitewash.

Two coats are better than one coat that is too thick. Strain the mix through three layers of cheesecloth before using a spray gun.

SOURCE: U.S. Department of Agriculture Bulletin 222.

11.8.3 Aggregate and natural stone   Aggregate is a completed architectural surface. If it is to be coated, be sure to find the original problem before making any attempt to coat it. Alkalinity, surface smoothness, porosity, and moisture may be difficult factors in coating.

The exposed aggregate surface will generally indicate a concrete base that has been pressure-blasted, acid-etched, or chemically treated to expose the aggregate materials. The aggregate itself may be any of hundreds of materials—manufactured or salvaged materials or a combination of these. The resulting surface of exposed aggregate has a very wide and well-accepted use. The durability of the surface will depend upon the type and kind of aggregate. Do not use soft, porous, or easily destroyed materials in heavy-wear areas. Aggregate is exposed, mainly for decorative purposes, in walkways, pour-in-place walls, and tilt-up wall panels. While these surfaces are rarely painted, they are often given a water-retardant treatment and/or a glaze coating treatment. Usually aggregate is exposed by treating either the forms or the surface of new concrete with a reveal chemical which softens the surface to a controlled depth of the concrete. After setting, the chemical can be washed off to reveal the stone aggregate. Because reveal chemicals contain sugar, it is absolutely necessary that the surface of concrete subjected to this treatment be washed with clean running water until all residual chemicals have been completely removed. Surfaces must be clean and dry prior to further surface treatment.

11.8.4 Asbestos siding (glazed and unglazed)   Specifications for preparing unglazed asbestos siding for coating should include a wash-down with a mild detergent in warm water. After the surface has been rinsed and dried, it is coatable. An exception would be a glaze of urethane or silicone resinous material. Such a surface should be whip-blasted or power-sanded and then washed and dried prior to coating. Normally, glazed asbestos is not painted. Weathering can remove the glaze, making the surface paintable, but severe peeling will result if a coating is applied to areas where glaze is still present, as under eaves. Here, sandblasting or some glaze-removal method is required. Wear an approved respirator.

Asbestos cement is usually highly alkaline. Surface smoothness is ordinarily fine unless a visible texture is applied. Porosity and moisture can cause the biggest problems. The application of an impermeable coating on one side of an asbestos-cement board can cause bowing or warping in the panel. This development is generally due to the nonuniform expansion and contraction that takes place in conditions of high humidity and moisture.

Asbestos siding is a manufactured siding material composed of cement, asbestos fiber, resin, and filler. It is available with a glazed surface, which is usually resin-treated, or with an unglazed surface, which is more porous and not as water-resistant. This material may soon disappear from the market because it is a suspected carcinogen.

BIBLIOGRAPHY

"A First Approach to Choosing a Coating for Concrete," *Concrete Construction,* September 1977, pp. 477–479.

MASTERSPEC 2, Sec. 0900, "Painting," Sec. 09800, "Special Coatings," Sec. 09820, "Cementitious Coatings," 1980 series, Production Systems for Architects and Engineers, Inc. (PSAE), 1735 New York Avenue N.W., Washington, D.C. 20006.

Moilliet, John L.: *Waterproofing and Water-Repellency,* American Elsevier Publishing Company, Inc., New York, 1963.

"Paints and Coatings for Concrete," *Concrete Construction,* April 1971, pp. 120–125.

Thomassen, Ivar P.: "Coatings for Masonry and Related Surfaces," *Proceedings of Sixteenth Coatings Course for Painting Contractors, Maintenance Engineers and Architects,* University of Missouri, Rolla, Mo., 1970.

Chapter **12**

Roof Coatings

ERIC S. WORMSER
KENNETH K. KAISER

Gibson-Homans Company

12.1 Evolution of Coatings into a System   The roof of a building is one of the most complex subsystems in a construction project. The various components arrive at the site from different manufacturers and are assembled by skilled mechanics, using a variety of equipment. Possible variations in the assembly of roof systems are almost infinite, and some of these variations can lead to premature failure of a roof.

When roof systems fail, there is need for materials which make quick repair possible even in unfavorable weather conditions. The roof-coating industry was born out of this need. At first, maintenance crews relied primarily on patching materials. Then came coatings for the entire roof, coatings supported by membranes and roofing sheets, cold-applied flashing systems, sophisticated preventive maintenance programs, and, finally, fluid-applied roofing systems for new construction.

Many roof manufacturers now recommend roof coatings on their hot-applied roofs to prevent premature failure and to extend the life of the original installation. Numerous manufacturers of hot-applied and prepared roofing materials now also offer fluid-applied roofing for new construction and maintenance. Until about a dozen years ago most such fluid-applied systems were bituminous.

When free-form structures with irregular roof surfaces became popular, new waterproofing methods had to be developed. The cost of materials became less of a consideration as the aesthetics of the system became more important, particularly when higher cost of materials could be balanced against lower labor cost of installation. The coating industry met this need by turning to the technically more sophisticated and expensive rubber and plastic materials.

12.2   Function of Roof Coatings   Roof coatings are used for the following reasons:

1. *Repair.* Leaks which have developed in existing roofs must be repaired. Much roof coating is still used for this purpose.
2. *Renewing old roofs.* A roof which has started to fail may be brought back to its functional waterproof condition with roof coating.

12.3   Maintaining Existing Roofs   Well-constructed roofs may be maintained indefinitely by periodic coating with quality roof coatings if such maintenance is started before the roof felts begin to deteriorate.

12.4   Installation of New Roof Systems   Conventional bitumen-and-felt roofs may be installed with coating materials. This is usually done when the roof is inaccessible, a melting kettle for hot bitumen is not available, the hazard of a hot-bitumen kettle cannot be allowed, or air pollution regulations preclude use of a kettle. There are now numerous elastomeric or plastic cold-applied roof systems. These new roof-coating systems offer many advantages over the old standby bitumen-and-felt roofs. They are light in weight and usually decorative in appearance, conform to free-form structures, have greater capacity to accept movement, and usually are very resistant to chemical attack.

Whatever the reason for using a coating (there may be more than one; see Table 12.1), selection should be made by a person knowledgeable in roof coatings and roof conditions after thorough evaluation of the surface and the exposure.

TABLE 12.1   Roof-Coating Selection Checklist

1. *Function.* Reroofing, new construction, stopping leaks, preventing corrosion, etc.
2. *Type of substrate.* Concrete, steel, plywood, tongue-and-groove wood, etc.
3. *Condition of substrate.* Alligatored, blistered, smooth, covered with gravel, porous, etc.
4. *Budget.* Dollars per square foot that can be spent.
5. *Aesthetics.* Is black acceptable? Are colors needed?
6. *Heat reduction or insulation value needed.*
7. *Traffic to be placed on roof.*

12.5   Objectives of This Chapter   Various surfaces and suitable coatings are described in this chapter. Topcoatings for existing roofs, fluid-applied roof systems for bare roof decks, and general roof coatings are three examples of roofing materials. Table 12.2 shows various roof surfaces covered in this chapter and indicates which coatings are suitable for each surface. Table 12.3 shows various types of roof decks and the fluid-applied roof systems suitable for use on each type.

Architects and specifying engineers should note that the Environ-

TABLE 12.2   Roof Coatings

| | Coating type | | |
| | Cutback asphalt | | |
Substrate	Nonfibered	Asbestos-fibered	Glass-fibered
Built-up asphalt	Used as primer before topcoating	Excellent coating on primed roof with moderate alligatoring or cracking	Excellent coating on moderately to severely alligatored roof which has been primed
Asphalt and gravel†	Used as primer before topcoating	Good coating on primed roof	Excellent on primed roof with all gravel removed; do not apply over gravel
Tar and gravel†	Not recommended	May be used if tar is very old and dry; tar roof coating is preferable	May be used if tar is very old and dry; all gravel must be removed
Asphalt shingle	Thin nonfibered asphalt may be used before asphalt aluminum paint	Suitable for waterproofing; appearance often not acceptable	Not recommended
Asbestos cement	Thin nonfibered asphalt used as primer before topcoating; more than one coat may be needed	Excellent on thoroughly primed roof	Excellent on thoroughly primed roof
Metal‡	Good coating to prevent corrosion; joints and seams must be treated with plastic roof cement or caulk to stop leaks	Excellent coating to prevent corrosion	Excellent to prevent corrosion; best coating for corrugated metal
Slate	Thin nonfibered asphalt may be used as primer before topcoating	Good coating on primed roof; black appearance often not acceptable	Not recommended
Typical formula	*Percent*	*Percent*	*Percent*
	Asphalt (175°F, softening point; 20 penetration) 65 Mineral spirits 35 100	Nonfibrated-asphalt roof coating 93 Canadian asbestos fiber, 7-D 7 100	Nonfibrated-asphalt roof coating 71 Polybutene (150 standard Saybolt units at 100°F) 6 Canadian asbestos fiber, 7-D 6 Glass fiber, ¼ in long 3 Kaolin clay 10 Mineral spirits 4 100

* When primer or application of aluminum over coating is recommended, allow curing time of the primer or previous coating in accordance with the recommendations of the manufacturer of the aluminum roof coating.

† On gravel roofs all *loose* gravel must be removed before coating unless complete removal is indicated in this table.

‡ When roof coatings are recommended for metal surfaces, they serve to prevent rust and corrosion, but they do not react to eliminate existing rust or corrosion. A rust-reactive primer may be used for this purpose under any of the roof coatings recommended. New galvanized metal should be permitted to weather 1 year before coating.

	Coating type		
Cutback asphalt (*Continued*)	Asphalt emulsion	Refined coal tar	
Mastic and granules	Fibered and nonfibered	Nonfibered	Asbestos-fibered
Good system when color is desired	Should be used only on roof with good drainage after priming; good only for lightly alligatored roofs	Not recommended	Not recommended
Good system if all gravel is removed first	Should not be used unless roof has good drainage, which is rarely the case; surface must be primed	Not recommended	Not recommended
Not recommended	Not recommended	Excellent primer after loose gravel has been removed	Excellent coating after loose gravel has been removed
Excellent system	Not recommended	Not recommended	Not recommended
Good system on thoroughly primed roof	Excellent on thoroughly primed roof	Not recommended	Not recommended
Good for metal when color is desired	Good on primed metal	Not recommended	Not recommended
Good system when color is desired	Good on thoroughly primed roof	Not recommended	Not recommended

Percent		*Percent*		*Percent*		*Percent*	
Nonfibered asphalt roof coating	78	Nonfibrated:		Refined coal tar	96	Nonfibrated-tar roof coating	94
Canadian asbestos fiber, 7-R	12	Asphalt (110°F, softening point)	50	AWPA creosote	3	Canadian asbestos fiber, 7-D	6
Kaolin clay	5	Bentonite clay	2	Phosphoric acid	1		
Mineral spirits	5	Water	48		100		100
	100		100				
		Asbestos-fibrated:					
		Nonfibrated-asphalt emulsion	97				
		Canadian asbestos fiber, 7-R	3				
			100				
		Glass-fibered:					
		Non-fibrated asphalt emulsion	96				
		Canadian asbestos fiber, 7-R	2				
		Glass fiber, ¼ in long	2				
			100				

TABLE 12.2   Roof Coatings (Continued)

Coating type				
Aluminum-pigmented asphalt*			Colored coatings	
Nonfibered	Asbestos-fibered	Glass-fibered	Solvent-based	Emulsion
Excellent on new or recently coated black roofs for appearance and heat reflection	Excellent on primed roof with moderate alligatoring	Excellent on primed roof with moderate to severe alligatoring	Excellent on roof at least 1 year old; primer not recommended	Excellent if roof has good drainage; primer not recommended
Recommended only when good drainage is provided, which is rarely the case	Good coating on primed roof when there is good drainage	Excellent on primed roof	Excellent	Excellent if roof has proper drainage, which is rarely the case
Not recommended	Not recommended	Not recommended	Not recommended	Not recommended
Good if first thoroughly primed	May be used; shingle-tab curling can be a problem	May be used	Excellent; shingle-tab curling can be a problem	Not recommended
Excellent on thoroughly primed or coated roof	Excellent on thoroughly primed roof	Excellent on thoroughly primed roof	Excellent; two coats, the first coat thinned, are recommended	Excellent if surface is relatively new and is not primed
Excellent when applied on a sound surface	Excellent when applied to a sound surface	Excellent when applied to a sound surface	Excellent when applied to a sound surface	Not recommended
Good when applied to a thoroughly primed surface	Good when applied to a thoroughly primed surface	Excellent when applied to a thoroughly primed surface	Excellent; two coats, the first coat thinned, are recommended	Not recommended
Percent	_Percent_	_Percent_	_Percent_	_Percent_
Nonfibrated-asphalt roof coating 61	Nonfibrated-asphalt roof coating 48	Nonfibrated-asphalt roof coating 43	Long-oil alkyd resin 34	Acrylic latex emulsion (55 percent nonvolatile) 40
Xylol 20	Xylol 17	Polybutene (150 standard Saybolt units at 100°F) 5	Polybutene (150 standard Saybolt units at 100°F) 5	Dibutyl phthalate 5
Aluminum paste (65 percent solids) 18	Aluminum paste (65 percent solids) 18	Xylol 17	Nonleafing aluminum paste (65 percent solids) 4	Talc, calcium carbonate 35
Desiccant 1 — 100	Canadian asbestos fiber, 7-D 10	Aluminum paste (65 percent solids) 18	Coloring pigment 5	Titanium dioxide 10
	Talc 5	Canadian asbestos fiber, 7-D 7	Mineral spirits 35	Additives (thickeners, antifoam, fungicide, surfactants) 2
	Desiccant 2 — 100	Glass fiber 3	Talc, kaolin clay, calcium carbonate 16	Water 8 — 100
		Talc 5	Additives (driers, antiskin, etc.) 1 — 100	
		Desiccant 2 — 100		

mental Protection Agency is seeking to curtail the use of asbestos under the provisions of the Toxic Substances Control Act. The EPA intends to regulate all "nonessential" uses of the mineral. While only proposed regulations on floor sheeting and friction are imminent, an eventual phase-out would affect roof-coating formulations. As substitute formulations are developed, the new coatings would be substituted in the tables or included as a separate category.

Selection of a topcoating depends upon the type of substrate, its condition, its exposure, and aesthetic considerations. Various substrates and topcoatings as well as the types of deterioration that are commonly found are described below.

12.6   Types of Roofing Substrates

12.6.1   Asphalt built-up roofs   These are the most common roofs on industrial and commercial buildings. Built-up roofs are suitable only for decks having good drainage. They are constructed of field-applied, asphalt-saturated roofing felts (normally 3 to 5 plies), plus hot-melt asphalt between each ply and as the top surface. The asphalt is the waterproofing agent, and the felts give strength to absorb normal building movement. The asphalt is oxidized over a period of time by ultraviolet rays and high temperatures from the sun. This oxidation causes shrinkage and embrittlement, which leads to alligatoring. Alligatoring becomes a problem when it is deep enough to expose the underlying felts. Asphalt built-up roofs allowed to deteriorate to this point usually show one or more of the following failure symptoms:

1. *Brittleness.* The oxidized topcoat and saturating asphalt become brittle. The roof is now susceptible to cracks from normal building movement and fractures from impacts.

2. *Loss of strength and loss of ability to absorb movement.* After the oxidation of the saturant the exposed felts are subject to decay of the organic materials used in their construction. This leads to cracking of the felts.

3. *Shrinkage and edge curling.* The exposed felts start to absorb and wick water. As the water dries out of the felts, they shrink, curling at the edges and exposing more of the next ply of felt.

This entire deterioration process can be prevented by periodic topcoating of the built-up roofs with cold-applied, fiber-reinforced roof coating. The first application should be made before the top layer of hot-applied asphalt alligators. Depending on exposure conditions, the surface should be recoated every 3 to 5 years.

12.6.2   Asphalt-and-gravel roofs   These roofs are constructed similarly to asphalt built-up roofs. However, this system is used for roofs with a pitch of less than ½ in/ft (41.7 mm/m). The top layer of asphalt is soft,

TABLE 12.3 Coating Systems for Decks

System substrate	Neoprene-Hypalon*	Silicone	Urethane	Low-molecular butyl
Asphalt built-up roof	Excellent if roof is at least 5 years old. Only neoprene is used	Excellent; roof must be over 1 year old	Excellent; roof must be at least 6 months old	Not recommended
Concrete deck	Excellent; concrete must be dry and smooth	Excellent; concrete must be dry and smooth	Excellent; concrete must be dry but need not be as smooth as for neoprene-Hypalon* or silicone	Excellent; surface must be sandblasted; Hypalon* topcoat can be used for light colors
Wood deck	Excellent after proper joint preparation	Excellent after proper joint preparation	Excellent after proper joint preparation	Excellent after proper joint preparation; may be coated with Hypalon* for light colors
Insulation-board deck	Excellent if board is firmly anchored and joints are prepared; not compatible with all-board stock	Excellent on some board stock; board must be firmly anchored and joints properly prepared	Excellent on almost every board stock; joints must be prepared	Excellent over selected board stock; joints must be prepared
Sprayed-on urethane-foam insulation	Excellent if foam is reasonably smooth and free-draining	Excellent if foam is reasonably smooth and free-draining	Excellent; fewer limitations than with neoprene-Hypalon* or silicone	Excellent; may be covered with Hypalon* for light colors
Metal deck†	Excellent if joint deflection is kept to a minimum and surface is sound	Excellent if joint deflection is kept to a minimum and surface is sound	Excellent if joint deflection is kept to minimum and surface is sound; primer must be used	Excellent; surface must be sandblasted; may be coated with Hypalon* for light colors

Typical formula	Neoprene: *Percent*
	 Type W neoprene 15
	 Phenolic resin 5
	 Carbon black 15
	 Curing agents 3
	 Xylol 62
	   100
	 Hypalon*:
	 Hypalon* No. 30 15
	 Dibutyl sebacate 3
		Actual formulations are not available‡	Component A: *Percent*
			 High-molecular-weight triol 16.5
			 Chlorinated paraffin 15.5
			 Mineral spirits 4.0
			 Pyrogenic silica 2.5
			 Calcium carbonate 59.5
			 Refined coal tar 2.5
			 Catalyst 0.5
			 Component B:
			 High-molecular-weight diol 19.0
			   100.0
				Percent
				 Low-molecular butyl rubber (75 percent solids) 50
				 Nonleafing aluminum paste (65 percent nonvolatile) 12
				 Colorant 2
				 Canadian asbestos fiber, 7-M 7
				 Additives 1
				 Mineral spirits 18
				 Hexane 10
				   100

Titanium dioxide	12
Talc, calcium carbonate	9
Curing agents	3
Cellosolve	8
Xylol	50
	$\overline{100}$

* Hypalon is a trademark of E. I. du Pont de Nemours & Co.
† New galavanized metal must be permitted to weather 1 year or undergo special surface treatment before any fluid-applied system may be installed over it.
‡ Uncured silicone materials have the following properties:

Property	Value	Test method
Solids content, percent by volume	64–68	ASTM D-2697
Solids content, percent by weight	78–82	ASTM D-2697
Flash point, °F	100, minimum	Tag closed cup

The cured silicone membrane shall have the following properties:

Property	Value	Test method
Tensile strength, psi	500–600	ASTM D-412
Elongation, percent	100–150	ASTM D-412
Permanent set, percent	<1.0	ASTM D-412
Heat aging, after 26 weeks at 180°F	None	ASTM D-412
Ultraviolet exposure, 4000 h in Atlas Xenon Weather-Ometer	No cracking, checking, or significant discoloration	
Ozone resistance, 150 ppm, 14-day exposure	No cracking	
Water absorption, 168 h, 75°F, weight percentage	0.5, maximum	ASTM D-570

usually with a 135 to 150°F (57 to 66°C) softening point. This layer is covered with gravel, which protects it from the ravages of exposure to direct sunlight.

The mechanics of failure of this system are similar to those of asphalt built-up roofs, with three exceptions:

1. The topcoat of asphalt does not alligator until much later in the aging process. Being low-melt, it exhibits cold flow during the early years and does not alligator until it becomes brittle because of oxidation.

2. Felt becomes exposed. The cold-flow characteristic of soft asphalt causes this material to flow off the high points on the roof, exposing felts.

3. The deterioration process is retarded by the gravel. These roofs must be watched closely. Because of their low pitch, they pond water, causing severe leakage and potentially much interior building damage as the system deteriorates. Leaks are difficult to find since the failures are often hidden by the gravel. For proper repair the loose gravel must be removed and reapplied after repair.

12.6.3 Tar-and-gravel roofs   These roofs are constructed of tar-saturated felts (usually 3 to 5 plies) and coal-tar pitch. The pitch usually has a low softening point (between 140 and 155°F, or between 60 and 68°C). Therefore, these roofs are limited to decks which have a slope of ½ in/ft (41.7 mm/m) or less. The top layer of pitch is covered with gravel. The coal-tar pitch is the waterproofing agent, and the felts give strength to the system. The gravel reduces the softening of tar from the heat of the sun (tar softens more readily than asphalt). The gravel allows people to walk on the roof without their feet sticking to the relatively soft coal-tar pitch. It also protects the coal tar from the ultraviolet rays of the sun.

Tar-and-gravel roofs fail similarly to asphalt-and-gravel roofs, except that tar normally does not alligator. When the roof is young, cracks that develop during cold weather heal in the hot sun. With age, the tar oxidizes and becomes brittle, so that self-healing can no longer occur. If the embrittlement of the tar is left unchecked, gravel is blown off the roof, and the felts are exposed and absorb water. To repair the roof, the loose gravel must be removed and reapplied after repair.

12.6.4 Asphalt shingles   Asphalt shingles consist of roofing felt saturated and coated with hot asphalt. Small granules are embedded in the top layer of asphalt. The complete shingle is factory-prepared and is installed at the jobsite with nails or staples. Shingle roofs function well when installed on roofs with a pitch of at least 3 in/ft (250 mm/m), since they must shed water to be effective.

In the case of shingle roofs also, the asphalt becomes brittle as it oxidizes from the sun; granules are lost, and felt is eventually exposed to the elements. As granules are lost and the weight of the shingle tabs is reduced, the shingle becomes subject to wind uplift and breakage unless the tabs are tightly adhered or constructed to be self-locking. Increasingly, shingles are produced with an adhesive strip on the bottom of the tab to prevent this problem.

12.6.5 Asbestos cement roofs   Asbestos cement is cast into both shingles and corrugated sheets. Both forms are factory-made and are erected on the jobsite. This type of roofing depends upon water runoff to remain watertight; to be effective, therefore, it must be put on roofs with a minimum slope of 2 in/ft (166.7 mm/m). Asbestos cement has good fire and chemical resistance.

As the binder weathers, the roof deteriorates. Typical evidence of deterioration is craze cracking and dusting or chalking of the surface. As the craze cracking becomes severe, the surface becomes porous, sometimes almost spongy. These conditions can be prevented by topcoating before deterioration begins.

12.6.6 Metal roofs   Although metal does not leak, the seams and holes for the fastening system do. Metal roofs may be of galvanized iron, aluminum, or copper, among other materials. Because of the high coefficient of thermal expansion with temperature changes, metal roofs experience much movement. This movement results in elongated fastener holes, fasteners which are sheared off or pulled loose, and loose seams. Many metals corrode under atmospheric conditions, and corrosion may be accelerated by chemicals in the air. Proper coatings prevent corrosion. Roof-repair materials are used to stop leaks.

12.6.7 Slate shingles   Slate is a naturally occurring stone which is cut at quarries into shingle fashion and erected on the jobsite to form the roof. The use of slate shingles is limited to steeply pitched roofs since they depend on their water-shedding characteristics to be effective. The life of this type of installation is usually limited to the life of the fastening devices. If the nails are put in too tightly during installation, the shingles may crack; if the installation is too loose, the shingles may come loose.

The deterioration of the slate itself starts with the surface becoming porous. As the slate absorbs water, under freeze-thaw conditions, spalling, craze cracking, or severe cracking is likely to occur. Patching materials are useful to stop leaks in these roofs; coatings are used to prevent deterioration of the shingles.

12.7 Types of Sidewalls   Roof coatings are sometimes used for the sidewalls of buildings. Selection of coatings must be determined after evaluating the substrate, its condition, exposure, and, of greater importance than with most roofs, the aesthetics desired. The more common substrates and the types of deterioration encountered are described below.

12.7.1 Concrete   Concrete is a mixture of portland cement, sand, and gravel. The portland cement acts as a binder for the sand and gravel. The cured mixture is somewhat porous.

In many cases too much water is added to the mix for easy handling. When this water evaporates, small voids which increase the porosity of the cast are formed. During the curing process some shrinkage occurs, but the rigidity of the cast allows no flexibility. Any movement of the wall, therefore, is likely to cause cracking. These conditions result in potential areas of water infiltration. The water can cause leakage, corrosion of the reinforcing rods, and spalling of the surface. For these reasons, concrete walls are often coated with heavy-bodied coatings.

12.7.2 Concrete block and brick   Concrete block and brick are made of cementitious binder with aggregate filler that is cast in a mold and cured quickly under uniform conditions. The resulting unit is uniform in finish and dimensionally stable. Concrete blocks and bricks are erected on the jobsite by binding them together with a cementitious mortar. The resultant structures are very rigid. This rigidity often results in cracks, usually in the mortar joints, caused by building movement from temperature change and building settlement.

The quick curing, lean mixes, and coarse aggregates used in blocks and bricks can result in rough surfaces with high degrees of porosity. A driving rain against porous walls often causes moisture seepage to the inside. Moreover, the rough surfaces become dirty in areas with air pollution problems. Porous bricks are also subject to spalling problems. For these reasons, concrete-block and brick walls are frequently coated with heavy-bodied coatings.

12.7.3 Stucco   Stucco is a cementitious binder with fine aggregate which usually is troweled over metal lath, brick, or cement block. The mixture is usually tinted and given an aesthetically pleasing, rough surface.

This substrate has many of the problems associated with concrete block and brick. It is subject to cracks and dirt collection, and weathering can cause bleaching of the color and increasing porosity. Once moisture has infiltrated this substrate, the finish can spall under freeze-thaw conditions.

12.7.4 Metal   Metal walls are subject to the same conditions to which metal roofs are subject, namely, movement, elongated fastener holes, fasteners shearing off or pulling loose, loose seams, and corrosion.

12.8 Functions of Roof Coatings   A coating should be selected after considering the condition of the substrate and evaluating the functions that the coating is expected to perform. Four basic functions may be required of a coating. Since in many cases more than one of these functions must be fulfilled, priorities must be set. Sometimes it is difficult to obtain all the desired results satisfactorily. The four primary functions are as follows:

12.8.1 Rejuvenation of existing surface   This function is preventive in nature. The coating is applied before any serious deterioration of the surface takes place, and it should renew the substrate insofar as possible to its original condition. A classic example is the coating of an asphalt built-up roof which has started to alligator but does not yet leak.

12.8.2 Stopping leaks   This requirement is corrective in nature. With proper selection, however, the measures taken can be both corrective and preventive by stopping additional leaks from occurring. A simple roof patch is corrective. Roofs are often patched and then coated. The coating not only completes the patch but also is designed to prevent additional leaks.

12.8.3 Improving appearance   Very few surfaces remain aesthetically pleasing for an indefinite period. Often changes in appearance are desired, possibly because of a change in building occupancy. A rusted metal wall may be coated with a colored coating to protect the metal from further deterioration and, at the same time, to improve the appearance of the building.

12.8.4 Protection against corrosion and/or spalling   Corrosion of metal and spalling of masonry are common problems. Some heavy-bodied coatings provide an excellent answer to these problems by preventing the infiltration of moisture, which is the usual cause. A good example is the coating of a concrete wall with a good colored roof coating.

12.9 Criteria in Selecting Coating Systems   Coating systems, as opposed to topcoatings, are usually intended to constitute the entire waterproofing protection for roof decks. The selection of a coating system depends upon the type of substrate, its condition, exposure, and aesthetic considerations. One of the many advantages of the modern coating systems is their light weight. Therefore, the support structure can be designed in a more economical manner than is necessary for the heavier built-up roofs. Roof designs that are readily visible from the ground can be employed because of the decorative characteristics of the coating system. With proper construction of the deck and selection of the coating system, it is possible to create a decorative pedestrian deck that adds very little weight to the building structure.

12.9.1 Types of substrate
Asphalt built-up roofs   Asphalt built-up roofs were described above. Sometimes they are used as the substrate for a coating system rather than as a topcoating. This selection is usually made because the asphalt built-

up roof has deteriorated. Treatment with a coating system is more economical than the removal of the roof covering down to the supporting deck and the subsequent installation of a system on the roof deck.

Concrete decks   Thin shell concrete is probably the most common substrate for roof-coating systems. Decks of this material are often constructed in irregular shapes, which lend themselves to a coating rather than a sheet system. These shapes normally are visible from the ground and so call for the attractive colors and seamless construction possible with a coating system.

Concrete is subject to hairline cracking, but elastomeric-type coating systems have the ability to span hairline cracks. Concrete is more or less porous, depending upon the precise mixture used. Coating systems generally have the ability to prevent water penetration into and through the concrete. Proper protection of the concrete also prevents spalling. Spalling results from absorption of moisture into the concrete, which freezes. The frozen moisture then expands within the concrete and causes a flaking action.

Tongue-and-groove or plywood wood decks   Wood decks must be waterproofed, and the wood must be protected from deterioration. For successful application of coating systems, the most important factor is the fastening devices that hold the deck to the substrate in a dimensionally stable manner. The fasteners must not be able to back out of the deck and puncture the system. The joints must be properly prepared to allow for the movement that takes place in them and to enable the coating to waterproof them. In the case of plywood, it is extremely important that the specification include a directive to seal properly all edges of plywood sheets. This precaution will prevent many moisture problems that often are blamed incorrectly on the coating system.

Insulation-board stock   Insulation-board stock is composed of a variety of materials, which include wood pulp that is glued together, fiber glass, cork, and plastic materials with a cellular structure. These boards are sold in various thicknesses from ½ to 2 in (from 12.7 to 50.8 mm) and in sheets from 2 by 4 ft (0.6 by 1.2 m) up to 4 by 8 ft (1.2 by 2.4 m) in size. Almost all insulation boards must be protected by a roof system. Even if a board itself is waterproof and resistant to sunlight, joints between boards require waterproofing. Care should be exercised in the selection of both the insulation and the coating system to make sure that they are compatible with each other. For instance, Styrofoam is dissolved by many common solvents used in coatings.

Fastening insulation to the deck is critical in using a coating system. Adhesives, nails, or clips are commonly used. When adhesives are used, care must be taken that evaporation of solvents or water from the coating, which causes shrinkage, does not curl the corners of the board. Nails and clips must not back out, puncturing the coating. The vertical differential and deflection between boards must be kept to a minimum for the success of the system. This is a function not only of proper fastening but also of the correct supporting structure under the insulation boards.

Sprayed-on urethane foam   Coating systems are the only practical means of protecting sprayed-on polyurethane foam. Urethane foam is applied by spraying two chemicals at the same time from a two-component spray gun. The two components react into a foam, expanding to 20 to 30 times their original volume within about 60 s after combining. No urethane foam should be applied on the exterior of a structure without protecting it from the elements. Urethane foam is especially susceptible to deterioration from the ultraviolet rays of the sun and must be protected as soon as possible after installation.

There are three important factors to consider in selecting a coating system for foam:

1. *Finish of the foam.* A wide variety of urethane-foam finishes are available. A slight orange-peel finish is the one recommended by most coating manufacturers.

2. *Drainage.* All applications of urethane foam should be free-draining to obtain the desired performance of the coating system.

3. *Compatibility.* The urethane foam and the coating system must be compatible. Consult the foam and coating manufacturers.

Metal decks   A metal-deck coating by itself cannot stop leaks which are caused by elongated bolt holes, missing fasteners, or opened-up laps, but it can prevent corrosion and improve the appearance of a metal deck. Bolt holes and laps require a sheet or mat material to be used in conjunction with the fluid roofing.

12.9.2   Types of exposure   Exposure is a very important consideration in selecting a coating system. Common exposures are discussed below.

Interior   Interior coatings are used for two reasons: for protection of the substrate and for color. Interior surfaces require the protection given by a roof coating when there is a possibility of leakage to the floor below, foot traffic, moisture vapor, chemical attack or corrosion of the substrate, or leakage of inert atmospheres.

Exterior   Coatings applied to the exterior of a structure must withstand the normal elements. On roofs the primary attack is from the ultraviolet rays of the sun, ozone, and heat absorption from the sun's rays. Water in its three forms (liquid, vapor, and ice) is the enemy of every roof. It does its greatest damage as it changes from one form to another. Both dirt in the atmosphere and chemical pollutants in the air attack roof-coating systems. Use caution when these enemies of roofs or walls are present, and closely match the specification of the coating system to environmental conditions.

Between-slab   The between-slab application of a liquid waterproofing membrane is used to prevent water migration through the total deck system. Typical examples are plazas, roof-parking areas, elevated swimming pools, bathroom floors, the floors of chemical, food, and beverage plants, and similar applications. Such liquid membranes must have the ability to keep their integrity underwater. They must remain waterproof, and they must not crack or otherwise deteriorate with age. Often overlooked, they must have the ability to withstand some traffic, since the membranes must be walked on to place the wearing surface on top of them.

Below-grade   Below-grade waterproofing of exterior building walls has many of the requirements of between-slab waterproofing. Coating systems must be able to keep their waterproof integrity underwater in spite of normal building movement, and they must not deteriorate because of microorganisms in the soil. They must also withstand a head of pressure, since in many cases they will be subjected to water pressure in below-grade areas. Because these are vertical applications, coatings must be nonsagging when applied in required thicknesses.

Pedestrian traffic   Before the development of the newer coating systems, the most common method of waterproofing a promenade deck or balcony employed the between-slab type of construction. Newer systems give building designers a new freedom because many of these systems perform the dual function of waterproofing and of offering wear resistance to pedestrian traffic. This is accomplished without adding weight to the building structure, weight that is often needed when the between-slab type of construction is used. The most common substrates for these new systems are concrete and plywood.

Vehicular traffic   Vehicular-traffic decks are similar to pedestrian decks, except that they must resist the ravages of vehicular traffic. The substrate is almost always concrete.

Drainage   Good drainage is important to the success of most systems described in this chapter. Most manufacturers will not warrant their materials if they are used under ponded water. A good rule of thumb is to specify a minimum drop of ¼ in/ft (20.8 mm/m) so that the system is free-draining under ideal conditions.

12.10   Types of Topcoating

12.10.1   Asphalt cutback

Nonfibered   These coatings are used primarily to restore old, dried-out felt roofs and to prime-concrete roof decks. They are also employed on metal roofs and gutters. Nonfibered coatings are usually made from air-blown asphalt plus solvent, a combination called "asphalt cutback." The solvent, which is used to obtain a viscosity that permits cold application, assists penetration so that the asphalt is carried into the roofing felt. The use of nonfibered roof coatings as primers under fibrated coatings extends the life of the topcoating.

Asbestos-fibered   Asbestos fiber is often added to the blown-asphalt and solvent mixture of nonfibered roof coatings. It serves two purposes. First, asbestos fiber acts as a bodying agent which allows a thicker film build to be placed in one application without running or sagging. Second, it serves as a reinforcing agent which helps to extend the life of the coating.

Asbestos-fibered roof coatings should be applied as soon as possible after the installation of felt or built-up roofs to protect the top layer of asphalt on the roofs. Well-constructed built-up roofs can be maintained almost indefinitely by frequent coating with quality asbestos-fibered roof coatings.

Asbestos-fibered roof coatings are also used to rejuvenate built-up or felt roofs which have alligatored or otherwise deteriorated. They can be expected to stop a few leaks, but leaks resulting from fish mouths, cracks, and other defects must be repaired with plastic roof cements and often with supporting membranes. The life of asbestos-fibered roof coatings applied to weathered roofs will always be extended if the roofs are first primed with a nonfibered roof coating.

Because cutback roof coatings do not permit water to pass through in either direction, the substrate to which they are applied must be dry. Water-soaked roofs must be dried out before coating, or they may be vented with vents installed approximately on 30-ft (9.1-m) centers. Unless these precautions are taken, cutback-asphalt coatings applied to wet roofs may be expected to blister. On a typical built-up roof requiring rejuvenation, nonfibered roof coating is applied at a rate of 1 gal/square (100 ft²), or 0.41 l/m², as the primer, and asbestos-fibered coating at 3 gal/square (1.22 l/m²).

Glass-fiber–reinforced   These coatings are similar in composition to asbestos-fibered–asphalt roof coatings, except that they are reinforced with chopped–fiber-glass strands. The fiber glass permits formulation on a lower–softening-point base without the danger of running or sagging on steep surfaces. It also adds tensile strength and elongation properties to coatings. In addition, fiber glass increases the ability of coatings to bridge cracks, voids, and alligators on roof surfaces. Whereas the successful use of asbestos-fibered roof coating without supporting membranes or felts is limited to roofs which are not yet severely alligatored and suffer from only relatively few cracks, glass-fibered roof coatings may be used successfully on roofs without supporting membranes that are considerably alligatored and cracked. There are obvious economies in material and labor when it is possible to treat such roofs without felts and membranes. Glass-fibered coatings are usually applied in minimum thicknesses of 3 gal/square (1.22 l/m²).

Mastics and granules A technique employed for rejuvenating built-up roofs and masonry sidewalls is to use heavy-bodied–asphalt mastics covered with colored roofing granules. The heavy-asphalt mastics rejuvenate and waterproof the surface, while the roofing granules lend attractive colors to the finished job and protect the mastics from deterioration. The granules also contribute to cooler temperatures in buildings, since they absorb less heat from the sun than do black-asphalt preparations.

Typically, such installations consist of a spray application of asphalt mastic at the rate of 3 to 5 gal/100 ft² (1.2 to 2 l/m²), covered with roofing granules at the rate of 100 lb/100 ft² (4.9 kg/m²). The granules may be broadcast with a shovel or, preferably, blown into the tacky mastic with sandblasting equipment at low air pressure.

12.10.2 Asphalt emulsions There are two basic types of asphalt emulsions: chemical emulsions and the mineral-colloid type. For roofing purposes the mineral-colloid type usually is employed. To produce asphalt-emulsion roof coatings, the asphalt is broken down into fine particles and suspended in water. The mineral colloid, usually bentonite clay, is used to surround each particle of asphalt in order to keep it in suspension. The resultant product is a thixotropic, buttery, brushable, and pumpable coating. Because asphalt emulsions are static, the applied coatings result in smooth-appearing roof surfaces. These coatings have good resistance to alligatoring and blistering.

By their nature emulsions are not waterproof, only water-repellent. They adhere well only to properly prepared surfaces and should be applied only at temperatures of 50°F (10°C) or over so that they can coalesce properly. For these reasons, surfaces to be treated with asphalt-emulsion roof coatings must be primed with cutback primer and have excellent drainage (normally a roof slope of 2 in/ft, or 166.7 mm/m, or more). For satisfactory results roof emulsions must be applied in thicknesses approximately double those of the asphalt cutbacks. Emulsions should not be applied when rain is imminent, since they may then be washed off during the first few hours after application.

Fibered and nonfibered There is little difference between fibered and nonfibered emulsions. Most roof emulsions contain a small amount of asbestos fiber. These emulsions are designed to be used in conjunction with glass mat in order to build sufficient film thicknesses to achieve a sufficiently water-repellent surface. Typically, a worn roof is treated with cutback primer at 1 gal/square (0.41 l/m²), followed by emulsion at 3 gal/square (1.22 l/m²). Then glass mat is set in place, and another coat of emulsion is applied at a rate of 3 gal/square.

Glass-reinforced   Glass-reinforced emulsions have chopped–fiber-glass strands mixed into them. This built-in fiber glass eliminates the need for a separate glass mat on most roofs. Glass strands increase tensile strength and help to bridge the checking and alligatoring typical of asphalt built-up roofs.

12.10.3 Refined coal tar   Refined coal tar, a by-product of the coking of coal, is an excellent waterproofing material for flat or nearly flat roofs. In its natural state it has greater penetrating power than asphalt, and it can be applied to damp roofs. Because it is heavier than water, it will displace the water. Refined coal tar softens in the hot sun and becomes self-healing, but it becomes more brittle than asphalt in cold temperatures and therefore is more subject to cracking. Most tar roofs are covered with gravel. All loose gravel should be removed before coating. Because tar coatings are self-healing, a reinforcing mat is not needed, but a very heavy application (up to 10 gal/square, or 4.1 l/m^2) is recommended to take advantage of the self-healing feature. Specifications can call for replacing the gravel on a roof repaired with a coal-tar coating, but such gravel replacement is optional. The gravel does protect the coating from the rays of the sun and reduces wind uplift of the roof. However, it interferes with the self-healing properties of the tar and makes leaks difficult to find.

12.10.4 Aluminum-pigmented asphalt   The use of aluminum-pigmented–asphalt roof coatings has steadily increased. Finely ground leafing-aluminum pigment is added to a cutback-asphalt vehicle, asbestos fiber, and other mineral pigments. The applied coating presents a reflective aluminum surface. The aluminum pigment reflects the heat from the sun, thus reducing the surface temperature of the roof. The lower temperature maintained on the roof surface, in turn, reduces the heat absorbed into the building, thus increasing comfort and reducing air-conditioning loads. The reduced temperature of the roof mat prolongs the life of the roof, which is subject to less heat. The aluminum pigment also protects the asphalt from the destructive rays of the sun. However, aluminum roof coatings are subject to discoloration when good drainage is not provided.

Nonfibered   Nonfibered aluminums are topcoats that are applied to roofs after all repairs and rejuvenation have been completed. Their sole function is to add an aluminum topcoat to the roof and thus gain the advantages mentioned above.

Asbestos-fibered   Aluminum-pigmented   asbestos-fibered–asphalt coatings are a variation of cutback-asphalt asbestos-fibered coatings. The vehicle content is modified to obtain good leafing of the aluminum pigment and yet maintain the protective qualities of an asbestos-fibered roof coating. These coatings combine the advantages of a coating, a restorer, and an aluminum surface in one package.

Glass-fibered   Aluminum-pigmented glass-fibered–asphalt coatings are a variation of glass-reinforced asphalt-cutback coatings. The vehicle is modified to obtain good leafing of the aluminum pigment. These coatings combine the advantages of higher tensile strength, ability to take more movement than asbestos-reinforced–asphalt coatings can, greater hiding of surface imperfections, and an aluminum surface in one package.

12.10.5   Colored coatings   Colored roof coatings are used for the same reason that aluminum roof coatings are, with the added feature of attractive colors. They lower the temperature of the roof, protect it from ultraviolet rays, and reduce the heat absorbed into the building. In addition, colored coatings make possible color coordination of the building components. There are two types of colored roof coatings: solvent-based coatings and emulsions.

Solvent-based   Solvent-based coatings are made from resins that are thinned with solvents and pigmented for color and body. They must have these characteristics: good color stability, good adhesion to the substrate, quick drying, and good hiding and coverage rates. These coatings are excellent for beautifying, waterproofing, and protecting built-up or felt roofs, masonry walls, and sprayed-on urethane foam. Built-up roofs must be at least 1 year old to avoid discoloration of the coatings caused by asphalt bleed-through. Surfaces must be properly patched to fill cracks and other voids before application. Solvent-based coatings are also recommended for restoring and beautifying asphalt-shingle roofs. Many such coatings can cause shingle tabs to curl. If the shingle tabs are not of the locking type or self-adhering, they should therefore be cemented down before application of the coatings.

Emulsions   Emulsion coatings are made from resin emulsions. They may be used successfully only on surfaces with excellent drainage, with a slope of at least 3 in/ft (250 mm/m). Most of these coatings have the following characteristics: excellent hiding of the surface; chalking, which makes them self-cleaning; and good adhesion to damp surfaces. Check weather conditions before applying emulsion coatings because they can be washed off by rain immediately after application, in the period before they set. They should be applied at temperatures of 50°F (10°C) or higher to obtain good adhesion. They must not be used on asphalt shingles. As a rule, emulsion coatings are not as durable as solvent-based colored coatings. Many of these coatings set up to a rigid film, causing cracking problems.

12.11   Coatings for Applying Cap Sheets   Each of the coatings described so far is limited to roofs which are still in good to fair condition, depending on the coating. Aluminum and colored coatings should be applied only to roofs which are still in reasonably good condition, while glass-fibered and granule-mastic coatings will correct worse conditions than can be corrected by any of the other coatings described. However, there comes a time when the condition of a roof can no longer be corrected with coating alone and a cap sheet must be applied. A cap sheet is basically a new roof that is installed directly over the old one.

12.11.1   Hot-applied versus cold-applied cap sheets   Cap sheets may be applied with hot bitumen just as most original roofs are, or they may be applied with cold adhesives and coatings. The type of reinforcing sheet used in a cap-sheet system depends upon the adhesive and the coating.

Hot-applied cap sheets normally are applied by professional roofers, since handling hot bitumen requires considerable skill. Cold-applied cap sheets also are applied by professional roofers, but they can readily be installed by plant maintenance crews and other nonprofessionals.

Cold-applied cap sheets normally outlast hot-applied sheets, since cold-applied adhesives and topcoatings usually are more flexible and durable than hot bitumens are. The cold-applied sheets normally are more highly plasticized, and their durability is extended by their asbestos-fiber reinforcement.

With hot-applied systems there is no waiting period for adhesives and coatings to dry; therefore, a larger roof area can be installed in a day. However, person-hours of labor expended per 100 ft² (9.29 m²) are about the same for both systems, since normally fewer persons are required to cover a given area with cold-applied systems. The following section covers only cold-applied cap-sheet systems.

12.11.2   Cold-applied roofing felts   Conventional roofing felts should not be used with cold-process roof adhesives. These felts are only saturated with bitumen and are not coated. Therefore conventional felts are very porous and would absorb too much of an adhesive. Cap sheets for cold application are double-coated roofing sheets. When adhesive is applied to them, it softens the asphalt coating and welds the two sheets together. Typical cold-process roofing felts are 45-lb and 55-lb roll roofing. All such roofing is "powdered" so that the rolls can be unrolled. For cold-application felts, sand is the preferred powder, and mica the least desirable.

12.11.3   Glass roofing sheets   The best cap sheets for use with cold-applied adhesives and coatings are made from glass fibers. Glass obviously is less subject to deterioration than are the vegetable fibers in ordinary roofing felts. Glass sheets normally used in cold-applied methods also have the advantage of greater flexibility and, therefore, of readier conformance to roof irregularities. The most successful type of cap sheet used with cutback-asphalt coatings is one of random fiber-glass construction held together with a resin and a light coating of asphalt on each side. When held up to the light, this cap sheet is found to be porous. The cap sheet is applied to the roof with an asphalt adhesive, care being taken to work the sheet into the adhesive, thus filling the pores. The cap sheet is then heavily coated with either asbestos-fibered or glass-fibered asphalt. Glass mats of random construction held together only with resin binder as well as glass membranes may also be used with cutback-asphalt coatings. The membranes are fiber-glass strands woven in a regular pattern. The more open the construction, the more care that must be taken to fill all the voids with sufficient asphalt coating for waterproofing. Also the lighter the sheet, the more difficult it is to handle on a windy roof.

With emulsion coatings a glass mat of random construction with an asphalt binder is normally used. Typically, such mats are of heavier fiber-glass construction than those used with cutback coatings; also typically, such mats are of more open construction. Emulsions are static, as previously described, and this characteristic makes penetration more difficult. The emulsion cannot work its way through the denser construction normally used with cutback coatings. When emulsions are used, the roof is primed, a heavy coating of emulsion applied, the mat laid into the wet emulsion, and another heavy coat of emulsion applied over the mat.

The cap-sheet systems described may also be used as original roofs on decks of concrete, wood, insulation board, and other materials.

12.12   Roof Cements   Roof cements (Table 12.4) are trowel-applied patching cements. They are used for patching cracks and other openings in roofs before such roofs are coated with rejuvenating coatings. Roof cements are used in both new and remedial construction, for installing flashings around projections coming through roofs, on parapet walls, along gutters, and wherever roof surfaces are interrupted. Until recently, all flashing cements had to be applied with a reinforcing sheet. Some flashings which need no separate reinforcement are now available. The most popular reinforcing material for roof cements is glass membrane, but many other materials, such as glass mats, roofing felts, aluminum foil, and even rags, are used.

12.12.1   Asphalt

Regular asbestos-fibered roof cements   These are the most commonly used roof cements. They are made from asphalt cutback (similar to that used in nonfibered and asbestos-fibered roof coatings) and are heavily pigmented with asbestos fiber and other mineral fillers to achieve a consistency which can readily be applied with a trowel. These roof cements should always be used in conjunction with a reinforcement. They are first applied to the imperfection or flashing to be treated, the reinforcement is

TABLE 12.4 Typical Formulas for Roofing Cements

	Percent
1. *Regular asbestos-fibered roof cement*	
Nonfibrated-asphalt roof coating	75
Canadian asbestos fiber, 7-D	15
Slate flour	10
	100
2. *Wet-surface roof cement*	
Regular asbestos-fibered roof cement	97
Wet-surface additive	3
	100
3. *Glass-fibered roof cement*	
Nonfibrated-asphalt roof coating	70
Polybutene (150 standard Saybolt units at 100°F)	10
Canadian asbestos fiber, 7-D	15
Glass fiber, ¼ in long	5
	100
4. *Coal-tar roof cement*	
Nonfibrated-tar roof coating	75
Canadian asbestos fiber, 7-D	15
Slate flour	10
	100
5. *Neoprene flashing cement*	
Type W neoprene	18
Phenolic resin	4
Carbon black	10
Canadian asbestos fiber, 7-D	4
Curing agents	3
Chlorinated paraffin	10
Mica	5
Xylol	46
	100

pressed into the plastic cement, and more of the cement is then applied over the reinforcement, care being taken to cover the reinforcement thoroughly with a heavy film. Typically 1 gal (3.785 l) of plastic roof cement so used will cover approximately 10 to 15 ft² (0.93 to 1.39 m²) of surface.

Wet-surface roof cements   By employing proper additives it is possible to use some asphalt roof cements on damp and even on wet surfaces. A material with this capability is in great demand to fix leaks during inclement weather.

Glass-fiber–reinforced roof cements   The addition of chopped–fiber-glass strands to plastic roof cement increases its tensile strength and allows it to take much more movement before rupturing than is possible with regular asbestos-fibered cement. The asphalt vehicle must be formulated with sufficient plasticity so that the glass fibers can move. Materials with this capability are used to repair breaks without additional membrane reinforcement, thus reducing the cost of patching and flashing. They also allow patching to be made on irregular surfaces that are very difficult to waterproof when a reinforcing sheet must be employed.

12.12.2 Coal-tar roof cements   Refined coal tar is used as a base for roof cement. When asbestos fiber and mineral fillers are added to make a trowelable material, these tar-based plastic cements are suitable for patching and for installing flashings on tar roofs. Since tar softens severely in the hot sun, use caution when employing these cements on vertical surfaces and specify supported glass-membrane or other reinforcing sheets. In many cases the use of asphalt plastic cement is preferable on vertical flashings even though there may be a problem of incompatibility between the asphalt cement and the tar with which it comes into contact.

12.12.3 Neoprene flashing cements   Neoprene is a synthetic rubber having outstanding weather resistance. It is compounded into trowelable flashing roof cements that cure into excellent rubbers within a few days after application. These elastomeric cements are completely compatible with built-up and most other roof systems and are recommended for high-movement areas where conventional materials fail. They are used without any separate reinforcement, since any reinforcing sheet would serve only to interfere with the natural elasticity of the neoprene. Typical applications include use in the areas around hangers for gutters or over gravel stops and in the installation of expansion joints.

12.13 Types of Liquid Roof Systems

12.13.1 Fluid-applied roofs   Liquid-applied roof systems are designed to be the sole waterproofing cover for such decks as concrete, plywood, and insulation board. These systems normally are used only where their attractive appearance is of benefit. Inherent in the application of liquid-applied roof systems is the need to treat all cracks and joints with either a trowel-consistency compound or a reinforcing sheet; otherwise the liquids would disappear in such cracks or joints. Such systems also require careful attention to flashing details. On the other hand, on many free-form types of roofs there is no practical way to waterproof the surface with a sheet system. On irregular surfaces, applying liquids rather than cutting and fitting sheets saves labor costs. Liquid systems also have the advantage of being free of all seams upon curing. With any sheet system the roof is waterproof only to the degree that the seams between the sheets remain permanently watertight.

A variety of fluid-applied roof systems are available. A description of these systems by generic groups follows.

Neoprene-Hypalon   Neoprene-Hypalon systems are based on the two synthetic rubbers so named (Hypalon is a registered trademark of E. I. du Pont de Nemours & Co.). The base coats are compounded from neoprene, and the topcoats from Hypalon. Neoprene coatings generally have better elastomeric properties than do Hypalon coatings. However, Hypalon coatings have much better color stability. They also are more expensive than neoprene coatings, and the two types therefore are normally used in combination.

Typically these coatings are approximately 25 percent solids and are applied in multiple coats. Therefore, three coats, each applied at the rate of 1½ gal/100 ft² (0.61 l/m²), produce a finished dry mil thickness of only 18 mils (0.4572 mm). A typical application consists of a neoprene primer, two coats of neoprene coating at 1½ gal/100 ft² each, and two coats of Hypalon coating at 1 gal/100 ft² (0.41 l/m²) each. Neoprene-Hypalon systems, being made from two very durable elastomers, have a reputation for excellent weathering characteristics and ability to bridge hairline cracks in the substrate. Hypalon coatings are available in white and a variety of colors.

Silicone   Silicone coatings are two-part synthetic-rubber coatings that exhibit excellent weather and chemical resistance. The two-component material has a high solids content so that thick films are possible in one coat. With proper priming, good adhesion is obtained to most common substrates. Silicone systems are difficult to patch because of poor adhesion, and they also tend to attract dirt. Typically, silicone roofing systems consist of a primer, a base coat of approximately 1 gal/square (0.41 l/m²), and a topcoat of approximately 1 gal/square. The base coat and the topcoat are two-part coatings which must be mixed on the job. The base coat is approximately 85 percent solids, while the topcoat consists of the same material thinned with 1 qt of solvent per 2 gal of coating (1 l of solvent per 8 l of coating).

Urethane   Urethane liquid roofing systems are relatively new in the marketplace, and extensive performance experience is not yet available. Urethane elastomers are highly weather-resistant. Urethane coatings are available as one-part, ready to use on the job, or two-part, to be mixed on the job immediately before use. Two-part coatings have the advantage of being 100 percent or nearly 100 percent solids, so that film thicknesses of 50 mils (1.27 mm) are readily obtained in one application. This feature not only places a thick protective film on the roof but also reduces the amount of crack, joint, and flashing preparation that is necessary with coating systems containing solvents (and therefore experiencing shrinkage upon curing). One-part urethane coatings have the advantage of being ready to use on the jobsite. On the other hand, they contain solvent to secure a consistency that is practical to apply. Because these coatings cure through the absorption of moisture from the air, they cure at unpredictable rates that depend upon ambient-moisture conditions. Also, stability of the product in the container can be a problem, since any moisture in the container before the product is used can cause a curing condition prior to use. Urethanes are usually applied to roofs in one or two coats for a total dry film thickness of approximately 40 mils (1.016 mm).

Low-molecular butyl   Low–molecular-weight butyls are relatively new on the marketplace, and further evaluation is needed to determine their effectiveness. Butyls inherently have good water resistance and low vapor transmission. Low–molecular-weight–butyl coatings are two-component materials which are mixed prior to application but still are only approximately 50 percent solids by volume. Therefore, they suffer from the disadvantages of both one- and two-part urethanes. In addition, they cannot be produced in white or pastel colors. A topcoat of Hypalon coating is recommended when light colors are desired. Low-molecular butyls normally are applied to roof decks in one to two coats for a total of 2 to 3 gal/square (0.81 to 1.22 l/m²), or a total dry film thickness of 15 to 22 mils (0.127 to 0.3048 mm).

Polyvinyl chloride   Polyvinyl chloride coating systems are similar to the cocoon type of coatings used during and after World War II for protecting equipment over a long period of time. They consist of polyvinyl chloride resins dissolved in solvents and plasticizers and then pigmented to a wide range of colors. These coatings have been used on masonry more extensively than on other surfaces. They are usually employed in conjunction with a glass-mat reinforcement. Like neoprene-Hypalon coatings, polyvinyl chloride coatings are relatively low in solids percentages; therefore, multiple coats are necessary to achieve sufficient film thickness. These coatings have the additional disadvantage of low elongation values. Because of plasticizer migration, they tend to become brittle over a period of time. Because of these serious disadvantages, polyvinyl chloride coatings hold a lower share of the market than they did a decade ago.

Asphalt emulsions   When liquid-applied coating systems were in their infancy, many roofs were covered with asphalt emulsions and a glass mat, topcoated with a colored emulsion roof paint. As more and more elastomeric-type liquid roofing systems have been perfected, fewer asphalt-emulsion systems are being applied as roof systems to bare decks. Asphalt-emulsion systems lack the ability to move with building movement, and they also lack the durability that is available in the newer elastomeric systems. (Discoloration of the topcoat due to bitumen migration from the asphalt emulsion has been a problem on many installations.)

12.13.2 Pedestrian and vehicular traffic decks   On many types of modern buildings there is a demand for waterproofing systems that maintain their integrity even when subjected to pedestrian or vehicular traffic. A variety of liquid-applied systems have been developed to meet this demand. By and large they are designed to withstand either pedestrian or vehicular traffic, but not both. Wear from vehicles is much more severe than that from foot traffic, so that a typical pedestrian deck system is not suitable for vehicular traffic.

Neoprene-Hypalon   Pedestrian and vehicular traffic decks made from neoprene and Hypalon normally are not a combination of the two coatings, as roof systems are. Traffic decks usually are made from either neoprene or Hypalon. They normally consist of a neoprene or Hypalon coating, with glass mat or glass roving embedded in the coating, a traffic surface of aggregate mixed with the coating, and, in many instances, a topcoat over the aggregate coat. Most neoprene or Hypalon traffic deck systems consist of from three to seven layers of material. The multiple coats are required to achieve the film thickness that is considered necessary for resistance to cracking and traffic wear.

Urethanes   Urethane traffic systems come in one-part and two-part versions, just as urethane roof systems do. They generally consist of one or two coats. In most urethane systems, rather than mixing the aggregate with the coating, the aggregate is sprinkled onto the applied coating before it cures. Even a one-part urethane system can produce a higher solids level than neoprene-Hypalon coatings can, thus reducing the number of coats and the labor required to install a satisfactory traffic deck.

Silicone   Silicone traffic deck systems are available for pedestrian traffic only. They have the advantage of being made from a relatively high-solids coating (approximately 85 percent), so that the deck can be laid with a prime coat plus one aggregate coat of 3 to 7 gal/100 ft² (1.22 to 2.85 l/m²), depending on the type of traffic. However, silicone systems are two-part systems which must be mixed on the job. Also, adhesion of subsequent coats is difficult, and problems thus may arise when repair is necessary.

12.13.3 Between-slab and below-grade moisture barriers   Coating systems are widely used in between-slab and below-grade waterproofing, for which they offer the advantages of quick application, quick cure, and good elastomeric properties. The various coatings supplied for this use are described by generic groups as follows:

Neoprene   Neoprene coatings similar to roofing solutions are used for between-slab and below-grade moisture barriers. These coatings are selected for their low vapor permeability and resistance to chemical attack.

At least two coats are usually required to attain a minimum film thickness of 20 mils (0.508 mm).

Urethane   While urethane moisture barriers are relatively new, they are already taking a greater share of the market than are neoprene coatings. Urethane coatings for between-slab and below-grade use come as either a one-part or a two-part system, just as urethane roof coatings do. The coatings used for this type of application normally are modified with bitumen to reduce their cost as compared with urethane roof coatings. The faster and more reliable cure of the two-part system is of special importance for between-slab and below-grade waterproofing, since on almost every construction project the fastest possible cure in these areas is wanted. When a urethane coating is used as between-slab waterproofing, contractors want to be able to walk over the surface, if, indeed, not to pour the final wear surface. On below-grade installations, contractors usually are in a rush to backfill. These coatings typically are applied in dry film thicknesses of 25 to 60 mils (0.635 to 1.524 mm).

Polysulfide   Polysulfide coatings are very successful moisture barriers. They have characteristics similar to those that can be formulated with urethanes, but at a higher price.

Asphalt-cutback and -emulsion coatings   Asphalt-cutback and -emulsion coatings similar to those used on roofs have been employed on below-grade foundations almost as long as they have on roofs. Since they lack the elastomeric properties of neoprenes, urethanes, and polysulfides, asphalt products are not recommended for critical installations. Because between-slab waterproofing is relatively new and elastomeric products were already available when it came into vogue, asphalt products are not extensively used for this purpose. Still, the minimum treatment for most below-grade foundations is coating with a nonfibered-cutback material similar to nonfibered roof coating. Such a thin coat, unreinforced, cannot resist even the slightest amount of cracking that may occur in the foundation. Therefore, it is of limited value.

Some foundations are coated with heavy-bodied asphalt mastics. For more dependable long-term below-grade waterproofing, however, a combination of asphalt mastics and supporting membranes or sheets, similar to those used on roofs, is recommended. A good below-grade waterproofing job consists of at least three coats of mastic and two layers of reinforcing sheet.

Although asphalt emulsions are used on below-grade foundations, their use is not recommended. Under the constant-moisture conditions often found below grade, emulsions have a tendency to reemulsify. Also, since every particle of asphalt is surrounded by clay in producing the emulsion, this coating is not waterproof and should be used only on applications where drainage is good.

Coal-tar coatings   Coal-tar coatings are used on below-grade foundations in a manner similar to those described for asphalt cutbacks. Coal tar is better for this application than asphalt because of its greater resistance to microorganisms in the soil. Because of the cold-flow characteristics of coal tar, however, it is difficult to obtain the required film thickness for good waterproofing on the vertical surfaces of foundations. When coal-tar products are used for this purpose, it is important to backfill promptly to avoid exposure to the hot sun.

12.14   Bonds and Guarantees   In many specifications architects call for a guarantee or bond of a liquid-applied roof system. There is a difference between these two types of instruments.

12.14.1   Roof bonds   These are called bonds because they are backed by an insurance company. The company steps into the picture only if the manufacturer who has issued the bond is financially unable to perform under it. The most common roof bond is issued for 20 years. Under the common roof bond, the manufacturer and, frequently, the applicator are obligated to make any repairs to the roof necessary to keep it waterproof because of a failure in material or workmanship. The normal roof bond contains many exclusions, sometimes even exempting the repair of flashings. The most restrictive factor of the roof bond is that the total obligation of the manufacturer and contractor normally is limited to a certain dollar amount, which usually is a small fraction of the original contract. It is not at all unusual for the entire obligation of the manufacturer and contractor to be used up within the first few years after installation of the roof, so that the owner has no further warranty protection.

12.14.2   Roof guarantees   Roof guarantees are normally given for a period of from 1 to 5 years or sometimes for as long as 10 years. The roof guarantee normally is a combination guarantee between the manufacturer and the applicator. There usually is no dollar limit. Therefore, during the life of the guarantee the manufacturer and sometimes the contractor are obligated to make whatever repairs are necessary to keep the roof waterproof regardless of the cost. The roof guarantee normally is not backed by an insurance company.

Since problems with roofs normally show up within the first 2 or 3 years after installation, the roof guarantee is considered to be a more valuable instrument than the normal roof bond, provided it is issued by a reputable and financially sound manufacturer.

BIBLIOGRAPHY

Abraham, Herbert: *Asphalts and Allied Substances,* D. Van Nostrand Company, Inc., Princeton, N.J., 1960.
Hoiberg, Arnold J.: *Bituminous Materials,* Interscience Publishers, New York, 1964.

Chapter 13

Fireproof Coatings

SEYMOUR I. KAWALLER

TSI, Inc.

13.1 Introduction   Until recently, fireproofing applications were not part of the painter's craft. However, the development of chemically reactive coatings and mastics and their growing acceptance by both code and insurance authorities created the need for specifications in this coatings area.

One must distinguish between *fire-inert* paints, which simply use specially formulated resins as binders and/or pigments or impregnants that do not support combustion, and *fire-retardant* coatings. When exposed to heat, the latter undergo physical and chemical changes to delay destruction of the substrate by fire.

While the latter part of this chapter touches on *fire-retardant* paints, the central focus is on chemically reactive coatings, including guidelines and specifications. The goal is to achieve both flame-spread protection and *fire-resistive* characteristics. With regulations and test procedures changing rapidly, specification writers must base their final specifications on currently applicable codes to meet fire-protection requirements. Specifiers should become totally familiar with the rules and ratings of insurance organizations (Table 13.1) and appropriate building and fire codes (Table 13.2).

TABLE 13.1   Insurance and Other Organizations

1. *American Insurance Association (AIA)*. Formerly the National Board of Fire Underwriters (NBFU), the AIA is supported by the insurance industry and provides local rating bureaus with rate-making guidance which influences virtually all building construction. The organization sponsors the national building and fire codes but has no enforcing authority.

2. *Industrial Risk Insurers (IRI)*. The IRI is an outgrowth of a recent merger of the Oil Insurance Association (OIA) and the Factory Insurance Association (FIA). It now provides engineering and inspection services to a major group of capital-stock insurance companies covering preferred-risk accounts. These are primarily multiplant companies in high-risk industries such as those producing and refining chemicals, petroleum, and petrochemicals.

3. *Underwriters Laboratories (UL)*. UL provides recognized testing facilities and reports as well as the UL label and listing services to identify tested materials. It is supported by fees paid by manufacturers whose products are tested by the organization.

NOTE: The UL label is *not* an indication of official approval but simply assurance that a product has been tested in accordance with a specified standard and manufactured under UL controls to conform to the standards of the materials used in the tests.

4. *Factory Mutuals (FM)*. FM provides services comparable to those of the IRI for a major group of underwriters in the mutual-insurance field. It also supports an engineering division and a research corporation which provides test facilities and reports to meet FM requirements.

5. *State rating bureaus*. These bureaus enjoy quasi-official status in most areas and offer liaison between owners, insurance underwriters, regulatory agencies, and officials. They establish rate schedules and provide inspection and engineering services to underwriters and their clients.

6. *American National Standards Institute (ANSI)*. This organization consists of industrial firms, trade associations, technical societies, consumer organizations, and government agencies. It serves as a clearinghouse for nationally coordinated voluntary safety, engineering, and industrial standards.

7. *National Fire Protection Association (NFiPA)*. This is a consensus standards-making body that publishes, among other manuals, the *Life Safety Code No. 101*, which is adopted by reference in many codes and covers standards governing the construction of areas of exit, entry, and public assembly.

TABLE 13.2   Building and Fire Codes*

1. *The Uniform Building Code.* Sponsored by the International Conference of Building Officials (ICBO), this code exerts a major influence from the West Coast through the Midwest.

2. *The Basic Building Code.* Sponsored by the Building Officials and Code Administrators International (BOCA), this code is strongest in the East and Midwest.

3. *The Standard Building Code.* Sponsored by the Southern Building Code Congress (SBCC), this code is used principally in the Southeast and Gulf Coast states.

4. *The National Building Code.* Sponsored by the American Insurance Association, this code exerts national influence. It should not be confused with a proposed *Federal Building Code.* It is the only model code that does not utilize a democratic code-changing procedure open to public comment and adoption in public voting sessions.

* Local codes are generally based on one of these "model codes."

The distinction between the terms "fire-inert," "fire-retardant," and "fire-resistive" (or "fire-resistant") needs specific clarification. A *fire-inert* material is nonburning; that is, it won't support combustion. A *fire-retardant* material slows burning across a surface, given a combustible substrate. A *fire-resistive* material delays heat penetration through a substrate. It is important to avoid any suggestion that intumescent *paints* are fire-resistive; they are fire-retardant. Intumescent *mastics,* however, *are* fire-resistant. The critical difference is based on the method by which the materials are tested and rated. Paints are tested in accordance with ASTM E-84; mastics, in accordance with ASTM E-119. While *subliming* paints use a different mechanism of protection, in most other characteristics they are similar to, and can be classified with, intumescent coatings—either paints or mastics, since they are available in both forms.

Specification writers should understand that there is no relationship between fire-endurance (ASTM E-119) and flame-spread (ASTM E-84) ratings. Many local codes require a flame-spread rating of 0 to 25 in crucial escape areas such as corridors and exits, while a flame-spread rating of 26 to 75 might be acceptable for other areas. Check local codes before specifying a coating.

Fire-endurance ratings apply more specifically to property protection by indicating a period of time before failure. Flame-spread ratings indicate a degree of danger to life by measuring how a flame will spread over a given surface. You must also include smoke-development criteria in setting specifications for these coatings.

13.1.1   Building officials and inspectors   These are the enforcing authorities for locally adopted building codes. They are responsible for code interpretations, code approvals, and enforcement.

13.1.2   Fire marshal   The fire marshal is usually the head of the local fire-prevention bureau (a division of the fire department) or other organization charged with inspection and enforcement of fire-prevention regulations and investigations. Generally the building department inspects plans and enforces codes relating to construction which requires a building permit, while the fire-prevention bureau inspects existing facilities to assure continued compliance with code provisions or regulations, particularly in licensed occupancies. However, the fire-prevention bureau often must approve plans before a permit is issued by the building department. No hard-and-fast rule applies to all situations in this field.

13.1.3 Definitions   Fire-protection work is complicated by semantics. To assure clarification, specifiers should be familiar with the following definitions and identifications:

- *Flame-spread rating.* A comparative measurement of the speed of flame travel across a given surface in accordance with ASTM E-84 (the 10-min Steiner tunnel test). It is generally identified with the evaluation of fire hazards of interior finishes. It should not be confused with fire endurance.
- *Fire-endurance (fire-resistance) rating.* A determination of the time required for a given material or assembly of building materials to reach a failure point, based on ASTM E-119 testing. Door assemblies are tested in accordance with ASTM E-152, which also evaluates the effective seal around a door within the time frame of the test exposure.
- *Fire-retardant paint.* A coating which retards flame spread on a combustible substrate by means of a chemical reaction in which the coating is converted to protective thermal insulation when exposed to heat. Intumescence and sublimation are two of the more common protective mechanisms.
- *Fire-retardant lumber.* Wood impregnated with chemicals and tested in an extended tunnel test (ASTM E-84) for 30 min, with the result of a flame-spread value of not more than 25 and no evidence of significant progressive combustion. Surface application of fire-retardant paint providing equal UL-rated protection also is often acceptable to meet this requirement.
- *Fireproofing.* Materials applied to protect other materials or assemblies against heat penetration, thus delaying the time in which the failure point is reached in ASTME E-119 tests (also identified as UL No. 263, ANSI A2.1, and NFPA No. 251).

13.2 Fire-Resistive Coatings   When exposed to the heat of a fire, these finishes convert to chemically reactive by-products which retard heat penetration. Formulations are designed to contain fire within an enclosed area or to offer protective cladding to structural steel and so to help prohibit stress failure at high temperatures (Fig. 13.1) through this chemical reaction.

For example, a mixture of ammonium phosphate and melamine and pentaerythritol results in a tremendous proportional increase in volume, with a thickness of 0.1 mm swelling to over 1 cm. Some intumescent paints are water-sensitive and thus are unsuitable for outdoor use. However, isano oil has proved to be quite successful as an exterior coating in White Sands, New Mexico, where a nuclear fire caused by a nuclear test nearby was considerably deterred through the use of this intumescent coating. Use of intumescent paints on wood basement ceilings in New York State tenements is credited with a lifesaving performance in a number of cases.

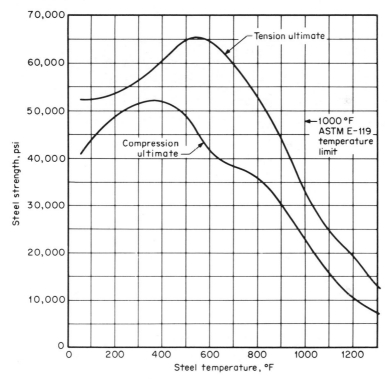

Fig. 13.1   Strength of structural steel under fire exposure. Note that ultimate strength in compression at 1000°F (538°C) approaches the usual design stress for steel. [*Carboline, Fireproofing Products Division*]

Chemically reactive coatings include fire-resistive formulations based on epoxies, vinyls, resorcinol resins, sublimation compounds, and tar-based materials. Ratings, which are established by testing in accordance with the approved standard, are listed as the lowest time segment during which the tested unit withstands fire exposure prior to reaching the failure point. Ratings are published for 1 h, 1½ h, 2 h, 3 h, and 4 h. Typically, fire-exposure tests are conducted in an enclosed (oven) environment under controlled conditions (Fig. 13.2).

13.2.1   Standard Fire Test: ASTM E-119   Ratings are based on two separate test procedures using the same fire exposures.

- *Structural members.* Columns and beam assemblies are tested for a specific time in this fire exposure. Fireproofings are installed to prevent the fire heat from being transmitted into the steel and causing premature collapse. For beams an average temperature reading of 1100°F (593°C) or a single-point temperature rise above 1300°F (704°C) has been established as the failure point in this test. Beams are rated for use in both restrained and unrestrained installations. Column-failure temperatures are set at 1000°F (538°C) on the average or at 1200°F (649°C) at a single point.

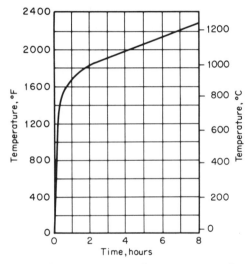

Fig. 13.2   Temperature curve for ASTM E-119 fire test:
1000°F (538°C) at 5 min
1300°F (704°C) at 10 min
1550°F (843°C) at 30 min
1700°F (927°C) at 1 h
1850°F (1010°C) at 2 h
2000°F (1093°C) at 4 h
2300°F (1260°C) at 8 h or over

- *Wall and floor assemblies.* These assemblies are tested to measure the fire-containing capability of a design. In this exposure the failure point has been established as the time at which heat transmission through the assembly of materials in the design reaches 250°F (139°C) above ambient room temperature, as read by thermocouples placed against the back of the tested section.

What the Standard Fire Test does   ASTM E-119 provides a fire exposure predicated on what would be expected from a fire occurring in a "typical" room or building containing "normal" quantities of combustibles. Since it is unlikely that any specific project will conform to the typical or the norm, this test merely provides a frame of reference by which all materials and designs can be compared.

What the Standard Fire Test does not do   Results of the ASTM E-119 test do not provide assurance that the specified rated protection will be obtained in any individual case. Fire-exposure differences related to fuel load and to area variations and environments are some of the considerations that will affect the actual performance of a design. These modifying influences should be considered in establishing the fire rating required in any specific case.

Example: If there is a need of actual protection for 1 h in a facility where fire-load exposures may be more severe than the standard ASTM E-119 time-temperature curve, a fire rating in excess of 1 h may be specified to help compensate for the more severe exposure.

13.2.2 Who runs the Standard Fire Test?   There are many independent, qualified testing agencies whose reports will be acceptable to approving authorities for most nonfire test requirements. However, only a select few have the facilities and capability of complying with ASTM E-119 standards. Underwriters Laboratories has unquestioned authority in this field. This organization not only tests but lists the test results in a manual called *The Fire Resistance Index*. Factory Mutual Engineering Laboratories, Ohio State University, Southwest Research Institute, United States Testing Laboratory, Innovative Fire Technology Laboratories, and others also conduct acceptable tests. However, to date only UL, Warnocke-Hersey, and FM provide the follow-up which assures the user that the product delivered to the project is identical to the product which was tested. The testing agency's label is applied to each package of the fireproofing material before it is shipped to the jobsite as assurance that it has been manufactured under controlled conditions and subject to the supervision and inspection of the testing organization.

NOTE: There is a popular misconception that the UL label itself indicates approval of a fire-retardant coating. *This is not true.* UL is simply a nonprofit testing organization whose function is to test and to list the results of its tests. Evaluation by a code or insurance authority whose responsibility includes approval is then required.

13.2.3 Fire-protective mechanisms of chemically reactive coatings   Low thermal conductivity, or heat-transfer coefficient, provides the basis for fire resistivity. Most of the physical characteristics of chemically reactive coatings are self-evident, but "paint-on" fireproofing materials provide fire protection through all or a combination of some of the following chemical reactions, which are dormant until triggered by heat from a fire.

Calcination   Heat breaks down chemical compounds of the stable material into new compounds or elements, leaving a powder residue. The process uses up British thermal units, which therefore are unavailable to affect the steel.

Ablation   Friction causes the peeling of exhausted surface layers of insulation to expose new protective layers. This reaction is commonly identified with the protection of nose cones on missiles during the reentry stages of flight, when friction heat is countered with an ablative shield.

Intumescence   Heat triggers a multifaceted reaction, converting a thin coating into a thick insulating barrier of highly reflective multicellular carbonaceous foam insulation.

Thermal hydrogeneration   This occurs during some calcinating reactions when the chemical breakdown releases H_2O as water vapor. Gypsum plasters and most cementitious materials utilize this reaction for fireproofing. A proprietary magnesium oxychloride–based plaster offers about 2½ times the water content of gypsum. In addition, the water-forming temperature for this material is about 572°F (300°C), compared with 180°F (83°C) for gypsum. As a result, fireproofing values are greatly increased. However, corrosion on steel substrates requires special priming protection.

Sublimation   Sublimation develops when heat triggers a reaction in which a solid is converted into a gaseous state without passing through a liquid stage. Large quantities of heat are necessary for sublimation, leaving less heat available to affect the substrate, and the outflowing gases block heat penetration from the fire and inhibit combustion.

Transpiration   This is a mode of fireproofing which is utilized by all porous materials in facilitating the release and dispersion of heat and cooling gases during a chemical reaction in the coating.

Reflection   Reflection helps to prevent penetration of some of the heat input which attacks a surface, thus aiding the fireproofing process. The greater the reflectivity, the more effective the result. Subliming mastics, for example, utilize this supplementary effect since the coating turns white-hot under fire exposure during the chemical reaction.

Film thickness required   A single application in a thickness of $^3/_{16}$ in (4.762 mm) can provide a 1-h fire rating on steel columns based on the ASTM E-119 fire test for subliming mastics. Urban-area codes and insurance companies more frequently require 2 h or more of fire-rated protection for most occupancies; thus the 1-h rating is not very meaningful. Unfortunately, to achieve a 2-h rating, some of the economic advantages of using mastics are negated by the complicated and costly procedure required to install the necessary fiber-glass reinforcing.

Fire protection can be achieved to some degree by almost any insulative or reflective coating. However, sophisticated, high-cost materials can often be justified by the economics of a thin application which saves not only material but the time and labor costs of installing a thick coating, especially if there is no requirement for costly embedment of reinforcing lath or fiber glass. Weigh installation economics carefully.

example: There is a maximum thickness (approximately $3/16$ to $3/8$ in, or 4.762 to 9.525 mm) beyond which intumescent mastics may be self-defeating. Thicker applications develop such heavy insulation that there is a danger that it may fall off of its own weight. This could result in less protection than a lighter coating would provide. As a preventive measure, fiber-glass reinforcement has been used to slow down and add cohesiveness to the intumescence. The addition of enough fiber glass to be effective interferes with the basic intumescent action. Therefore, the fiber glass must be introduced as an intermediate cloth reinforcement, embedded between layers of intumescent mastic. Such an application procedure is slow and costly. Still, this system represents an attractive alternative for installations where the possibility of physical abuse makes softer materials inappropriate, especially for applications requiring the fire protection of columns. Galvanized-metal reinforcements such as hardware cloth or chicken wire also are used to support certain intumescent coatings during application and to provide resistance to fire-hose streams in case of a fire. This support also is required for most cementitious fireproofings, including concrete encasement installations.

13.2.4   Cementitious plasters

While poured-in-place concrete offers fire protection in new construction, there is also a need to protect existing structures or those where concrete is not poured. Magnesium oxychloride cements (which contain 35 percent water of hydration plus some absorbed water) can be considered for these applications.

Magnesium oxychloride plasters rely on thermal hydrogeneration to provide water-cooling protection for the structural steel. At a temperature of about 600°F (316°C) water vapor is released. The result is akin to the effect of water in a pot over a direct flame. As long as the water can absorb the heat, the pot is unaffected. In the same way, as long as the plaster continues to react and actively releases water vapor, fire protection is effective.

Cementitious plasters decompose at the rate of 0.04 in (1.016 mm) during the first hour of exposure in a 2000°F (1093°C) fire. Because magnesium oxide is a very white material, it also acts as an insulating and reflective shield. This residue of the chemical reaction tends to reduce the rate of heat activity transmitted to the fireproofing, thus prolonging the protective process of thermal hydrogeneration. In contrast to the intumescent mastics, there is no theoretical limitation on the thickness of these coatings, as it is possible to develop fire protection of any required duration with the application of a sufficient thickness.

Architects and engineers should realize, however, that magnesium oxychlorides often are corrosive and that substrates must be protected with appropriate primers. If the corrosion affects the reinforcing lath, there may be some loss of fire protection. Also, the presence of free chlorides can damage stainless steel. For stainless-steel applications, specification writers can specify new chloride-free cementitious products to avoid the corrosion problem.

13.2.5 Choosing the proper chemically reactive coating   All building codes make similar fireproofing requirements based on such considerations as building size, height, location, and occupancy classification:

- Identify the use or type of occupancy for your building under "Use Group" in the *Code Manual*.
- For the identified occupancy, find the listing of the square footage and height limitations of the building.
- These data will identify the type of construction which is required.

Table 13.3 shows the specific fire ratings required for each structural element of a building, keyed to the type of construction and using *The Basic Building Code* criteria.

Using Table 13.3, note the requirements for columns, beams, trusses, girders, etc. Under each type of construction, minimum fire-resistance ratings are listed for these structural members. All codes have similar requirements. Since it is possible to cover most wall or roof deck and

FLAME-SPREAD AND OTHER DATA ON RESORCINOL-ALDOL-PLUS-ALCOHOL (RAA) TWO-COMPONENT FIRE RETARDANT		
Test	Current specification	RAA results
UL-required Steiner 25-ft tunnel test		
Flame-spread rate	25	5
Smoke density	50	15
Ohio State University		
Heat-release rate	500 Btu/min/ft^2	15 Btu
Smoke release (particles per minute per square foot)	150 ppm	10 ppm
U.S. Bureau of Standards smoke chamber: determinations under direct flaming and also in nonflaming condition		
Average results for binder PVC and acrylonitrile-butadiene-styrene	600–1100	
RAA over Nomex paper		36
RAA over fiber glass		4
Low-oxygen index tests: minimum amount of O$_2$ needed to maintain combustion		
Red-oak wood (the standard)	20 percent	
RAA over Nomex paper		32 percent
RAA over fiber glass		64 percent
University of Utah toxicity test		
No toxic fumes found on pyrolysis of RAA		

beam requirements with a rated interior finish (Sheetrock or acoustical ceiling), chemically reactive fireproofings may not be competitive if this type of construction is to be used.

In general, limit specifications for chemically reactive fireproofings to projects where steel will be exposed to view or to abuse (Table 13.4). If the interior finish (membrane ceiling) will be breached by air-conditioning ducts, lighting units, or similar openings, it may be more economical to fireproof hidden steel behind the finish than to provide a rated membrane system which must include rated lighting fixtures and air-conditioning ducts.

13.2.6 Additional potential for chemically reactive fireproofings

Upgrading existing masonry materials   The Portland Cement Association (PCA) has published data indicating that the addition of a $^3/_{16}$-in (4.762-mm) thickness of intumescent mastic to the underside of a concrete slab will add 1 h to the rated value of the concrete. It seems reasonable to assume that similar improvement can be achieved with other coatings, such as magnesium oxychloride plasters, at lower cost.

In addition, new ambient-temperature–curing two-component resins are becoming available. One innovation that dramatically reduces smoke propagation is based on a resorcinol-aldol resin combined with an alcohol catalyst.

The UL *Fire Resistance Index* indicates that the addition of ¾ in (19.05 mm) of portland cement or gypsum plaster will add 30 min to the rating of any masonry fire wall. Since paint-on coatings have superior fire-resistance qualities, a good case can be made for substituting these materials at reduced thicknesses.

Because chemically reactive coatings have been fire-tested to determine the specific rate of heat transmission through the coating, it is possible to relate this rate to ASTM E-119. By measuring the heat-transmission rate, a fire-protection engineer can develop a basis for the required film thickness of a fireproof coating in order to upgrade an existing substrate to meet any code requirement. Chemically reactive fire-retardant coatings could provide the basis for thinner coverage and, at the same time, a suitable interior finish, thus justifying higher material costs.

Protection of urethane-foam interior finishes   It has been determined that the ASTM E-84 flame-spread test method is not applicable to foam plastics because they decompose at low temperatures and release highly toxic and flammable gases. Fire-insurance underwriters have established new test standards and criteria for the acceptance of such materials. An approved method of compliance is to cover the foam with a coating which will provide a "15-min finish rating" while remaining in place during the exposure.

TABLE 13.3 Fire-Resistance Ratings of Structural Elements (in Hours)

Structural element[a]		Type of construction, Sec. 214.0									
		Type 1 Sec. 215.0 Fireproof		Type 2 Sec. 216.0 Noncombustible[b]			Type 3 Sec. 217.0 Exterior masonry walls			Type 4 Sec. 218.0 Frame	
				Protected		Unprotected	Heavy timbers (mill)	Ordinary			
								Protected	Unprotected	Protected	Unprotected
		1A	1B	2A	2B	2C	3A	3B	3C	4A	4B
Exterior walls (Sec. 906.0)[c]											
On street lot lines or with fire separation of 30 ft or more from interior lot lines or any building	Bearing	4	3	2	1	0	2	2	2	1	0
	Nonbearing	0	0	0	0	0	0	0	0	0	0
1. On interior lot lines or less than 6 ft therefrom or from any building	Bearing	4	3	2	1½	1	2	2	2	1	1[d]
	Nonbearing	2	2	1½	1	1	2	2	2	1	1[d]
6 ft or more but less than 11 ft	Bearing	4	3	2	1	0	2	2	2	1	0
	Nonbearing	2	2	1½	1	0	2	2	2	1	0
11 ft or more but less than 30 ft	Bearing	4	3	2	1	0	2	2	2	1	0
	Nonbearing	1½	1½	1½	1	0	See Sec. 217.0	1½	1½	1	0

No.	Construction element	4	3	2	1	0	2	1	0	1	0
2.	Interior bearing walls and partitions	4	3	2	1	0	2	1	0	1	0
3.	Fire walls and party walls (Sec. 907.0)	4	3	2	2	2	2	2	2	2	2
4.	Fire-separation assemblies[e]	Not less than fire grading of use group →									
		Fire-resistance rating corresponding to fire grading of use group →									
5.	Fire enclosure of exitways, exitway hallways, and stairways (Sec. 909.8)[f]	2	2	2	2	2	2	2	2	1	1
6.	Shafts other than exitways, elevator hoistways (Sec. 910.0)	2	2	2	2	2	2	2	2	1	1
	← Noncombustible →										
7.	Exitway access corridors and vertical separation of tenant spaces (Sec. 909.0)	1	1	1	←— Footnote g —→		1	1	0	1	0
	Other nonbearing partitions	0	0	0	←— Footnote g —→		0	0	0	0	0
8.	Columns, girders, trusses (other than roof trusses), and framing (Sec. 911.0) — Supporting more than one floor	4	3	2	1	0	See Sec. 217.0	1	0	1	0
	Supporting one floor only	3	2	1½	1	0	See Sec. 217.0	1	0	1	0
	Supporting a roof only	3	2	1½	1	0	See Sec. 217.0	1	0	1	0
9.	Structural members supporting wall (Sec. 911.0)	3	2	1½	1	0	1	Not less than fire-resistance rating of wall supported →		1	0
10.	Floor construction including beams (Sec. 912.0)[h]	3	2	1½	1	0	See Sec. 217.0	1	0	1	0

TABLE 13.3 Fire-Resistance Ratings of Structural Elements (in Hours) (Continued)

| Structural element[a] | Type 1 Sec. 215.0 Fireproof | | Type 2 Sec. 216.0 Noncombustible[b] | | | Type 3 Sec. 217.0 Exterior masonry walls | | | Type 4 Sec. 218.0 Frame | |
| | | | Protected | | Unprotected | Heavy timbers (mill) | Ordinary Protected | Ordinary Unprotected | Protected | Unprotected |
	1A	1B	2A	2B	2C	3A	3B	3C	4A	4B
11. Roof construction including beams, trusses, and framing arches and roof deck (Section 912.0)[h,i]										
15 ft or less in height to lowest member	2	1½	1	←— Footnote g —→	0	See Sec. 217.0	1	0	1	0
More than 15 ft but less than 20 ft in height to lowest member	1	1	1	0	0	See Sec. 217.0	0	0	1	0
			——— Footnote g ——→							
20 ft or more in height to lowest member	0	0	0	0	0	See Sec. 217.0	0	0	0	0
			——— Footnote g ——→							

SOURCE: *The Basic Building Code*, 1975. Sections noted refer to this code.

[a] For special high-hazard uses involving a higher degree of fire severity and higher concentration of combustible contents, the fire-resistance rating requirements for structural elements shall be increased accordingly (see Sec. 400.3).

[b] Protected exteriors shall be required within the fire limits in Type 2 construction as follows: high-hazard uses, 2-h fire resistance with fire separation up to 11 ft.

c The fire separation or fire exposure in feet as herein limited applies to the distance from other buildings on the site, from an interior lot line, or from the opposite side of a street or other public space not less than 30 ft wide to the building wall (see definitions, Sec. 201.3).

d See Sec. 303.2.

e See Secs. 213.0, 909.0, and 912.0.

f In all buildings of Types 3 or 4 construction, the stairways and their enclosures may be constructed of wood or other approved materials of similar characteristics and of adequate strength, except in school buildings (see Sec. 616.9.3).

g Fire-retardant treated wood, complying with Sec. 903.6.1, may be used as provided in Sec. 903.6.2.

h In Type 3A construction, members which are of material other than heavy timber shall have a fire-resistance rating of not less than 1 h (see Sec. 853.2).

i If the omission of fire protection from roof trusses, roof framing, and decking is permitted, the horizontal or sloping roofs in Type 1 and Type 2 buildings immediately above such members shall be constructed of noncombustible materials of the required strength without a specified fire-resistance rating or of mill-type construction in buildings not over five stories or 65 ft in height (see Sec. 913.3).

TABLE 13.4   Pros and Cons of Commonly Used Fireproofing Materials

Type	Advantages	Disadvantages
Concrete	Hard, durable; meets requirements up to 4 h; may be used in place of structural steel	Heavy; costs more than other types of fireproofing for structural steel; requires expensive labor and longer installation time
Gunite	Hard, durable; quickly installed	Heavy; requires steel reinforcing; messy spray application
Masonry	Durable; attractive exterior finish	Slow, costly installation
Troweled plasters	Hard, durable; attractive interior finish	High-cost multistage installation; easily damaged unless reinforced with corner beads
Sprayed plasters	Low-cost, lightweight	Messy, dusty installation that cures to soft, friable insulation; easily damaged; may dust or flake
Magnesium oxychloride plaster	Hard, durable, lightweight "paint-on" fireproofing; suitable for interior or exterior use	Higher-cost than soft spray-on materials
Gypsum board	Interior finish for large surface areas such as walls or ceilings	Costly installation for beams and columns; not suitable for exteriors
Intumescent	Quickly installed, hard, durable "paint-on" coating; best for short-term column protection	Installations requiring more than 1-h rated protection require costly, complicated installation; some brands use hazardous "red-label" toxic solvents which are not compatible with many primers
Sprayed mineral fibers	Lightweight; lowest-cost fireproofing for hidden installations	Susceptible to damage from physical contact, vibration, or air movement; dust may be a health hazard both during and after application

13.2.7 The fireproofing specification   Many variables affect the specification for fireproofing. For example, if concrete encasement is selected in place of sprayed fireproofing, the steel frame may require beefing up to carry this load. The cost of fireproofing a lighter steel frame with more sophisticated coatings may be justified. However, for specifying purposes it is important to indicate the degree of fireproofing durability that will be acceptable in order to obtain the lowest possible bid. For example, if a soft cementitious material will meet the requirements of the project, a lightweight plaster or fibrous fireproofing can be selected. A chemically reactive plaster or a subliming coating will not be competitive, but the latter will provide a more durable, wear-resistant finish.

For cases in which chemical or physical abuse is anticipated or for weathering exposures, the selection of a hard-finish product is indicated. Acceptance of a soft material as an "equal" penalizes the owner and all bidders who have tried to provide the specified characteristics. No specifier should request a superior grade of material, only to accept a subcontractor's substitution of a softer material after bids have been accepted. Yet this situation frequently occurs because of inadequate understanding of the characteristics of various fireproofing materials.

Warning: Never accept a UL-listed fire rating as the sole criterion for the selection of a fireproofing product.

Another warning: Consider the project in its entirety. A material which may meet the requirements of 90 percent of a structure may be inappropriate for selected areas which may be subject to unusual service conditions. For example, in a multistory commercial building with parking facilities below the principal occupancies, one of the fiber or cementitious fireproofings will serve adequately for areas where membrane ceilings will protect against dusting or flaking, but a hard, weather-resistant product should be selected for the parking levels and the service areas in the boiler room, elevator shafts, and steel supporting the air-conditioning equipment, printing presses, or similar operating equipment. Proper specification may call for more than one fireproofing product.

Although virtually every building requires this discriminating approach to fireproofing selection, few structures benefit from such specifications. Invariably, the specifier indicates a fire-rating requirement and leaves it to the applicator to provide a suitable product. Naturally, the cheapest material will be installed, and fire protection can be virtually nonexistent soon after the building has been occupied.

Selection of the right fireproofing product for a specific requirement must take into consideration conditions of actual use. A product which might be suitable for use in a library, for example, would certainly not meet the requirements of a printing plant. Anticipated exposure to physical or chemical abuse might easily determine the suitability of one material versus another. Available clearance space, temperature conditions, and design and aesthetic considerations all must be evaluated. Use Table 13.5 to select correct fireproofing materials.

Before writing the specification, be able to answer these typical questions:

■ *What is the nature of the fire-protection requirement?* Commercial structure? Light industrial structure? Petrochemical structure? Is the ASTM E-119 fire test meaningful for the requirement?

■ *To what environment will the fireproofing be exposed?* Interior or exterior? Is a weatherable fireproofing material required? Will it be subject to physical abuse, vibration, impact, or chemical attack?

■ *What are the corrosion considerations?* Is the fireproofing expected to provide corrosion protection? Is there anything in the fireproofing which can cause corrosion of the substrate or adjoining surfaces? Is corrosion resistance a requirement of the material selected?

■ *Costs?* What are the first costs? What are the expected maintenance costs?

Keep in mind that it is the designer's responsibility to establish a credible design based on the test data available and to obtain the prior approval of the responsible code or insurance authority for the designated

TABLE 13.5   Material Installation Features and Problems

	Typical thickness and weight	Installation features	Major problems faced by contractors
High-density cementitious materials (concrete, gunite)	2 in 24 lb/ft²	Requires heavy steel reinforcement. Thickness controlled by forms or edge guides. Surface preparation minimal	High-cost, time-consuming installation; very difficult in congested areas
Low-density cementitious materials (plasters, fiber- and vermiculite-loaded cementitious compounds)	½ to 2 in 2 to 12 lb/ft²	Requires lighter-gauge reinforcement; anchoring very important. Surface preparation required (critical in some cases). Sealing a problem. Thickness control important	Sealing difficult; some constructions fragile. Maintenance sometimes required. Not well suited to congested areas
Spray-on mastic coatings (generally intumescent or subliming compounds)	3/16 to ½ in 1 to 3 lb/ft²	Reinforcement not required in some materials. Surface preparation important. Most coatings seal well. Thickness control extremely important in these thin coatings	Thickness control and application in proper weather conditions. With one exception, maintenance required

Source: G. K. Castle, Castle Fireproofing & Coating, Inc., Baton Rouge, La., "A Contractor Views Fireproofing," *Hydrocarbon Processing*, August 1979, p. 180.

use. (Substituting one material for another has definite legal ramifications if the substitution is made by the applicator.)

13.2.8 Fireproofing costs   Typical guidelines indicate that the cost of concrete encasement would be 10 times the cost of spray-on fireproofing and that gunite or conventional plaster fireproofing would be applied for about 7 times the cost of spray-on fireproofing. More durable intumescent or subliming coatings would also cost more than soft conventional spray-ons.

The range of fireproofing possibilities and cost comparisons is endless, but comparison figures can point the specifier in the right direction in deciding on the type of use in the specification and the range of pricing to be expected. Job conditions play an important part in sprayed-fireproofing costs, and yardsticks apply only to projects with ready access to the steel. Outdoor industrial applications involving complicated scaffolding and staging will be far more costly. For lowest cost be sure to specify that fireproofing should be installed after all steel is in place, but before hangers, ducts, equipment, and similar obstructions have been installed.

Finish requirements are another cost factor. If extra finishing is required, identify the minimum acceptable level. For example, smoothing or leveling sprayed finishes will require extra work and should be specified only if desired. Normal spray applications will reflect the spraying technique in a dimpled or irregular surface finish. An applied sample should be required for approval before proceeding with the installation. Failure to insist on this procedure is a major cause of controversy in fireproofing installations.

As a rule of thumb, intumescent or subliming coatings provide a protected-beam and metal-deck application at a budget of $3 per square foot of surface area to be covered with a thickness of ¼ in (6.35 mm) for a 1-h rating. Ratings of 2 h on either beams or columns require additional labor to install reinforcing–glass-fiber embedment, plus greater thicknesses, almost to triple the cost of 1-h applications. In such cases, subliming mastics are economical, especially if reinforcing can be omitted.

Sprayed cementitious fireproofing is competitive with fibrous materials; it provides reduced acoustical values but slightly more durability at 5 to 10 cents per square foot over fibrous materials. For magnesium oxychloride plaster a 3-h fire-rated beam application should cost about $3.50 per square foot; column applications require greater thicknesses. Applications of 1 to 4 h on columns are possible at installed costs of $6.50 to $10.50 per square foot.

13.3 Fire-Retardant Paints   While the majority of paint binders are flammable, binders containing nitrogen, bromine, or chlorine are much less flammable (exceptions are nitrocellulose lacquers, particularly if they are combined with phosphorus, or silicon). In general, water-based paints and those using inorganic binders such as lime, cement paints, or paints based on the silicates as well as such pigments as zinc borate and antimony oxide are fire-inert (they will not add fuel to a fire). However, they do not protect the substrate.

Fire retardants may be either impregnants or paints (Table 13.6). The theory behind the success of the impregnating fluid is that it tends to combine with the free radicals of the cellulose in wood to prevent the formation of other flammable breakdown products. Another theory is that the acids of the impregnant, such as hydrobromic or hydrochloric acid, act as dehydrating agents and remove all the water from the burning medium, leading to rapid carbonization and lowering the heat transfer, with the ultimate result of reducing flame spread.

TABLE 13.6   Fire Retardance of Coatings

Binder	Burning rate (inches per minute)
Polyvinyl chloride	Slow to self-extinguishing
Cellulose acetobutyrate	Very slow
Polyethylene	1.04 (slow)
Nylon	Self-extinguishing
Chlorinated polyether	Self-extinguishing
Fluorocarbon	None
Polyester	Self-extinguishing
Epoxy	Slow
Polypropylene	Slow

The success of fire-retardant paints depends on one of the following:

- Ability to produce nonflammable gases such as hydrochloric or hydrobromic acids, carbon dioxide, ammonia, or water vapor
- Formation of glasslike melts from silicates, borates, or phosphates by the development of the porous, spongy layer characteristic of intumescent coatings
- Absorption or utilization of heat in the process of sublimation to retard heat penetration to the substrate

These points have certain implications (Table 13.7). The specifier should assure that any off gases produced will not harm human life under emergency conditions.

13.3.1   Pigments for fire-retardant paints   Most inorganic pigments will retard the spread of fire. The most effective are calcium carbonate (whiting), antimony oxide, and silicate-containing pigments. Organic pigments, on the other hand, are generally flammable.

TABLE 13.7 Typical Applications for Coatings with Fire-Resistant Characteristics

Surface	Coating or system	Comments
Any combustible surface, including painted and unpainted wood, cardboard, fiber, and pressed boards	Intumescent or subliming paints	Fire protection is designed to provide protection against flame propagation along a combustible surface. The rate of application coverage is based on testing in accordance with ASTM E-84 and is rated by flame spread, fuel contribution, and smoke development
Steel	Chemically reactive mastics or plaster coatings	Protection criteria are based on testing in accordance with ASTM E-119. Formulations are designed to retard heat penetration to or through substrates. Hourly ratings are assigned on the basis of thickness and design details of installation
Concrete	Chemically reactive mastics or plaster coatings	Application is limited to requirements for upgrading fire-resistive capability. The degree of improvement is affected by the thickness applied and the temperature of effective chemical reaction. The synergistic value of the coating plus substrate usually is greater than mathematically calculated values

13.3.2 Impregnants used for fire-retardant purposes   Impregnants are substances which are applied to surfaces to penetrate the pores of wood or wallboard in order to retard the spread of fire. Such substances are ammonium salts such as ammonium sulfate, ammonium bromide, and ammonium dihydrogen phosphate. Under conditions of fire, the last-named compound decomposes into ammonia, water, and phosphoric acid. Impregnants should be applied only to unsealed porous surfaces. Since these materials are soluble in water, the treated surface should be sealed with a compatible paint if further exposure to water is expected.

13.3.3 Types of binders for coatings
Binders which produce nonflammable gases
Those Producing Hydrochloric Acid

- Chlorinated rubber (Parlon)
- Polyvinyl chloride
- Polyvinylidene chloride
- Chlorinated plasticizers (chlorinated paraffin and chlorinated diphenyl)
- Tetrachlorophthalic anhydride used in alkyds

Those Producing Hydrobromic Acid

- Alkyds using tetrabromophthalic anhydride
- Epoxy resins which include tetrabromobisphenol in their manufacture

Binders which form glasslike melts These contain inorganic compounds like phosphates, borates, or silicates. The theory is that the melt formation absorbs heat and that this tends to reduce the temperature and at the same time introduces a new barrier between the air and the substrate, lowering the rate of combustion.

At one time silicate paints were the only fire retardant paints of importance. The incombustible product, which is water-soluble, forms a porous, heat-insulating layer when heated. A shortcoming is that it tends to react slowly with carbon dioxide in the air; when heated, it could flake off the surface.

13.3.4 Other fire-retardancy test methods

Fire retardancy of paints (cabinet method: 1360-58) This method determines the fire-retardancy properties of a coating by determining its comparative weight loss and char volume as compared with a standard. A 6- by 12- by $\frac{1}{4}$-in (152.4- by 304.8- by 6.35-mm) coated panel is weighed and placed at 45° from the vertical in a sheet-metal cabinet. In a container placed 1 in (25.4 mm) from the panel, 5 ml of ethyl alcohol is permitted to burn completely. The panel is then cooled and weighed again, and the weight loss determined. The volume of char is determined by cutting along the lines of maximum length and width of attack on the panel. The maximum depth is multiplied by the other two dimensions to get the char volume.

Fire retardancy of paints (stick-and-wick method: D1361-58) This method determines the percentage of weight loss, the average rate of flame spread, and the average char height. The test piece is a coated and weighted stick, measuring 1 by 1 by 16 in (25.4 by 25.4 by 406.4 mm) and having a folded piece of gauze wrapped around one end, which is placed vertically, gauze end down, in a fire-shield box measuring 8 by 11 by 30 in (203.2 by 279.4 by 762 mm). The gauze is saturated with 4 ml of absolute alcohol and ignited. The average height of the flame (flame spread) is taken from 10 readings (a reading is made every 10 sec). After the flame is extinguished with CO_2 gas, the weight of the stick and the height of the char are determined. The percentage of weight loss (W), rate of flame spread (F), and height of char (C) are determined by comparison with an uncoated burned stick and combined to compute the burn index from the following formula:

$$\text{Burn index} = \frac{W + F + C}{3}$$

Schulz fire-resistance tester (Gardner)   An 0.5-cm gas-burner flame is displaced toward and away from the test surface by a hand-driven cam. At its closest point the nozzle is 1.0 cm away from the panel. With the flame approaching the panel every 5 sec, the number of times the panel is heated before it ignites determines its fire resistance.

Truax fire-tube apparatus (Forest Products Laboratory)   A 40-in (1016-mm) specimen of coated wood is suspended on one arm of a balance through a fire tube 50 in (1270 mm) long. A gas flame is applied to the lower end of the test strip for 4 min. The flame is controlled to give a temperature of 1832°F (1000°C) at the hottest of the flame and of 356°F (180°C) 1 in (25.4 mm) above the center of the top of the tube without a specimen in place. Loss of weight and maximum temperature at the top of the tube determine the fire resistance of impregnated wood.

Fire Underwriters Laboratories tunnel (Steiner)   The tunnel, which measures 20 in by 18 in by 25 ft (508 mm by 457.2 mm by 7.62 m), is heated at one end by two gas burners placed 12 in (304.8 mm) from the fire end of the tunnel. A sample of coated lumber measuring 20 in by 1 in by 25 ft (508 mm by 25.4 mm by 7.62 m) is placed at the top of the tunnel, forming its ceiling. A slight suction at the vent end of the tunnel carries the flame 4½ ft (1.37 m) from the burners. A photoelectric cell at the vent end indicates the density of smoke, and thermocouples indicate the temperature near the downstream end of the sample.

Three characteristics of the coating under test are determined.

- *Flame-spread value.* With the flame-spread value of oak taken as 100, the coating must reduce the value to 70 to qualify as a fire-retardant coating. If codes call for a noncombustible interior finish, a flame-spread rating of 0 to 25 is required.
- *Fuel-contributed factor.* The temperature of the furnace at a point 23 ft (7 m) from the burners is 1200°F (649°C) when using uncoated oak. The fuel-contributed factor is the proportion of heat produced by the burning of the coating itself.
- *Smoke-developed factor.* The amount of smoke developed is determined by the reading on the photoelectric cell. The ratio of smoke developed by the coated oak panel to that of the uncoated panel defines the smoke-developed factor.

Forest Products Laboratory tunnel   A sample measuring 13¾ in by 8 ft (349.25 mm by 2.4384 m) is supported under the top of the tunnel, which is a triangular combustion chamber 14 in (355.6 mm) wide on one side that slopes from end to end at an angle of 6°. Photoelectric cells and thermocouples are located in the stack to measure smoke density and temperatures.

In this apparatus, the flame spread of red oak will be 87 in (2209.8 mm) in 18.4 min. The flame-spread index of a fire-retardant coating will be determined by the ratio

$$\text{Flame-spread index} = \frac{\text{distance reached on specimen in 18.4 min}}{87}$$

Schlyter fire test   Two panels, each measuring 12 by 31 in (304.8 by 787.4 mm), are supported vertically 2 in (50.8 mm) apart, forming a sort of chimney. The flame is placed between the panels, and the height of flame is observed for 15 min at 15-sec intervals. The test may be either mild or severe. For the mild test, a bunsen burner with a fishtail tip is placed at right angles to the panels. For the more severe test, a Meker burner having a T head 5 in (127 mm) long with 12 holes drilled in it is placed between the panels.

Radiant-panel test for flame spread (162-60T)   A porous ceramic material measuring 12 by 18 in (304.8 by 457.2 mm) has radiant heat (from a gas heater) maintained at 1238°F (670°C). A test specimen measuring 6 by 18 in (152.4 by 457.2 mm) is placed at an angle of 30° and 4¾ in (120.65 mm) from the radiant heat. The amount of smoke is determined by weighing a sample collected on glass-fiber filter paper. The flame spread is recorded for the entire length by noting the time for each 3-in (76.2-mm) interval.

British box test   The sample is used to build a scale-model room 7½ in (190.5 mm) square. The room, containing furniture of comparable size, is burned and compared with the standard.

13.4   Design Approach to Fire-Safe Structural Steel   Requirements for fire-protective cladding on structural steel *outside* a building have rested on standard furnace tests (ASTM E-119) that approximate fire conditions *inside* a building. For external building elements, architects, engineers, and other building professionals are considering an alternative design method, *FS3: Fire-Safe Structural Steel,* sponsored by the American Iron and Steel Institute (AISI). In time, this development may result in shifts in building codes related to alternative test methods.

The FS3 method was developed for AISI by a firm of British engineers, Ove Arup & Partners. FS3 uses heat-transfer principles to calculate the temperature that steel structural members will reach in a building fire. The critical temperature for load-carrying safety is pegged at 1000°F (538°C) for columns and 1100°F (593°C) for beams. The FS3 method tests a building design to see where that temperature may be exceeded, then lets the designer try various alternatives for keeping the steel structure out of danger.

One set of alternatives involves moving the steel away from the windows or changing the window design to accomplish the same thing. In other cases, beams and columns can be partly surrounded by sheet-steel or chemically reactive "flame shields." In all cases, the goal is to allow the use of either bare steel or steel with a minimum of applied protection.

In recent years, about 35 buildings around the world have gone through often-complicated ad hoc code approvals to use bare exterior structural steel. Among them are United States Steel's Pittsburgh headquarters, a 45-story tower with liquid-filled columns; the Centre Pompidou, in Paris; and a savings and loan institution in California. In the last-named case rust runoff has accumulated and stained the sidewalks around the building.

AISI has prepared the 96-page *Design Guide for Fire-Safe Structural Steel,* which, when used in conjunction with accepted practices, offers architects and design engineers additional safety alternatives in construction practices.

ACKNOWLEDGMENT: The author expresses gratitude to William Lawrence for contributing Sec. 13.3 to this chapter.

BIBLIOGRAPHY

"Fire Retardant Coatings," *Masterspec 2,* sec. 09841, Production Systems for Architects and Engineers, Inc., New York, 1980.

Kawaller, Seymour I.: "Fireproofing: Considering the Alternatives," *Construction Specifier,* April 1980, p. 20 ff.

———: "Specifying Fireproofing for Structural Steel," *The Specifier,* January 1977.

Ranney, Maurice W.: *Fire Retardant Building Products and Coatings,* Noyes Data Corp., Park Ridge, N.J., 1970.

Williams, Alec: *Flame Retardant Coatings and Building Materials,* Noyes Data Corp., Park Ridge, N.J., 1974.

Yehaskel, Albert: *Fire and Flame Retardant Polymers,* Noyes Data Corp., Park Ridge, N.J., 1979.

Chapter **14**

Marine Paints

RICHARD J. DICK
Battelle Columbus Laboratories

14.1 Introduction   This chapter provides a practical, current, and concise guide to the use of marine coatings for protecting structures near, on, or in seawater. All good general coating practices carry over to marine technology, but unfortunately they constitute only a small portion of what must be considered to operate effectively in a marine environment.

When marine coatings are first encountered, one outstanding fact is apparent: as opposed to the selection of a single coating for a single application, *complete systems of coatings are required*. The specification must tailor a complete coatings system from individual primers, anticorrosive midcoats, tie coats, and antifouling topcoats. Each component is selected to perform a specific function to be compatible within the total system.

The marine environment contains such destructive forces as salt water, constant washings, sunlight, and fouling. These are complicated by high condensing humidity, chemical pollution in ports, extreme ranges in service temperatures, and severe mechanical abuse. Within this extremely hostile environment, marine coatings systems are required to protect vessels, offshore structures, and dock, terminal, and storage facilities.

Vessels must be subdivided by class (commercial, government, pleasure) and by area (exterior, interior, deck, bottom, etc.). Further distinctions must be made for substrate (steel, aluminum, fiber glass, wood) and for service and size (cargo versus tanker, tug, barge, workboat, motor launch, hydrofoil). In considering offshore structures, it is important to differentiate between the conventional rigid and the new compliant stationary structures (such as drilling platforms) and anchored structures (such as navigational aids). Again, the specific area of the structure and the type and geography of the service must be considered. Dock facilities include everything from the type of service (e.g., tanks) to supportive pilings. Terminals and storage facilities, including pumping equipment, require interior and exterior pipe coatings.

To select a marine system intelligently it is essential to have a working knowledge of (1) organic-coatings technology, (2) marine corrosion, (3) marine fouling, (4) the function of each general component coating within a selected system, and (5) the economic considerations which overshadow any coating selection. Good organic-coatings practice is the general subject of this *Paint Handbook,* and additional detailed comments on organic-coatings technology are not necessary here. For those desiring additional background information, an excellent current treatment of marine coatings was published in October 1974 by the Maritime Administration of the U.S. Department of Commerce in cooperation with the Quincy Shipbuilding Division of General Dynamics.[1]

This chapter is organized in three sections. First, a brief general introduction to marine surface-preparation and application techniques concentrates on pleasure boats and the needs of the small-boat owner. The main (second) section is devoted to coatings systems and is broken down into units on primer, anticorrosive, and antifouling coatings. Introductory remarks on corrosion and fouling preface the appropriate units. The third section presents a comprehensive listing of surface requirements, coatings systems, etc., for each area of each major marine application.

---

[1] *National Shipbuilding Research Program, Project SP-2-10: Prefailure Evaluation Techniques for Marine Coatings.* Inquiries for copies can be addressed to General Dynamics, 97 East Howard Street, Quincy, Mass. 02169.

In addition to the topics covered in these three sections, certain economic considerations are unique to marine coatings and must be considered before intelligent coatings selections can be made. In theory, it is possible to determine the cost of applying and maintaining (both under way and in dry dock) a complete coatings system. A theoretical estimate can also be made of the expected service life of most coatings systems and the subsequent loss-of-services time in dry dock. However, marine painting has, by necessity, been approached under the most adverse conditions of climate and time, which make even the most general estimates unreliable. Much of this adversity reflects the low priority given to painting even though it is generally known today that "premium coatings properly applied" are the best investment of the shipowner.

A first consideration for marine economics is a definition of "expected service life," i.e., how many months of protection the shipowner desires or is willing to pay for. This can range from purely aesthetic considerations (e.g., exterior appearance for a pleasure-boat owner) to basic, practical considerations for commercial owners. When considering the painting of vessels, it must be remembered that the greatest costs are incurred when the shipowner is faced with tremendous daily dry-dock charges together with even greater daily costs arising from loss of services. Therefore, only after the expected service life has been defined, and a realistic time between dry-dockings has been programmed, can preliminary cost estimates be calculated.

Users of this *Handbook* should strive for (1) selection of proper materials, (2) rigid use of adequate surface preparation, (3) employment of satisfactory application techniques, and (4) insistence on sufficient time schedules to implement the program.

14.2 Surface Preparation and Application   For marine applications, certain minimum requirements are essential, and these vary according to substrate. Emphasis is placed on pleasure boats in this section because government surface-preparation procedures are completely covered by specifications, while commercial interests are, more likely than not, closely tied to the yard procedures of choice. Therefore, the persons who are in greatest need of being impressed with the importance of surface preparation and those who also have the greatest difficulty in obtaining adequate information are the owners of pleasure craft.

Even the small-boat owner should be aware of the impact of the Environmental Protection Agency (EPA) on marine-coatings procedures, for example, in the elimination of open blasting and in restrictions on solvents. Again, commercial yards must conform to current Occupational Safety and Health Administration (OSHA) regulations. The small-boat owner can become acquainted with the flavor of these regulations by ordering the U.S. Department of Labor brochure *Safety and Health Regulations for Ship Repairing.*[2]

Further to assist the small-boat owner, the following section is devoted to information specific to surface preparation and application of marine coatings on a do-it-yourself basis.

14.2.1 Pleasure boats   Pleasure boats require as much surface preparation, area for area before painting, as do the largest vessels afloat. Since the advice of paint contractors and the expertise of yards are not available to the small-boat owner, this type of information must be obtained independently.

For any application it is advisable to have all the necessary equipment at the jobsite before starting. Even for the smallest painting tasks this may include wire brushes, scrapers, a power sander, sandpaper and blocks, tack cloths, steel wool, a heat source to blister off old coatings (blowtorch, infrared lamp, or flatiron), putty knives, caulking guns, paint buckets, brushes, a spray gun, paint strainers, varnish cups, gauze face masks, cleaning rags, and solvent. These materials vary greatly depending on the task at hand, and a little preplanning will pay unexpected dividends in time and effort.

It is well to remember that a boat is traditionally coated or refinished from the top to the bottom (inside spars, cabin top and sides, deck, brightwork, all topsides, hull boot top, and then bottom). Generally, a new boat is completely primed before any finish work is started. An alternative procedure allows performing much of this work while the boat is afloat after the hull has been made seaworthy. It is assumed that the first task in hull preparation is sealing all scuppers and overboard drains to maintain optimum dry working surfaces.

Wood boats   Surface preparation begins with bedding, seam and glazing tasks, and sanding, sanding, sanding. Many commercial oil- and rubber-based materials (e.g., underwater-seam compounds) are specific for each job. Caulking and puttying are extremely important because they are the first line of defense against dry rot above the waterline, and they ensure a watertight bottom below the waterline. As a general rule, all new bare wood must be primed before these materials are applied (there are exceptions). Most suppliers provide truly helpful hints with their materials, such as the desirability of finishing caulk seams in new wood with a V notch by running a spatula point down the wet seam. This practice will prevent surface roughness resulting from seam protrusion when the new wood swells.

---

[2] The Bureau of Labor Standards publishes and updates this brochure as an excerpt from the voluminous *Code of Federal Regulations, Title 29.* Amendments are published in the *Federal Register.*

Old, cracked caulk must be completely routed out, and all blistered and peeled paint must be removed. After all this work has been finished, the surface should be completely smooth, and no scratches or dents (old work) or screw and nail heads (new work) should be visible. A smoothness profile of 10 mils (0.254 mm) is desirable, and any surface roughness over 20 mils (0.508 mm) must be considered significant. After the desired smoothness profile has been obtained by sanding, perserverance, and more sanding, nothing remains but painting. In areas not to be painted or varnished (e.g., forepeaks, below the level of floors), wood should be given a generous brush coat of wood preservative. Many preservatives are covered by military specifications, for example, MIL-W-18142 B. A listing of some commercial products may be found in the *OPD* (oil, paint, and drug) *Chemical Buyers Directory.*[3]

Fiber-glass boats   It is necessary to obtain a clean, oil-free surface. Light scuff sanding is beneficial in providing an anchor pattern to promote physical adhesion. For hulls not previously painted, it is essential to remove all parting (mold-release) compounds used by the manufacturer. These oily and/or waxy films are almost impossible to detect, and a thorough solvent scrubbing should routinely be used to remove them. Minor spot repairs should be done quickly with touch-up–kit materials and should never be postponed to the following season's maintenance list.

Aluminum boats   For aluminum boats, an oil-free surface is usually obtained by copious solvent washings. Sand dusting will provide a desirable anchor pattern, but severe grit blasting should be avoided because of the possibility of shot becoming embedded in the soft metal. In new construction, service can be greatly improved by specifying pretreatments such as anodizing. This is an economic decision which must be weighed independently for each application.

Ferrocement boats   The problems of maintenance for ferrocement boats are few, but little "dock information" is available because the small-boat owner has only recently discovered the merits of these boats. The use of steel-reinforced concrete to form the hull and sometimes the deck of a vessel is ideally suited to the do-it-yourself approach to marine construction. This material is inexpensive, does not require a high level of construction skill, is maintenance-free, and has an initially good structural strength which improves with aging.

The usual method of finishing the wet concrete is hand troweling. This leaves an open and somewhat rough surface that is an ideal anchor pattern for promoting the physical adhesion of coatings. Before concrete can be painted, it must be aged and then acid-etched to remove the alkaline deposits which form on the dry surface. Any very dilute acid will

[3] Schnell Publishing Company, Inc., New York.

serve this purpose. There are no known restrictions for selecting a resin system to finish a ferrocement hull.

Small boats For small boats (rowboats, duckboats, dinghies, canoes, "kit" boats, etc.), surface-preparation requirements remain unchanged. Varnish on wood and old canvas coatings can be removed with most commercial paint and varnish removers; other old and worn coatings can be removed by blowtorch or hand sanding. The user of paint-stripping materials should be cautioned against any wax-containing removers. The wax is frequently left on the stripped surface and presents a problem as serious as that before stripping. Solvent removal of the wax may be necessary.

Chipped, peeled, and cracked coatings must be removed to the bare substrate, but tightly adhering old coatings may need only scuff sanding to obtain a clean and dry surface before spot-priming. The entire surface is then sanded smooth in preparation for painting.

14.2.2 Application data for small-boat owners   Coatings suppliers help the weekend painter in many ways. Catalyzed coatings frequently change color to ensure that the catalyst has been added properly. Contrasting colors of the same formulation allow easy application of a high-build system from multiple coats. Some coatings have low hiding power, so adequate masking is not obtained until the desired film thickness is present. Mildewcides are commonly added to finishes that might mildew in service (it is advisable to double-check for their presence).

The relationship between temperature and the pot life of catalyzed coatings should be understood, since temperature variations of as little as 20°F (11°C) from an average of 70°F (21°C) may double pot life (at 50°F, or 10°C) or half pot life (at 90°F, or 32°C). On a hot day the benefits of applying paints taken directly from cool storage areas are obvious. It should also be noted that catalyzed systems produce heat, and a 5-gal (18.925-l) bucket of paint will rise to a higher temperature and set up much more quickly than will a 1-gal (3.785-l) can of the same material or will a small amount in a large container. Good agitation will dissipate some of this heat.

If application of the recommended system produces a lower total dry film thickness than is desired, it is necessary to add coats. *Corrosion resistance is directly related to film thickness,* and thinner films will not be as serviceable as heavier ones. However, do not apply excessively thick films (beyond the manufacturer's recommendations), which would result in solvent entrapment or pinholing. Excessive thinning of material will produce comparably thinner dry films. Wet film thickness estimates can be made by gauges pressed into the coating. Many nondestructive dry-thickness gauges are available for measuring this extremely important parameter of marine coatings.

Each manufacturer formulates coatings so that a recommended total dry film thickness can be obtained in a specific number of applications. The manufacturer's directions must be followed explicitly, as one heavy coat will never equal the performance of three thinner ones. It is impossible to list specific paint recommendations of dry film thickness for each part of a boat because of the wide differences in formulation among commercial products (use of puffing agents, extender pigments, etc.). The safest course is to select only quality materials of known performance and rigidly adhere to the manufacturer's directions.

Above the waterline, good application technique includes such items as (1) avoiding early-morning and late-afternoon dampness and all cold-weather painting (varnishes are particularly susceptible to moisture and cold) and (2) allowing sufficient time for drying before scuff-sanding for the following coat. Below the waterline, it is important to consider the unusually heavy weight of the toxicant in antifouling bottom paints. Adequate mixing and agitation are essential. Red-lead–oil primers are seldom used on wood bottoms because of incompatibility with many bottom coatings. Two thin coats are many times better than one heavy coat, and it is impractical to skimp on the labor of application.

Special problems in coatings procedure may arise regardless of the type of surface or boat being finished. For example, the *centerboard trunk* of a small boat may be almost inaccessible. To prepare the surface and apply new paint, it may be necessary to improvise and perform each operation at the end of a broom handle. *Brightwork* may require special attention if it is badly discolored. The discoloration can usually be removed by applying an acid bleach to the bare wood and then neutralizing the bleach with a borax solution. To avoid a spotty appearance, the area to which the bleach is applied should be feathered to a fine outside edge. Typical solutions can be made from (1) oxalic acid in water or (2) borax in water (each is prepared at about 1 to 8 parts by volume in hot water, with slow addition and good agitation to obtain a satisfactory solution). Most marine-paint suppliers have developed *coating systems for specific areas* of a small boat (e.g., metal fittings, interior enamels). These coatings should not be interchanged for those formulated for other applications. An engine paint developed for optimum adhesion to metal and a coating designed to protect a wood engine compartment may be based on the same vehicle, but their formulations may vary considerably.

Canvas covers for decks and cabin tops are frequently overlooked in planning general-maintenance programs. If they are new, it is advisable to prime and then paint them with materials specifically recommended by reputable marine dealers. Canvas that is old and in disrepair must be stripped, primed, and repainted. Any sanding technique that avoids gouging is acceptable. Sanding is particularly important in guarding against adding film thickness and weight with each year's yard work.

Federal standard and safety colors   These colors, which are covered in *Federal Standard 595*, should be understood by all boat owners for purposes of general marine intelligence. The standard includes such colors as the following:

Color	Number
International orange	12197
White	17886
Black	17038
Dark red	10076
Buff	10371
Blue gray	16099
Green	24432
Light gray	16251
Medium gray	16187

Items not to paint   It is extremely important to know which items should *not* be painted. These include all threaded parts, springs, oil holes, grease cups, activating mechanisms of electrical safety devices, identification plates, and joint faces of gaskets and packing surfaces. A complete list would be lengthy. It should be sufficient to remember that it is important to know how to paint properly and also what properly not to paint.

Estimating material needs   It is usually difficult for the small-boat owner to estimate the amount of coating needed for each area. All such calculations are based on the fact that 1 United States gallon of coating material (231 in^3, or 3.785 l) will cover 1604 ft^2 (149 m^2) of area at a wet film thickness of 1 mil (0.0254 mm). This can be adjusted to practical spreading rates by reducing the coverage estimate by more than 25 percent for the first coat and by less than 25 percent for the following coats. Many factors account for this 25 percent reduction (surface roughness, material losses such as overspray, applicator's skill in brushing or rolling).

It is also necessary to consider the percentage of solids since a 12-mil (0.3048-mm) wet film at 60 percent solids will deposit only a 7-mil (0.1778-mm) dry film (12 mils × 0.60 = 7 mils).

On the basis of practical experience, Woolsey[4] has organized the following remarks and formulas which illustrate the application of these facts to actual-use situations. In estimating the amount of material needed for a specific job, you may assume that 1 gal (3.785 l) of paint or enamel will cover 500 ft^2 (46.5 m^2) for one coat on the average painted surface. Over new wood, figure 325 ft^2/gal (7.9 m^2/l). A gallon of varnish will cover an average of 750 ft^2 (69.7 m^2) on recoat work and 500 ft^2 on

---

[4] Reproduced with permission of the Woolsey Paint and Color Company, Marine Paint Division.

new wood. As paint and varnish remover may require several applications, it can be expected to soften only about 200 to 250 ft²/gal (4.9 to 6.1 m²/l).

For *varnished spars,* multiply the greatest diameter (in feet) by the length (in feet), and multiply the result by 2.5. For new wood, divide the result by 500, and for previously finished wood divide by 750 to obtain the gallonage required. For example, suppose that you have a new spar 8 in (203.2 mm) in diameter and 40 ft (12.2 m) long. Then $\frac{^8/_{12} \times 40 \times 2.5}{500} = \frac{67}{500}$, or approximately ⅛ gal (1 pt, or 0.47 l) for the priming coat. For refinishing work, a pint is enough for about 1½ coats. To determine the requirements for painted spars, change the coverage factor to 325 for new work and to 500 for previously painted wood.

For *cabins* or *deckhouses,* multiply the height of the deckhouse (in feet) by the girth (in feet). Deduct the area of any large openings such as windows and doors. If the deckhouse is to be painted, divide the result by 325 for the priming coat and 500 for each finishing coat. If it is to be varnished, divide by 500 for the first coat and 750 for the following coats.

For *decks,* multiply the length of the boat (in feet) by its greatest beam (in

Estimating material to be varnished	Formula	Approximate required gallonage (1 gal = 3.785 l)
Mast (5-in diameter, 40 ft long)	$\frac{^5/_{12} \times 40 \times 2.5}{750} \times 2$	= 0.112
Boom (4-in diameter, 20 ft long)	$\frac{^4/_{12} \times 20 \times 2.5}{750} \times 2$	= 0.044
Cabin sides (25 ft²)	$\frac{25 \times 2}{750}$	= 0.067
Brightwork (30 ft²)	$\frac{30 \times 2}{750}$	$= \frac{0.080}{0.303}$ (3 pt)

Estimating material to be painted		
Deck and cabin top	$\frac{30 \times 9 \times 0.75}{500} \times 2$	= 0.81 (1 gal)
Topsides (freeboard, 2 ft 6 in)	$\frac{30 \times 2½ \times 1.5}{500} \times 2$	= 0.45 (2 qt)
Bottom	$\frac{22 \times 5 \times 3.5}{400} \times 2$	= 1.92 (2 gal)
Boot top		1 pt
Interior trim (150 ft²)	$\frac{150}{500}$	= 0.30 (3 pt)
Cabin floor		1 pt
Engine		½ pt
Metal fittings		1 pt

feet), and then multiply the result by 0.75. From this deduct the areas of cabin houses, hatches, etc. Divide the remainder by 325 to obtain gallons required for the priming coat and by 500 for each finishing coat of color. If the deck is to be coated with varnish, divide the figure by 500 and 750 respectively.

For *topsides,* multiply the length overall (in feet) by the greatest freeboard (in feet). Multiply the result by 1.5. Divide by 325 for new work and by 500 for old work to obtain the required gallonage.

For *bottoms,* multiply the waterline length by the draft (in feet). For a keelboat multiply by 3.5, and for a centerboard boat multiply by 3. Divide the final result by 300 for priming new work and by 400 for subsequent coats to give the required gallonage.

An example of estimating material for a typical 30-ft (9.1-m) *auxiliary* (general dimensions: length overall, 30 ft, or 9.1 m; length at waterline, 22 ft, or 6.7 m; beam, 9 ft, or 2.7 m; draft, 5 ft, or 1.5 m) is shown on page 14-9. The calculations are for previously finished surfaces, with two coats to be applied on the exterior and one coat on interior work.

14.2.3 Large structures and vessels   Large structures require much shop priming and finishing. On-site coatings-application techniques vary with the area of the structure (legs and substructure, tanks, superstructure). Again, the general rules apply: (1) desirability of a near-white blast if possible and (2) acceptability of a commercial blast above the waterline as an alternative procedure. Appropriate surface pretreatments and recommended coatings systems are listed in Sec. 14.4.

Vessels can conveniently be divided into government and commercial vessels. It is equally convenient to subdivide commercial vessels into cargo ships and tankers because of their different operating schedules and service requirements.

Government requirements represent standards that are justifiably high on performance and, from the standpoint of the commercial owner, equally low on considerations of economy. Current specifications are contained in *Naval Ships Technical Manual.*[5] The advantages of becoming conversant with these specifications are obvious, since they generally represent a close-to-ideal situation. On steel substrates, most government specifications for new work prior to fabrication require blasting to near-white metal. However, Chapter 9190 of the *Manual* still recognizes pickling as an acceptable method of removing rust and mill scale. The acid descaling (pickling) removes only the surface oxides, and the rather smooth resulting surface has little anchor pattern to promote the mechanical adhesion of a coating. After-fabrication techniques, including touch-up and repair, usually require blast cleaning.

[5] Chap. 9190, "Preservation of Ships in Service (Paints and Cathodic Protection)," Naval Ships System Command, Washington, D.C. 20360, January 1970.

Most commercial applications require either pickling or blasting to near-white metal. They respect the proven performance records of coatings systems applied over a blast-to-white-metal surface treatment. All rough welds, burrs, and sharp surface projections must be ground smooth, and all weld spatter must be removed prior to blast cleaning. For maintenance and repair applications below the waterline, a blast to white metal is required, and at least a commercial blast is specified above the waterline. The pretreatment or primer coating must be applied on the same day as the blasting and certainly before any visible rusting occurs.

Power and hand-tool cleaning can be acceptable if (1) a conscientious job is done and (2) it is realized that the resultant service life of the coatings will not compare with that obtained from pickled or blasted surfaces. Some areas such as machinery space bilges are so nearly inaccessible for maintenance and repair that surface preparation is a serious problem. Chemical cleaning may be the most effective method of surface preparation, and even in combination with it the use of high-performance coatings is not indicated. It is always difficult to justify a high-cost material when it cannot perform near its maximum efficiency. At the extreme minimum end of acceptable surface preparation is the use of high-pressure freshwater hoses, stiff brooms, and scrapers to remove loose rust, oil, salt, marine growth, solvent, dirt, and moisture. This practice should be considered only as an emergency pretreatment. If pits, plate edges, rivet heads, and other surface irregularities must be contended with, it is advisable to hand-brush a prime coat of the first specified material before the coating program begins. Although the following caution appears to question the intelligence of offending parties, many workers must be reminded that no marine finish is available for application over ice. (Under hockey ice use PPG 97-812 acrylic urethane.)

Of all the application techniques available today, spraying is a widely used and convenient method because it is quick and uniform and permits faster recoating. If overspray is a serious problem, it can be substantially reduced by adopting "hot spray." This technique is used effectively for applying extremely viscous materials that thin upon heating. The result is heavier films for each coat and considerable savings in labor costs.

Airless spray, which is usually available today, is generally recognized as a significant improvement in application technique for large surfaces such as ship hulls. Because no air is used, there is no paint scatter, and a uniform, wet, and heavy film results. One of the big advantages of airless spray is the ease of handling and applying coatings at normal film thicknesses during low temperatures. The proper nozzle tip (orifice size), particle size of the paint (fineness of grind), and pressure must be synchronized to obtain maximum benefit. If an oversize nozzle orifice is used, too much paint will be applied per unit area, and flooding, running, and sagging will result. Too small an orifice will prohibit the application of a wet uniform film. Marine-paint suppliers will readily furnish appropriate information for their products.

When less desirable application conditions are encountered, it should be remembered that most paints that can be sprayed can also be brushed or rolled. If labor is inexpensive, rolling is well suited to large flat surfaces, and film thickness can be somewhat controlled by cross-rolling. Touch-up brushing is required in crevices and blind areas to complete the painting. For surfaces that have been blasted to white metal and then found to be badly pitted (e.g., a plate is close to its maximum corrosion allowance), cross-rolling can be a desirable application technique. When the owner is faced with an alternative of complete plate replacement, the economics of applying high-build anticorrosive coatings becomes extremely attractive.

Brushing is more costly than rolling because more labor is involved, but it is superior to rolling for rough surfaces because slower solvents can be used. These solvents hold the paint open for longer periods of time, and greater flow and leveling are obtained. It is impractical to attempt brushing paints with fast solvents because the quick set time will not allow uniform film thickness. Over and above these considerations, it is assumed that the marine "chip and paint" tradition will continue as a general-maintenance program when under way. Accordingly, most superstructure coatings, exclusive of new work, are supplied at either brush or roller viscosity. It should be recognized that concern about environmental pollution and energy conservation is having a tremendous impact on conventional application techniques. Such innovations as water-based and solventless coatings require special application techniques.

14.3 Coating Systems   To simplify the task of discussing all component coatings used in marine coating systems, this section is divided into (1) primer coatings for metal, (2) anticorrosive coatings for metal, and (3) antifouling coatings.

14.3.1 Primer coatings for metal   Primer coatings are used mainly to provide an adhesive bond between the substrate and the complete coating system. If inhibitive pigments such as zinc are present, it is also possible to obtain anticorrosive properties. Primers gain adhesion to a good clean surface both mechanically and chemically because of their resin matrix. Proper selection of solvents improves the ability of the surface to become wet and permits the adhesive bond to strengthen as the solvents are released during cure. Primers commonly used today are listed below.

Shop-plate primers   Shop-plate primers are often called "after-blast primers" because they are applied by the plate manufacturer during the normal production schedule immediately after blasting. The plate is thus afforded corrosion resistance for long periods of time, possibly until entire sections of a ship's hull or structure have been completed. This assembly-line coating operation has been pioneered in Europe and Japan, and its attractive economics is now recognized in most parts of the United States. One of the main advantages of shop-plate primers is their cost. Blasting of new construction on an assembly-line basis may cost as little as one-fifth of the cost of manual blasting in a yard. Thus, a significant amount of the cost of a primer coating job may be saved by using factory blasting and primers. Among other advantages are (1) factory control over all pretreatment conditions (application, film thickness, and cure), (2) added coating strength gained by oven curing if baking-type materials are selected, and (3) cleaner yard working areas because of the elimination of rusty materials which previously were being blasted at the jobsite.

Epoxy resins are prime candidates for shop-plate primers because they have excellent physical properties and because the cured coatings are compatible with a wide range of topcoat vehicles. Most formulations contain inhibitive pigments. All zinc-rich paints are widely used, but these may require a tie coat to gain adhesion for the subsequent anticorrosive coatings. In practice, many vehicles, with almost unlimited combinations of anticorrosive pigments, are applied on the production line before the plate is fabricated. Little remains for the yard except a visual check for the quality of the coating application and film thickness and a nominal touch-up program.

The plate manufacturer selects primers that will (1) cure quickly, (2) not emit toxic or noxious fumes during cure, (3) not interfere with the flame cutting used for edge beveling or the large amounts of water used for cooling, and (4) tolerate automatic welding equipment. As a general statement, the development of shop-plate primers that may be welded through (see section "Weld-Through Primers" below) is one of the major breakthroughs in marine coatings which have influenced the economic feasibility of constructing large tankers (over 100,000 tons).

Wash primers   Wash primers are the materials most commonly used at the jobsite. They are so called because little application technique is required and they can literally be washed on. Because they are applied at extremely thin film thicknesses (0.3 to 0.4 mil, or 0.000763 to 0.001016 mm), they are often called "pretreatment coatings" or "metal-conditioning primers." They are pigmented with zinc chromate, are based on a polyvinyl butyral resin cut in alcohol, and are catalyzed with phosphoric acid. This formulation results in a phosphatizing action which is the same as passivating the metal, thus making it impervious to corrosion. Advantages of the wash primer include (1) possible use over damp, newly blasted surfaces because of the compatibility of alcohol with moisture; (2) ease in handling and application; (3) excellent adhesive bond with minimum air-cure time, which then allows immediate application of the next coating; (4) economy; and (5) ideal touch-up capabilities, even in applications that require spot blasting.

The phosphoric acid catalyst is added by volume in the specified ratio with the resin component (e.g., 1 : 4). Two directions must be observed. To allow for maximum benefit of the catalyst, the primer should not be applied for 30 min after mixing. Also, regardless of appearance, the solution is not acceptable for use after it has been mixed for 8 h. If the material is being applied in continuous shifts, it is advisable to label the mixing time and caution the painters. Unlike most catalyzed coatings, which become too viscous for application, the wash primer does not change appearance or consistency with time.

Most of the myriad commercial formulations available today may be considered offshoots of specification Formula 117, *MIL-P-15328 B, Metal Treatment.* The basic formulation is as follows:

Resin component	*Pounds (1 lb = 0.4536 kg)*
Polyvinyl butyral	56.0
Basic zinc chromate	54.0
Talc	8.0
Lampblack	0.6
Butyl alcohol	125.0
Isopropyl alcohol (99 percent)	353.0
Water	15.0
Acid component	
Phosphoric acid	28.0
Water	25.0
Isopropyl alcohol (99 percent)	99.0
Total weight, pounds	763.6
Total yield, United States gallons	100 (378.5 l)

As a general consideration in painting metal surfaces (i.e., steel, aluminum, galvanized metal, magnesium), a wash primer of this quality or better should be specified unless contrary information is supplied by the manufacturer of the subsequent coatings to be used in the system. There is some latitude in the quality of the surface that will accept the wash primer. For premium applications, however, the only acceptable condition is a near-white blast cleaning. However, when ideal conditions are not available or are not economically feasible, the wash primer may be applied directly over tightly adherent mill scale and even over very small amounts of rust. In these situations, the expected service life of the entire coating system must be downgraded accordingly.

Wash primers are also commonly used to seal the rough open surfaces of zinc-rich primers before additional topcoats are applied.

Weld-through primers   These primers permit a satisfactory weld to be made through them without compromising the integrity of the weld or the coating. Unless a particular technique is specified (e.g., hand welding), it should be assumed that the coatings must withstand twin-arc machine welding. Most of today's shop-plate primers emphasize this property and are formulated accordingly.

Primers based on specific vehicles   Marine primers include formulations based on specific materials such as vinyl, chlorinated rubber, epoxy, and epoxy–coal tar. Most of these materials are also used as the resin basis for successful, high-performance anticorrosive and antifouling coatings. Since all pertinent remarks are equally applicable to anticorrosives, these primers are discussed generally by resin type in the section "Anticorrosive Coatings for Metals." Notable exceptions are the zinc-rich materials, which are used mainly as primers and are employed exclusively as such in immersion situations where they must be topcoated. In atmospheric applications the thin films (under 3.0 mils, or 0.0762 mm, dry) might classify them as anticorrosive primers even if they are not subsequently topcoated. Regardless of classification, zinc-rich materials are discussed below as primers.

Zinc-rich primers   The marine chemist should know that these paints were developed to extend the known advantages of hot-dip galvanizing to materials difficult to galvanize. A single coat provides galvanic protection of steel in atmospheric exposures, but a topcoat is generally required for long exposure and more severe service such as immersion or strong acid or alkaline exposures. Zinc-rich primers are used extensively as shop primers and as weld primers in new work, and they enjoy wide service in the repair yard. The group is somewhat difficult to understand because of its many varieties. Basically, these varieties fall into two major types according to vehicle: organic (e.g., epoxy) and inorganic (e.g., silicates).

Premium-quality anticorrosive primer coatings can be made from each type. When subsequent topcoats (vinyl, epoxy, etc.) are applied over inorganic zinc-rich primers, the adhesive bond is mostly mechanical and polar. However, this bond is more than adequate because of the irregular nature of the cured zinc-rich film. By contrast, organic zinc-rich materials are readily softened by the solvents in topcoats such as vinyls and epoxies, and the coatings fuse to produce an almost ideal adhesive bond.

Organic zinc-rich coatings can tolerate poorer surface preparation (at least a commercial-grade blast) than can the inorganics (usually a near-white blast). Both types are easily applied by conventional techniques in a range of 2 to 3 mils (0.0508 to 0.0762 mm) dry, although some current formulations specify only 1 mil (0.0254 mm) for the organic type.

The inorganics are considerably more fire-resistant than the organics. The organics are well suited to spot repair over intact old paint and are more economical because of lower applied costs. Because of the extremely high pigment loadings used with both types, their open surface is not suited to immersion service until adequate sealer topcoats are applied. It is customary to apply a wash coat or a topcoat thinned with extra diluent over the zinc-rich primer before topcoating.

Inorganic Zinc Primers. Sometimes called "zinc silicates," these can be subdivided into three types. The *post-cured type* is based on silicates or lead silicates and requires an acid curing agent. *Self-cured types* are classed as either the *alcohol type* (based on hydrolyzed ethyl silicates) or the *water type* (based on water-soluble silicates). The alcohol type could be classed as an organic zinc-rich primer when it is in the liquid state. In the dry state, however, it is inorganic and therefore is classed as a self-cured inorganic zinc-rich primer. Because of the low flash point (55 to 70°F, or 13 to 21°C) of the alcohol type, the industry has developed cellosolve silicates which flash above 110°F (43°C; Tag closed cup).

All the inorganic types have excellent abrasion resistance, hardness, and toughness, but not flexibility. One coat is sufficient to obtain galvanic protection of steel exposed to the atmosphere or to clean petroleum cargoes. In general, zinc loadings exceed 75 percent by weight and are a mixture of superfine zinc dust and high-metallic zinc produced by a distillation process. Red lead, lead peroxide and chromate, and sulfides are added to extend pot life. The lead compounds also produce harder-cured coatings.

If these paints are being prepared from old materials, it should be recognized that moisture and CO_2 will cause a film to form over the zinc-dust particles. This will result in a very limited pot life of the mixed coating. All these primers must be applied at temperatures above 40°F (4°C). When applying organic topcoats over inorganic zincs, special care must be taken because of reported serious problems of blistering and adhesion. If the topcoat barrier is penetrated by seawater, the zinc will sacrifice to the steel substrate and zinc salts will be deposited on the surface. It is also possible that the acid postcure of the inorganic zinc may adversely influence adhesion and produce blistering. Because of these problems, which may actually accelerate corrosion, the tendency has been not to recommend inorganics for use below the waterline. This tendency probably only reflects the need for an intimate knowledge of these materials before they can be used to optimum advantage.

The inorganic zinc primers are two-package materials which are somewhat difficult to mix because of the extreme weight of the zinc pigment and of the water-viscosity of the silicate resin. Pot life is usually about 8 h, and good cleanup technique is an absolute requirement. Because of the weight of the catalyzed material, it is difficult to pump more than several feet in a vertical direction, and it is important to have the pot at the same approximate height as the gun while spraying. An agitator pot is essential.

One-pot products which require no mixing and have no problems of pot life (for example, Amercoat's Dimetcote E-Z) have recently been introduced. These materials may appeal to the small-boat owner who prefers the handling properties of an "ordinary paint."

Among other vehicles used successfully for the inorganic type are lithium silicate, phosphates, silicate esters, zinc-lead silicates, and many combinations of these.

Organic Zinc-Rich Primers. These materials are based on organic binders and contain at least 85 percent by weight (45 to 55 percent by volume) of zinc dust. The principal binders used today are vinyl, epoxy, epoxy-polyamide, and chlorinated rubber; but successful primers and anticorrosive coatings have been based on styrene, polyesters, acrylics, urethanes, and silicones. Considerable work has been done by Bakelite (Union Carbide) in developing formulations based on its phenoxy resins. These resins are somewhat similar to conventional epoxies but are purported to have superior properties.

Vinyl zinc-rich primers topcoated with vinyl anticorrosives are well suited to freshwater immersion applications. Epoxy zinc-rich coatings, especially shop primers, are now widely recommended, and this recommendation supports the general trend toward greater use of the organics and less use of the inorganics. Again, the basic formulation is simply epoxy resin and zinc dust, in a ratio ranging from 20 : 80 to 5 : 95 parts by weight, which is cut in a usable solvent blend such as methyl isobutyl ketone (MIBK)–xylene (70 : 30) or methyl ethyl ketone (MEK)–xylene-cellosolve acetate (45 : 45 : 10). Silane additives have been recommended to promote adhesion further. If galvanic protection is desired, the coatings must contain at least 90 percent zinc by weight. The organic zincs are less conductive than the inorganics and sacrifice at a lower rate because the polymer resin is present to retard the rate of sacrifice. This characteristic results in longer service lives than those observed with inorganic zincs.

The organics should *not* be used directly under any vehicles containing oil (alkyds, modified alkyds, epoxy esters) when considered for service in chemical (tank) or salt (ballast and hull) atmospheres. Zinc and water in the presence of salt produce sodium hydroxide.

Miscellaneous primers Countless private formulations of individual marine-paint companies are available. The small-boat owner is more likely to encounter these materials. From considerations of economy, desired performance, and available surface-preparation and application facilities, it is reasonable to assume that most of these primers will provide acceptable service if the manufacturer's recommendations are followed. The rationale behind all these materials is the selection of a vehicle-pigment combination that will support a certain level of performance and economics. Typical examples are phenolics and oil- or phenolic-modified alkyds. These resins are filled with inhibitive pigments to meet individual requirements such as a specific service condition (immersion, atmosphere), type of work (new, touch-up), available surface-preparation facilities (desirable, less than adequate), and compatibility with specific topcoats. Again, no specific product recommendations are possible because of the wide variations found in formulation. In selecting these materials, emphasis must be placed on the integrity of the supplier.

14.3.2 Anticorrosive coatings for metal   Anticorrosive, or barrier, coatings present a physical barrier between moisture (salt spray, seawater, fresh water) and metal. Their effectiveness, *which is directly related to film thickness,* is measured generally by the water-permeability rate. For immersion applications, electrolytic resistance becomes equally important. For all applications, the barrier coating must have outstanding abrasion and impact resistances to maintain a critical film thickness. Also in immersion applications, the barrier must be compatible with the impressed current used for cathodic protection, and when used under copper-bearing antifoulants, the anticorrosive must insulate the copper from the metal substrate. In considering the protection of vessels, it is essential to recognize the relationship between friction resistance (loss of operating speeds) and bottom-surface roughness. The value of maintaining a desirable surface smoothness is well documented, and typical studies report shaft-horsepower savings up to one-third by reducing roughness from 30 to 10 mils (from 0.762 to 0.254 mm). Obtaining this quality is closely tied to proper surface preparation (e.g., removal of mill scale) and choice of primer. The largest contribution to surface smoothness is made by proper selection and application of the anticorrosive material. The small-boat racer must consider drag reduction and understand the relationship between speed and bottom smoothness. Other pleasure-boat owners and all commercial-boat owners view this property from the standpoint of more economical performance. With the advent of the supertanker, renewed emphasis has been placed on bottom smoothness, and this concern is being reflected throughout the marine community.

Anyone involved with marine finishes is deeply involved with corrosion, and all users of the *Paint Handbook* are admonished to pursue the subject in the depth befitting their special circumstances. The marine-corrosion information summarized below should serve as an introduction to this study.

Some aspects of marine corrosion   The pleasure-boat owner as well as the marine-coatings chemist should possess a working knowledge of three general concepts to understand and protect effectively against marine corrosion. In brief, these are (1) the extreme complexity of marine environments; (2) the major mechanisms of marine corrosion, particularly that caused by galvanic coupling; and (3) the performance of

TABLE 14.1   Factors in Seawater Environment*

Chemical	Physical	Biological
Dissolved gases[a]	Velocity[b]	Biofouling[c]
Oxygen	Air bubbles	Hard-shell types
Carbon dioxide	Suspended silt	Types without hard shells
		Mobile and semimobile types
Chemical equilibrium[d]	Temperature[e]	Plant life
Salinity		Oxygen generation
pH		Carbon dioxide consumption
Carbonate solubility		
	Pressure[f]	Animal life
		Oxygen consumption
		Carbon dioxide generation

* With iron used as the reference, the following trends are typical:
[a] Oxygen is a major factor in promoting corrosion.
[b] Increasing velocity tends to promote corrosion, especially if entrained matter also is present.
[c] Biofouling can reduce attack or promote local-corrosion cells.
[d] The tendency to form protective scale (carbonate type) increases with high pH.
[e] Temperature increase tends to accelerate attack.
[f] Pressure may affect corrosion.

individual metals in each phase of a specific marine environment. A current text by F. W. Fink and W. K. Boyd[6] is recommended as a cogent, concise, and practical reference. The following remarks outlining the three concepts are summarized, with permission, from this reference.

The corrosivity of seawater is affected by the interaction of many chemical, physical, and biological factors. The diversity of these factors, as summarized in Table 14.1, points to the necessity of weighing the corrosion aspects of each marine-service climate independently.

Also, a specification should consider corrosion situations such as cavitation, pitting, impingement, and crevice attack. Of special concern is a thorough understanding of galvanic attack. Because seawater is an excellent electrolyte, a corrosion cell is instantly formed as soon as two dissimilar metals are joined and placed in the sea. This hazard is unavoidable because of the diverse structural requirements of vessels.

One coupled metal (e.g., a bronze propeller) immediately becomes the anode, and the other (e.g., a steel-hulled vessel) becomes the cathode in the seawater corrosion cell even if the dissimilar metals are as distant as 100 feet (30.48 m) from each other. Any coating holiday or defect which exposes bare steel to seawater will, in effect, couple the entire steel cathode to any small anode, and severe corrosion will quickly result. In

[6] *The Corrosion of Metals in Marine Environments,* Bayer and Co., Inc., Columbus, Ohio, 1970.

the atmosphere, in contrast, corrosion from galvanic couples is limited to the immediate area of the joint.

The rate of corrosive attack is governed partly by the position of the metals in the galvanic series in seawater (selected examples are listed in Table 14.2). The metal which is sacrificed (the anode) is the one nearest the top of this list; as a general rule, the rate of anode corrosion increases as the metals become farther apart on the series. The ratio of anode to cathode area is also important, as, for example, in an increased rate of

TABLE 14.2   Galvanic Series in Seawater Flowing at 13 Ft/S (3.96 M/S); Temperature, about 77°F (25°C)

Material	Steady-state electrode potential, volts (saturated-calomel half cell)
Zinc	−1.03
Aluminum 3003-(H)	−0.79
Aluminum 6061-(T)	−0.76
Cast iron	−0.61
Carbon steel	−0.61
Stainless steel, Type 430, active	−0.57
Stainless steel, Type 304, active	−0.53
Stainless steel, Type 410, active	−0.52
Naval rolled brass	−0.40
Copper	−0.36
Red brass	−0.33
Bronze, composition G	−0.31
Admiralty brass	−0.29
90 Cu–10 Ni, 0.82 Fe	−0.28
70 Cu–30 Ni, 0.47 Fe	−0.25
Stainless steel, Type 430, passive	−0.22
Bronze, composition M	−0.23
Nickel	−0.20
Stainless steel, Type 410, passive	−0.15
Titanium*	−0.15
Silver	−0.13
Titanium†	−0.10
Hastelloy C	−0.08
Monel 400	−0.08
Stainless steel, Type 304, passive	−0.08
Stainless steel, Type 316, passive	−0.05
Zirconium‡	−0.04
Platinum	+0.15

* Prepared by powder-metallurgy techniques. Sheath-compacted powder, hot-rolled, sheath removed, cold-rolled in air.
† Prepared by iodide process.
‡ From other sources.

corrosion resulting from a small steel anode coupled to a large copper cathode. The performance of some typical metal couples is listed in Table 14.3.

In addition to these general considerations, it is equally important to know the corrosion behavior of an individual metal in a marine environment because its performance in each environmental zone (atmosphere, splash, tide, immersed, mud, deep ocean) will undoubtedly vary. Only a few metals (titanium and tantalum) and alloys (Hastelloy C) are "completely resistant" to all zones of marine corrosion. The most common metal in marine service is plain carbon steel even though it readily corrodes and must be protected. Splash-zone corrosion is significantly worse than that encountered in a tidal zone and several times worse than that in a zone of complete immersion.

Total protection includes using the proper substrates (new work), anticorrosive barrier coatings, sacrificial anodes, and cathodic protection. Anticorrosive barrier coatings (Table 14.4) are the first line of defense against corrosion in any program and are the only tool available to the small-boat owner. The marine aspects of sacrificial anodes and cathodic protection are briefly discussed on pages 14-27 and 14-28.

TABLE 14.3   Examples of Galvanic Couples in Seawater

Metal A	Metal B	Comments
Couples that usually give rise to undesirable results on one or both metals		
Magnesium	Low-alloy steel	Accelerated attack on Metal A; danger of hydrogen damage on Metal B
Aluminum	Copper	Accelerated pitting on Metal A; ions from Metal B attack Metal A. Reduced corrosion on Metal B may result in biofouling on Metal B
Bronze	Stainless steel	Increased pitting on Metal A
Borderline couples that may work, but with uncertain results		
Copper	Solder	Soldered joint may be attacked but may have useful life
Graphite	Titanium or Hastelloy C	
Monel 400	Type 316 SS	Both metals may pit
Generally compatible		
Titanium	Inconel 625	
Lead	Cupronickel	

TABLE 14.4 Typical Coatings and Service and Maintenance Requirements for Specific Marine Areas

Area	Prime service requirements	General maintenance requirements; surface-preparation requirements	Application requirements	Typical coatings
Bottom (vessels), mudline to splash zone (offshore structures), pilings (docks)	Resistance to abrasion, impact, salt water, pitting, fouling, cathodic protection; smoothness (vessels)	Adhere to blasted touch-up repair spots; recoatable after good surface cleaning	Wide temperature range; accept some dampness; applicable with all techniques	1. Wash primer–bituminous AC; hot or cold plastic AF 2. Wash primer–coal-tar AC; hot or cold plastic AF 3. Vinyl primer plus AC; rosin-vinyl AF 4. Epoxy, inorganic zinc, or organic zinc-rich primer; epoxy–coal-tar AC plus AF (any AF) 5. Inorganic zinc primer; epoxy-polyamide AC (any AF) 6. Chlorinated-rubber primer; AC plus AF (any AF)
Boot top (vessels), splash zone (legs, substructure, first platform in offshore structures)	Flexibility; resistance to abrasion, impact, thermal shock, salt water plus salt water spray, oxygen, ultraviolet light, cathodic protection, undercutting; smoothness (vessels)	Adhere to blasted or power-tool–cleaned touch-up repair spots; recoatable after good surface cleaning	Wide temperature range; accept some dampness; applicable with all techniques	1. Inorganic zinc primer; epoxy-polyamide AC, epoxy–coal-tar AC, or tie coat plus vinyl-alkyd AC 2. Organic zinc-rich primer; high-build–epoxy AC or tie coat plus vinyl-alkyd AC 3. Epoxy–coal-tar primer plus AC 4. Chlorinated-rubber primer plus AC 5. Vinyl primer plus AC 6. Polyester glass 7. Modified alkyd primer plus AC
Freeboard (vessels), sides plus rakes	Resistance to abrasion, thermal shock, under-	Adhere to blasted or power-tool–cleaned	Fast drying time; easy application	1. Epoxy or epoxy–coal-tar primer; high-build– 4. Inorganic zinc primer; high-build–

Location	Properties	Surface preparation	Drying	Recommended coating systems
(barges), deep-load line to top of gunnel (workboats, tugs, motor vessels)	cutting, salt spray; gloss and color retention	touch-up repair spots; recoatable after minimum surface cleaning		epoxy or epoxy–coal-tar AC. 2. Organic zinc-rich primer; high-build–epoxy or vinyl tie coat plus vinyl-alkyd. 3. Vinyl primer; vinyl AC plus vinyl-alkyd topcoat. epoxy or vinyl, epoxy-polyamide, modified vinyl tie coat plus vinyl- plus alkyd or alkyd AC. 5. Chlorinated-rubber primer plus AC. 6. Modified alkyd primer; various topcoats
Weather working decks, heliports, forecastle decks, catwalks (all vessels and structures)	Resistance to abrasion, impact, thermal shock, salt spray; gloss and color retention; nonskid properties	Adhere to power-tool–cleaned surfaces; recoatable after minimum surface cleaning	Fast drying time; accept some dampness; underway application	1. Modified alkyd primer plus AC. 2. Inorganic zinc primer; epoxy AC. 3. Vinyl primer plus AC; vinyl-alkyd topcoat. 4. Epoxy, organic zinc, or inorganic zinc primer; tie coat plus vinyl-alkyd, high-build–epoxy, or epoxy–coal-tar plus epoxy AC
Superstructure (masts, stacks, fittings for all vessels; derricks, towers, exposed machinery for offshore structures)	Resistance to under-cutting, edge corrosion, salt spray; gloss and color retention	Adhere to power-tool–cleaned surfaces; recoatable after minimum surface cleaning	Fast drying time; accepts some dampness; underway application	1. Wash primer; vinyl-alkyd AC plus topcoat (vinyl-alkyd or various modified resins). 2. Organic or inorganic zinc primer; tie coat plus alkyd, tie coat plus vinyl-alkyd, high-build–epoxy or vinyl, epoxy-polyamide, or vinyl plus modified acrylic topcoat, modified vinyl plus oil-based–topcoat AC. 3. For aluminum: epoxy primer (not inorganic zinc); high-build–epoxy AC

TABLE 14.4 Typical Coatings and Service and Maintenance Requirements for Specific Marine Areas (Continued)

Area	Prime service requirements	General maintenance requirements; surface-preparation requirements	Application requirements	Typical coatings
Heat exchanger, hot-stack exteriors	Resistance to temperatures of 1000°F (538°C), corrosion, salt spray; gloss and color retention	Adhere to blasted touch-up repair spots; recoatable after good surface cleaning	Applicable with all techniques in wide temperature range	1. Inorganic zinc primer; modified silicone AC 2. Stainless-steel–pigmented epoxy resin
Working and living areas	Color and gloss retention appearance; easy maintenance	Recoatable after minimum surface cleaning	Underway application (roller and brush)	1. Alkyd; chlorinated-alkyd 2. Epoxy primer; urethane, high-build–epoxy, or vinyl topcoat 3. Organic zinc primer; epoxy-polyamide, high-build–epoxy, or vinyl topcoat 4. Blasted surface: inorganic zinc primer; epoxy-polyamide, high-build–epoxy, or vinyl topcoat 5. Wire-brushed or solvent-cleaned surface; zinc chromate primer; alkyd topcoat
Dry cargo space, dry stores	Resistance to impact, abrasion, corrosion	Recoatable after minimum surface cleaning; easy spot repair	Extremely fast drying time; high-flash solvents only	1. Wash primer; alkyd topcoat 2. Epoxy or vinyl primer; high-build–epoxy or vinyl topcoat 3. Organic zinc primer; high-build–epoxy topcoat 4. Inorganic zinc primer; tie coat plus alkyd, tie coat plus vinyl-alkyd, chlorinated-rubber topcoat

Application/Location	Property 1	Property 2	Property 3	Coating system 1	Coating system 2
Above-bilge spaces, voids, cofferdams, wet bilges	Resistance to general corrosion	Recoatable after minimum surface cleaning; easy spot repair	Extremely fast drying time; high-flash solvents only	1. Epoxy or inorganic zinc primer; high-build–epoxy AC	2. Epoxy–coal-tar primer plus AC
Tanks (cargo and ballast)	Resistance to salt water; impervious to cargo (chemical resistance)	Easy spot repair; recoatable after minimum surface cleaning	Extremely fast drying time; high-flash solvents only	1. Inorganic or organic zinc primer; high-build–eopxy AC 2. Epoxy–coal tar	3. Polyester-glass (ballast only)
Tanks (forepeak and mud)	Resistance to general corrosion	Easy spot repair; recoatable after minimum surface cleaning	Extremely fast drying time; high-flash solvents only	1. Epoxy–coal tar	2. Organic zinc or epoxy primer; high-build–epoxy AC
Deep tanks (chemicals, edible oils)	Impervious to cargo; resistance to general corrosion	Easy spot repair; recoatable after minimum surface cleaning	Extremely fast drying time; high-flash solvents only	1. Generally high-build epoxy but cargo-dependent	
Tanks (potable water)	Resistance to general corrosion; inert (low extractables) in fresh water	Easy spot repair; recoatable after minimum surface cleaning	Extremely fast drying time; high-flash solvents only	1. Wash primer; vinyl AC	2. Inorganic zinc alone or as primer; for epoxy–coal-tar or epoxy-polyamide AC
Tanks (sweet plus sour crude oil)	Resistance to general corrosion, specific cargo; inert (non-contaminating) to cargo	Easy spot repair; recoatable after minimum surface cleaning	Extremely fast drying time; high-flash solvents only	1. Organic zinc primer; epoxy–coal-tar, high-aluminum epoxy–coal-tar, or high-build–epoxy AC	2. Service over 160°F (71°C): wash primer; modified phenolic AC

TABLE 14.4 Typical Coatings and Service and Maintenance Requirements for Specific Marine Areas *(Continued)*

Area	Prime service requirements	General maintenance requirements; surface-preparation requirements	Application requirements	Typical coatings	
Tanks (clean petroleum, fuel, aviation gas, JP-5 fuel)	Resistance to general corrosion, cargo fuel; inert (noncontaminating) to fuel	Easy spot repair; recoatable after minimum surface cleaning	Extremely fast drying time; high-flash solvents only	1. Organic zinc	2. Epoxy primer; high-build–epoxy AC
Tanks (solvents)	Resistance to general corrosion, specific cargo	Easy spot repair; recoatable after minimum surface cleaning	Extremely fast drying time; high-flash solvents only	1. Inorganic zinc	

NOTE: AC = anticorrosive; AF = antifouling.

Cathodic protection   This is an electrochemical technique used to protect metals in a marine environment by impressing a current which stops the current flow between anode and cathode. It is commonly referred to as "polarization" or simply as "polarizing the cathode." Two methods are used to obtain this condition: sacrificial anodes and impressed current.

Sacrificial Anodes. These are so named because they are less noble metals which are selected for sacrifice in protecting more noble structural metals. In this manner, aluminum, zinc, or magnesium bars are attached as anodes to an immersed object and protect it (the cathode) by eliminating local anodes. Two disadvantages are present: the necessity of periodic inspection and replacement and the burden of added weight. On large steel vessels and structures, anodes are added on the basis of square feet of underwater area. Both the exterior hull and the interior bilges of vessels are included. For example, 1 ft^2 (0.09 m^2) of zinc is commonly used to protect 200 ft^2 (18.58 m^2) of coated metal or 50 ft^2 (4.65 m^2) of bare metal. In stagnant areas, the quantity can be reduced to 1 ft^2 of zinc per 400 ft^2 (37.16 m^2) of painted area. Aluminum hulls and structures are usually protected by zinc anodes, but magnesium or mercury-bearing aluminum anodes may not be used. Wood and plastic hulls and structures do not require anodes for themselves, but anodes are commonly specified for each 25 ft^2 (2.32 m^2) of immersed metal area and also to protect the metal rudder, struts, sea chests, etc. In general, anodes are commonly employed close to junctions of dissimilar metals in equipment which is used in saltwater service (i.e., valves, bilge-pump strainers).

In both fresh and brackish waters, magnesium anodes are usually selected because of the higher driving power of the metal. Magnesium is not used to protect aluminum in seawater or when extended protection (several years) is desired. It is also necessary to paint the area beneath, and for several feet around, a magnesium anode array. An impervious coating such as a tank lining (MIL-P-23236, as discussed below) is recommended. Zinc anodes are commonly used to afford long-term protection to painted steel structures. They may not be attached to bare metal, but any anticorrosive coating is satisfactory to insulate the anode from direct contact with the immersed structure.

Aluminum anodes continue to find wider use, although to date no consensus has been reached as to their relative merits over zinc or magnesium. Most aluminum anodes contain some mercury, and wide use is now being made of aluminum-indium as an acceptable replacement.

The marine-coatings chemist must always keep several factors in mind. Before ship painting is begun, all anodes must be masked, and although this admonition appears unnecessary, the mask must be removed to expose the bare anode after painting has been completed. *Under no circumstances are anodes painted or abrasive-blasted.* Since it is common practice to bolt or strap these anodes to a structure to facilitate replacement, special attention must be paid to the attachment mechanism to ensure that it has been completely painted. Frequently a welder will blister the coating on a bolt or strap and thereby seriously impair its service life.

Impressed Current. This is a system of corrosion protection which introduces direct low-voltage current from an external power source into seawater. The current is introduced at an appropriate distance from the object to be protected by using insoluble anodes which are fastened directly to the hull or structure. The anodes (graphite, silicon iron, lead-silver, platinum, platinized titanium or tantalum) are driven at higher voltages than the galvanic type. This heightens the importance of the shield that protects the anode array from the structure at the point where it is secured. The outstanding advantage of impressed current is the ability to adjust the current to meet the needs of the moment (e.g., after damage to an anticorrosive coating). This type of system is commonly employed as one component of an anticorrosive program designed to protect today's large vessels and offshore structures.

Formulations   Anticorrosive needs are served by a wide range of coatings. A very general usage pattern can be identified by noting that today's high-performance materials (e.g., zinc-rich, vinyl, epoxy, epoxy-polyamide, epoxy–coal-tar, chlorinated-rubber, and polyester-glass coatings) are selected for splash-zone and immersion applications. The older (bituminous and oleoresinous) coal-tar materials still find considerable use, although the trend since World War II has been toward high-performance coatings. Above the waterline and in salt-spray applications on shore, corrosion requirements are less severe, and such materials as phenolic–tung oil, alkyds and modified alkyds, acrylics, epoxy esters, polyurethanes, and polyesters are common choices.

As vessels and structures become larger and service requirements are tightened, economics becomes even more important. For vessels, this importance reflects an urgent need to extend significantly the time between dry-dockings. For structures, the additional service time postpones not only repair but actual replacement of metal. This need has led to the use of hybrid anticorrosive systems composed of almost every conceivable combination of surface treatments, primers, anticorrosive barrier coats, and special topcoats. A general guide to the compatibility of various primers and topcoats is presented in Tables 14.5 and 14.6.

The maritime specification at the top of page 14-31 is typical of conventional coal-tar resin-based anticorrosive materials. It is a good illustration of the complex mixture of rust-inhibitive pigments selected to impart corrosion resistance to a barrier coating.

This coating dries by solvent evaporation and has good adhesion and corrosion resistance, but its abrasion resistance is only fair. Many com-

TABLE 14.5 Compatibility of Some Coating Materials with Various Primers

Primers	Topcoats											
	Alkyd	Alkyd-phenolic	Vinyl-alkyd	Vinyl	Vinyl-acrylic	Epoxy catalyzed	Epoxy ester	Coal-tar epoxy	Chlorinated-rubber	Phenolic oleoresinous	Poly-urethane	Polyester flake or glass
Alkyd	R	R	NR	NR	NR	NR	NR	NR	NR	R	NR	NR
Bituminous (aluminum)[a]	NR	NR	NR	NR	NR	NR	NR	NR	NR	NR	NR	NR
Vinyl-alkyd[b]	R	R	R*	R*	R*	NR	R	NR	NR	NR	NR	NR
Vinyl[b]	R	R	R	R	R	NR	NR	NR	R	NR	NR	NR
Epoxy ester	R	R	NR	NR	NR	NR	R	NR	X	R	NR	NR
Epoxy catalyzed[c]	NR	NR	R*	R*[d]	R*	R*	R*	R*	X	NR	R*	R*
Epoxy noncatalyzed[c]	R	R	R	X	X	R	R	X	X	R	NR	R*
Epoxy–organic zinc[c]	NR	NR	NR	R*	NR	R	R	R	R	R	NR	NR
Phenolic oleoresinous	R	R	R	NR	NR	NR	R	NR	R	R	NR	NR
Vinyl-phenolic	R	R	R	R	R	NR	R	NR	R	R	X	NR
Coal-tar epoxy	NR	NR	NR	NR[d]	NR	R*	R*	R*	NR	NR	NR	NR
Inorganic zinc, postcure[e]	NR	NR	NR	R*	NR	R*	R*	R*	R*	NR	NR	NR
Inorganic zinc, self-cure[c]; water-base[e]	NR	NR	NR	R*	NR	R*	R*	R*	R*	NR	NR	NR
Inorganic zinc, self-cure[c]; solvent-base[e]	NR	NR	NR	R*	NR	R*	R*	R*	R*	NR	NR	NR
Chlorinated-rubber	R	R	NR	NR	R	NR	R	NR	R	R	NR	NR

NOTE: R = known compatibility; normal practice. R* = known compatibility with special surface preparation and/or application. NR = not recommended. This is defined as meaning that it is not common practice to apply this topcoat over the specified primer, although certain products, if properly formulated, may be compatible. Attention is called also to the fact that certain combinations marked NR may be used, provided a suitable tie coat is applied between the two. X = not recommended because of insufficient data.

[a] Topcoated with itself or with an antifouling coating.
[b] Vinyl wash primer required.
[c] May be used as an after-blast primer.
[d] Vinyl antifouling coating such as MIL-P-15931 may be applied.
[e] May be used without topcoat.

14-29

TABLE 14.6 Compatibility of New Coatings with Previously Applied Coatings

New topcoat (excluding antifouling)

Existing topcoat (aged a minimum of 6 months)	Alkyd	Silicone-alkyd	Alkyd-phenolic	Vinyl-alkyd	Vinyl	Vinyl-acrylic	Acrylic (solvent type)	Epoxy catalyzed	Epoxy ester	Coal-tar epoxy	Chlorinated-rubber	Phenolic oleoresinous	Poly-urethane	Polyester flake or glass
Alkyd	R	R	R	NR	NR	NR	NR	NR	NR	NR	NR	R	NR	NR
Silicone-alkyd	R	R	R	NR	NR	NR	NR	NR	NR	NR	NR	R	NR	NR
Alkyd-phenolic	R	R	R	R	NR	NR	NR	NR	X	NR	NR	R	NR	NR
Vinyl-alkyd	R	R	R	R	R	R	R	NR	R	NR	R	R	NR	NR
Vinyl	R	NR	R	R	R	R	R	NR	R	NR	R	NR	NR	NR
Vinyl-acrylic	X	NR	X	R	R	R	R	NR	X	NR	R	NR	NR	NR
Acrylic (solvent type)	NR	NR	X	R	R	R	R	NR	X	NR	X	NR	NR	NR
Epoxy catalyzed	NR	NR	NR	R*	R*	R*	R*	R*	R*	R*	NR	NR	R*	R*
Epoxy ester	R	R	R	R	R	R	R	NR	R	NR	R	R	NR	NR
Coal-tar epoxy	NR	NR	NR	NR	NR	NR	X	R*	R*	R*	NR	NR	NR	NR
Chlorinated-rubber	R	R	R	R	NR	NR	NR	NR	R	NR	R	R	NR	NR
Phenolic oleoresinous	R	R	R	NR	NR	NR	NR	R*	R	R*	NR	R	NR	NR
Polyurethane	NR	NR	NR	NR	NR	NR	NR	R*	NR	NR	NR	NR	R*	R*
Polyester flake or glass	NR	NR	NR	NR	NR	NR	NR	R*	NR	NR	NR	NR	R*	R*

NOTE: R = known compatibility; normal practice. R* = known compatibility with special surface preparation and/or application. NR = not recommended. This is defined as meaning that it is not normal practice to apply the new topcoats over the existing topcoats. However, certain products, if properly formulated, may be compatible. Attention is called also to the fact that certain combinations marked NR may be used, provided a suitable tie coat is applied between the two. The age of the existing coat may also be a factor in determining compatibility. X = not recommended because of insufficient data.

Specification TT-P-00118 (Formerly 52-MA-401b)

Ingredient	Parts by weight
Zinc oxide	170
Zinc yellow	115
Mica	45
Indian red and bright red	15
Magnesium silicate	185
Aluminum stearate	10
Hard coumarone-indene resin	200
Soft coumarone-indene resin	51
Coal tar	56
Chlorinated rubber	29
Naphtha	210
Petroleum spirits	130

mercial variations have substituted petroleum resins for the coal-tar and coumarone-indene resins. The formulation is economical and serviceable, but it is not a premium, high-performance anticorrosive coating.

The Navy coumarone-indene resin-based anticorrosive shows a radically different pigmentation which cures to a hard, adhesive, and water-impermeable coating:

Specification MIL-P-19453 (Formula 14N)

Ingredient	Parts by weight
Zinc chromate	297
Venetian red	50
Mica	75
Silica (diatomaceous)	75
Varnish*	640
Lead linoleate	1
Xylene	20
Total weight, pounds	1158 (525.3 kg)
Total yield, United States gallons	100 (378.5 l)

*Varnish	Pounds
Phenolic resin	106
Cumar resin A	106
Cumar resin B	73
Tung oil	66
Alkali-refined linseed oil	33
Xylene	192
Turpentine	64
Total weight, pounds	640 (290.3 kg)

Heat the phenolic, Cumar resin A, tung oil, and linseed oil to 460°F (238°C) in 45 to 50 min. Hold for 20 min. Remove heat, and add Cumar resin B. Cool to 275°F (135°C), and add solvents.

Today's major anticorrosive coatings are discussed below according to the resin binder upon which the formulation is based. The remarks are general, for there is considerable latitude in developing a formulation based on a specific resin. The type of resin is the major consideration for meeting specific service requirements, but performance can be significantly varied by other formulation variables such as type and amount of pigmentation, additives, and modifiers. For example, the "classic" anticorrosive pigments such as red lead, zinc chromate, and zinc dust are seldom formulated into the topcoat. However, rust inhibition can be extended into the topcoat by using lead silicochromate. This pigment possesses good anticorrosive properties and has a low tinting strength which allows green, gray, and brown topcoats to be prepared. One of the most novel pigments used in anticorrosive formulations is stainless steel. Epoxy resins pigmented with stainless steel provide good service up to 450°F (232°C) and are particularly impervious to edge and crevice corrosion. No reasons are apparent for this good performance.

Vinyl Coatings.   These coatings evolved as the first really premium coatings as a result of performance during and after World War II. Outstanding primer, anticorrosive, and antifouling formulations are now available. In addition to these solvent-type finishes, vinyl latex coatings are widely used for interior and some exterior surfaces above the waterline. Still other vinyl organosol and plastisol formulations are used to protect many items whose size permits factory finishing (structures, equipment, small tools, etc.).

Vinyl-solution coatings are the primary concern of this section, and their performance is directly related to the effort spent in surface preparation. The unusually strong, impervious barrier that they present to the marine atmosphere is due to outstanding cohesive strength. Because this force of cohesion is greater than that of adhesion, a common failure mechanism is blistering or simply a sheeting off of coating from an improperly prepared surface. Abrasive blasting to white metal is a minimum surface-preparation requirement, and wash primers should be used if at all possible. Wash primers used as a tie coat over zinc primers are also recommended with vinyl barrier coatings. Special care must be exercised when applying vinyls to obtain the desired quality of cured film. The coatings should (1) be applied at low viscosity, or about 300 centipoises (60 KUs, or 0.3 Pa · s) for spray application; (2) contain a slow-evaporating solvent system such as methyl isobutyl ketone–xylene (80 : 20), in preference to a fast solvent such as methyl ethyl ketone; and (3) be hesitation-sprayed, allowing up to 1 min (depending on film thickness) between passes for solvent evaporation.

A significant advantage of the vinyls is the short curing time required before they may be placed in service. As compared with the much longer time required for an epoxy–coal tar, this is an important factor for drydock consideration.

A typical example of a vinyl resin-based anticorrosive coating is Navy Formula 119:

Specification MIL-P-15929B, Type 119

Ingredient	Parts by weight
Red lead (97 percent)	250
PVC-anticorrosive copolymer (alcohol-modified)	165
Tricresyl phosphate	17
Aluminum stearate	1
Methyl isobutyl ketone	290
Toluene	270
Total weight, pounds	993 (450.4 kg)
Total yield, United States gallons	100 (378.5 l)

This type of anticorrosive is frequently recommended for severe service situations such as tropical immersion. The excellent service is due partly to the high loading of red-lead pigment and also to the choice of resin binder (VAGH, Union Carbide). A similar vinyl resin-based anticorrosive primer coating is pigmented with zinc chromate in place of red lead. This coating is covered in MIL-P-15930.

Vinyl coatings can be applied by conventional, airless, or hot spray, and a system of four or five coats of 1.5 mils (0.0381 mm) each is usually specified to obtain a dry film thickness of 6 to 8 mils (0.1524 to 0.2032 mm). The entire system requires considerable labor, but it can be applied in 1 day. The recent introduction of high-build vinyls may eliminate some of this labor. Today's vinyl coatings are impervious to the cathodic protection measures required for additional corrosion protection of large vessels and structures. Some of the few applications for which vinyl anticorrosives are not recommended are areas of high cavitation and erosion, as on rudders and struts. The U.S. Navy has shown its confidence in vinyl marine finishes by incorporating them into many of its specifications. A typical complete vinyl system for ship bottoms is covered by the specifications in Table 14.7.

Epoxy Coatings. These coatings are used for premium marine applications. The most common types are amine-cured epoxy, epoxy-polyamide, epoxy ester, and epoxy–coal tar. Epoxy zinc-rich coatings, another distinct class, have been discussed in the section "Zinc-Rich Primers."

High-performance primers and anticorrosive coatings can be formulated from each type, but epoxy esters are usually limited to service above the waterline. Antifouling topcoats based on epoxy resins are limited to the epoxy–coal-tar type.

Amine-cured and polyamide-epoxy coatings are extremely durable, tough, and smooth coatings with excellent adhesion and resistance to

TABLE 14.7   Vinyl Systems for Ship Bottoms

Component	Specification	Number of coats	Film thickness, mils (mm)
Metal pretreatment primer	Formula 117, MIL-P-15328 B	1	0.5 (0.0127)
Anticorrosive coating	Formula 119, MIL-P-15929 B	4	6.0 (0.1524)
Antifouling topcoat	Type 121, MIL-P-15931 A	2	4.0 (0.1016)
Total		7	10.5 (0.2667)

blistering, peeling, and undercutting. They possess excellent resistance to salt air and water and to most chemicals. Some latitude is available in surface preparation, since a blast to white metal is not an absolute requirement. The tolerance of these coatings for surface contamination is considerably greater than that of vinyls and chlorinated rubbers. This property is one of the main reasons that epoxies have taken over so much of the marine-coatings field. However, it is strongly recommended that the best possible surface preparation be rigidly maintained regardless of the broader tolerance.

Epoxy resins cure by a chemical reaction. After the solvent has evaporated, the curing time depends mostly on temperature; other factors such as humidity and airflow are not significant. Film thickness is of minor importance to curing unless the heat of reaction in thicker films may hasten the ultimate cure. Complete cure is usually effected in about 7 days at 70°F (21°C), with longer times for colder weather and shorter times for hotter weather. It is common practice to consider a solvent-free epoxy film as serviceable even for immersion applications. However, complete cure is extremely important for tank coatings, and several extra days at 70°F are usually allotted before service.

From the standpoint of the applicator, epoxy paints are ideal because any conventional equipment can be used, pot life is excellent, drying times are relatively short, and touch-up is simple. Special formulations which will not lift an old but intact paint film are available. Naturally, all touch-up work is best approached on a trial-spot basis before the entire job is undertaken. Because of the epoxies' good adhesion and wetting, they can be applied at any temperatures above 45°F (7°C). If the surface is dry, 20°F (−7°C) outside temperatures are permissible, provided hot spray is used.

The small-boat owner should know that no harm will result from water spotting (rain or splash) before the epoxy film is hard. For appearance, however, a smooth, unspotted topcoat is always desirable. The epoxies also are used successfully as mastics to fill in pits and areas around rivet heads, etc., in order to obtain a smooth bottom before the coating system is applied.

As compared with vinyls, epoxies are economical from two standpoints. First, application costs are considerably lower because epoxies can be applied in a few high-build coats to obtain a desired thickness (i.e., 10 to 12 mils, or 0.254 to 0.3048 mm, dry). Vinyls, exclusive of the recently introduced high-build types, must be built up from many separate thin coats. Second, epoxies, which are supplied at high solids, are generally less expensive at equal dry film thicknesses than vinyls, which are supplied at lower solids.

Epoxies usually are not applied more heavily than 6 mils (0.1524 mm) per coat so that the danger of entrapped solvent is avoided. It is well to remember that each coat must be hard to the fingernail before the subsequent one is applied.

Epoxy-polyamide coatings are premium anticorrosive materials that possess outstanding water resistance. Because of their exceptionally good chemical resistance, they find wide usage as interior tank coatings. The Navy has introduced a new anticorrosive-coating specification, MIL-P-24441, which can be considered a performance standard for premium materials.

Epoxy-polyamide coatings are frequently used for offshore immersion applications such as splash zones and well jackets. Because of the high viscosity of these two-part coatings, they find considerable use as anticorrosive mastics. They are also well suited to new application techniques such as high-pressure two-nozzle spraying at 100 percent solids and elevated temperatures. They are so tolerant of moisture that many formulations can be applied underwater by trowel after an underwater sandblast. Coatings applied in this manner are surprisingly serviceable, but they certainly cannot be considered premium, high-performance materials.

Epoxy–ester coatings find limited use as field maintenance primers, marine interior enamels, and other applications for which salt water, splash, and spray are not significant factors. The vehicles are prepared by modifying the basic epoxy resin with drying oils. These modified resins, or epoxy esters, dry by oxidation.

Epoxy–coal-tar coatings combine the advantages of coal-tar pitch (excellent adhesion and chemical resistance, low moisture absorption) with those of epoxy resins (outstanding hardness, toughness, durability, and chemical resistance). The resultant properties are ideally suited to the complete range of severe anticorrosive-service uses. They include (1) outstanding adhesion to all metal and even wood substrates; (2) superior chemical resistance, which makes them candidate coatings for tank linings (their resistance is considered equal to that of conventional epoxy coatings); (3) impact resistance sufficient to protect sheet piling against abrasive washings in the mud zone and to resist the shock of pile driving during installation; and (4) an unusually wide range of temperature service, reportedly as diverse as 400°F (204°C; dry service only) and 250°F (121°C; thermal-shock threshhold) to −30°F (−34°C).

Catalyzed coatings are usually applied directly to blasted steel in two heavy coats of 8 to 10 mils (0.2032 to 0.254 mm) each (dry) for a total dry thickness of 16 to 20 mils (0.4064 to 0.508 mm). Over an intact old film, surface preparation usually includes dusting with fine sand to roughen the old film. These coatings are applied at a rate of about 100 ft²/gal (2.454 m²/l) and in a manner allowing 50 percent overlap per pass. To ensure uniform coverage, the second pass is applied at right angles to the first. The materials set up quickly, and close attention must be paid to spray equipment once application has started. Cleaning solvent should be readily available and routinely used if spraying is interrupted for even short periods of time. Epoxy–coal tars cure by both chemical reaction and solvent evaporation. The second coat is applied as soon as the first coat begins to harden, but the total package is not serviceable for 24 h (tank-lining applications require a final cure of several days).

Epoxy–coal-tar resins are commonly cured by either amines or polyamides. The polyamide type of cure is preferred for immersion service because of its outstanding water resistance. Both types have greater tolerance for damp surfaces (moisture) during application than have vinyls or epoxies, but their curing rate is dramatically reduced by cold weather. Thus, English shipyards, which usually are quite damp, prefer epoxy–coal tars to vinyls, but yards in the cold Scandinavian countries prefer vinyls.

Because epoxy–coal tars are more abrasion-resistant than vinyls, it is logical to assume that there will be less spot touch-up repair. This fact, together with reduced labor for application (two coats for epoxy–coal tar versus four or five coats for vinyl), brightens the long-term economic picture.

Conventional application techniques can be used, but applicators who balk at handling high-viscosity catalyzed materials complain of the problems of mixing, applying, and cleanup. On the other side of the ledger, epoxy–coal tars offer a formidable barrier to corrosion and a unique approach to fouling protection, since the same material may be used for both applications (see "Antifouling Coatings," Sec. 14.3.4). They must be considered premium high-performance materials for such applications as offshore structures, bottom and boot-top coatings, tanks, buoys, and power-plant intake and discharge lines for seawater.

Chlorinated-Rubber Coatings. These are sophisticated, high-performance coatings based on plasticized chlorinated rubber and pigmented to meet a desired service application. Entire chlorinated-rubber systems of primer, tie coat, and anticorrosive and antifoul coatings can be formulated, or specific components (e.g., tie coat and anticorrosive over an inorganic zinc primer) can be selected. Thin-film primers usually contain only anticorrosive pigments such as aluminum, red lead, and lead silicochromate. High-build anticorrosive coatings also contain talc and extender pigments (e.g., TiO_2 and barites).

These high-performance materials have drawn the interest of the industry since the late 1960s, and they should not be confused with the older low-build chlorinated-rubber systems that have been available since the middle 1930s. The newer materials meet all the general anticorrosive criteria of being practical, versatile, and economical. They have outstanding (low) water-vapor–permeability rates and possess the required chemical resistance to withstand both the alkaline conditions of cathodic protection and the overall acid and alkali conditions encountered with the electrolytic aspects of corrosion. As might be expected, they are durable and possess some measure of flexibility. Because of their high chlorine content, these paints will not support combustion and may be considered nonflammable.

Chlorinated-rubber paints are of the lacquer type which dries simply by solvent release. They can be applied by any technique available. Airless spray is commonly used in a shipyard because of economy, and coverage rates of 300 to 400 yd^2/h (250.8 to 334.5 m^2/h) are typical. Atmospheric conditions during application are not extremely critical because there is some tolerance for high humidity and temperature. If it is possible to work with the paint supplier, the paints can be applied at most temperatures above freezing by adjusting the amount of thixotropic agent, which is added to most high-build coatings to prevent sagging and running on vertical surfaces. Some viscosity adjustment is possible through solvent addition; xylene is recommended.

Complete chlorinated-rubber systems (primer, anticorrosive, and antifoul) have been formulated for both new and repair applications. For new work, high-build coatings are applied directly to bright metal or over wash or zinc-rich shop-plate primers. Dry film thicknesses of at least 8 mils (0.2032 mm) are recommended in two coats of 4 to 6 mils (0.1016 to 0.1524 mm). Recoating is possible after 8 h above 70°F (21°C) and after proportionally longer times at lower temperatures. This is required because the paints dry only by solvent evaporation and not by chemical reaction. Premium applications now specify an initial coat of chlorinated-rubber primer (1 to 1.5 mils, or 0.0254 to 0.0381 mm, dry) before applying the 8 mils (0.2032 mm) of high-build anticorrosive. The thinner primer may be recoated after 2 to 3 h.

If surface preparation is less than desirable, the chlorinated-rubber primer must be used to obtain adequate adhesion. The primer is identified by pigmentation (only anticorrosive pigments with no high-build talc-extender types) and film thickness (1 to 1.5 mils, or 0.0254 to 0.0381 mm, versus high-build films of 4 to 6 mils, or 0.1016 to 0.1524 mm, per coat). For applications over older coatings, the best adhesion is obtained over vinyls and epoxies. The adhesion obtained over other paints (phenolics, alkyds, etc.) depends upon the completeness of cure of the older paint. Intercoat adhesion will be lost if the rubber is lifted by residual solvents being released from the older paint or, for example, by bleeding from an incompletely cured epoxy–coal tar.

To complete the system, chlorinated-rubber–based topcoats are used both above and below the waterline. Above the waterline the topcoats are attractive because of their excellent ultraviolet-light resistance, mold resistance, and overall physical properties. Below the waterline, the topcoats definitely appear to be of premium quality, but insufficient data are available further to define their potential as vehicles for antifouling paints. The can stability of chlorinated-rubber–based antifouling paints can be improved by added magnesium oxide as 10 percent of the total pigment.

Polyester-Glass Coatings. These represent a relatively new concept in barrier anticorrosive coatings which is based on extremely thick films of catalyzed-polyester resins loaded with minute (>0.005-in, or 0.127-mm) glass fibers or flakes. On the basis of limited experience, they appear to be extremely desirable for premium applications. The cured coatings are tough and water-impermeable, and they maintain a desirable bottom-smoothness profile for extended periods. Their initial high cost appears to be more than offset by the additional service life obtained.

Glass loading permits high-build application and restricts film loss in high-abrasion areas. Normal failures in these areas usually leave serious undercutting that results either in high underfilm corrosion or in difficult touch-up repair. Typical of the excellent service of these coatings is the outstanding resistance they have shown to ice erosion on the hulls of icebreakers.

A near-white blast is the required surface pretreatment, and wash or epoxy-resin primers are suggested. Polyester-glass coatings are easily applied with conventional spray equipment when the temperature is above 60°F (16°C). The solution is almost 100 percent solids, and the wet and dry film thicknesses therefore will be almost equal. One cross pass produces the recommended dry film thickness of 30 to 35 mil (0.762 to 0.889 mm). The cured coatings have demonstrated good compatibility with many premium topcoats (e.g., vinyl, vinyl-alkyd, epoxy).

Since the solvent is styrene, many safety precautions must be observed. Yard operators may be familiar with these, but the small-boat owner tends to minimize their importance. For example, the flash point (Tag open cup) is 90°F (32°C), and application areas must be separated from contact with heat, sparks, and open flames. Special attention must be given to grounding all nearby equipment and to the use of explosion-proof motors. Styrene is toxic, and even spray-mist contact can harm skin and eyes. Vapors are equally harmful, and adequate ventilation is required to reduce the danger of breathing the vapors and the possibility of forming explosive concentrations with air. As a further precaution, CO_2 or dry chemical (Class B) fire extinguishers should be readily accessible.

The catalyst is a peroxide and is as combustible as the styrene resin. It must not be stored or used around steam pipes, sparks, etc. Many small-boat owners are tinkerers who take pride in embellishing the manufacturer's recommendations. These people especially must be made aware of the violent explosions that result from peroxide catalysts in combination with other accelerators such as paint driers (metal salts such as cobalt naphthenate). Accidental spills should be absorbed into vermiculite and burned in an open area. The cleaning solvent is methyl ethyl ketone, and high-flammability precautions similar to those for the resin and catalyst are required. Because of MEK's fast evaporation rate, there is no serious problem with spills as long as adequate ventilation is maintained.

When the outstanding service properties obtainable over extended periods are considered, polyester-glass materials may offer one of the most economical choices of anticorrosive coating available for premium applications. The use of these coatings is certainly within the ability of the intelligent weekend applicator.

Urethane Coatings (Polyurethanes). These resins are relative newcomers to the field of marine coatings. Since the early 1950s they have been known for their toughness, durability, and resistance to abrasion, chemicals, and temperature extremes. However, early formulations were subject to unacceptable yellowing above the waterline and were not suited to complete-immersion service below the waterline. Later formulations have proven performance records of good gloss and color retention without sacrifice of toughness and durability. Today, cost-competitive urethane coatings are ideal choices for use above the waterline. Not all formulations possess good light stability, and it is essential to specify this quality if decorative properties are required. Resin suppliers suggest some formulations such as coal-tar urethanes below the waterline. These have not received wide market acceptance to date.

Urethane resins were introduced to the marine market through the small-boat owner, who found them excellent wood coatings because of low curing temperatures (around 32°F, or 0°C), fast drying and sanding properties, and ease in cleaning and maintenance (although nonskid coatings prepared by sprinkling sand over a wet resin will gather and hold dirt). More extensive marine uses will be seen as the coatings industry takes advantage of the wide range of formulations available. Basically, a urethane resin is a product of two components, one containing an isocyanate group (NCO) and the other a hydroxyl (OH). The ratio of these components (commonly varying from 1.0 to 1.4) influences the final properties of the coating. For example, primers are usually formulated at a low NCO/OH of about 1.0 to obtain maximum adhesion and flexibility. Many isocyanates are available for formulation. Tolylene diisocyanate (TDI) and its adducts, such as the Mondur resins from Mobay Chemical Company, are typical sources. Hydroxyl can be supplied by polyesters, polyethers, or many mixed polymers (for example, the Multron and Desmophen resins of Mobay Chemical Company). All these combinations, prepared at selected NCO/OH ratios, can be modified with other resins to obtain a specific mix of physical properties. Polyvinyl butyral resins are frequently added to primer formulations, and many resins such as epoxy and vinyl may be added to topcoats.

Two-component urethane formulations are now seeing service as coatings for bridge and deck housings, funnel scoops, and rough-service areas such as working alleyways and anchor decks. One-component moisture-curing coatings are also available, but they do not have all the proven outstanding properties of two-component coatings.

14.3.3 Decorative and functional topcoats

Special topcoats are frequently specified when the appearance or performance of the anticorrosive alone is not satisfactory for the application. Topcoats generally selected are modifications of the base coat, such as a chlorinated-rubber enamel-type color-coded topcoat over a chlorinated-rubber anticorrosive coating.

An example of a special topcoat for appearance reasons might be a topcoat in a light-service area such as a ship's living quarters. Here a less expensive topcoat such as an alkyd or a modified alkyd may be applied directly over a primer. An example of a special topcoat for performance reasons might be an antifouling coating to prevent or retard marine growth in a total-immersion application. The technology of antifouling coatings is sufficiently unique and complex to justify separate treatment.

14.3.4 Antifouling coatings

Several thousand marine plants and animals will attach to and grow on surfaces in either fresh or salt water. When an immersed object has accumulated a noticeable growth of these marine plants and animals, it is said to be fouled. If this growth consists predominantly of barnacles or other hard-shelled animals, fouling can cut through the protective coatings, expose bare metal, and allow severe corrosion of a metallic surface. If the growth on a ship bottom is long streamers of grass (algae), the ship will require substantially more fuel to maintain speed and will be comparably more expensive to operate. From these concerns materials have been developed that retard the growth of marine fouling by releasing a substance which is toxic to fouling organisms. As long as the toxicant is present (released) in sufficient quantities, the materials are effective as antifoulants.

In economic magnitude, the problem of marine fouling is second only to that of corrosion itself. Because of this and of the problems of introducing another scientific discipline (biology) for consideration and also because of the radically different nature of antifouling coatings themselves, it is important to possess basic information in each of these areas. Some fundamental remarks on the nature of fouling are presented below. Remarks pertinent to antifouling coatings are included in the section "Formulations."

Some aspects of marine fouling One outstanding work, *Marine Fouling and Its Prevention,*[7] is suggested for an in-depth treatment of marine fouling. The overall nature of fouling is so complex that no procedure for becoming an overnight expert can be recommended. The following remarks are offered to whet the appetite of those who have the need to know.

The several thousand known marine fouling plants and animals can be divided into microscopic and macroscopic organisms, and discussion can be substantially reduced by considering only the major species. Microscopic organisms are important, and their effects are well known to all boat owners because they produce the slime film found on all immersed objects. Slime is composed of bacteria, single-celled plants, and other larger microscopic organisms (Protozoa and rotifers). The early slime film is important because it seals off the antifouling paint from functioning properly and because it provides both a site of attachment and a nutrient broth for the macroscopic fouling organisms. Theoretically, as long as a surface can be kept free of slime, it presumably will not foul.

Macroscopic fouling organisms can be divided into those with and without shells. Grass, or algae, is typical of soft fouling (without shells), and today it is considered a major problem on ship bottoms. This attitude reflects the effectiveness of the huge strides taken since World War II to combat hard fouling and the recognition of the high fuel penalties paid for grass-fouled bottoms. Grass growth is reported for most marine environments and is usually more of an annoyance than a serious problem to the pleasure-boat owner. However, because of high ultraviolet light and increased oxygen in waters while a ship is under way, grass growth reaches luxuriant proportions and becomes a significant problem for the commercial-boat owner. Freshwater (sweet-water) fouling is almost exclusively soft, and inland sailors concern themselves primarily with problems of grass.

Fouling organisms with shells include (1) barnacles (the widest-known and most troublesome species), (2) annelids (worms with coiled and twisted shells), (3) Bryozoa (flat, spreading limy patches), and (4) Mollusca (mussels, oysters, and clams with paired shells). Acorn barnacles are the most common, but even these vary widely in species (a California

---

[7] Prepared by the Woods Hole Oceanographic Institution for the United States Naval Institute, Annapolis, Md., 1952.

species, *Tintinnabulum,* grows to 1.58 to 1.73 in, or 40 to 44 mm). All barnacle embryos undergo several free-swimming metamorphoses and a "walking" stage (cyprid) that culminates in attachment. This is accomplished by depositing an adhesive through antennules and, somewhat later, by depositing additional shell material as cement after the shell has formed. It is the embryo barnacle that must be repelled by the toxicant in the antifouling coating since the cyprid does not require constant feeding. After attachment has taken place, other cyprids are attracted to the slime covering on the first barnacle, and layer upon layer of hard fouling is thus deposited.

If this were the extent of the problem, it might not be so formidable. In actual fact, a fouling community will vary year to year and seasonally within any given year, responding to such factors as water temperature, light, oxygen content, salinity, water depth and current, location, and the amount of organic nutrients present. To the ship operator, the largest single factor affecting hard fouling attachment is ship speed, since no hard fouling generally occurs at speeds over 1 to 2 knots. This fact is exploited by large vessels, which, because of their fast turnaround times, rely only on a premium anticorrosive coating system and periodic hull scrubbings with underwater brushes to remove grass. This procedure may be satisfactory in certain situations, but the case for antifouling coatings is well established and cannot be ignored.

Antifouling coatings function by controlled release of a toxic substance that is soluble in seawater. If the leaching rate is too high, the antifouling effectiveness will be short-lived; if the leaching rate is too low, the coating will be ineffective. The rate varies with the toxicant; for example, cuprous oxide is the most common toxicant, and an effective leaching rate is known to be 10 μg of copper per 1 cm² of surface area per day. Older paints using mercury as a toxicant required a leaching rate of only 2 μg/cm²/day because of the greater toxicity of mercury.

This picture is further complicated by other factors which directly affect the leaching rate, such as (1) the temperature, salinity, and pH of seawater; (2) the speed of water flow; (3) the type of primer coating and the age of the anticorrosive system; (4) the effect of slime; and (5) dry film thickness. As an example of the importance of these factors, water temperature is so critical that each degree Fahrenheit (0.55°C) of rise or fall can increase or decrease the leaching rate by 5 percent. This explains the rapid failure of paints in tropical waters (accelerated leaching rate due to warmer water temperatures) even though identical coatings perform well in temperate waters. Another example is the adverse effect of the rate of water flow on film erosion. Rapid paint failures are encountered on high-speed vessels because the toxicant is depleted quickly if the cruising speed of the vessel has not been considered and the formulator has not added tougheners to retard erosion.

Formulations Today's antifouling coatings use two general leaching mechanisms, depending on the type of resin matrix selected, soluble or insoluble. The insoluble-matrix type leaves a resinous skeleton intact as the toxicant particles are removed by dissolving into solution in seawater. This is also called the contact type because it depends upon the toxicant migrating to the surface and entering solution by making contact with seawater. Since the resins are somewhat water-permeable, the toxic particles may diffuse through the semipermeable coating, and as one particle dissolves, another is exposed to seawater. The contact type contains several times more toxicant than the soluble type. The resultant thicker films of toxicant provide a longer service life to the antifouling topcoats.

Vinyl-Rosin Antifouling Coatings. These coatings are typical of insoluble-matrix contact-type paints. Type 121 vinyl standard finds wide usage today. It is applied in two coats of 1.5 to 2.0 mils (0.0381 to 0.0508 mm) each (dry thickness). Even in hot-spray applications, it should never be heated over 120°F (49°C).

MIL-P-15931 A (Type 121) Insoluble-Matrix Antifouling Coating

Ingredient	Pounds (1 lb = 0.4536 kg)	Gallons (1 gal = 3.785 l)
Cuprous oxide	800	16.4
Vinyl resin	80	7.1
Rosin	80	9.0
Tricresyl phosphate	30	3.1
Total solids	990	35.6
Methyl isobutyl ketone	375	41.1
Xylene	185	25.5

In this coating, the vinyl-rosin matrix has been plasticized with tricresyl phosphate, and the Cu_2O toxicant is added at a level of 8 lb/gal (958.6 kg/m^3). In temperate waters, this topcoat should afford 18 to 24 months' protection against hard fouling. Cuprous oxide is not particularly effective against soft fouling.

A recent modification of the Type 121 coating is the Type 121/63 vinyl–high-rosin formulation. On the basis of resin solids, the two formulations are essentially equal except for solvent and the volume ratio of vinyl resin to rosin. A service life of about 2 years in temperate waters can be expected from this formulation.

As a general rule, the insoluble-matrix type of paint does not contain an extender pigment, and the geometry of the dry film requires high-toxicant loadings (52 to 74 percent by volume) to ensure that the Cu_2O particles will be in continuous contact with each other. Below the level of

MIL-P-15931 (Type 121/63) Insoluble-Matrix Antifouling Coating

Ingredient	Pounds (1 lb = 0.4536 kg)	Gallons (1 gal = 3.785 l)
Cuprous oxide	1440	29.6
Vinyl resin	55	4.9
Rosin	215	24.2
Tricresyl phosphate	50	5.1
Total solids	1760	63.8
Methyl isobutyl ketone	165	24.6
Xylene	115	15.9

cubic packing (52 percent) the resin will encase the Cu_2O particles and prevent solution; above the level of hexagonal packing (74 percent) the coating will be too resin-poor to maintain film integrity. These figures may vary somewhat in actual practice, and it is common to adjust the leaching rate and the effective range of toxicant loading (e.g., by the addition of rosin or other natural resins). In commercial practice, both natural resins and extender pigments are frequently used. When high levels of rosin are used and high erosion might be expected, tougheners such as ester gum, ethyl cellulose, and modified rubbers are added.

Soluble-Matrix Antifouling Coatings. These coatings function by simultaneous removal of both toxicant and matrix (resinous vehicle). The matrix dissolves by mechanical erosion or by bacterial degradation when a biodegradable matrix is used. The owners of merchant ships have favored these paints, which expose new toxicant to seawater as the resin binder slowly dissolves.

Two typical soluble-matrix paints are the rosin–fish-oil type and the rosin–coal-tar type. The rosin–fish-oil or oleoresinous cold plastic dries by solvent evaporation and represents a significant improvement in application technique over the older hot-plastic type. The hot-plastic type

MIL-P-19451A (Type 105) Cold-Plastic Oleoresinous Antifouling Coating

Ingredient	Parts by weight
Cuprous oxide	58.9
Zinc oxide	16.1
Talc	5.6
Zinc stearate	1.8
Gum rosin	27.7
Blown fish oil	11.8
Total	121.9
Solvent (naphtha)	as required

TABLE 14.8   Complete Coating System for Ship Bottoms

Component	Specification	Number of coats	Film thickness, mils (mm)
Metal pretreatment primer	Formula 117, MIL-P-15328 B	1	0.5 (0.0127)
Anticorrosive coating	Formula 14N, MIL-P-19453	4	6.0 (0.1524)
Oleoresinous cold-plastic antifouling topcoat	Formula 105, MIL-P-19451	2	4.0 (0.1016)
Total		7	10.5 (0.2667)

(MIL-P-19452), which still finds occasional use, has a sag temperature of 140°F (60°C), and in hot weather the vessel must be undocked as soon as possible after application or given a coat of whitewash to retard sagging.

A typical complete coating system for protecting ship bottoms by using the cold-plastic antifouling topcoat is covered by the military specifications in Table 14.8.

Rosin–Coal-Tar Antifouling Coatings. These coatings are used extensively in maritime service. This type of topcoat is usually applied over the TT-P-00118 anticorrosive formulation. Because the vehicle is soft, zinc oxide is added as a toughener. Iron oxide is a good rust inhibitor, and magnesium silicate is used as a high-performance extender pigment.

TT-P-00117 (Formerly 52-MA 403c) Rosin–Coal-Tar Antifouling Coating

Ingredient	Parts by weight
Cuprous oxide	42.5
Mercuric oxide*	2.1
Zinc oxide (lead-free)	21.0
Indian and bright red	8.0
Magnesium silicate	8.0
Gum rosin	26.5
Coal tar (80 percent nonvolatile matter)	8.0
Pine oil	4.2
Total	120.3
Solvent	as required

* United States regulations now prohibit mercury.

Extender pigments are used to reduce the amount of toxicant required in soluble-matrix coatings. This is possible because the entire group of extender pigments, toxicant, and soluble plasticizers is added in a volume range of 74 to 80 percent. Cuprous oxide can be added as a very low percentage of this group. Actually the percent-by-volume range of Cu_2O in soluble-matrix paints is only 6 to 14, as compared with a range of 14 to 40 percent for insoluble-matrix coatings that do not contain extenders. When the volume percentage of total soluble material exceeds 86 percent, the paint fails quickly by excessive chalking.

All the above copper-bearing antifouling topcoats are effective, but they are handicapped by two shortcomings: (1) heavy barrier coats are required to insulate the copper toxicant from the metal substrate and prevent accelerated corrosion, and (2) performance records in protecting against soft fouling are less than desirable.

Organometallic Toxicants in Antifouling Paints. These toxicants do not have the disadvantages listed above, and successful commercial materials based on vinyl, chlorinated-rubber, and epoxy–coal-tar resins are available. Advantages commonly cited include the following: (1) Stable white and light-colored coatings are possible (epoxy–coal tar is an exception). (2) Galvanic corrosion is not accelerated when these toxicants are applied directly over aluminum (for severe service an anticorrosive coating is still recommended). (3) Organotin toxicants are more effective against soft fouling than are copper compounds (not all are as effective against hard fouling as are copper or organolead toxicants).

The first commercially successful organometallic toxicant was tributyl tin oxide (TBTO). It still finds wide usage today, often in combination with a more chemically active toxicant such as tributyl tin fluoride (TBTF), as, for example, in U.S. Navy experimental Formula No. 1020A. TBTF is more toxic than TBTO and is probably the most popular single organometallic toxicant used in the United States today. Triphenyl tin fluoride (TPTF), which is more toxic than TBTF, finds usage in many parts of the world but is not approved for use in the United States. Organolead toxicants also have not been approved for use in the United States. Typical formulations are as follows.

Navy Formula No. 1020A

	Parts by weight
Vinyl resin (Gelva C-5, V-16, Monsanto)	17.5
TBTO	4.2
TBTF	18.1
Carbon black	2.1
Titanium dioxide	0.8
Ethylene glycol monoethyl ether acetate	3.0
n-Propanol	11.1
n-Butyl acetate	43.2
Total	100.0

Vinyl Antifouling

	Parts by weight
Red iron oxide	15.12
Talc	11.22
Zinc oxide	7.08
Vinyl resin (VAGH, Union Carbide)	11.16
Rosin	3.73
Methyl isobutyl ketone	20.31
Xylene	18.84
Bentone 27 } pregel	0.51
Methanol 95 percent }	0.17
TBTF	11.86
Total	100.00

It must be emphasized that the toxicity and environmental acceptability of all organometallic marine coatings are still being assessed.

Chlorinated-Rubber Antifouling

	Parts by weight
Red iron oxide	20.0
Talc	8.5
Zinc oxide	9.4
Bentone 27 } pregel	0.7
Methanol 95 percent }	0.2
Chlorinated rubber (Parlon S-20, 50 percent in xylene, Hercules)	13.8
Rosin, water-white gum, 60 percent in xylene	23.0
Xylene	8.7
TBTF	15.7
Total	100.0

Of interest to pleasure-boat owners is the fact that these paints can be applied months before launching, thereby permitting leisurely hull painting during the late winter. Because of the toxicity of these materials, however, special handling conditions must be observed during application and removal. Commercial painters must wear complete protective apparel including respirators. The paint is applied by brush or roller and *should not be sprayed.* The brushes or rollers should be disposed of by burying, not by burning. Steel which is to be welded should not be painted with organometallics. For the repair of old coatings, high-pressure water or wet-sand washing must be used in preference to dry blasting. The final verdict on the toxicity of these paints has not been reached.

An interesting use of the organotin toxicant is its incorporation into an elastomeric sheet which is then bonded to a surface with a marine-grade adhesive. The product No Foul, of B. F. Goodrich Company, appears extremely attractive for buoys and other small-area applications, but the economics and logistics problems of bonding the sheet to an entire vessel's hull remain controversial. The toxicant diffusion (leaching rate) has been shown to be temperature-dependent and to be adversely influenced by water velocity. However, when this toxicant is specified for the proper application, an effective antifouling service life of up to 10 years can be expected.

New Antifouling Materials. These materials deserve mention even though no service-life recommendations are possible at this time. Two 100 percent solids materials are of interest because their application does not cause any air pollution. The first, Vitron (Southern Imperial), is a solventless epoxy–coal-tar coating with reported superior antifouling properties. The second, Zebron (Xenex), is a nonsolvent polyurethane coating which is basically an anticorrosive material. It appears to have modest antifouling properties by virtue of its high-density surface. Another new material, Hydron (Hemple Marine), is a clear hydrophilic coating which purportedly reduces drag resistance and retards the release of the toxicant while the ship is under way.

A new generation of antifouling paints may emerge from current studies of polymers containing chemically bound toxicants. By incorporating the toxicant into the polymer molecule, toxicity problems may be reduced. Solubility in seawater (leaching rate) is decreased, thereby significantly extending the antifouling service life. Further, the addition of several different chemically bound toxic moieties onto a common resin backbone broadens the scope of the paint's antifouling activity. Encouraging exposure data are reported for unpigmented polymers by Naval Ship Research and Development Center personnel at Annapolis, Maryland. Pigmentation studies are now in progress. Environmental acceptability is still unknown.

14.4   Application of Coating Systems to Specific Service Areas   Selecting coatings systems for specific service areas of a particular application is a formidable task. Countless systems and intercombinations of systems could be listed for each area. The few systems noted in this chapter have been selected as examples of combinations meeting the criteria of corrosion resistance, fouling resistance, protection of cargo, and appearance. However, for true economy (obtained through significantly longer service lives) the coatings must be applied properly over adequately prepared surfaces. These examples are intended mostly for maintenance and repair systems. New construction work may specify the same materials but has more options available, for example, that of using shop primers such as organic or inorganic zinc.

14.4.1 Coatings for service areas   Much information is required before coatings are selected. If a seagoing vessel is to be adequately protected, it is necessary to know its purpose, itinerary, cruising speed, turnaround time, dry-docking time schedule and locale, facilities for underway maintenance, and plans for cathodic protection. Even if the vessel is a small tug, barge, or workboat, it is equally important to know its operating conditions (i.e., fresh water, polluted harbor, climate, maintenance schedule). Coatings systems for offshore structures must meet all the high-performance criteria of surface vessels except, occasionally, that of fouling resistance. The same type of information that is needed for vessels is required for these structures.

After this information is in hand, one more significant factor must be considered. Widely different resin materials are commonly used as components to build a complete coatings system, and the final success of the system will be only as good as the weakest adhesive bond between coats. Great care must be taken to select and then use components in such a manner that they are completely compatible with one another. Factors that influence intercoat adhesion have been touched upon throughout this chapter. When designing complete coating systems, it will be valuable to recall some of these factors, among them the following:

1. Wash primers can tolerate some surface moisture when being applied (epoxy tolerance is more severely limited than the tolerance of wash primers).

2. Vinyl coatings have no tolerance for surface moisture, and chlorinated-rubber coatings have very little.

3. Tie coats are, more often than not, used over the inorganic zinc-rich primer (no surface moisture is tolerable when this primer is applied).

4. Coatings that cure by solvent evaporation (chlorinated rubber, vinyl, etc.) develop an adhesive bond by softening the previous coating and anchoring to it.

5. Coatings that cure by a chemical reaction should be applied before the previous coat is completely cured if the maximum adhesive bond is to be obtained.

6. Coatings that cure by solvent evaporation can be effectively applied at lower temperatures than can coatings that cure by chemical reaction.

The following remarks are organized by service area in an attempt to narrow the task of selecting the proper materials for a particular application.

Immersion coatings   Immersion applications include the bottoms of vessels, dock pilings, and the area from the mud line to the splash zone on structures. Premium materials with known performance records are usually selected.

In addition to corrosion resistance, immersion applications for vessels must include protection from fouling. Antifouling coatings have been discussed separately above.

Boot-top and splash-zone coatings   The boot-top area of a vessel is that part of the hull between the light- and deep-load lines. The splash zone of a structure is that area extending between several feet above the high-tide line to several feet below the low-tide line. This area is the most severe of all marine environments, and special attention must be given to coatings systems selected for it. Of particular importance on stationary structures is the area extending from the point of mean low water to a depth of 2 ft (0.6 m) below this line. This area acts anodically to the splash zone and, because of its aggressive character, must be considered part of the maximum-service zone. Special attention should be paid to metal sheathing which may influence this anodic area.

Boot-top requirements are quite severe because the mobility of a vessel will expose the boot top to a wider variety of conditions, among them, brackish and other polluted waters. In general, outstanding abrasion and impact resistances are required to protect against wave slap and mechanical damage from tugs, piers, pilings, and, in northern waters, ice. The splash zone contains an unusually high degree of oxygen, which, in combination with the ultraviolet-light energy from sunshine, accelerates the corrosion process. Resistance to ultraviolet light is extremely important because ultraviolet degradation will effect complete loss of film integrity in an alarmingly short time. Boot-top and freeboard coatings are exposed to rapid changes in temperature and hence must possess outstanding resistance to thermal shock. This property is formulated into the coating by combining good adhesion with flexibility so that the material is able to expand and contract with the metal. Some studies have shown that the problem can be further minimized by selecting light colors in preference to dark.

Special application and handling properties such as manageable pot life, fast drying times, and excellent air-drying adhesion are required to permit field application. Because of these requirements, vinyls and chlorinated rubbers have appeared most attractive. Epoxy–coal tars are sometimes chosen but are not recommended for application above the light-load line because of color limitations. Many vessels are painted without delineating a boot top, and the bottom paint is simply extended to the deep-load line. In this circumstance, special attention must be given to the selection of a bottom coating which meets all the rigorous boot-top requirements.

Metal claddings, sheathings, and flame-sprayed metals used either alone or in combination with premium coating systems are mentioned below in the section "Dock and Support-Piling Coatings."

Appearance problems arising from rust bleeding at damaged areas may be overcome by using an inorganic zinc primer. This procedure may be practical only with new work. Pigment and filler selection for boot-top coatings is as broad as that for all anticorrosive paints. Iron oxide is widely used because of economy and good anticorrosive properties. When used in a fine-particle grade, it provides good packing, which in turn reduces water-vapor transmission through the film. A high ohms-resistant type of barium sulfate (barites) can be selected along with typical inexpensive extenders such as magnesium silicate or diatomaceous silica. Both of these extenders are completely inert and have high oil-absorption values.

Strut and rudder coatings   In addition to the abrasion and impact resistances required in all immersion situations, struts and rudders need specific resistance to cavitation and impingement. Epoxy–coal tar and polyester-glass are ideal candidate materials; vinyls are not recommended. It can be assumed that polyester-glass will provide a longer service life than epoxy–coal tar even though substantiating data are not yet available.

Superstructure coatings   Superstructure requirements are less rigorous than those for splash or immersion zones. All primers are acceptable for steel, but an inorganic zinc primer is preferred in new construction. Aluminum should be anodized, primed, or treated with a wash primer or a chromate primer. Topcoats are selected for gloss and color retention, cost, and maintenance. The traditional oleoresinous and alkyd finishes are being replaced with silicone-alkyds because of their superior gloss and color retention. Since silicone-alkyds are more expensive than vinyls and vinyl-alkyds, however, the latter are often selected as a compromise. Epoxy coatings have poorer gloss and color retention but are well known for durability. Vinyls and alkyds are more easily maintained. However, most studies of total costs (including application costs) have shown alkyds to be less expensive than epoxies, and vinyls to be the most expensive.

As can be seen in Table 14.4, specific superstructure areas such as weather decks and hot stacks require totally different coating systems from those mentioned above. Urethane resins make excellent deck coatings because of their superior wear resistance. Sand sprinkled over the uncured-resin surface provides a good nonslip surface, but pleasure-boat owners may be unhappy with the dirt that these finishes collect. Most nonskid coatings have a tendency to be somewhat water-permeable because of the porosity resulting from the sand content. When steel surfaces must be protected, it is customary to apply the nonskid topcoat over a premium-grade primer. A military specification for a nonskid surface is listed in MIL-D-23003, Type II. Hot stacks are usually protected with a silicone coating system which meets the effluent-temperature requirements.

Tank coatings   The merits of tank coatings or linings were well established in the early 1950s, when the U.S. Navy specified the Saran resin-based Formula 113 for jet-fuel tanks. The vinylidene chloride–acrylonitrile copolymer formulation has a flash point of only 30°F (-1°C), and other materials were quickly developed. Today the Saran specification has been deleted, and the product is not available. The entire field of tank coatings has become quite complex because of the many cargoes which must be considered. The best introduction to tank coatings is familiarity with MIL-P-23236: (*Ships*), *Paint Coating Systems, Steel Ship Tank, Fuel and Salt Water Ballast.* This is divided into Type I (new and complete application) and Type II (maintenance), with each type subdivided into several classes such as Class 1 (epoxy) and Class 2 (epoxy–coal tar). Chapter 9190 now lists at least 60 different materials such as epoxies and inorganic zincs which meet these specifications, exclusive of potable-water applications.

Epoxy, epoxy–coal tar, or urethanes are commonly used in combination with thermosetting topcoats, and vinyls are usually employed in combination with thermoplastic-resin topcoats. Epoxy and urethane resins are applied in three to five coats to a dry thickness of 6 to 8 mils (0.1524 to 0.2032 mm), whereas epoxy–coal tars are applied in two coats of 8 to 10 mils (0.2032 to 0.254 mm) each. The low solids of vinyls generally require at least five coats for proper build (6 to 8 mils dry), although some applications specify vinyl mastics that do not require as many coats.

For supertankers which can carry as much as 60 million gal (287,125 m³) of liquid cargo, it is planned to segment each compartment further with several rubber bladder tanks in preference to relying solely on coatings on compartment bulkheads for protection. Floating rubber-fabric tanks have been suggested for both marine storage and transportation of liquids. The new liquid-natural-gas (LNG) tank carriers are expected to proliferate during the next decade because of the vast needs for transporting gas (methane, ethylene, butane, etc.) at cryogenic temperatures. To date, no serviceable marine coatings have been developed for LNG applications, and operators are relying solely on such materials as stainless-steel membranes.

Each type of tank may be considered a separate application, and this factor is reflected in the requirements outlined in Table 14.4. For example, the outside of mud tanks on offshore structures is treated in the same way as the superstructure (e.g., with inorganic zinc topcoated with an epoxy-polyamide), but the inside is coated with an epoxy–coal-tar material.

The most rigid, best-established specifications have been written for potable-water tanks. Early coating systems included hot bituminous and phenolic varnish pigmented with red lead, as covered in MIL-P-15145 A or TT-P-86a. Today's coatings are formulated to a high pigment-volume concentration (PVC), which gives a low-sheen finish without compromising adhesion. The pigmentation is low on TiO_2 and high on extenders such as silicates and barites to give adequate hiding and economy. Solvent selection is extremely important for potable-water–tank coatings because of the need for a completely inert surface free from outgassing. High-flash solvents are always selected to provide a flash point (Pensky Martin cup) over 100°F (38°C). Because of these rigid requirements, Chapter 9190.171 lists only five materials currently acceptable in both the wet and the cured states. In addition to Sec. 9190, potable-water–tank coatings must meet MIL-P-23236 and Food and Drug Administration (FDA) standards. Vinyl, epoxy, inorganic zinc, and inorganic zinc plus an aluminum-pigmented phenolic tung-linseed binder are being specified today. Recently the trend is not to specify an inorganic zinc primer without a topcoat. The criticisms raised include a suggested sensitivity of inorganic zinc to hydrogen sulfide and a tendency to pick up heavy metals. If these problems are verified, the inorganic zinc could not be used for jet fuels or for potable water.

The relative merits and disadvantages of high-performance materials as potable-water–tank coatings are similar to those for most marine applications. Vinyls have long service records behind them and are easily patched, but their low solids require many coats and entail high labor costs. Also, highly volatile, flammable solvents are required. Epoxy systems are less expensive because they are easier to apply and fewer coats are required. If chlorinated solvents are added to an epoxy resin for nonflammability during application, the cost rises accordingly. Inorganic zinc is inexpensive, only a single coat is required, and no flammable solvents are needed. However, strict attention must be paid to climate (low humidity) during surface-preparation blasting, and as noted above, a topcoat may be required. Any of these premium systems may be selected with equal confidence in their performance. The ultimate choice is usually made on the basis of secondary considerations such as time schedules and yard procedures.

Surface-effects–ship coatings   Surface-effects ships (SES) must be protected from corrosion and fouling similarly to all conventional vessels. In addition, these ships are subjected to erosion from high-velocity water and to cavitation. Ships operating under 20 knots are considered hull-borne vehicles, and neoprene anticorrosive coatings have performed satisfactorily. Those operating at speeds between 20 and 80 knots are cushion-borne and are usually protected by metal or metal claddings because little research has been done with organic coatings.

One suggested system for 45-knot hydrofoils is a wash primer (Type 117) plus a urethane topcoat filled with Teflon and polypropylene. Supercavitating applications (any speed over 60 knots) are proliferating, and satisfactory marine finishes must be developed to meet these needs.

Dock and support-piling coatings   All coating materials for docks are selected similarly to those for other marine application (e.g., decks, superstructure, etc.) with one outstanding exception. The exception is wood pilings which must be protected from borer attack as well as from abrasive bottom washings. The traditional approach to the protection of wood pilings is by impregnation with creosote, creosote–coal tar, and creosote plus toxicants (chromated copper arsenate, Malathion, pentachlorophenol, etc.).

In general, it is common to protect metals cathodically below the water-line and to use claddings, sheathings, and coatings in the splash zone and above. Flame-sprayed metals are receiving considerable attention because of increased service requirements arising from support pilings for offshore structures. It is assumed that approximately 1 year of service life can be obtained from each mil (0.0254 mm) of zinc coating. Thus, a system of 5 to 7 mils (0.127 to 0.1778 mm) of flame-sprayed zinc plus a premium high-build epoxy topcoat should provide outstanding service. Other metals such as aluminum are also easily applied by flame spray. The use of flame-sprayed metals to mitigate corrosion is not limited to pilings. Entire vessels such as tugs and workboats have been successfully protected by a system including 3 to 4 mils (0.0762 to 0.1016 mm) of flame-sprayed aluminum, a wash primer to seal the aluminum, a thin vinyl coat to prevent leaching of chromate from the wash primer, and an antifouling topcoat.

A suggested new approach to obtaining splash-zone corrosion resistance is the use of Monel Metal sheathing. Substantial improvements in service life are reported for a cost per square foot of approximately 40 percent more than that of conventional coatings. Fouling resistance is not a factor in these applications except to the extent that fouling may cut into the anticorrosive system and initiate corrosion. Epoxy–coal-tar and epoxy-polyamide anticorrosive coatings are tough enough to withstand this undercutting even though they become heavily fouled in a short time. Several anticorrosive and antifouling wrapping materials have been introduced to the market for the protection of pilings, but they are too new for their purported outstanding performance records to be confirmed.

Tie coats   Any coating used to promote adhesion between two otherwise-incompatible materials is considered a tie coat. Several circumstances have contributed to the increased use of tie coats: (1) So much emphasis is placed on the weld-through shop primer that adhesion of subsequent materials has become a secondary consideration. (2) Since so many types of coatings can be interchanged in one complete system, serious adhesion problems can be encountered. (3) For touch-up repair over intact old coatings, any premium coating is likely to be chosen without regard to compatibility.

Tie coats can be selected from any material, but they are usually based on premium resins such as those used in high-build vinyl, epoxy, or chlorinated-rubber formulations. Thin coatings applied at reduced viscosity may also be considered tie coats. They may be used to bond primers to anticorrosives, one anticorrosive to a second anticorrosive of a different type, or anticorrosives to antifouling coatings. The most often cited example of a tie coat in this chapter is the wash primer required to obtain adhesion of almost all coatings to the inorganic zinc-rich primer. The wash primer is an excellent tie coat for this application because it has good alkali resistance and because its water-thin viscosity will not trap moisture or air in the porous zinc silicate surface.

14.4.2 Critique   Two approaches to tabulating the information required to match coatings systems to service areas are presented below. First, the general list of service areas, application and maintenance requirements, and selected coating systems discussed above is outlined in Table 14.4, which is intended as an overview to tie major marine application areas into a single package. Second, reference is made to the Society of Naval Architects and Marine Engineers, which has attempted to resolve the complexity of understanding marine coatings by tabulating basic information on ship-bottom, boot-top, freeboard, weather-deck, and superstructure coatings systems.[8]

Finally, note that a new marine coatings standard is being drafted by the California Air Resources Board (CARB); it will undoubtedly be considered a model for all states. The standard, which may significantly alter current formulation and application practice, is slated for implementation in 1982.

BIBLIOGRAPHY

Cleveland, John L.: "California Air Resources Board Gives Chief Exec Power to Perfect Model Rule for Marine Coatings," *American Paint Journal,* July 17, 1978, p. 7 ff.

Matanzo, F.: "Service Life Performance of Marine Coatings and Paint Systems," *Journal of Coatings Technology,* June 1980, pp. 55–63.

"Organometallic Polymers Targeted at Barnacles," *Chemical and Engineering News,* Jan. 6, 1975, pp. 16–19.

"Performance of Selected Marine Coatings," *Journal of Coatings Technology,* February 1980, pp. 35–45.

Williams, Alec.: *Antifouling Marine Coatings,* Noyes Data Corp., Park Ridge, N.J., 1973.

[8] *Coating Systems Guide for Hull, Deck, and Superstructure,* Technical and Research Bulletin 4-10, Society of Naval Architects and Marine Engineers, 77 Trinity Place, New York, N.Y. 10006, 1973.

Chapter **15**

Specification Products

SIDNEY B. LEVINSON
D/L Laboratories

15.1   Introduction   Coatings can be purchased either by brand name or by specification. Purchasing by brand name relies on the paint manufacturer's name and reputation. This is a nebulous base on which to determine quality and anticipate good performance. Fortunately, most companies guard their reputations carefully and honestly attempt to sell quality merchandise at a competitive price. However, sometimes a company sells strictly on price and does not use a brand name, giving a product a name which is not even associated with the company. Some wholesalers and dealers, especially those who do a large volume of business, often have their paints made by a paint manufacturer under their own private labels. This practice is not necessarily a sign of inferior quality, but it does lack the support of the paint manufacturer's brand name. Also, many reputable paint manufacturers and dealers offer more than one line of coatings, in some cases as many as three or four. These coatings vary in quality and price. Sometimes a manufacturer may put several different labels on the same batch of paint. Each label may have a different selling price.

Consequently, it is apparent that purchasing by brand name must rely on the reputation of the supplier rather than on any specific data or parameters by which quality of the product can be determined. There is one exception to this statement. Most national paint manufacturers of trade sales paints do supply analyses of their products on the can labels. While these help to determine the type of vehicle used, the amount of color or hiding pigment, and total solids, they cannot delineate the anticipated performance of the paint. Specifications supply this information. They often include analytical data and always include paint performance requirements.

15.2   Purpose of Specifications   Specifications are used to delineate the type of paint for the anticipated end use. Although there is a trend toward omission of analytical data, paint specifications do describe the general makeup of the paint to be used and, of major importance, the anticipated properties with definite minimum and/or maximum values. Specifications also include the actual test methods to be used to determine those parameters or refer to these test methods elsewhere.

Specifications are of special value to large purchasers since they enable them to obtain bids on a definite product with definite parameters of performance. Thus they can obtain a quality product at a competitive price instead of relying on brand names. In fact, many paint manufacturers prefer not to bid closely on their brand-name products since this practice may affect their normal pricing arrangements, especially if bids are made through dealers.

Specifications do have limitations, and these should be understood before purchasers rely on them completely. Paint manufacturers differ in their views of which properties should be emphasized in a specific type of coating. Some stress ease of application; others stress ease of stain removal, scrubbability, etc.; and still others stress appearance, such as complete opacity in one coat or high leveling. Such differences are often reflected in reports by a major consumer testing organization. This circumstance demonstrates the fallacy of specifying ". . . or equal." However, specifications often tend to embody minimum acceptable values for most of these properties. This tendency is particularly true of paints manufactured to meet government specifications. Some first-line brand-name products are superior to the equivalent specification in certain properties, and a specification should therefore be used only as a minimum standard of performance.

In view of this consideration, there is an ongoing effort to establish performance criteria for the government procurement of commercial off-the-shelf paint and coatings. ASTM has a special subcommittee, DO1.41, to deliberate and select tests actually to be used for the federal procurement of these coatings and to assign numbers to these tests.

Another problem in accepting bids from all companies is that some manufacturers specialize in borderline performance and thus are able to quote lower prices than reputable companies. The federal government unfortunately must buy from anyone, but private companies can limit bid purchasing to suppliers whose reputation is known.

One fact is often overlooked by many paint purchasers. Specifications are of little value without systematic inspection. Products should at least be spot-checked before and after purchase. If products are not checked, suppliers can become careless and disregard specifications completely. A few systematic spot checks will keep them "honest" and on their toes.

Specification data are often submitted by paint manufacturers, especially in the architectural, marine, and maintenance fields. These data are invaluable. Not only do they explain the use of the product, but they also give definite performance information and recommended procedures for surface preparation, application, film thickness, drying time before use, etc. The latter information is omitted from most government specifications. Use these data; they are much better than the usually limited instructions on labels.

15.3 Specification Composition   The makeup of specifications varies among United States federal, Canadian, state or provincial, and municipal government agencies, but all contain most of the items listed below. A typical federal specification can be used as an example of the composition of an ideal document. It includes the following sections:

- Title
- Scope and classification

- Type and colors
- Applicable documents
- Requirements
- Sampling, inspection, and testing procedures
- Delivery and use

15.3.1 Title   The title describes the product, its type, finish, and color, and its general area of use. It also contains a designation showing whether the product has been modified and the date on which it was issued.

Example: TT-P-29J: *Paint, Latex Base, Interior, Flat, White and Tints,* dated August 27, 1976 (Fig. 15.1). The letter *J* designates the tenth modification since the original specification was issued, and the date is the date on which this modification was issued.

15.3.2 Scope and classification   This section of the specification explains the essential purpose of the product, its general makeup, and the places where it should be used, for example, "a lead-free ready-mixed primer paint for previously unpainted exterior woodwork or for exterior surfaces previously painted with house paint."

15.3.3 Type and colors   Often more than one type of the product is available, and, of course, many products are offered in a variety of colors or, alternatively, may be available only as a clear finish such as a varnish or a glaze. A typical example is TT-C-598 caulking compound, which has two types: Type I for gun application and Type II for knife application. Another is TT-P-641 primer, which has three types: Type I, oil; Type II, alkyd; and Type III, phenolic.

15.3.4 Applicable documents   All applicable references, such as specifications for ingredients, color cards (e.g., *Federal Standard No. 595*), packaging, and test methods, are listed.

15.3.5 Requirements   This section of the specification is of greatest interest to the architect, contractor, specifier, or maintenance engineer. It lists and describes the properties of the products. Usually it is divided into three subsections, as described below.

General requirements   This subsection describes what is generally required of the product, for example, that it should be well ground, easily mixed for use, and free of toxic materials (e.g., a lead-free paint).

Qualitative requirements   Requirements which cannot actually be measured but must be checked subjectively are described in this subsection. Examples are condition in the container, working properties, color match, and flexibility.

Quantitative requirements   Requirements which can be measured are listed in this subsection with actual values for (*continued on page 15-17*)

```
                                        TT-P-29J
                                        August 27, 1976
                                        SUPERSEDING
                                        Fed. Spec. TT-P-29H
                                        March 26, 1974
```

FEDERAL SPECIFICATION

PAINT, LATEX BASE, INTERIOR, FLAT, WHITE AND TINTS

This specification was approved by the Commissioner, Federal Supply Service, General Services Administration, for the use of all Federal agencies.

1. SCOPE AND CLASSIFICATION

1.1 Scope. This specification covers a ready-mixed, latex-base paint for interior walls and ceilings. This specification provides two types of flat paint.

1.2 Classification. The paint shall be of the following types.

Type I   - Tints (pastel) and whites (colors 37875 and 37778) specified
             by reference to Fed. Std. No. 595.
Type II - A high-hiding white (no color number), suitable for use as
             is or as a tint-base (see 3.4.3 and 6.5).

2. APPLICABLE DOCUMENTS

2.1 The following documents, of the issues in effect on date of invitation for bids or request for proposal, form a part of this specification to the extent specified herein.

Federal Specifications:

```
H-B-420     -  Brush, Paint, Flat, Metal-bound.
H-R-550     -  Roller, Kit, Paint.
L-S-626     -  Sponges, Synthetic.
SS-L-30     -  Lath, Sheathing, and Wallboard, Gypsum.
TT-E-545    -  Enamel, Odorless, Alkyd, Interior Undercoat, Tints and White.
TT-P-650    -  Primer, Coating, Latex Base, Interior, White (For Gypsum
                 Wallboard).
TT-S-179    -  Sealer Surface: Pigmented Oil, Plaster and Wallboard.
TT-T-390    -  Tinting Medium, Concentrate General-Purpose.
PPP-P-1892  -  Paint, Varnish, Lacquer, and Related Materials; Packaging,
                 Packing, and Marking of.
PPP-T-60    -  Tape, Packaging, Waterproof.
```

Federal Standards:

```
Fed. Test Method Std. No. 141  -  Paint, Varnish, Lacquer, and Related
                                    Materials, Methods of Inspection,
                                    Sampling and Testing.
Fed. Std. No. 595  -  Colors.
```

(Activities outside the Federal Government may obtain copies of Federal Specifications, Standards, and Handbooks as outlined under General Information in the Index of Federal Specifications and Standards and at the prices indicated in the Index. The Index, which includes cumulative monthly supplements as issued, is for sale on a subscription basis by the Superintendent of Documents, U.S. Government Printing Office, Washington, DC 20402.

FSC 8010

Fig. 15.1

TT-P-29J

(Single copies of this specification and other Federal Specifications required by activities outside the Federal Government for bidding purposes are available without charge from Business Service Centers at the General Services Administration Regional Offices in Boston, New York, Washington, DC, Atlanta, Chicago, Kansas City, MO, Fort Worth, Denver, San Francisco, Los Angeles, and Seattle, WA.

(Federal Government activities may obtain copies of Federal Specifications, Standards, and Handbooks and the Index of Federal Specifications and Standards from established distribution points in their agencies.)

Military Standard:

MIL-STD-105 - Sampling Procedures and Tables for Inspection by Attributes.

(Copies of Military Specifications and Standards required by suppliers in connection with specific procurement functions should be obtained from the procuring activity or as directed by the contracting officer.)

2.2 Other publications. The following documents form a part of this specification to the extent specified herein. Unless a specific issue is identified, the issue in effect on date of invitation for bids or request for proposal shall apply.

American Society for Testing and Materials (ASTM) Standards:

D 476	-	Titanium Dioxide Pigments.
D 562	-	Consistency of Paints Using The Stormer Viscosimeter.
D 1210	-	Fineness of Dispersion of Pigments in Vehicle System.
D 1296	-	Odor of Volatile Solvents and Diluents.
D 2244	-	Instrumental Evaluation of Color Difference of Opaque Materials.
D 2369	-	Volatile Content of Paints
D 2486	-	Scrub Resistance of Interior Latex Flat Wall Paints.
D 3273	-	Resistance to Growth of Mold on the Surface of Interior Coatings in an Environmental Chamber.
D 3274	-	Evaluating Degree of Surface Disfigurement of Paint Films by Fungal Growth or Soil and Dirt Accumulation.
D 3335	-	Low Concentrations of Lead in Paint by Atomic Absorption Spectroscopy.
E 97	-	Daylight Directional Reflectance.

(Application for copies should be addressed to the American Society for Testing and Materials, 1916 Race Street, Philadelphia, PA 19103.)

3. REQUIREMENTS

3.1 Qualification. All paint supplied under this specification shall be a product which has been tested and approved for listing on the applicable Qualified Products List (QPL) maintained by the Federal Supply Service (FMBP), General Services Administration, Washington, DC 20406. Hereinafter the term "qualifying activity" shall mean GSA-FMBP. All inquiries regarding qualification shall be directed to this address.

3.1.1 Qualification samples. Qualification testing is limited to these colors for type I: blue, color no. 35526; green, color no. 34554; and yellow, color no. 33717. Type II also requires qualification. Qualification for each color will include extension of approval to related colors or shades (see 4.2.1.1 and 6.6). This carryover of qualification does not in any way waive the responsibility of the manufacturer to fully meet the specification requirements for all shades and colors supplied. Full details on sample size, method of payment for tests, testing schedules, and report of test results are available from the qualifying activity.

3.1.2 Qualification testing. Paint for test may be submitted by prospective suppliers to the General Services Administration. Should the submitted product(s) fail to meet all requirements of the specification as defined herein, the qualifying activity reserves the right to refuse to accept such products for additional qualification tests until satisfactory data and test results have been submitted indicating correction of product deficiencies. The qualifying activity reserves the right to levy a charge to cover the cost of testing for qualification of products.

Fig. 15.1 (Continued)

TT-P-29J

3.1.3 <u>Qualified products listing</u>. The Qualified Products List shall consist of products which have been tested and have passed all qualification tests specified herein (see 4.2.1). Qualification and listing in the Qualified Products List does not guarantee acceptance of the products in any future procurement, nor constitute a waiver of the requirements of the specification as to acceptance, inspection, test, or other provisions of any contract involving these products. Different plants of the same manufacturer must be qualified individually in order to be listed on the Qualified Products List.

3.1.4 <u>Formulation change</u>. Qualification of a supplier's paint under this specification, once established, applies only to that paint manufactured according to the specific formulation in use at the time of qualification. Any change in formulation or manufacturing procedures may result in removal of the product from the Qualified Products List. If material supplied by a manufacturer on the Qualified Products List to any agency of the U.S. Government, or for U.S. Government end use, is found to deviate from the originally qualified formula, or not to meet all the requirements of the specification, this shall be considered cause for possible removal from the Qualified Products List (see 3.1.5). All proposed formulation changes shall be reported to the qualifying activity, with a statement by the supplier as to the general nature of the changes, and the extent and effect of such changes on the delivered product. Changes will be evaluated by the qualifying activity for possible removal of the product from the Qualified Products List.

3.1.5 <u>Qualification withdrawal</u>. A supplier's product may be removed from the Qualified Products List in accordance with "Provisions Governing Application by Manufacturers for Inclusion on Federal Qualified Products Lists (QPL's)" (see 6.4).

3.1.6 <u>Requalification</u>. A supplier's product, once removed from the Qualified Products List, shall not be accepted for requalification until satisfactory data and test results have been submitted to the qualifying activity by the supplier indicating correction of the product fault(s). The qualifying activity reserves the right to levy a charge to cover the cost of testing for requalification of product(s).

3.2 <u>Materials requirements</u>. The paint covered by this specification shall consist of the pigments and vehicle specified, so combined as to produce a ready-to-use paint meeting all the requirements of this specification.

3.2.1 <u>Vehicle</u>. The vehicle shall be of the latex type, i.e., a stable aqueous dispersion of synthetic resin particles prepared by emulsion polymerization. Small additions (not in excess of 10 percent) of emulsified modifying resins may be made, provided the finished product meets all the requirements specified herein.

3.2.2 <u>Pigments</u>. The prime pigments shall consist of non-chalking titanium dioxide conforming to ASTM D 476, type III or IV. Suitable extender pigment or pigments may be used provided the paint meets all the requirements specified herein. Tinting pigments may be used when necessary to match the color required, provided these pigments have good color permanence. Pigments and any combination thereof shall be lightfast, alkali-resistant, and good commercial quality.

3.3 <u>Quantitative requirements</u>.

3.3.1 <u>General quantitative requirements</u>. The general quantitative requirements of the paint shall be as specified in table I.

TABLE I. Quantitative requirements, types I and II

Characteristic	Minimum	Maximum
Consistency, Krebs-Stormer, shearing rate 200 rpm		
Grams	200	475
Equivalent Krebs Units	82	110
Nonvolatile, percent by weight of paint	50	----
Dry hard, minutes	---	60
85° Specular Gloss	3	10
Fineness of grind	3	----
Lead content, percent by weight of total nonvolatile (as Pb metal)	---	0.5
Yellowness index difference (after accelerated yellowing), (type I, only color 37875; and type II)	---	0.07

Fig. 15.1 (*Continued*)

TT-P-29J

3.3.2 Quantitative requirements for type I only. The dry and rewetted opacity for type I paints, applied at a spreading rate of 630 square feet per gallon, shall be as specified in table II. The maximum difference between dry and rewetted opacity, for all values of the directional reflectance, shall not exceed 0.02.

TABLE II. Minimum opacity, type I only

Directional Reflectance	Opacity Dry	Rewetted
80 and above	0.95	0.93
79 - 76	0.96	0.94
75 - 72	0.97	0.95
71 - 68	0.98	0.96
67 and below	0.99	0.97

3.3.3 Quantitative requirements for type II only. Requirements for type II paint only, applied at a spreading rate of 630 square feet per gallon, are given in table III.

TABLE III. Quantitative requirements, type II only

Characteristic	Minimum	Maximum
Directional Reflectance	90	----
Opacity		
Dry	0.95	----
Rewetted	0.93	----
Difference between dry and rewetted	-----	0.02

3.4 Qualitative requirements.

3.4.1 Condition in container. The paint as received shall be ready-mixed and shall show no evidence of biological growth, livering, skinning, putrefaction, corrosion of the container, or hard settling of the pigment. Any settled pigment shall be readily dispersible in the liquid portion by stirring with a paddle to form a smooth homogeneous paint, free from persistent foam.

3.4.2 Storage stability.

3.4.2.1 Partially full container. The paint shall show no skinning when tested as in 4.4.2. After storage as specified in 4.4.2.1, the paint shall show no livering, curdling, hard caking, or gummy sediment. It shall mix readily to a smooth homogeneous state.

3.4.2.2 Full container. The paint shall show no skinning, livering, curdling, hard or dry caking, or tough gummy sediment when tested as in 4.4.2.2. After storage for 12 months as specified in 4.4.2.2, the paint shall remix readily to a smooth homogeneous state, there shall be no change in drying time, and the viscosity shall be within the range of 77 to 115 K.U.

3.4.3 Color. The color of all type I paint specified in the contract or order (see 6.2) shall match that of the standard color chip in Fed. Std. No. 595 when tested as specified in 4.4.3. Type II (tint-base) shall meet the directional reflectance specified in table III when tested as specified in table V.

3.4.4 Flexibility. When tested as described in 4.4.4, the paint shall show no evidence of cracking, chipping, or flaking.

3.4.5 Working properties. The paint shall be easily applied by brush, roller, and spray equipment when tested as specified in 4.4.5. The paint shall show no foaming, spattering, or pigment separation during application.

Fig. 15.1 (Continued)

TT-P-29J

3.4.6 _Appearance of the dried paint_. When the paint is applied by brushing or rolling as specified in 4.4.6, the film shall dry to a smooth, uniform finish free from craters and other defects caused by bubble retention. There shall be no shiners or flashing, no streaking, and no conspicuous laps or objectionable brush marks on the dried film. Between any two of the eight films, the difference in 85° specular gloss shall not be more than 2 units, and the difference in reflectance shall not be more than 1 percent (absolute).

3.4.7 _Anchorage_. A film of the paint, when tested as specified in 4.4.7, shall show no removal or loosening of the paint beyond 1.5 mm (one-sixteenth inch) on either side of the score line.

3.4.8 _Scrubbability_. The film shall not be worn through to the panel in fewer than 400 cycles (800 separate strokes), when tested as specified in 4.4.8.

3.4.9 _Washability_. When painted panels are tested as specified in 4.4.9, the soil shall be substantially removed without any exposure of the undercoat. The reflectance of the cleaned area shall be not less than 95 percent of the value measured on the unsoiled area before the test; the 85° specular gloss of the washed area shall be not greater than 20. The color of the film in the washed area shall not be different from that in the unwashed area.

3.4.10 _Freeze-thaw resistance_. The paint as received shall withstand the freeze-thaw test performed as specified in 4.4.10, and the viscosity shall not change more than 5 K.U. After completion of this test, the paint shall dry to a smooth uniform finish when applied to a wallboard panel.

3.4.11 _Water resistance_. The film shall show no wrinkling, re-emulsification, or other changes when tested as specified in 4.4.11.

3.4.12 _Alkali resistance_. When the paint film is tested as specified in 4.4.12, the film shall show no change in hue, lightness, or 85° gloss.

3.4.13 _Resistance to biological growth_. When tested as specified in 4.4.13, the paint film shall have a surface disfigurement rating of 5 or greater. All biological growth shall be included in the evaluation of disfigurement.

3.4.14 _Resistance to reflectance variation_. When tested as in 4.4.14, the general appearance of the panel, including the color of tinted paints, shall be substantially uniform. Any variation in reflectance of the film between the coated sealed and coated unsealed areas shall not exceed 1.0 percent (absolute).

3.4.15 _Compatibility (type II only)_. When tested as specified in 4.4.15, the dried film shall show uniform color, an 85° gloss between 3 and 10, and no streaks, craters, or pigment floating.

3.4.16 _Recoating properties_. The paint shall produce no lifting, softening, or other film irregularities upon recoating of a previously painted surface when tested as specified in 4.4.5.

3.4.17 _Odor_. When tested as specified in table V, the odor of the paint shall not be putrid or otherwise offensive or irritating before, during, and after application. There shall be no residual odor after 24 hours of drying.

4. QUALITY ASSURANCE PROVISIONS

4.1 _Responsibility for inspection_. Unless otherwise specified in the contract or purchase order, the supplier is responsible for the performance of all inspection requirements as specified herein. Except as otherwise specified in the contract or order, the supplier may use his own or any other facilities suitable for the performance of the inspection requirements specified herein, unless disapproved by the Government. The Government reserves the right to perform any of the inspections set forth in the specification where such inspections are deemed necessary to assure that supplies and services conform to prescribed requirements.

Fig. 15.1 (Continued)

TT-P-29J

4.2 <u>Classification of tests</u>. Inspections specified herein are classified as follows:

(a) Qualification testing (see 4.2.1).
(b) Contractor's production testing (see 4.2.2).
(c) Acceptance testing (see 4.2.3 and 4.3.5).
(d) Inspection of preparation for delivery (see 4.3.4).

4.2.1 <u>Qualification testing</u>. Qualification testing shall be conducted at a laboratory designated by the qualifying activity (see 3.1) on paint submitted in accordance with 3.1.1. Qualification inspection shall consist of tests for all requirements in section 3. The results of each test shall be compared with the applicable requirement in section 3. Failure to conform to any requirement shall be counted as a defect, and the paint represented by the sample tested shall not be approved for inclusion on the Federal Qualified Products List (QPL) under this specification.

4.2.1.1 Qualification, once established, is extended to related colors and shades (see 3.1.1 and 6.6).

4.2.1.2 Different plants of the same manufacturer must be qualified individually in order to be listed on the Qualified Products List.

4.2.2 <u>Contractor's production testing</u>. The contractor shall perform production inspections on all paint submitted for delivery. Production inspections shall be performed in accordance with test procedures of section 4 to determine conformance to requirements specified in section 3. All necessary test equipment and facilities required to conduct the subject inspections shall be furnished by the contractor.

4.2.3 <u>Acceptance testing</u>. Testing for acceptance of individual lots shall consist of tests specified in section 4, for all requirements specified in section 3, with the exception of storage stability (see 3.4.2.2 and 4.4.2.2) and resistance to biological growth (see 3.4.13 and 4.4.13).

4.3 <u>Sampling and inspection for acceptance</u>.

4.3.1 <u>Lot</u>. For the purpose of sampling, a lot of the paint shall consist of a manufacturer's batch. A batch is defined as the end product of all raw materials mixed, blended, or processed in a single operation.

4.3.2 <u>Sampling for inspection of containers</u>. A random sample of containers offered for delivery shall be selected in accordance with MIL-STD-105 at inspection level I and acceptable quality level (AQL) = 2.5 percent defective to verify compliance with this specification regarding fill, closure, marking, and other requirements not involving tests.

4.3.3 <u>Inspection of containers</u>. Each container randomly selected from the lot offered for inspection shall be examined for defects of construction of the container and the closure, for evidence of leakage, and for unsatisfactory markings; each container selected shall also be tested to determine the amount of contents. Any container in the sample having one or more defects or under required fill shall be rejected, and if the number of defective containers in any sample exceeds the acceptance number for the appropriate sampling plan of MIL-STD-105, the lot represented by the sample shall be rejected.

4.3.4 <u>Inspection of preparation for delivery</u>. An inspection shall be made in accordance with PPP-P-1892, to determine that the packaging, packing, and marking comply with the requirements of section 5 of this specification. Defects shall be scored in accordance with table IV. For examination of interior packaging, the sample unit shall be one shipping container, fully prepared for delivery, and selected at random just prior to the closing operations. Sampling shall be in accordance with MIL-STD-105. Defects of closure listed shall be examined on shipping containers fully prepared for delivery. The lot size shall be the number of shipping containers in the end item inspection lot. The inspection level shall be S-2 and the AQL shall be 4.0 defects per hundred units.

Fig. 15.1 (Continued)

TT-P-29J

TABLE IV. Classification of preparation for delivery defects

Examine	Defect
Markings (exterior and interior)	Omitted; improper size, location, sequence, or method of application.
Materials	Any component missing or damaged.
Workmanship	Inadequate application of components such as incomplete closure of container flaps, loose strapping, inadequate stapling. Bulging or distortion of container.

4.3.5 Testing of the end item. The methods of testing specified in 4.4 shall be followed. For purposes of sampling, the lot shall be expressed in units of gallons of paint. The sample unit for testing shall be 1 gallon of paint, randomly selected from containers in the lot. The sample size shall be as follows:

Lot size (gallons)	Sample size (gallons)
800 or less	2
801 up to and including 22,000	3
22,001 or more	5

All test reports shall contain the individual values utilized in expressing the final result. The lot shall be unacceptable if any test result is not in conformance with the corresponding requirement in section 3.

4.4 Test procedures. Samples shall be tested as specified in table V. Alternate test techniques and equipment may be used by the Government, but in case of dispute, the test methods specified here shall prevail. Unless otherwise specified, standard test conditions are $23 \pm 1°C$ ($73 \pm 2°F$) and a relative humidity of 50 ± 5 percent.

TABLE V. Index

Characteristics	Requirement reference	Test Method Fed. Test Method Std. No. 141	ASTM	Para. ref.
Condition in container	3.4.1	3011	------	4.4.1
Storage stability	3.4.2	3021, 3022	------	4.4.2
Color	3.4.3	4250	------	4.4.3
Flexibility	3.4.4	6221	------	4.4.4
Working properties	3.4.5	----	------	4.4.5
Appearance	3.4.6	----	------	4.4.6
Anchorage	3.4.7	----	------	4.4.7
Scrubbability	3.4.8	----	D 2486	4.4.8
Washability	3.4.9	6141	------	4.4.9
Freeze-thaw resistance	3.4.10	----	------	4.4.10
Water resistance	3.4.11	----	------	4.4.11
Alkali resistance	3.4.12	----	------	4.4.12
Resistance to biological growth	3.4.13	----	D 3273	4.4.13
			D 3274	------
Resistance to reflectance variation	3.4.14	----	------	4.4.14
Compatibility	3.4.15	----	------	4.4.15
Recoating properties	3.4.16	----	------	4.4.5
Odor	3.4.17	----	D 1296	------
Consistency	Table I	----	D 562	------
Nonvolatile	Table I	----	D 2369*	------
Dry hard time	Table I	4061	------	------
Reflectance	Table III	----	E 97	------
Opacity, dry	Table II and III	4121, Proc. B	------	------
Opacity, rewetted	Table II and III	----	------	4.4.16
85° specular gloss	Table I	6103	------	------
Yellowness, accelerated	Table I	6132	------	------
Lead content	Table I	----	D 3335	------
Fineness of grind	Table I	----	D 1210	------

* Distilled water shall be used in place of toluene.

Fig. 15.1 (Continued)

TT-P-29J

4.4.1 <u>Condition in container</u>. Examine the paint as received in accordance with method 3011 of Fed. Test Method Std. No. 141, and evaluate for compliance with 3.4.1.

4.4.2 <u>Storage stability</u>.

4.4.2.1 <u>Partially full container</u>. Determine skinning after 48 hours in accordance with method 3021 of Fed. Test Method Std. No. 141, except use a 3/4 filled 1/2 pint, multiple friction top can. Reseal and store for 14 days at 49°C (120°F), and evaluate for compliance with 3.4.2.1.

4.4.2.2 <u>Full container</u>. In accordance with method 3022 of Fed. Test Method Std. No. 141, allow a full standard quart can of the paint to stand undisturbed at standard conditions for 12 months, and then examine the contents. Evaluate pigment settling or caking, then agitate the can for 5 minutes on a paint shaker and re-evaluate. Make other applicable tests, and evaluate for compliance with 3.4.2.2.

4.4.3 <u>Color</u>. The film shall be applied on a smooth, flat chart in successive coats, each having a dry-film thickness of 0.076 mm (0.003 inch), until complete hiding is achieved, and shall be allowed to dry for 24 hours at standard conditions. Determine the color of the dried film in accordance with method 4250 of Fed. Test Method Std. No. 141, and evaluate for compliance with 3.4.3.

4.4.4 <u>Flexibility</u>. Prepare the test panel in accordance with method 2012 of Fed. Test Method Std. No. 141. Supplement the test panel cleaning procedure with an additional cleaning with an abrasive soap so that the entire surface of the panel is water-wet. Apply the paint in accordance with method 2162 of Fed. Test Method Std. No. 141, on the clean, dry panel to a dry-film thickness of 0.025 + 0.003 mm (0.001 + 0.0001 inch). Air dry for 18 hours at standard conditions, bake for 3 hours at 105 + 2°C. Cool for 1/2 hour, bend over a 6.35 mm (1/4-inch) mandrel, and examine the film in accordance with method 6221 of Fed. Test Method Std. No. 141. Evaluate for compliance with 3.4.4.

4.4.5 <u>Working properties</u>. Prepare four 1-foot square panels of gypsum wallboard conforming to SS-L-30. Leave panel 1 bare. On panel 2 apply one coat of latex priming coat conforming to TT-P-650, at a spreading rate of 450 square feet per gallon. On panel 3, apply one coat of primer coating conforming to TT-S-179, at a spreading rate of 450 square feet per gallon. On panel 4, saw one groove and apply joint cement of low absorption or of the same type used to fill joints on plasterboard so that a 6-inch wide strip 1/16 inch high in the crater with feathered edges is produced. Sand the rough spots after joint cement has dried. Allow the panels to dry for 24 hours, and then apply the paint at a spreading rate of 530 square feet per gallon over the entire surface of the four panels using a brush conforming to H-B-420, grade AA. Allow the panels to dry for 3 hours; then apply by brush, a second coat over the upper half of the panels at a spreading rate of 530 square feet per gallon. Cover the lower half of the four panels using a roller conforming to H-R-550, class II. While applying the second coat, observe leveling, application characteristics, and other film irregularities such as softening or lifting. After a 24-hour drying period, obtain gloss and reflectance readings in accordance with methods 6103 and 6121 of Fed. Test Method Std. No. 141 respectively, on the roller-coated and brushed surfaces for a total of eight different readings. Compare the eight readings with each other, and evaluate for compliance with 3.4.6. After examining for deficiencies as specified in 3.4.5 and 3.4.6 (see 4.4.6), apply a spray coat over a portion of a panel to determine sprayability of the paint.

4.4.6 <u>Appearance of dried film</u>. The dried film of the paint as applied in 4.4.5 shall be examined for compliance with 3.4.6.

4.4.7 <u>Anchorage</u>. Prepare a panel as in 4.4.4. Score a line through to the metal across the width of the film using a sharp pointed knife. The film shall then be taped perpendicular to and across the score line with waterproof, pressure sensitive tape 19 mm (3/4 inch) wide conforming to PPP-T-60, type IV. Press the tape with two passes of a 2 kg (4-1/2 pound) rubber-covered roller approximately 89 mm (3-1/2 inch diameter by 45 mm (1-3/4 inches) wide. The surface of the roller shall have a Duro-meter hardness value of 70 to 80. Allow approximately 60 seconds for the test area to return to room temperature. Grasp a free end of the tape, and then strip it at a rapid speed from the film by pulling the tape back upon itself at an angle of about 180°. Observe for compliance with 3.4.7.

Fig. 15.1 (Continued)

TT-P-29J

4.4.8 <u>Scrub resistance</u>. Determine the scrub resistance of the paint film in accordance with ASTM D 2486, except:

1. Use a sponge 1/ conforming to L-S-626, type II, porosity B, with dimensions of 95 x 73 x 38 mm (3-3/4 x 2-7/8 x 1-1/2 inches) when wet. The direction of least compressibility shall be in the 1-1/2 inch dimension. Soak the sponge for 30 minutes in distilled water at ambient laboratory temperatures, squeeze dry with maximum hand pressure, and evenly distribute 50 ml of distilled water over the surface of the sponge. Do not wet the panel with additional water. Spread 10 g of the specified scrub medium evenly over the wearing surface of the sponge. Recharge the sponge with 10 g of scrub medium after each 100 strokes.

2. Use a holder weighing 454 g (1.0 lb) suitable for holding the sponge. The apparatus used shall have a stroke length of 380 mm (15.0 inches). Evaluate for compliance with 3.4.8.

4.4.9 <u>Washability</u>. Prepare panels in accordance with method 6142 of Federal Test Method Standard 141, except use enamel conforming to TT-E-545 as an undercoat. Using method 6141 of Federal Test Method Standard 141, perform the washability test on the prepared panels, except:

1. Lava hand soap or equal may be used.

2. Use a sponge 1/ conforming to L-S-626, type II, porosity B, with dimensions of 95 x 73 x 38 mm (3-3/4 x 2-7/8 x 1-1/2 inches) when wet, and a suitable sponge holder weighing 454 g (1.0 lb). The direction of least compressibility of the sponge shall be in the 1-1/2 inch dimension.

3. The stroke length of the tester shall be 380 mm (15 inches).

4. Recharge the sponge after every 25 cycles until a total of 100 cycles has been run.

5. Measure 85° gloss in accordance with method 6103 of Federal Test Method Standard 141, instead of 60° specular gloss as described in the method.

6. Measure the reflectance according to ASTM E 97.

Evaluate for compliance with 3.4.9.

4.4.10 <u>Freeze-thaw resistance</u>. Fill a 1-pint resin-lined friction-top can two-thirds full with the paint, as received, and close the can tightly. Expose the can and contents three times to the following temperature cycle:

(a) Low temperature of -9 $\pm$ 2°C (+15 $\pm$ 3°F) for 16 hours.

(b) High temperature of 25 $\pm$ 3°C (77 $\pm$ 5°F) for 8 hours.

At the completion of the exposure test, measure the consistency of the paint using ASTM D 562, and compare with the original consistency to determine compliance with the requirements of 3.4.10. Brush the paint on a composition or gypsum wallboard panel and observe while brushing and after drying whether the paint is normal and usable in all respects. Compare with the unexposed paint with respect to freedom from coagulation, agglomeration, speckiness, and change in sheen or color.

4.4.11 <u>Water resistance</u>. Prepare two glass panels in accordance with method 2021 of Fed. Test Method Std. No. 141. Apply the paint to a dry-film thickness of 0.076 mm (0.003 inch), and allow the film to air dry at standard conditions for 120 hours. On one panel, place on three different spots approximately an inch apart 1 ml of distilled water, and immediately cover the area with a 50-mm watchglass. After 4 hours remove the watchglass and note any change in the appearance of the exposed areas. Wipe off the water, and gently rub the film with cheesecloth. Allow 2 hours at standard conditions for recovery; then examine the film and evaluate for compliance with 3.4.11.

1/ Suppliers: (1) Reeves Bros., P.O. Box 188, Cornelius, NC 28031; (2) Bunell Foam Products, 344 Bulcan, Buffalo, NY; (3) American Sanitary Products Co., 2301 Blake, Denver, CO 80205; (4) O'Cell-O Co., 205 Fawyer Ave., Tonawanda, NY

Fig. 15.1 (*Continued*)

TT-P-29J

4.4.12 <u>Alkali resistance</u>. On the other glass panel prepared in 4.5.11, place 1 ml of an 0.5 percent solution of reagent-grade sodium hydroxide in distilled water on three different spots and immediately cover the areas with a 50-mm watch glass. Allow 4 hours contact time, and remove the watch glasses. Wash off the alkali solution in running water and allow 2 hours recovery, then examine the film and evaluate for compliance with 3.4.12.

4.4.13 <u>Resistance to biological growth</u>. Determine biological growth on the surface of the paint film in accordance with ASTM D 3273. Evaluate the extent of surface disfigurement in accordance with ASTM D 3274. All fungal mycelium and spores, slime, and dirt and soil accumulation, whether opaque or transparent, shall be considered to be disfiguring agents in the evaluation. Evaluate the rating obtained for compliance with 3.4.13.

4.4.14 <u>Resistance of reflectance variation</u>. Apply the paint to a penetration chart to a dry-film thickness of 0.076 mm (0.003 inch) and width of 89 mm (3-1/2 inches). Allow the material to dry, in a horizontal position, 24 hours at standard conditions. After the drying period, obtain 45° directional reflectance measurements in accordance with ASTM method E 97 on the sealed and unsealed portions of the penetration chart, and report the difference. Evaluate for compliance with 3.4.14.

4.4.15 <u>Compatibility test (type II only)</u>. In a beaker containing approximately 100 ml of type II paint, place 2.0 g of tinting medium concentrate conforming to TT-T-390. Stir thoroughly until the tinting concentrate is evenly dispersed to a homogeneous mixture. Allow the mixture to stand undisturbed for 5 minutes. On one clear plate-glass panel, prepared in accordance with method 2021 of Fed. Test Method Std. No. 141, brush a coat of the mixture to approximately 0.025 mm (0.001 inch) dry film thickness and allow to dry at room temperature in a vertical position for 24 hours. While brushing, observe for streaks and pigment separation. On another panel prepared in the same way, draw down a 0.051 mm (0.002 inch) wet film thickness of the mixture. While the paint is still wet, rub-up an area using the index finger in circular motion and continue for a minimum of 20 revolutions. Exert light pressure of the finger while rubbing so as not to rub off the film. Allow the paint film to dry at standard conditions for 24 hours. Examine the dried film, and compare the rubbed-up area against the unrubbed-up area. A difference in color, 85° gloss (tested in accordance with method 6103 of Fed. Test Method Std. No. 141A), or texture of the dried film between these areas shall constitute incompatibility. Evaluate for comformance with 3.4.15.

4.4.16 <u>Rewetted opacity</u>. Apply a coat of water-white mineral oil (U.S.P. Liquid Petrolatum, Heavy) to a wet-film thickness of 0.037 mm (0.0015 inch) over each of the dried test panels prepared for determining dry opacity (method 4121 of Fed. Test Method Std. No. 141, procedure B). Allow the panels to stand horizontally in a dust-free atmosphere for 10 minutes at standard conditions, blot the excess oil, and perform the test. Evaluate for compliance with table II (type I only), or table III (type II only).

5. PREPARATION FOR DELIVERY

5.1 <u>Packaging, packing, and marking</u>. The paint shall be packaged, packed, and marked in accordance with PPP-P-1892. The level of packaging shall be level A, B, or C, and the level of packing shall be level A, B, or C, as specified (see 6.2). The paint shall be furnished in 1-gallon cans, or 5-gallon pails as specified (see 6.2). All containers shall be lined to prevent corrosion.

6. NOTES

6.1 <u>Intended use</u>. The latex-base paint in whites and pastel colors covered by this specification is intended for use on such interior wall and ceiling surfaces as wallboard and plaster. It may be applied to previously painted wood, plaster, or drywall surfaces. Gloss finishes should be dulled either by sanding or washing with a solvent-type cleaner before application of the paint. Where flat, deep-tone colors are desired, latex-base paint conforming to TT-P-1728 should be used.

Fig. 15.1 (*Continued*)

TT-P-29J

6.2 <u>Ordering data</u>. Purchasers should select the preferred options permitted herein, and include the following information in procurement documents.

    (a) Title, number, and date of this specification.
    (b) Type required (see 1.2).
    (c) Color required (see 3.4.3).
    (d) Level of packaging and level of packing required (see 5.1).
    (e) Quantities to be packed (see 5.1).

6.3 With respect to products requiring qualification, awards will be made only for such products as have, prior to the time set for opening of bid, been tested and approved for inclusion in Qualified Products List QPL-TT-P-29, whether or not such products have actually been listed by that date. The attention of suppliers is called to this requirement and manufacturers are urged to arrange to have the products that they propose to offer to the Federal Government tested for qualification in order that they may be eligible to be awarded contracts or orders for the products covered by this specification. The activity responsible for the Qualified Products List is the Federal Supply Service (FMBP), General Services Administration, Washington, DC 20406, and information pertaining to qualification of products may be obtained from that activity. Application for qualification tests shall be made in accordance with "Provisions Governing Application by Manufacturers for Inclusion on Federal Qualified Products Lists (QPL's), (see 6.4).

6.4 Copies of "Provisions Governing Application by Manufacturers for Inclusion on Federal Qualified Products Lists (QPL's)" (published as Appendix IV-A to the Federal Standardization Handbook), may be obtained upon application to Director, Standards Control and Support Division, Federal Supply Service, General Services Administration, Washington, DC 20406.

6.5 The type II tint-base white paint is a high-hiding white paint which can be tinted to the desired light color before application. This paint can also be used directly as a regular white paint.

6.6 All colors specified in table VI may be procured by the General Services Administration. Qualification testing is limited to the three colors and tint-base (type II) shown in the left column. Upon successful completion of qualification testing of each sample, qualification will be extended to the colors listed in the right column (see 3.1.1 and 4.2.1.1).

TABLE VI. Colors approved by extension of qualification

Color qualified		Additional colors to which approval is extended	
Color	Color No. Fed. Std. No. 595	Color	Color No. Fed. Std. No. 595
Yellow	33717	Brown	30233
		Red	31643
		Red	31667
		Red	31668
		Red	31670
		Orange	32630
		Orange	32648
		Yellow	33617
		Buff	33690
		Yellow	33695
		Yellow	33711
		Yellow	33727
		Ivory	37855
Green	34554	Green	34277
		Green	34300
		Green	34373
		Green	34424
		Green	34491
		Green	34516
		Green	34558
		Green	34670
		Green	34672

Fig. 15.1 (Continued)

TT-P-29J

TABLE VI. Colors approved by extension of qualification (con.)

Color qualified		Additional colors to which approval is extended	
Color	Color No. Fed. Std. No. 595	Color	Color No. Fed. Std. No. 595
Blue	35526	Blue	35622
		Gray	36293
		Gray	36357
		Gray	36492
		Gray	36521
		Gray	36555
White	Type II (no color number)	White (type I)	37778
		White (type I)	37875

6.7 If other colors are required by the General Services Administration, approval will be extended as follows: For browns (color numbers x0xxx), reds (color numbers x1xxx), oranges (color numbers x2xxx), and yellows (color numbers x3xxx) approval will be extended following qualification of yellow 33717. For greens (color numbers x4xxx), approval will be extended following qualification of green 34554. For blues (color numbers x5xxx) and grays (color numbers x6xxx), approval will be extended following qualification of blue 35526.

MILITARY CUSTODIANS:

Navy - YD
Air Force - 84

Review Activities.

Army - CE, MR

User Activities:

Army - MD
Navy - SH

Military Coordinating Activity:

Naval Facilities Engineering Command - YD

CIVIL AGENCIES COORDINATING ACTIVITIES:

Commerce - NBS
DOT - CG, RDS
D.C. GOV'T - DGS
GSA - FSS, PBO
HEW - NIH
HUD - HHE
INTERIOR - BOR
POSTAL - POS

PREPARING ACTIVITY: GSA - FSS

GSA DC-01901288

Orders for this publication are to be placed with General Services Administration, acting as an agent for the Superintendent of Documents. See Section 2 of this specification to obtain extra copies and other documents referenced herein.

Fig. 15.1 (Continued)

minimum or maximum limits, or both, as applicable. Examples are weight per gallon, pigment concentration, vehicle solids, gloss, consistency, drying time, and fineness of grind.

15.3.6 Sampling, inspection, and testing procedures   This section covers procedures required to check performance. It contains two subsections.

Sampling and inspection   This subsection states who is responsible for inspection and how sampling shall be carried out.

Test procedures   The tests to be conducted to determine performance are listed or described in this subsection. Whenever possible, standard methods are used, as described in *Federal Test Method Standard No. 141* or *ASTM Standards.*

15.3.7 Delivery and use   This section describes the packaging and use of the product. It is divided into two subsections.

Packaging and marking   The type of packaging desired is described in this subsection: for example, 1-gal (3.785-1) lined cans, 5-gal (18-925-1) lug-cover steel pails, or 55-gal (208.18-1) drums. Labeling should include the name and number of the product, the name of the manufacturer, and a code to designate the batch and date of manufacture. Other special packing instructions, such as for overseas shipment, may be included. The common federal reference is Specification TT-P-143.

Directions for use   Any special instructions for use are covered in this section. This is a very important part of the specification that is often overlooked, with consequent misuse of the product. A good reference work is the United States military painting manual *Paints and Protective Coatings*. Others are the Canadian government 1-GP-72: *Guide to the Selection of Paint Specifications on Use Basis* and NRC No. 2111: *Selection and Use of Paints* (NRC No. 2420 in French).

Notes   The rest of the specification describes the intended use in greater detail: for example, "The primer is especially useful on wood surfaces of residential housing, to which children may be commonly exposed." It also includes purchasing information.

15.4 Types of Specifications   Paint specifications are used by all United States federal and Canadian agencies and by most state, provincial, and municipal agencies. They are also used by some industry- and government-related associations, by large private agencies that purchase in sizable quantities, and by architects, specifiers, and engineers who design or specify materials for large projects. The most common government and industry association specifications of interest are the following:

15.4.1 American Institute of Steel Construction (AISC)
    1221 Avenue of the Americas
    New York, New York 10020

AISC provides guidelines to the shop painting of structural steel, including paint systems and specifications.

15.4.2 American Iron and Steel Institute (AISI)
    1000 Sixteenth Street N.W.
    Washington, D.C. 20036

AISI publishes technical bulletins related to paint specifications. Of particular interest is the result of 9-year tests entitled *Paintability of Galvanized Steel.*

15.4.3 American National Standards Institute (ANSI)
1430 Broadway
New York, New York 10018

ANSI provides indexes of standards for thousands of manufactured items including paint. It covers both United States and international standards.

15.4.4 American Nuclear Society (ANS)
555 North Kensington Avenue
La Grange, Illinois 60525

Besides furnishing specifications for painting nuclear-reactor facilities (N5.9-1967, N101.2-1972, et al.), ANS provides specifications that cover finishes for walls and other surfaces.

15.4.5 American Plywood Association (APA)
1119 A Street
Tacoma, Washington 98401

APA does not issue specific paint specifications but does have a "qualified coatings" program to assure that proper, durable coatings are used for exterior plywood.

15.4.6 American Society for Testing and Materials (ASTM)
1916 Race Street
Philadelphia, Pennsylvania 19103

ASTM standards do not list formulations for specific applications. ASTM does cover:

Part 27: Paint tests for formulated products and applied coatings
Part 28: Paint pigments, resins, and polymers
Part 29: Paint fatty oils and acids, solvents, and other raw materials

ASTM Subcommittee D01.41 is investigating criteria for government procurement of commercial paint and coatings. This investigation could clear the way toward federal purchases based on performance characteristics and a set of standard tests to assure compliance.

15.4.7 American Society of Civil Engineers (ASCE)
345 East 47th Street
New York, New York 10017

ASCE does not provide actual specification data, but at sessions of society meetings it does cover painting, including corrosion and methods of protection of structural steel.

15.4.8 Association for Finishing Processes (AFP)
20501 Ford Road
Dearborn, Michigan 48128

AFP of the Society of Manufacturing Engineers (SME) mainly serves the manufacturing industries, but in some cases its surface preparation and application procedures are applicable to architectural and engineering problems.

15.4.9   Construction Specifications Institute (CSI)
1150 Seventeenth Street N.W.
Washington, D.C. 20036

CSI provides specification documents including an outline for an actual specification, Document No. 09900. This is one of the more important documents to be included in a specification writer's file.

15.4.10   Cooling Tower Institute (CTI)
9030 1H-45 North
Houston, Texas 77037

CTI recommends wood maintenance and preservative treatments and provides standard specifications in these fields.

15.4.11   Federation of Societies for Coatings Technology (FSCT; formerly Federation of Societies for Paint Technology)
1315 Walnut Street
Philadelphia, Pennsylvania 19107

FSCT does not publish paint specifications per se, but its technical committees are excellent sources of coatings information.

15.4.12   Gypsum Association (GA)
1603 Orrington Avenue
Evanston, Illinois 60201

GA has recommendations on the application of certain coatings to wallboard (GA 202-72), including specific practices covering textured finishes and veneer plastic.

15.4.13   Hardwood Plywood Manufacturers Association (HPMA)
2310 South Walter Reed Drive
Arlington, Virginia 22206

HPMA provides purchasing data on hardwood plywood and veneer.

15.4.14   Maple Flooring Manufacturers Association (MFMA)
1800 Pickwick Avenue
Glenview, Illinois 6002J

MFMA publishes specifications and recommendations for finishing and maintaining maple, beech, and birch flooring including gymnasium finishes.

15.4.15   National Association of Corrosion Engineers (NACE)
P.O. Box 986
Katy, Texas 77450

NACE publishes various bulletins on surface preparation and coating systems, one of the most important being *Industrial Maintenance Painting.*

15.4.16   National Association of Pipe Coating Applicators (NAPCA)
2504 Flounoy-Lucas Road
Shreveport, Louisiana 71108

NAPCA recommends certain coating specifications and provides pocket editions of inspection procedures.

15.4.17   National Coil Coaters Association (NCCA)
1900 Arch Street
Philadelphia, Pennsylvania 19103

NCCA does not prepare coatings specifications but does provide detailed performance data (Technical Bulletin IV) for various coatings systems that can prove valuable in specification writing.

15.4.18   National Paint and Coatings Association (NPCA; formerly National Paint, Varnish and Lacquer Association)
1500 Rhode Island Avenue N.W.
Washington, D.C. 20005

NPCA does not issue paint specifications but does publish a paint raw-material index that lists properties. The NPCA Scientific Committee has printed a 26-page bulletin, *The Selection of Paint.* NPCA publishes the excellent *Guide to U.S. Government Paint Specifications,* which contains a summary of all federal, military, and American Association of State Highway and Transportation Officials specification paint products. It can be obtained only by members.

15.4.19   Painting and Decorating Contractors of America (PDCA)
7223 Lee Highway
Falls Church, Virginia 22046

The PDCA specification document *Masters' Guide to Paint Specifications* is out of print, but the New Orleans section of PDCA is asking that it be updated.

15.4.20   Production Systems for Architects and Engineers (PSAE)
1735 New York Avenue N.W.
Washington, D.C. 20006

Owned by the American Institute of Architects (AIA), PSAE has master specifications for painting concentrated in the 9T series of *Masterspec,* a specification published and copyrighted by PSAE.

15.4.21   Red Cedar Shingle and Handsplit Shake Bureau (RCSHSB)
515 116th Avenue N.E.
Bellevue, Washington 98004

The bureau does not set paint or stain specifications, but in cooperation with NPCA it has published data on application procedures and problems.

15.4.22   Society of Automotive Engineers (SAE)
400 Commonwealth Drive
Warrendale, Pennsylvania 15096

SAE provides specification data covering corrosion problems (Bulletin HSJ 447a), including chemical treatments for surface preparation.

15.4.23   Steel Structures Painting Council (SSPC)
4400 Fifth Avenue
Pittsburgh, Pennsylvania 15213

SSPC publishes detailed technical information for coatings used to protect steel from corrosion. The most important data appear in two volumes, Volume I, *Good Painting Practice,* and Volume II, *Systems and Specifications.*

15.4.24   Water Pollution Control Federation (WPCF)
2626 Pennsylvania Avenue
Washington, D.C. 20037

WPCF Manual No. 17 is titled: *Paints and Protective Coatings for Waste-Water Treatment Facilities.*

15.4.25   Western Wood Products Association (WWPA)
1500 Yeon Building
Portland, Oregon 97204

WWPA provides technical guides for painting wood siding and interior paneling and actual specifications for painting over knots.

15.4.26   Wood and Synthetic Flooring Institute (WSFI)
1800 Pickwick Avenue
Glenview, Illinois 60025

WSFI publishes detailed practices on construction, care, and preservation of flooring, including stage floors, cushioned floors, mosaics, and others.

15.4.27   Wood Moulding and Millwork Producers (WMMP)
P.O. Box 25278
Portland, Oregon 97225

WMMP sets some specifications for priming and inspection.

15.4.28   Zinc Institute (ZI)
292 Madison Avenue
New York, New York 10017

ZI, in cooperation with the International Lead-Zinc Research Organization (ILZRO), provides painting guidelines for coatings systems for galvanized steel.

15.4.29   State governments
Traffic paints   The line marking of highways is a major function of state governments and the American Association of State Highway and

Transportation Officials (AASHTO; see Table 15.1). Most road marking in the United States is done by the various states. Each state usually has its own specifications, although many are similar and some duplicate or use federal specifications. AASHTO is a good overall source for specification information. Its specifications are included in two volumes entitled *Standard Specifications for Highway Materials* and *Methods of Sampling and Testing,* which may be obtained from

American Association of State Highway and Transportation Officials
444 North Capitol Street
Washington, D.C. 20001

TABLE 15.1   State Highway Departments*

Alabama	State of Alabama Highway Department, State Highway Building, 11 South Union Street, Montgomery, Alabama 36104
Alaska	Department of Highways, Third Street, Douglas, Alaska 98824. Mailing address: P.O. Box 1467, Juneau, Alaska 99801
Arizona	Arizona Highway Department, 206 South Seventeenth Avenue, Phoenix, Arizona 85007
Arkansas	Arkansas State Highway Department, State Highway Department Building, 9500 New Benton Highway, P.O. Box 2261, Little Rock, Arkansas 72203
California	Department of Transportation, Public Works Building, 1120 N Street, P.O. Box 1499, Sacramento, California 95814
Colorado	State Department of Highways, 4201 East Arkansas Avenue, Denver, Colorado 80222
Connecticut	Department of Transportation, 24 Wolcott Hill Road, Wethersfield, Connecticut 06109. Mailing address: Department of Transportation, P.O. Drawer A, Wethersfield, Connecticut 06109
Delaware	Department of Highways and Transportation, Highway Department Administration Building, P.O. Box 778, Dover, Delaware 19901
District of Columbia	Department of Highways and Traffic, Room 508, Presidential Building, 415 Twelfth Street N.W., Washington, D.C. 20004
Florida	Florida Department of Transportation, Haydon Burns Building, 605 Suwannee Street, Tallahassee, Florida 23204
Georgia	Department of Transportation, No. 2 Capitol Square, Atlanta, Georgia 30334
Guam	Department of Public Works, Government of Guam, Agana, Guam 96911
Hawaii	Department of Transportation, 869 Punchbowl Street, Honolulu, Hawaii 96813
Idaho	Department of Highways, 3311 West State Street, P.O. Box 7129, Boise, Idaho 83707

TABLE 15.1   State Highway Departments*   (*Continued*)

Illinois	Department of Transportation, 2300 South 31st Street, Springfield, Illinois 62764
Indiana	State Highway Commission, 100 North Senate Avenue, State Office Building, Indianapolis, Indiana
Iowa	Iowa State Highway Commission, State Highway Commission Building, 826 Lincoln Way, Ames, Iowa 50010
Kansas	State Highway Commission of Kansas, State Office Building, Topeka, Kansas 66612
Kentucky	Kentucky Department of Transportation, State Office Building, High and Clinton Streets, Frankfort, Kentucky 40601
Louisiana	Department of Highways, Capitol Station, P.O. Box 44245, Baton Rouge, Louisiana 70804
Maine	Maine Department of Transportation, State House, Augusta, Maine 04330
Maryland	Maryland Department of Transportation, Office of the Secretary, P.O. Box 8755, Friendship International Airport, Maryland 21240
Massachusetts	Executive Office of Transportation and Construction, 18 Tremont Street, Twelfth Floor, Boston, Massachusetts 02108
Michigan	Michigan Department of State Highways, State Highway Building, 425 West Ottawa, P.O. Drawer K, Lansing, Michigan 48904
Minnesota	Department of Highways, State Highway Building, St. Paul, Minnesota 55155
Mississippi	State Highway Department, Woolfolk State Office Building, Northwest Street, P.O. Box 1850, Jackson, Mississippi 39205
Missouri	Missouri State Highway Commission, State Highway Building, 119 West Capitol Avenue, Jefferson City, Missouri 65101
Montana	Department of Highways, East Sixth Avenue and Roberts Street, Helena, Montana 59601
Nebraska	Department of Roads, Central Office Building, Room 212, South Junction, U.S. 77 and N-2, Lincoln, Nebraska. Mailing address: P.O. Box 94759, Statehouse Station, Lincoln, Nebraska 68509
Nevada	Nevada Department of Highways, Administration Building, Room 201, 1263 South Stewart Street, Carson City, Nevada 89701
New Hampshire	Department of Public Works and Highways, John O. Morton State Office Building, 85 Loudon Road, Concord, New Hampshire 03301
New Jersey	Department of Transportation, 1035 Parkway Avenue, Trenton, New Jersey 08625
New Mexico	New Mexico State Highway Department, 1120 Cerillos Road, P.O. Box 1149, Santa Fe, New Mexico 87501

New York	State Department of Transportation, State Campus Site, 1220 Washington Avenue, Albany, New York 12226
North Carolina	Department of Transportation and Highway Safety, Highway Building, 1 South Wilmington Street, Raleigh, North Carolina 27611
North Dakota	State Highway Department, State Highway Building, Capitol Grounds, Bismarck, North Dakota 58501
Ohio	Ohio Department of Transportation, Highway Department Building, 25 South Front Street, Columbus, Ohio 43215
Oklahoma	Oklahoma Department of Highways, Jim Thorpe Building, Lincoln Boulevard at Northeast 21st Street, Oklahoma City, Oklahoma 73105
Oregon	State Department of Transportation, State Highway Building, Room 307, Salem, Oregon 97310
Pennsylvania	Pennsylvania Department of Transportation, Transportation and Safety Building, Commonwealth and Forster Streets, Harrisburg, Pennsylvania 17120
Puerto Rico	Department of Transportation and Public Works, Box 3909, General Post Office, San Juan, Puerto Rico 00936
Rhode Island	Department of Transportation, State Office Building, Providence, Rhode Island 02903
Samoa (American)	Department of Public Works, Government of American Samoa, Pago Pago, American Samoa 96799
South Carolina	State Highway Department, State Highway Building, Columbia, South Carolina 29201
South Dakota	Department of Transportation, State Highway Building, East Broadway, Pierre, South Dakota 57501
Tennessee	Department of Transportation, Highway Building, corner of Sixth Avenue North and Deaderick Street, Nashville, Tennessee 37219
Texas	Texas Highway Department, State Highway Building, corner of Eleventh and Brazos Streets, Austin, Texas 78701
Utah	Utah State Department of Highways, 603 State Office Building, Salt Lake City, Utah 84114
Vermont	Department of Highways, State Administration Building, State Street, Montpelier, Vermont 05602
Virginia	Department of Highways, 1221 East Broad Street, Richmond, Virginia 23219
Virgin Islands	Department of Public Works, Charlotte Amalie, St. Thomas, Virgin Islands 00801
Washington	State Highway Commission, Highway Administration Building, Franklin at Maple Park, Olympia, Washington 98504

TABLE 15.1   State Highway Departments*   (Continued)

West Virginia	West Virginia Department of Highways, State Office Building, 1900 Washington Street, East Charleston, West Virginia 25305
Wisconsin	Department of Transportation, Hill Farm, 4802 Sheboygan Avenue, Madison, Wisconsin 53702
Wyoming	Wyoming Highway Department, State Highway Office Building, P.O. Box 1708, Cheyenne, Wyoming 82001

* Address all correspondence to Director of Highways.

Environmental specifications   The California Air Resources Board has established model rules regulating solvent emissions from industrial finishing operations. It is now turning its attention to general products, including solvent limitation for architectural and marine coatings. Paint specification must now take into account local regulations. New Jersey, for example, requires white paint on storage tanks with a capacity of 2000 gal (7570 l) or more which contain volatile organic substances with a vapor pressure of 0.02 psi (137.9 Pa) or greater (Regulation N.J.A.C. 7: 27-16, Sec. 16.2[a]). This regulation is designed to reduce pollution caused by evaporative losses.

15.4.30   United States federal government and Canadian specifications   For individual federal and military specifications and *Standard No. 595* colors (these colors are available as individual 3- by 5-in [76.2- by 127-mm] chips or as a set of 495 chips), get in touch with:

General Services Administration
Specification Activities Division, 3 FRI
Building 197, Navy Yard
Washington, D.C. 20407

For the following:

1. *Federal Test Method Standard No. 141b: Paint, Varnish, Lacquer, and Related Materials; Methods of Inspection, Sampling and Testing*
2. *DODIS: Department of Defense Index of Specification and Standards*

Part I: Alphabetical listing
Part II: Numerical listing (including user data)
Part III: Federal supply classification listings; bimonthly cumulative supplements issued for each

3. *Index of Federal Specifications, Standards and Handbooks* ($1.50)
4. *Paints and Protective Coatings,* Army TM5-618, Navfac MO-110, Air Force AFM 85-3

Get in touch with:

Superintendent of Documents
U.S. Government Printing Office
Washington, D.C. 20402

For individual specifications and U.S. Naval Facilities Engineering Command specifications, get in touch with:

Naval Publications and Forms Center (Code 1032)
5801 Tabor Avenue
Philadelphia, Pennsylvania 19120

Department of the Interior specifications may be obtained from:

Bureau of Reclamation
Department of the Interior
Engineering Research Center
Denver Federal Center
Denver, Colorado 80225

Canadian specifications and standards may be obtained from:

Canadian Government Specifications Board
Department of Supply and Services
Ottawa 4, Canada

Division of Building Research
National Research Council
Ottawa 7, Canada

The Occupational Safety and Health Administration (OSHA) also issues standards that affect paint specifications, including sandblasting requirements. The National Institute of Occupational Safety and Health (NIOSH), the research wing of the Department of Health, Education, and Welfare, openly advocates a standard that would outlaw conventional sandblasting. OSHA is also expected to issue standards on skid- and slip-resistant coatings.

In the United States, the major products specified for typical end uses are listed below, first by end use and then by source. Titles have been shortened but are complete in the source lists which follow. United States federal specifications are presented when possible since they are the most popular government specifications, but those from other United States government agencies are shown when they are not yet available as federal documents.

For the sake of simplicity, the specifications listed cover only products used for architectural and maintenance painting on site. Omitted specifications include those for raw materials, industrial finishes applied on manufactured items such as furniture and vehicles, and coatings specifically used on ships.

15.5 Products for Surface Repair   The following are the major products used to repair surfaces before painting. They are listed in numerical order (C = concrete or masonry; E = exterior; I = interior; M = metal; P = plaster; W = wood).

15.5.1 Federal

		Area	Surfaces
SS-P-450	Patching plaster	I	P
TT-C-598	Caulking compound	E	C, W
TT-F-340	Plastic wood filler	I	W
TT-G-410	Glazing compound	I, E	M sash
TT-P-791	Putty	I, E	W sash
TT-S-227	Sealant: two-component	E	C, M, W
TT-S-230	Sealant: one-component	E	C, M, W

15.5.2 Canadian

		Area	Surfaces
1-GP-103	Paste wood filler	I	W
1-GP-137	Chromate putty	E	M

15.6 Products for Surface Preparation   The following products, listed in numerical order, are of major importance in the proper preparation of surfaces for painting (C = concrete or masonry; Ch Pt = chalked paint; E = exterior; I = interior; M = metal; S = steel; VT/B = vinyl toluene–butadiene; W = wood).

15.6.1 Federal

		Area	Surfaces
TT-C490	Metal pretreatment	I, E	S
TT-F-336	Paste wood filler	I	W
TT-F-1098	VT/B block filler	I, E	C
TT-P-620	Masonry conditioner	I, E	C, Ch Pt
TT-R-251	Paint remover	I, E	All

15.6.2 Military

		Area	Surfaces
MIL-S-12935	Knot sealer	I, E	W
MIL-P-15328	Wash primer	I, E	M
MIL-R-46073	Remover (water rinse)	I, E	All

15.6.3 Canadian

		Area	Surfaces
1-GP-102	Alkyd sealer	E	C, W
1-GP-121	Vinyl wash primer	I, E	M
I-GP-126	Vinyl knot sealer	I, E	W
1-GP-142	Alkali-resistant sealer	I, E	C, Ch Pt
1-GP-188	Block filler	I, E	C

15.7 Products for Interior Use   The following specification products, listed in numerical order, are those most commonly used on interior surfaces except floors (Al = aluminum; B = wallboard, plasterboard, drywall; C = concrete, masonry, block; Cl = clear; F = flat; H = high; L = low; M = metal; NS = not specified; P = plaster; Pl = reinforced plastic; SG = semigloss; T = tints; V = various; W = wood; Wh = white; Y = yellow). An asterisk (*) indicates that the surface is primed unless previously painted.

15.7.1 Federal

		Color	Gloss	Surfaces
TT-C-535	Two-component epoxy	Wh, T	H	C, M, W
TT-C-540	Oil-modified urethane	Cl	H	C, M, W
TT-C-542	Moisture-cured urethane	Wh, Cl	H	C*, M, W
TT-C-555	Textured coating system	Wh, T	F	C
TT-E-489	Enamel: alkyd, gloss	V	H	M*, W*
TT-E-505	Enamel: alkyd, odorless, high gloss	Wh, T	H	M*, P*, W*
TT-E-506	Enamel: alkyd, gloss	Wh, T	H	M*, P*, W*
TT-E-508	Enamel: alkyd, semigloss	Wh, T	SG	M*, P*, W*
TT-E-509	Enamel: alkyd, odorless, semigloss	Wh, T	SG	M*, P*, W*
TT-E-529	Enamel: alkyd, semigloss	V	SG	M*, P*, W*
TT-E-543	Enamel undercoat: alkyd	Wh, T	F	C*, M*, P*, W
TT-E-545	Enamel undercoat; alkyd, odorless	Wh, T	F	C*, M*, P*, W
TT-P-29	Latex flat paint	Wh, T	F	B, C, P
TT-P-30	Alkyd flat paint, odorless	Wh, T	F	B*, C*, P*
TT-P-35	Cement-powder paint	V	F	C
TT-P-38	Phenolic aluminum paint	Al	SG	M*
TT-P-95	Rubber-based paint	Wh, T	H, SG, F	C, M, P
TT-P-645	Zinc chromate primer	Y	L	M
TT-P-650	Latex primer	Wh	SG	B, P
TT-P-659	Alkyd primer for TT-E-489	Wh, T	L	M, W
TT-P-1511	Latex gloss and semigloss paint	Wh, T	SG	B, P, W
TT-P-1728	Latex flat paint; deep colors	Deep	F	B, P
TT-S-179	Surface sealer	Wh, T	SG	B, P
TT-S-711	Oil stain	V	L	W
TT-V-85	Flat varnish	Cl	L	W
TT-V-109	Alkyd spar varnish	Cl	H	W
TT-V-119	Phenolic spar varnish	Cl	H	W

15.7.2 Canadian

		Color	Gloss	Surfaces
1-GP-36	Alkyd varnish	Cl	H, SG	W
1-GP-38	Enamel undercoat	Wh	L	C*, M*, P*, W
1-GP-57	Alkyd semigloss enamel	NS	SG	M*, P*, W*
1-GP-60	Alkyd gloss enamel	NS	H	M*, P*, W*
1-GP-68	Primer-sealer	Wh	SG	B, P
1-GP-69	Aluminum paint	Al	SG	M, W
1-GP-100	Latex flat paint	NS	F	B, C, P
1-GP-118	Alkyd flat	Wh, T	F	B, P, W*

		Color	Gloss	Surfaces
1-GP-119	Latex primer-sealer	Wh, T	SG	B, C, P
1-GP-145	Pigmented stain	V	F	W
1-GP-146	Epoxy coating	Cl, V	H	C, M, P, W
1-GP-153	Epoxy high-build coating	T		C, M, P, W
1-GP-175	Oil-modified urethane	Cl	H, L	W
1-GP-176	Moisture-cured urethane	Cl	H	C, M, P, W
1-GP-177	Nonyellowing urethane (two packages)	Cl, V	H	C, M, Pl, W
1-GP-180	Urethane (two packages)	Cl, V	H	C, M, Pl, W

15.8 Products for Exterior Use   The following specification products, listed in numerical order, are those generally recommended for use on exterior surfaces. Most are intended for use on surfaces other than floors. Al = aluminum; BC = blast-cleaned; BCT = blast-cleaned plus pretreatment; Bl = black; Br = brown; C = concrete, masonry, block; Cl = clear;   F = flat;   galv. = galvinized;   Gn = green;   Gr = gray; H = high; L = low; M = metal; NS = not specified; NV = not visible; Or = orange;   Pl = reinforced   plastic;   PTC = power-tool–cleaned; R = iron oxide red; S = steel; SG = semigloss; T = tints; V = various; W = wood; Wh = white; X = not required; Y = yellow; (*a*) = primer tinted; (*b*) = topcoat tinted; (*c*) = alternative colors. An asterisk (*) indicates that the surface is primed unless previously painted.

15.8.1 Federal

		Color	Gloss	Surfaces
TT-C-555	Textured coating system	V	F	C
TT-E-489	Alkyd enamel	Wh, V	H	M*, W*
TT-E-490	Silicone-alkyd enamel	Wh, V	SG	M*
TT-E-522	Phenolic enamel	V	F	M*, W*
TT-E-1593	Silicone-alkyd enamel	NS	H	M*
TT-P-19	Acrylic latex paint	Wh, V	F	C, W*
TT-P-25	House-paint primer	Wh, T	SG	W
TT-P-35	Cement-powder paint	W, V	F	C
TT-P-37	Alkyd trim paint; deep colors	V	H	W*
TT-P-38	Phenolic aluminum paint	Al	SG	M, W
TT-P-52	Alkyd-oil shingle paint	Wh, V	SG	W
TT-P-55	Polyvinyl acetate latex paint	Wh, T	F	C
TT-P-57	Zinc chromate primer	Y, R	SG	S
TT-P-81	Oil exterior paint	V	SG	S
TT-P-86	Red-lead primer	R	SG	S
TT-P-95	Rubber-based paint	Wh, V	H, SG, F	C
TT-P-97	Rubber-based paint (styrene-butadiene)	Wh	F	C
TT-P-105	Oil house paint, lead-free	Wh, T	SG	M*, W*
TT-P-615	Basic lead silicochromate primer	Or	SG	S
TT-P-641	Zinc dust–zinc oxide primer	Gr	F	Galv.

		Color	Gloss	Surfaces
TT-P-1046	Zinc-dust primer, rubber-based	Gr	F	M
TT-P-1181	Styrene acrylate masonry paint	V	F	C
TT-V-119	Phenolic spar varnish	Cl	H	M, W

15.8.2 Military

		Color	Gloss	Surfaces
MIL-P-12742	Phenolic primer	Br, R, Y	SG	M, W
MIL-P-15328	Wash primer (two packages)	Y	L	M
MIL-P-15929	Vinyl red-lead primer	Or	SG	M
MIL-P-16738	Vinyl enamel	Wh	H	M*
MIL-C-22750	Epoxy-polyamide coating	V	H, SG	M
MIL-P-26915	Zinc-dust primer (organic)	Gr	Fl	S
MIL-P-38336	Zinc-dust primer (inorganic)	Gr	Fl	S
MIL-P-38427	Topcoat for MIL-P-38336	Gn, Gr	SG	S
MIL-P-52192	Epoxy primer	Br, R	SG	M
MIL-P-52324	Blister-resistant alkyd exterior paint	Wh, T	SG	M*, W*

15.8.3 U.S. Bureau of Reclamation

		Color	Gloss	Surfaces
VR-3	Vinyl-resin paint	R, Wh, Gr, Br, Al	SG	C, S
VR-6	Vinyl-resin paint (four coats)	Gr, Al	SG	M

15.8.4 Canadian

		Color	Gloss	Surfaces
1-GP-14	Red-lead–oil primer	Or	SG	S
1-GP-28	House paint	Wh, T	H	W*
1-GP-40	Oil-alkyd primer	NS	SG	S
1-GP-41	Linseed-oil house paint	V	H	M*, W*
1-GP-43	Shingle stain	NS	L	W
1-GP-55	Wood primer	Wh	SG	W
1-GP-59	Alkyd enamel	V	H	M*, W*
1-GP-69	Aluminum paint	Al	SG	M, W
1-GP-93	Aluminum marine paint	Al	SG	M
1-GP-99	Phenolic varnish	Cl	H	W
1-GP-132	Zinc chromate primer	Y	L	M
1-GP-138	Latex paint	NS	F	C, W*
1-GP-140	Red-lead iron oxide primer	Br	L	S
1-GP-144	Alkyd masonry coating	V	L	C
1-GP-145	Pigmented stain	V	F	W
1-GP-162	Emulsion masonry paint	V	F	C
1-GP-163	Styrene-butadiene masonry paint	V	F	C
1-GP-165	Epoxy primer	Or, Y	L	M
1-GP-166	Lead silicochromate–oil primer	Or	SG	S
1-GP-167	Lead silicochromate–alkyd primer	NS	SG	M

		Color	*Gloss*	*Surfaces*
1-GP-171	Zinc-rich primer (inorganic)	Gr	F	S
1-GP-177	Nonyellowing urethane (two packages)	Cl	H	C, M, Pl, W
1-GP-178	Zinc–zinc oxide–alkyd primer	Gr	L	Galv.
1-GP-180	Urethane (two packages)	Cl, V	H	C, M, Pl, W
1-GP-181	Zinc-rich primer (organic)	Gr	F	S
1-GP-182	Vinyl paint	V	H	M*
1-GP-183	Epoxy zinc-rich paint	Gr	F	S
1-GP-189	Alkyd wood primer	Wh	L	W

15.8.5 American Association of State Highway and Transportation Officials

		Color	*Gloss*	*Surfaces*
M 69	Aluminum paint	Al	SG	S
M 70	Ready-mixed paint	W, T	SG	C, M, W
M 72	Red-lead primer	Or	SG	S
M 229	Lead silicochromate primer	Or (essentially same as TT-P-615

15.8.6 Steel Structures Painting Council
Paints

		Color	*Gloss*	*Surfaces*
1	Red-lead linseed-oil primer	Or	SG	S
2	Red-lead alkyd-oil primer	Br	SG	S
4	Extended red-lead oil primer	Or	SG	S
5	Zinc dust–zinc oxide phenolic primer	Gr	SG	S, galv.
6	Red-lead iron oxide phenolic primer	Br	SG	S
8	Aluminum vinyl paint	Al		S
9	Vinyl paint	Wh, V	SG	S
11	Zinc chromate primer	Br	SG	S
14	Red-lead iron oxide oil primer	Br	SG	S
101	Aluminum alkyd paint	Al	SG	S
104	Alkyd paint	Wh, T	SG	S

Paint systems

System specification	Primer	Second coat	Third coat	Topcoat	Finish	Surface
PS 1.01	SSPC No. 14	TT-P-86, II	X	SSPC No. 101, I, or 104	Al Wh, T	PTC S
PS 1.02	SSPC No. 14	→(a)	SSPC No. 101, II	SSPC No. 101, I or 104	Al Wh, T	PTC S
PS 1.03	AASHTO M72, I	AASHTO M72, III	X	AASHTO M67, II or SSPC No. 104	Gr Wh, T	PTC S
PS 1.04	TT-P-641, I	SSPC No. 104 or 101, I (b)	X	→ or →	Wh, T Al	Galv.

System specification	Primer	Second coat	Third coat	Topcoat	Finish	Surface
PS 1.05	SSPC No. 2	SSPC No. 101, II	X	SSPC No. 101, I or 104	Al Wh, T	PTC S
PS 1.06	TT-P-86, I	Al paint (b) or SSPC No. 104	X X	→ →	Al Wh, T	PTC S
PS 2.01	TT-P-86, III	→(a)	SSPC No. 101, II	SSPC No. 101, I or 104	Al Wh, T	BC S
PS 2.02	TT-P-645	→(a)	X	SSPC No. 101, I or 104	Al Wh, T	BC S
PS 2.03	TT-P-86, II	→(a)	X	SSPC No. 101, I or 104	Al Wh, T	BC S
PS 2.04	TT-P-57, II	SSPC No. 101, II	X	SSPC No. 101, I or 104	Al Wh, T	BC S
PS 2.05	TT-P-641, II	SSPC No. 104 (b) or AASHTO M70	X	→ →	Wh, T Wh, T	Galv.
PS 4.01	MIL-P-15929	→(a)	SSPC No. 9 or 8 (b)	→ →	Wh, V Al	BCT S
PS 4.02	VR-3	→(c)	→(c)	→(c) or VR-3	Wh, Gr Al	BC S
PS 4.03	MIL-P-15929	SSPC No. 9	X	SSPC No. 8 or 9	Al	BCT S
PS 4.04	SSPC No. 9	→(c)	→(c)	→(c) or No. 8	Wh, V Al	BC S
PS 4.05	MIL-P-15929	MIL-P-16738 (b)	X	→	Bl	BCT S

15.9   Products for Floors   The following specification products are those generally recommended for use on floors (C = concrete; Cl = clear; E = exterior; H = high; I = interior; L = low; M = metal; NS = not specified; Or = orange; P = plaster; Pl = reinforced plastic; R = iron oxide red; S = steel; SG = semigloss; V = various; W = wood; Wh = white; Y = yellow). An asterisk (*) indicates that the surface is primed unless previously painted.

15.9.1   Federal

		Color	Gloss	Area	Surfaces
TT-C-542	Moisture-cured urethane	Cl, Wh, V	H	I	C, M*, W
TT-E-487	Floor and deck enamel	V	H	I, E	C*, W
TT-P-57	Zinc chromate primer	Y, R	SG	E	S
TT-P-86	Red-lead primer (alkyd and phenolic only)	Or, R	SG	E	S
TT-P-91	Rubber-based paint	V	H	I	C
TT-P-615	Basic lead silicochromate primer (alkyd and phenolic only)	Or	SG	E	S
TT-S-176	Varnish-sealer	Cl	L	I	W
TT-V-71	Oleoresinous varnish	Cl	H	I	W
TT-V-119	Phenolic spar varnish	Cl	H	I, E	W

15.9.2   Military

		Color	Gloss	Area	Surfaces
MIL-W-5044	Nonslip walkway compound	V	SG	I, E	M*

15.9.3   Canadian

		Color	Gloss	Area	Surfaces
1-GP-36	General-purpose varnish		H, L	I	W
1-GP-66	Floor enamel	V	H	I	C
1-GP-73	Floor enamel	V		I, E	
1-GP-154	Latex floor paint	NS	L	I	C
1-GP-175	Oil-modified urethane	Cl	H, L	I	W
1-GP-176	Moisture-cured urethane	Cl	H	I	C, M, P, W
1-GP-177	Nonyellowing urethane (two packages)	Cl, V	H	I, E	C, M, Pl, W
1-GP-180	Urethane (two packages)	Cl	H	I, E	C, M, Pl, W

15.9.4   Maple Flooring Manufacturers Association

	Color	Gloss	Area	Surfaces
Heavy-duty finish	Cl	H	I	W

15.10   Products for Road Marking   The following specification products are those which are most commonly used for traffic striping and zone marking in parking lots in manufacturing and warehousing areas and also in institutional buildings (bit. = bituminous; Bl = black; C = concrete; F = flat; H = high; M = metal; V = various; W = wood; Wh = white; Y = yellow).

15.10.1   Federal

		Color	Gloss	Surfaces
TT-C-542	Moisture-cured urethane	Wh, V	H	C, M, W
TT-P-85	Traffic paint	Wh, Y		C, bit.
TT-P-110	Black traffic paint	Obliterate lines		C, bit.
TT-P-115	White and yellow traffic paint	Wh, Y	F	C, bit.

15.10.2   Canadian

		Color	Gloss	Surfaces
1-GP-74	Alkyd traffic paint	Wh, Y, Bl	F	C, bit.
1-GP-149	Reflectorized traffic paint	Wh, Y		C, bit.

15.11   Products for Special Uses or Surfaces   The following specification products are especially designed for the end use or surface listed (A = asbestos insulation; Al = aluminum; B = wallboard, drywall; bit. = bituminous; Bl = black; C = concrete, masonry, block; Ca = canvas; Cl = clear; E = exterior; Gr = gray; I = interior; M = metal; NS = not specified; Or = orange; P = plaster; S = steel; T = tints; V = various; W

= wood; Wh = white). An asterisk (*) indicates that the surface is primed unless previously painted.

15.11.1 Federal

		Color	Area	Surfaces
TT-C-492	Antisweat paint	NS	I	A, B, C, M, P
TT-C-555	Textured paint	NS	I, E	C
TT-E-496	Heat-resisting paint (below 400°F, or 204°C)	Bl	I	M
TT-P-26	Fire-retardant paint: interior	Wh	I	B, W
TT-P-28	Heat-resistant aluminum (1200°F, or 649°C)	Al	I	M
TT-P-34	Fire-retardant paint: exterior	Wh	E	W
TT-P-95	Swimming-pool rubber-based paint	Wh, T	I, E	C
TT-P-595	Canvas preservative	Wh, V	I, E	Ca

15.11.2 Military

		Color	Area	Surfaces
MIL-C 4556	Epoxy fuel-tank lining	Wh	I	S
MIL-P-15144	Fire-retardant antisweat paint	Wh	I, E	M*
MIL-P-15145	Zinc-dust water-tank paint	Gr	I, E	S
MIL-C-46081	Insulating intumescent coating	Wh, V	I, E	M, W

15.11.3 U.S. Naval Facilities Engineering Command

		Color	Area	Surfaces
TS-09801	Lining for concrete fuel tanks	Cl, Wh, Or	I	C

15.11.4 Canadian

		Color	Area	Surfaces
1-GP-76	Heat-resistant enamel	NS	I, E	M
1-GP-91	Anticondensation fire-retardant paint	Wh	I, E	M*
1-GP-93	Aluminum tank lining	Al	I	M
1-GP-127	Freshwater tank lining	Bl	I	S
1-GP-139	Multicolor coating	NS	I	B, C, M, P
1-GP-143	Heat-resistant aluminum (800°F, or 427°C)	Al	I, E	M
1-GP-151	Fire-retardant intumescent paint	V	I	S, W
1-GP-152	Overcoat for 1-GP-151	V	I	
1-GP-187	Anticondensation coating	NS	I	C, P, A, M, bit.

15.12 Miscellaneous Specifications
The following specifications and manuals are of value as reference works and guides in the choice of product, color, or test methods referred to in the product specifications.

15.12.1 Federal

Standard No. 595: Colors
Standard Test Method Standard No. 141: Paint, Varnish, Lacquer, and Related Materials; Methods of Inspection, Sampling and Testing

15.12.2 Military

Army TM5-618, NAVFAC MO-110, Air Force AFM 85-3: *Paints and Protective Coatings*   This is an excellent manual describing potential conditions and procedures for surface application and paint application. It recommends the products to be used on various surfaces and in different environments. The manual contains 19 reference tables, 128 illustrations, and a glossary of paint terms.

15.12.3 Canadian

1-GP-12: *Standard Paint Colors*
1-GP-71: *Methods of Testing Paints and Pigments*
1-GP-72: *Guide to the Selection of Paint Standards on Use Basis*   This includes an alphabetical and numerical list of all specifications as well as a classification by end use. It also includes a résumé of the scope and intended uses for each specification and a glossary of paint terms.

85-GP Series   These are standards for painting on various surfaces and structures and in a variety of environments. A manual on proper surface preparation and paint application, NRC No. 2111: *Selection and Use of Paints* (in French as No. 2420), is also available.

15.12.4 Steel Structures Painting Council

PS 1.00: *Guide to Oil-Base Systems*
PS 2.00: *Guide to Alkyd Paint Systems*
PS 3.00: *Guide to Phenolic Paint Systems*
PS 4.00: *Guide to Vinyl Paint Systems*
PS 12:00: *Guide to Zinc-Rich Paint Systems*

15.13 Numerical Lists

The product specifications described above are listed below in numerical order. The exact title is given, but no attempt has been made to note the latest modification, inasmuch as specifications are subject to change.

15.13.1 Federal specifications

The latest issue or modification is designated by either a capital or a lower-case letter at the end of the number. The letter advances in alphabetical order with each new change. The new issue also is dated. If the new specification or modification is a temporary issue subject to industry review, the number is preceded by 00, e.g., TT-P-0029J.

SS-P-450: *Plaster, Patching, Gypsum (Spackling)*
TT-C-492: *Coating Compound, Paint, Anti-sweat*
TT-C-535: *Coating, Epoxy, Two-Component, for Interior or Exterior Use on Metal, Wood, Concrete and Masonry*
TT-C-540: *Coating, Polyurethane, Clear, Linseed-Oil Modified*

TT-C-542: *Coating, Polyurethane, Oil-Free, Moisture-Curing*

TT-C-555: *Coating, Textured (for Interior and Exterior Masonry Surfaces)*

TT-C-598: *Caulking Compound, Oil and Resin Base Type (for Masonry and Other Structures)*

TT-E-487: *Enamel: Floor and Deck*

TT-E-489: *Enamel, Alkyd, Gloss (for Exterior and Interior Surfaces)*

TT-E-490: *Enamel, Silicone Alkyd Copolymer, Semi-Gloss, Exterior*

TT-E-496: *Enamel, Heat-Resisting (400°F), Black*

TT-E-505: *Enamel, Odorless Alkyd, Interior, High Gloss, White and Light Tints*

TT-E-506: *Enamel, Alkyd, Gloss, Tints and White (for Interior Use)*

TT-E-508: *Enamel, Interior, Semigloss, Tints and White*

TT-E-509: *Enamel, Odorless, Alkyd, Interior, Semigloss, White and Tints*

TT-E-522: *Enamel, Phenolic, Outside*

TT-E-529: *Enamel, Alkyd, Semigloss*

TT-E-543: *Enamel, Interior, Undercoat, Tints and White*

TT-E-545: *Enamel, Odorless, Alkyd Interior-Undercoat, Flat, Tints and White*

TT-E-1593: *Enamel, Silicone Alkyd Copolymer, Gloss (for Exterior and Interior Use)*

TT-F-336: *Filler, Wood, Paste*

TT-F-1098: *Filler, Block, Solvent-Thinned, for Porous Surfaces (Cinder Block, Concrete Block, Stucco, etc.)*

TT-G-410: *Glazing Compound, Sash (Metal) for Back Bedding and Face Glazing (Not for Channel or Stop Glazing)*

TT-P-19: *Paint, Acrylic Emulsion, Exterior*

TT-P-25: *Primer Coating, Exterior (Undercoat for Wood, Ready-Mixed, White and Tints)*

TT-P-26: *Paint, Interior, White and Tints, Fire Retardant*

TT-P-28: *Paint, Aluminum, Heat Resisting (1200°F)*

TT-P-29: *Paint, Latex Base, Interior, Flat, White and Tints*

TT-P-30: *Paint, Alkyd, Odorless, Interior, Flat White and Tints*

TT-P-34: *Paint, Exterior, Fire Retardant, White and Light Tints*

TT-P-35: *Paint, Cementitious, Powder, White and Colors (for Interior and Exterior Use)*

TT-P-37: *Paint, Alkyd Resin, Exterior Trim, Deep Colors*

TT-P-38: *Paint, Aluminum, Ready-Mixed*

TT-P-52: *Paint, Oil (Alkyd-Oil) Wood Shakes and Rough Siding*

TT-P-55: *Paint, Polyvinyl Acetate Emulsion, Exterior*

TT-P-57: *Paint, Zinc Yellow–Iron Oxide–Base, Ready-Mixed*

TT-P-81: *Paint, Oil: Ready-Mixed, Exterior, Medium Shades*

TT-P-85: *Paint, Traffic and Airfield Marking, Solvent*

TT-P-86: *Paint, Red-Lead–Base, Ready-Mixed*

TT-P-91: *Paint, Rubber-Base, for Interior Use (Concrete and Masonry Floors)*

TT-P-95: *Paint, Rubber, for Swimming Pools and Other Concrete and Masonry Surfaces*

TT-P-97: *Paint, Styrene-Butadiene, Solvent Type, White (for Exterior Masonry)*

TT-P-105: *Paint, Oil: Chalk-Resistant, Lead-Free, Exterior Ready-Mixed, White and Tints*

TT-P-110: *Paint, Traffic, Black (Non-Reflectorized)*

TT-P-115: *Paint, Traffic, Highway, White and Yellow*

TT-P-595: *Preservative Coating, Canvas*

TT-P-615: *Primer Coating: Basic Lead Silico Chromate, Ready-Mixed*

TT-P-620: *Primer Coating, Conditioner for Chalking Exterior Surfaces*

TT-P-641: *Primer Coating: Zinc Dust–Zinc Oxide (for Galvanized Surfaces)*

TT-P-645: *Primer, Paint, Zinc-Chromate, Alkyd Type*

TT-P-650: *Primer Coating, Latex Base, Interior, White (for Gypsum Wallboard)*

TT-P-659: *Primer Coating and Surfacer, Synthetic, Tints and White (for Metal and Wood Surfaces)*

TT-P-791: *Putty: Linseed Oil Type (for Wood Sash Glazing)*

TT-P-1046: *Primer Coating, Zinc Dust, Chlorinated Rubber (for Steel and Galvanized Surfaces)*

TT-P-1181: *Paint, Styrene-Acrylate Solvent Type, Tints and Deep Tones (for Exterior Masonry)*

TT-P-1511: *Paint, Latex Base, Gloss and Semigloss, Tints and White (for Interior Use)*

TT-P-1728: *Paint, Latex-Base, Interior, Flat, Deep-Tone*

TT-R-251: *Remover, Paint (Organic Solvent Type)*

TT-S-176: *Sealer, Surface, Varnish Type, Floor, Wood or Cork*

TT-S-179: *Sealer, Surface: Pigmented Oil, for Plaster and Wallboard*

TT-S-227: *Sealing Compound, Elastomeric Type, Two Component (for Caulking, Sealing, and Glazing in Buildings and Other Structures)*

TT-S-230: *Sealing Compound, Elastomeric Type, Single Component (for Caulking, Sealing, and Glazing in Buildings and Other Structures)*

TT-S-711: *Stain, Oil Type, Wood Interior*

TT-V-71: *Varnish, Interior, Floor and Trim*

TT-V-85: *Varnish, Oil, Low Sheen, Brush or Spray Application*

TT-V-109: *Varnish, Spar, Alkyd-Resin*

TT-V-119: *Varnish, Spar, Phenolic-Resin*

15.13.2 Military specifications   The latest issue is designated by a capital letter under the same procedure as that used with federal specifications.

MIL-C-4556: *Coating Kit, Epoxy for Interior of Steel Fuel Tanks*

MIL-W-5044: *Walkway Compound and Walkway Matting, Nonslip*

MIL-P-12742: *Primer Coating; Phenolic, Water Immersible*

MIL-S-12935: *Sealer, Surface; for Knots*

MIL-P-15144: *Paint, Formula No. 34 (Binder for Anti-Sweat Coating)*

MIL-E-15145: *Enamel, Zinc Dust Pigmented, Fresh Water Tank Protective (Formula No. 102)*

MIL-P-15328: *Primer, (Wash) Pretreatment (Formula No. 117 for Metals)*

MIL-P-15929: *Primer Coating, Shipboard, Vinyl–Red Lead (Formula No. 119 for Hot Spray)*

MIL-P-16738: *Enamel, Exterior, White, Vinyl-Alkyd (Formula No. 122-82)*

MIL-C-22750: *Coating, Epoxy-Polyamide*

MIL-P-26915: *Primer Coating, Zinc Dust Pigmented, for Steel Surfaces*

MIL-P-38336: *Primer Coating, Inorganic Zinc Dust Pigmented, Self-Curing, for Steel Surfaces*

MIL-C-38427: *Coating Kit, Topcoat, Pigmented, for Application over Inorganic Zinc Primer–Coated Steel Surfaces*

MIL-R-46073: *Remover, Paint; Organic Solvent Type*

MIL-C-46081: *Coating Compound, Thermal Insulating (Intumescent)*

MIL-P-52192: *Primer Coating, Epoxy*

MIL-P-52324: *Paint, Oil, Alkyd, Exterior, White and Light Tints*

15.13.3 Canadian specifications   All paint specifications are listed by number following 1-GP-. All new issues are marked by a lower-case letter at the end of the specification number; the series continues in alphabetical order with each new issue. The date of the issue also is included.

1-GP-12: *Standard Paint Colors*

1-GP-14: *Primer: Lead in Oil*

1-GP-28: *Paint: Exterior, House, White and Tints*

1-GP-36: *Varnish: General Purpose, Interior, Gloss and Satin*

1-GP-38: *Undercoater: Enamel, Interior*

1-GP-40: *Primer: Structural Steel, Oil Alkyd Type*

1-GP-41: *Paint: Exterior, Linseed Oil Type, Solid Colors*

1-GP-43: *Stain: Shingle, Drying Oil Type, for Sidewalls and Roofs*

1-GP-55: *Primer: Wood, Exterior*

1-GP-57: *Enamel: Interior, Semigloss, Alkyd Type*

1-GP-59: *Enamel: Exterior, Gloss, Alkyd Type*

1-GP-60: *Enamel: Interior, Gloss, Alkyd Type*

1-GP-66: *Enamel: Interior, for Concrete Floors*

1-GP-68: *Primer-Sealer: Wall, Interior, Solvent Type*

1-GP-69: *Aluminum Paint*

1-GP-71: *Methods of Testing Paints and Pigments*

1-GP-73: *Enamel: Exterior and Interior, for Floors*

1-GP-74: *Paint: Traffic, Alkyd*

1-GP-76: *Enamel: Heat Resistant, Interior and Exterior*

1-GP-91: *Paint: Binder, Marine, Anticondensation, Fire Retardant*

1-GP-93: *Paint: Aluminum, Marine*

1-GP-99: *Varnish: Phenolic Resin, Exterior and Marine*
1-GP-100: *Paint: Interior, Latex Type, Satin Finish*
1-GP-102: *Sealer: Clear, Alkyd Type*
1-GP-103: *Filler: Wood, Paste*
1-GP-118: *Finish: Interior, Alkyd, White and Tints, Flat*
1-GP-119: *Primer-Sealer: Wall, Interior, Latex Type*
1-GP-121: *Coating: Vinyl, Pretreatment, for Metals (Vinyl Wash Primer)*
1-GP-126: *Sealer: Vinyl, for Wood*
1-GP-127: *Coating: Internal, for Fresh-Water Storage Tanks*
1-GP-132: *Primer: Zinc Chromate, Low Moisture Sensitivity*
1-GP-137: *Putty: Chromate, Corrosion Inhibiting*
1-GP-138: *Paint: Flat, Exterior, Latex Type*
1-GP-139: *Coating: Multicolor, Dispersion Type*
1-GP-140: *Primer: Red Lead, Iron Oxide, Oil Alkyd Type*
1-GP-142: *Sealer: Clear, Alkali Resistant*
1-GP-143: *Paint: Aluminum, Heat Resistant, Silicone Alkyd*
1-GP-144: *Coating: Stucco and Masonry, Exterior and Interior, Alkyd Type*
1-GP-145: *Stain: Pigmented, Exterior and Interior*
1-GP-146: *Coating: Epoxy, Cold Cured, Gloss*
1-GP-149: *Paint: Traffic, Reflectorized, Alkyd, White and Yellow*
1-GP-151: *Paint: Interior, Fire Retardant*
1-GP-152: *Paint: Interior, Overcoat for Fire Retardant Paint*
1-GP-153: *Coating: Epoxy, High Build, Gloss*
1-GP-154: *Paint: Latex Type, for Concrete Floors*
1-GP-162: *Coating: Stucco and Masonry, Exterior and Interior, Emulsion Type*
1-GP-163: *Coating: Stucco and Masonry, Exterior and Interior, Styrene-Butadiene Copolymer, Solvent Type*
1-GP-165: *Primer: Epoxy, Cold Curing*
1-GP-166: *Primer: Basic Lead Silico-Chromate, Oil Alkyd Type*
1-GP-167: *Enamel: Exterior, Basic Lead Silico-Chromate, Alkyd Type*
1-GP-171: *Coatings: Inorganic Zinc*
1-GP-175: *Polyurethane Coatings: Oil Modified, Interior Clear, Gloss and Satin*
1-GP-176: *Polyurethane Coating: Moisture Cured*
1-GP-177: *Coating: Polyurethane, Two-Package, Interior and Exterior, Nonyellowing, Nonchalking*
1-GP-178: *Paint: Zinc Dust/Zinc Oxide, Alkyd (for Galvanized Surfaces)*
1-GP-180: *Coating: Polyurethane, Two-Package, Interior and Exterior, General Purpose*
1-GP-181: *Coating: Zinc Rich, Organic, Ready Mix*
1-GP-182: *Paint: Vinyl, Exterior*
1-GP-183: *Coating: Zinc Rich, Epoxy*
1-GP-187: *Coating Compound: Anticondensation*

1-GP-188: *Filler: Block, Emulsion Type*
1-GP-189: *Primer: Alkyd, Wood Exterior*

15.13.4   U.S. Bureau of Reclamation

VR-3: *Vinyl Resin Paint*
VR-6: *Vinyl Resin Paint*

15.13.5   U.S. Naval Facilities Engineering Command (NAVFAC)

TS-09801: *Protective Lining and Treatment for Concrete Storage Tanks for Petroleum Fuels*

15.13.6   American Association of State Highway and Transportation Officials (AASHTO)   The latest issue is designated by the year of issue, e.g., M-69-70.

M 69: *Aluminum Paint*
M 70: *White and Tinted Ready-Mixed Paint*
M 72: *Red Lead Ready-Mixed Paint*
M 229: *Basic Lead Silico Chromate, Ready-Mixed Primer*

15.13.7   Steel Structures Painting Council (SSPC)   The latest issue is designated by the year of issue. A T, meaning "tentative," is added until the specification achieves final acceptance: e.g., 1-64T.

1: *Red Lead and Raw Linseed Oil Primer*
2: *Red Lead, Iron Oxide, Raw Linseed and Alkyd Primer*
4: *Extended Red Lead, Raw and Bodied Linseed Oil Primer*
5: *Zinc Dust, Zinc Oxide and Phenolic Varnish Paint*
6: *Red Lead, Iron Oxide and Phenolic Varnish Paint*
8: *Aluminum Vinyl Finish Paint*
9: *White (or Colored) Vinyl Paint*
11: *Red Iron Oxide, Zinc Chromate, Raw Linseed and Alkyd Primer*
13: *Red or Brown One-Coat Shop Paint*
14: *Red Lead, Iron Oxide and Linseed Oil Primer*
101: *Aluminum Alkyd Paint*
104: *White or Tinted Alkyd Paint*
1.01: *Oil Base Paint System with Linseed Oil Primer and Alkyd Topcoat*
1.02: *Oil Base Paint System: Four Coat System with Linseed Oil Primer*
1.03: *Oil Base Paint System with AASHTO Linseed Oil and Alkyd Paints*
1.04: *Oil Base Paint System for Galvanized Steel*
1.05: *Oil Base Paint System with Linseed Oil and Alkyd Primer*
1.06: *Oil Base Paint System with Red Lead Linseed Oil Primer*
2.01: *Alkyd Paint System: Four Coat System with Red Lead Primer*
2.02: *Alkyd Paint System with Zinc Chromate Primer*
2.03: *Alkyd Paint System with Red Lead, Iron Oxide Primer*
2.04: *Alkyd Paint System with Zinc Chromate, Iron Oxide Primer*

2.05: *Alkyd Paint System for Unrusted Galvanized Steel*

4.01: *Vinyl Paint System for Salt Water or Chemical Use*

4.02: *Vinyl Paint System for Fresh Water, Chemical and Corrosive Atmospheres*

4.03: *Vinyl Paint System with Wash Primer for Salt Water and Weather Exposure*

4.04: *Vinyl Paint System with Four White or Colored Coats*

4.05: *Vinyl Paint System: Vinyl Alkyd for Atmospheric Exposure*

15.14 Proprietary Specifications   A significant number of medium to large paint manufacturers issue specification sheets on their products. This is especially true of those who sell to architects and maintenance engineers. Some manufacturers issue complete binders which include not only specification sheets but also color cards and recommended complete paint systems for a variety of substrates, conditions, and service requirements. Many of these items are covered in Sweet's Catalog File, *Products for General Building.*

BIBLIOGRAPHY

Cleveland, John L.: "Government Specs: An Updated Look," *American Paint Journal,* April 28, 1980, pp. 54–58.

"Storage of Volatile Organic Materials in White Tanks," Regulation NJAC 7:27-16, Sec. 16.2(a), New Jersey State Department of Environmental Protection, Trenton, N.J., 1978.

Sweet's Catalog File, *Products for General Building (GB),* McGraw-Hill Information Systems, 1980.

U.S. Bureau of Reclamation: *Paint Manual,* Washington, 1961.

Chapter 16

Clear Coatings

BENJAMIN FARBER
Consultant

16.1 Introduction A surface is painted with a clear coating to enhance the beauty of the substrate or to protect the surface itself. A clear coating also reduces the abrasion of a surface, making the surface easier to clean. In addition, it gives some measure of fire retardance and acts as a barrier against aggressive agents like soap, alcohol, and acid. When specifying a clear coating, give careful consideration to the desired results and the properties of the coating. Carefully consider surface preparation and application conditions so that dry film thickness and adhesion are optimized.

A coating may be applied by brush, roller, aerosol, dip, curtain, airless, or electrostatic-spray methods. Each method has its own optimum viscosity. To get a good finish, therefore, follow directions for thinning and use the solvents recommended. Optimum film thickness for clear coatings varies from a thin film that just seals the surface to prevent staining or moisture absorption to films approximately 3 to 5 mils (0.0762 to 0.127 mm) thick for long abrasion cycles. To complicate matters, sometimes a thin film will show less wear than a thick film. Follow the manufacturer's thinning directions carefully to get optimum film thickness.

Clear finishes dry by evaporation (lacquers), oxidation (air-drying varnishes), reaction with moisture in the air (moisture curing), or the combination of two ingredients just before use (the two-can, or two-component, system).

Air-drying varnishes may be classified by *oil length* as long-oil, medium-oil, and short-oil (oil length is the ratio of oil to resin). In general, long-oil varnishes are very flexible and are used mainly for exterior work. Medium-oil varnishes, which are usually harder, are used on cement and woodwork. The short-oil type, which is harder still and more brittle, is used for furniture finishes since it can be sanded without gumming the sandpaper.

Sealers are thin liquids which are applied to surfaces like wood to prevent absorption and act as good anchors for succeeding coats. They are also used on top of surfaces like plastic, where they serve as a barrier to the topcoating, which could soften the plastic. In such cases, the sealer acts as the adhesive bond between the plastic and the top finish.

Exterior clear coatings do not have the durability of pigmented coatings when used outdoors, even when the best transparent ultraviolet absorbers are put into varnishes. Failures occur by delamination, cracking, and peeling.

16.2 Application Suggestions   Because clear coatings show any discoloration of the substrate, the surface must be thoroughly cleaned. If it is sanded, all dust must be removed. If soap and water are used for cleaning, thorough rinsing with clear water is imperative so that no soap remains. The surface must be dry, and grease and wax must be completely removed; otherwise poor drying and peeling may result.

Varnishes may also be classified by *composition*. On this basis, recommended specifications are found in Table 16.1. The types of clear coatings discussed below are made and sold commercially.

16.3 Alkyd Resins   Air-drying alkyds are composed of a dibasic acid (like phthalic anhydride), a polyol (like glycerin or pentaerythritol), and a monobasic acid (like soybean-oil or linseed-oil fatty acids, which may be

TABLE 16.1   Recommended Specifications for Clear Finishes

Surface	Finish	Type	Typical finishing schedule*
Wood paneling; wood trim, doors, and cabinets	Gloss	Lacquer	1 coat lacquer sanding sealer 2 or more coats alkyd gloss lacquer
		Varnish	1 coat high-gloss varnish (thinned 10 percent) 1 or 2 coats high-gloss varnish
		Polyurethane varnish	1 coat urethane gloss (thinned 10 percent) 1 or 2 coats urethane gloss
	Semigloss	Lacquer	1 coat lacquer sanding sealer 1 coat alkyd gloss lacquer 2 or more coats alkyd semigloss lacquer
	Satin	Varnish	1 coat high-gloss varnish (thinned 10 percent) 1 or 2 coats satin varnish
		Polyurethane varnish	1 coat urethane satin (thinned 10 percent) 1 or 2 coats urethane satin
	Waxlike finish	Lacquer	1 coat lacquer sanding sealer 1 coat alkyd gloss lacquer 2 or more coats waxing lacquer
	Flat	Lacquer	1 coat lacquer sanding sealer 1 coat alkyd gloss lacquer 2 or more coats high-solids flat lacquer
	Flat	Varnish	1 coat high-gloss varnish (thinned 10 percent) 1 or 2 coats flat varnish
Wood floors	Gloss	Polyurethane	2 or more coats urethane gloss
Wood floors and heavy-use areas; gymnasium floors	Gloss	Phenolic varnish	1 coat gymnasium-floor finish (thinned 10 percent) 1 coat gymnasium-floor finish

* When a coating system is specified, make sure that topcoats are compatible with undercoats. Products of one manufacturer (e.g., gloss lacquers) may be incompatible with those of another manufacturer.

used in their forms as oil). Typical oils are linseed, soybean, safflower, and dehydrated castor oil.

A long-oil clear coating may be used on all types of exterior surfaces except those which are alkaline. The medium-oil type shows better durability on exterior metal surfaces. Because it dries harder than the long-oil type, it is also used for wood finishes including floor finishes. Soybean types have better nonyellowing qualities but poorer dry and poorer gloss retention than linseed types. Short-oil varnishes usually contain aromatic solvents like xylol instead of aliphatic solvents like mineral spirits. They are generally used in baking finishes.

REMOVING OIL AND WAX FROM A PREVIOUSLY VARNISHED SURFACE

Dissolve 4 oz (0.11 kg) of trisodium phosphate in 1 gal (3.785 l) of hot water, and apply the solution to a small area at a time. Let it lie for 15 min, scrub well with No. 2 steel wool dipped in the same solution, rinse well with clean water, and wipe dry. When the surface has dried thoroughly, wash small areas in succession with denatured alcohol. Use clean cloths to dry the surface thoroughly before applying a new finish. Should any wax deposits remain, use steel wool and mineral spirits to remove them. Be sure that the room is well ventilated while work is in progress. Remove every speck of dirt and loose steel wool by wiping with a lint-free cloth dampened with turpentine or mineral spirits. Allow the surface to dry thoroughly.

Another way to restore worn surfaces is to buff with steel wool dipped in mineral spirits. Again, be sure to remove all dirt and steel wool by wiping with a lint-free cloth dampened with turpentine or mineral spirits. Allow the surface to dry thoroughly. Then apply thin coats of varnish. They should be confined to renewed areas.

Whereas alkyds lack the durability of pure phenolic varnishes on sea exposure, they do well in less aggressive environments. A coating system which shows good durability consists of a pure phenol formaldehyde–tung-oil varnish as the primer and a linseed-oil alkyd as the topcoating. The varnish does not crack or peel easily, and the alkyd topcoating maintains its gloss and color longer.

Ultraviolet-ray absorbers help to protect clear coatings against the harmful rays of the sun. Under the sun's rays the clear finish darkens, embrittles, cracks, delaminates, and loses gloss. Even the wood underneath is affected adversely. Sunlight and moisture cause wood to decay in a process known as photochemical degradation.

Ultraviolet absorbers, or stabilizers, inhibit the destruction of the film by strongly absorbing the sun's rays of wavelengths of 2000 to 4000 Å, or

200 to 400 nm (a nanometer is a billionth of a meter). Most degradation of film and wood occurs in that range of wavelengths. The short wavelengths which are absorbed by these materials are reemitted as longer wavelengths and as heat. Usually an addition of 1 to 4% absorber is needed to achieve definitive results (Fig. 16.1). This ideal curve for an absorber shows that it absorbs sunlight in the ultraviolet region and transmits 100 percent of the wavelengths in the visible-light region. If it were to transmit less than 100 percent in the 4000-Å range, the finish would show up yellow because a portion of the blue light would be absorbed.

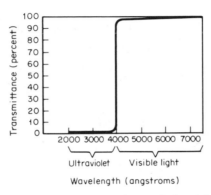

Fig. 16.1 Transmittance curve of a good ultraviolet absorber.

Clear finishes are subject to mildew problems just like any other type of coating. The mildew shows up as unsightly black or gray patches in areas where moisture is present. The right type and the right amount of fungicide (mildewcide) will inhibit mildew growth. The best mildewcides are mercury compounds, but the U.S. Environmental Protection Agency (EPA) prohibits using mercury in most cases. Mercury substitutes are being tested.

16.4 Urethanes   There are five types of urethane coatings (Table 16.2); they fall into two classes, nonreactive and reactive. Nonreactive coatings are essentially alkyds in which a diisocyanate (usually toluene diisocyanate, or TDI) replaces the difunctional acid or anhydride (e.g., phthalic anhydride). No free isocyanate (NCO) groups are present. By contrast, reactive urethanes contain free NCO groups (in a prepolymer) which upon application react with a variety of active hydrogen compounds to cross-link and thus to cure the coating. In these latter systems, coating properties show major dependence on the molecular weight and functionality of the polyhydroxy material used in the prepolymer or polyol component.

TABLE 16.2   Five Types of Urethane Coatings*

ASTM designation	One-component coatings			Two-component coatings	
	Type 1	Type 2	Type 3	Type 4	Type 5
Type	Oil-modified	Moisture-cured	Blocked-adduct	Catalyzed	polyol-cured
Curing agent	Oxygen	Water in air	Heat	Amine	Polyester and polyether epoxide
Pot life	Unlimited	Extended	Unlimited	2–5 days	1 day
Drying time (Hours to tack-free state)	¾–3	1–2	Temperature-dependent	½–2	2–8
Chemical resistance	Good	Very good	Excellent	Excellent	Excellent

ASTM Type 1. An oil-modified urethane coating can be applied like an ordinary varnish. It is used when greater hardness and abrasion resistance are desired. The surface must be free of wax and oil and be roughened to get good adhesion. If fillers are used, they should not contain stearate, which acts as a soap and can prevent good adhesion.

ASTM Type 2. A moisture-cured urethane coating can be applied like an ordinary varnish by brush, roller, or pad. Because the solvents are somewhat hazardous, the room must be well ventilated. This finish, which has outstanding abrasion resistance, is among the best finishes for floors exposed to heavy traffic. For best results the floor should be scraped to the original wood.

ASTM Types 3, 4, and 5. These coatings are used mainly in factory-applied finishes.

Polyurethane films are characterized by:

- Resistance to abrasion
- Solvent resistance
- Mar resistance
- Chemical resistance
- Fast curing rate
- Resistance to oxidation

By using aliphatic and cycloaliphatic isocyanates, polyurethane coatings of much greater durability and nonyellowing qualities under exposure can be obtained. In fact, these coatings are so good that they are used on airplanes, where they withstand quick temperature changes and exhibit abrasion resistance to the dust and dirt that impinge on them during flight.

Urethane alkyds are used when better color retention and lower costs are desired. However, increasing the ratio of alkyd to polyurethane lowers abrasion resistance and chemical resistance. There should be a minimum of 7 percent toluene diisocyanate in the urethane alkyd for its properties to approach those of a urethane oil.

One-package moisture-cured urethane coatings (ASTM Type 2) cure primarily through the reaction of moisture with the terminal isocyanate groups. In this case a low–molecular-weight resin polymerizes with moisture to give a high–molecular-weight firm film. These coatings perform well on gymnasium floors, decks of boats, and furniture exposed to the elements.

A moisture-cured polyurethane film yields considerably higher Sward or pencil hardness than the average nitrocellulose lacquer. Tensile strength and elongation also are superior. These coatings are the choice for concrete floors. Wheeled traffic can go over them with minimum wear, and they are not affected by cleaning agents. As their name indicates, they dry by reacting with the moisture, not the oxygen, in the air.

Moisture-cured polyurethanes have become popular in so-called seamless flooring. Besides the above-mentioned properties, they have the nonslip qualities desirable in floor coatings. Their ease of cleaning and stain resistance make them popular for such coatings. Decorative effects are obtained by binding colored thin flakes together in the matrix of the

Fig. 16.2 Sward hardness rocker. The horizontal pendulum responds to damping differences in resilient surfaces. [*Courtesy of Gardner Laboratory, Inc., Bethesda, Maryland*]

film. A final clear glaze coat is always used as a topcoating to protect these flakes.

Urethane sealers are used in high-grade furniture because they upgrade the appearance and general build of nitrocellulose lacquers. They also act as a barrier coat between dyestuff stains and polyester lacquers to prevent poisoning of the catalyst systems used. Polyurethane sealers are quick-drying, and relatively thin films can do a satisfactory job. Wood changes in dimensions with the absorption and desorption of water, depending on relative humidity and changing temperatures. Therefore, the applied film must adhere and not crack as the wood expands and contracts. Do not specify a clear-coating system that lacks proven compatibility between coats.

Two-component polyurethane finishes show exceptional durability on exterior wood surfaces. The polymerization occurs by chemical reaction, and no air is needed for drying. Urethane clear coatings based on castor oil are often more durable than standard-quality marine varnishes. In a test, three coats of clear varnishes were applied to mahogany, southern yellow pine, and redwood panels. The panels were exposed to a marine atmosphere on a 45° fence facing south at Miami, Florida. After 1 year gloss retention and film integrity were very good for the castor-oil urethanes, while the commercial spar varnishes and one long-oil varnish showed inferior gloss retention and moderate to severe cracking, checking, and flaking.

Rapid cure and film toughness make urethanes a natural choice for furniture finishing to replace the nitrocellulose finishes that have been used for many years. Nitrocellulose finishes do not equal urethanes in stain resistance or in solvent, heat, scratch-abrasion, and mar resistance. In addition, nitrocellulose film is highly flammable. (See Tables 16.3, 16.4, and 16.5.)

TABLE 16.3   Comparison of Nitrocellulose Lacquer and Polyurethane Finish

Coating	Resistance to harsh household chemicals	Scratch coefficient	Abrasion coefficient	Crack resistance	Moisture resistance
Nitrocellulose lacquer	Staining or softening	170 g	24 l/mil	Fail ⅛-in bend	Fail 6 cycles
Polyurethane finish	No damage	405 g	45 l/mil	Pass ⅛-in bend	Pass 20 cycles
	ASTM D-1308-57 Lipstick, alcohol, nail-polish remover, ball-point pen, shoe polish, cologne	Taber Shear Hardness Test Higher number indicates that greater pressure is required to damage film	ASTM D-968-51 Liters of falling sand required to damage film	ASTM D-1737-62 Films bent over ⅛-in mandrel	Wet Towel Test 1 cycle = 1 h at 100°F; 1 h at 70°F

TABLE 16.4   Household-Chemical Resistance of Some Clear Coatings

Staining agent	Comm.[a]	NC/EVA/SA[b]	UR[c]	UR/BU[d]	UR/NC[e]
Ball-point pen (1 min)	3 St	3 St	OK	OK	OK
Black shoe polish (1 h)	3 St	3 St	OK	1 St	1 St
Boiling water (2 h)	2 St	OK	OK	OK	OK
Catsup	OK	OK	OK	OK	OK
Cologne (1 min)	3 Ds	3 Ds	OK	OK	OK
Ethanol (50 percent)	3 Ds	1 Ds	OK	1 Ds	OK
Grape juice	2 St	OK	OK	OK	OK
Hot coffee (2 h)	2 St	OK	OK	OK	OK
Lipstick	3 St	2 St	OK	OK	OK
Merthiolate (1 min)	3 St	OK	OK	OK	OK
Mustard	3 St	2 St	1 St	2 St	2 St
Nail polish (1 min)	3 Ds	3 Ds	OK	OK	OK
Nail-polish remover (1 min)	3 Ds	3 Ds	OK	OK	OK
Vinegar	2 Ds	OK	OK	OK	OK

NOTE: Exposed for 6 h unless otherwise indicated. 1 = slight effect; 2 = moderate effect; 3 = considerable effect; St = Staining or whitening; Ds = dissolving, softening, or dulling.
[a] Commercial lacquer.
[b] Nitrocellulose, ethyl vinyl acetate, sucrose acetate.
[c] Air-dry polyurethane coating.
[d] Air-dry polyurethane coating combined with cellulose acetate butyrate.
[e] Air-dry polyurethane coating combined with nitrocellulose.

16.5   Epoxy Finishes   Epoxy varnishes are outstanding in their resistance to caustic substances and show excellent adhesion on most surfaces. On interior walls, cement, and cement block, an epoxy coating can be washed repeatedly with soap and water without losing its gloss or film

TABLE 16.5   Physical Properties of Some Clear Coatings

	Comm.[a]	NC/EVA/SA[b]	UR[c]	UR/BU[d]	UR/NC[e]
Sand abrasion coefficient (liters/mil; D-968-51)	15	24	45	40	38
Mandrel flexibility (inches; D-1737-62)	½	¼	⅛	⅛	⅛
Taber Shear Hardness Test (grams)	120	170	405	380	360
Cold check cycles (D-1211-60)	2	+20	+20	+20	+20
Print resistance (D-2091-67)	OK	OK	OK	OK	OK
Wet towel cycles[f]	2	6	+20	+20	+20

[a] Commercial lacquer.
[b] Nitrocellulose, ethyl vinyl acetate, sucrose acetate.
[c] Air-dry polyurethane coating.
[d] Air-dry polyurethane coating combined with cellulose acetate butyrate.
[e] Air-dry polyurethane coating combined with nitrocellulose.
[f] Test consists of placing a wet towel on film, cycling 1 h at 100°F (38°C), and observing for whitening, crazing, lifting, blistering, or other damage to the coating or substrate.

integrity. Epoxy clear finishes can be used as metal finishes and on substrates which must withstand corrosive atmospheres.

Two-can epoxies are mixed prior to use. They cure by chemical reaction. Properly formulated and applied, they are solvent- and chemical-resistant.

Epoxies have excellent impact resistance, good adhesion, and general detergent and chemical resistance. One important use is as a glaze on cement and cinder blocks. Because epoxies have properties normally associated with baked coatings, the amine- and polyamide-cured coatings resist heat, solvents, salt water, and chemicals. They are outstanding against mechanical shock, and they do not lose adhesion. Epoxies can be made highly flexible. Typical uses include coatings for corridors, laboratories, bathrooms, and ships, where the average coating would fail in a short time.

The specification writer should pay close attention to the pot life of epoxy products. After the two components have been mixed, the pot life (usable time for application) should be closely adhered to; otherwise problems can result. Many epoxy systems are 100 percent solids; i.e., no solvents are used. Other formulations are not 100 percent solids, and here proper application is required to avoid solvent entrapment that can lead to blistering.

Oil-modified epoxy ester varnishes dry by oxidation. They are lower in cost than the two-component epoxy systems. Properly formulated, they show excellent adhesion as primers for metal, concrete, and wood and better caustic resistance than pure phenol formaldehyde–tung-oil varnishes. Oil-modified epoxies are not suited to marine use.

16.6 Phenol Formaldehyde Varnishes Phenol formaldehyde varnishes are noted for all-around good resistance to weathering under marine conditions. A wood like mahogany with two or three coats of this type of varnish containing an ultraviolet-ray absorber and an antioxidant can preserve and maintain an attractive appearance for more than one season at sea. Because they are highly resistant to caustic material, soap, and acids, these varnishes (sometimes called Bakelite varnishes) can be washed with soap and water very often without impairing their finish. While the finish does have a tendency to yellow and lose its gloss on long exposure to the sun's rays, these varnishes rarely crack, chip, or peel.

16.7 Modified Phenolic Varnishes Although modified phenolics lack the resistance to weathering of pure phenolic varnishes, they are lower in cost. The films, which are hard and abrasion-resistant, are suitable for floors, furniture, and toys. They show good adhesion.

Modified phenolics also tend to yellow, but they may be used for furniture, floors, and dadoes. Their abrasion and reagent resistance is better than that of alkyds for interior use. Short-oil types may be used for rubbing varnishes.

16.8 Fluorocarbons   These are organic compounds in which percentages of hydrogen have been displaced by fluorine. This element in carbon compounds offers a measure of water repellency, depending on the percentage of fluorine in the compound. Fluorocarbons are also stable compounds against all types of reagents. They are very inert polymers with exceptional weathering resistance. (As an example, when a fluorocarbon coating is pigmented and applied to properly prepared surfaces, it will last for more than 20 years. This is 4 to 6 times the durability of the average outside paint.)

Fluorocarbon coatings are so thoroughly inert and show such exceptional resistance properties that they are best used where surfaces come in contact with severely acid and alkaline conditions. They provide long-term savings in labor, maintenance, and replacement of equipment. Their rate of erosion is low, and their gloss retention and fading resistance are exceptional.

16.9 Silicone Finishes   Silicone resins show high heat resistance, excellent outdoor durability, excellent water resistance, excellent electrical resistance, and high resistance to corrosive atmospheres. General practice includes baking because the material is rather soft when air-dried. Clear silicone coatings are used when high heat resistance is desired since silicones experience less decomposition under heat than do most polymers.

Silicone-acrylics have high gloss and extended protection against blistering, peeling, checking, and crazing. Pigmented, they will last from 10 to 15 years. As clear finishes, they are much less durable, but they have much better film properties than phenol-resin or alkyd spar varnishes.

Silicone-modified alkyds contribute markedly to exterior durability. Their gloss retention depends on the ratio of silicone to alkyd. Federal Specification TT-E-00490 describes a silicone-alkyd enamel containing 30 percent silicone-alkyd resin. Although its durability is vastly increased when pigmented, it can be applied as a clear coating. As a clear coating, it is second only to the fluorocarbons in durability.

Heat resistance is directly proportional to the ratio of silicone resin to alkyd. The best heat resistance is obtained with a pure silicone, and 75 percent silicone-polyesters will maintain color, hardness, and gloss up to 550°F (288°C). In silicone blends such as silicone-alkyds and silicone-polyesters, the silicone imparts weatherability and heat resistance. The organic alkyd imparts quick drying, abrasion resistance, and hardness.

Straight silicones have poor solvent resistance, but with the right type of alkyd this drawback is minimized. Moreover, modified silicones have better abrasion resistance than do pure silicones. Better film integrity, hardness, and toughness result if the finish can be cured at a high temperature.

Silicones have good adhesion to wood, but steel must be sandblasted and bonderized to get acceptable adhesion. A clear silicone may be added in small quantities to improve mar resistance and give slip to a coating. On the other hand, improper use of silicones can lead to fisheyes and adhesion problems when recoating.

Silicone finishes have outstanding heat stability, high corrosion resistance, excellent adhesion, color, and gloss, and very little degradation at 500°F (260°C). Some formulations are suited to service between 1200 and 1400°F (between 649 and 760°C). Silicone finishes do not normally blister as readily as straight organic finishes do under these high-heat conditions.

Silicone solutions containing from 2 to 20 percent nonvolatile material can be applied to masonry, brick, concrete, mortar, gypsum, cement and cinder block, porous stone like sandstone, and tile. The silicone is deposited in the pores of the substrate, leaving no continuous film on top. Water is thus prevented from entering. However, moisture vapor can still pass through, and it is common to say that the substrate can breathe. While keeping the surface clean, these sealants extend its useful life by increasing weather resistance and reducing spalling, cracking, and efflorescence. Efflorescence occurs when water penetrates a brick or masonry surface, dissolving salts (e.g., sodium sulfate) and leaching them to the surface. The water then evaporates, leaving an unsightly white salt deposit. Silicone sealants retard spalling and cracking because the water cannot penetrate, freeze, and then expand in the concrete, causing cracking and flaking. These sealants also prevent interior paint from peeling because water does not pass through to the interior walls. Their effectiveness depends on making the pores hydrophobic without plugging them. Water will not pass through under mild hydrostatic pressure, but air and water vapor are not stopped.

Water-repellent silicone sealers are available in either organic solvents or water emulsions. The water emulsions may take more than 24 h to achieve water repellency, whereas the organic-soluble types need only 2 h of drying after application to be effective.

16.10 Acrylic and Nitrocellulose Lacquers Clear acrylic coatings have become popular because their films are colorless, show minimal yellowing, and have gloss retention. They possess excellent water and stain resistance when used on metals, woodwork, plastics, brass, and textiles and as a paper coating. They are usually soluble in esters, ketones, and aromatic solvents, but not in aliphatic solvents. Acrylics show poor miscibility with other types of coating compositions. They are available in emulsion form as acrylic latices, retaining their nonyellowing and weathering resistance. Because of the surfactants needed for emulsification, however, detergent and general reagent resistance is lowered.

Nitrocellulose lacquers find wide application in tract housing construction, hotels, high-rise buildings and apartments, commercial jobs, and places where a quick dry is necessary. These coatings consist of nitrocellulose materials blended with plasticizers to produce a film that is tough yet flexible.

A good lacquer will run about 30 percent solids by weight and will dry quickly, eliminating film dust. Recoating time is usually about 2 h, and coverage runs about 300 ft²/gal (7.36 m²/l). Both acrylics and nitrocellulose lacquers use lacquer thinners based on true solvents such as esters, ketones, alcohols, and diluents. The diluent is usually a fast-evaporating naphtha or an aromatic solvent that will leave the film before the true solvents evaporate. Substituting a cheap lacquer thinner for one recommended by the manufacturer can cause the diluent to remain in the film for too long a time. Because it is not a solvent for the acrylic or nitrocellulose resins, it kicks them out of solution, causing blushing and white marks in the film and sometimes even seeding of the resin. A good craftsperson can sometimes correct the problem by mist-spraying a true solvent onto the surface and redissolving the resin. (This procedure can also work on water and alcohol ring marks on furniture, but it is not a job for an amateur.)

Depending on the architect's preference, the lacquer specification may call for a lacquer stain prior to the application of a lacquer finish coat. Stains are penetrating-type lacquer formulations of approximately 6 percent solids that cover about 600 ft²/gal (14.72 m²/l) to give the wood a tone or hue specified from a color chart.

It is common practice to test-stain the actual boards to be coated. A coating will slightly change the color of the wood, and you should approve and keep a sample of the test board for comparison with the actual job. When preparing this test, leave a portion of the board in its virgin state, spray stain over about three-fourths of the board, and when the stain is dry, spray one-half of the board with the topcoat. (If more than one coat is used, space the board swatches accordingly; cover the section or sections not to be painted with a piece of cardboard attached to the swatch board with masking tape. This is easily removed immediately after spraying.)

Choose a clear sealer to act as a base for succeeding coats of lacquer and to facilitate sanding by holding nibs and fibers of the wood stiffly enough so that they are removed by the sandpaper. Fill open-grain woods with a suitable paste wood filler (one that will not lift when strong lacquer solvents are applied over it). Presand, dust, and wipe all woodwork before applying the sealer. Resand after application of the sealer.

Most lacquer products now are formulated for either conventional or airless spraying. Among typical uses for high-solids lacquers are cabinets, doors, paneling, woodwork, and fixtures.

16.11 Miscellaneous Clear Finishes

16.11 Miscellaneous Clear Finishes   Alcohol-soluble or spirit varnishes such as shellac, manila gum, zein, maleic and fumaric resins, and nitrocellulose can be dissolved in alcohol either alone or in combination with other types of thinners. They can be applied on wooden surfaces such as floors, yielding a quick-drying, hard, nontacky finish. They do not discolor on aging, but their resistance to abrasion, water, and detergents is much poorer than that of other types of varnishes.

Styrene-modified alkyds (styrenated alkyds) are quick-drying clear finishes which normally are dissolved in aromatic solvents like toluol and xylol. They show good water and alkali resistance and maintain good color on aging, but they are not recommended for exterior use.

Vinyl toluene oils and alkyds are similar in properties to styrene-modified alkyds, except that they can be dissolved in low-solvency thinners like the aliphatics, of which mineral spirits are an example. They are satisfactory for use on wood and metal and show excellent detergent resistance.

Polyester finishes are two-component finishes which form a film by combining a polyester with styrene or diallyl phthalate. A peroxide and a promoter are added for curing since the two components are to be mixed on the job. Because the odor of one component like styrene is obnoxious, experienced painters should apply these finishes in areas where there is good ventilation. Polyester finishes are attractive, durable, high-gloss finishes for masonry, furniture, and plaster. Although adhesion on a smooth surface is a problem, the use of polyurethanes as part of the composition or employment of the right type of sealer vastly improves adhesion. Because there are no solvents evaporating and drying occurs by chemical reaction, through-drying films can be deposited in thicknesses up to 10 to 20 mils (0.254 to 0.508 mm) per coat. A solvent-thinned coat normally is up to 3 mils (0.0762 mm) thick.

The Fine Hardwoods Association, in conjunction with D. L. Laboratories in New York, has devised a benefit-rating system to compare clear coatings used for furniture. The system identifies protection values related to typical household hazards. The tests, which are applicable to wood, marble, slate laminates, and plastics, prove valuable in writing certain clear-coating specifications (Table 16.6).

16.12 Application Specifications

16.12 Application Specifications   While many of the clear coatings discussed in this chapter require strict adherence to the application instructions set forth by the coating manufacturer, there are some general guidelines. All wood surfaces require sanding. These rules apply:

TABLE 16.6   Family-Benefits Ratings of Finishes on Wood: Comparative Properties*

	Regular lacquer	Catalyzed lacquer	Catalyzed varnish	Wood foil pressed vinyl	Liquid plastics, heat-catalyzed	Printed plastic laminates
Chemical resistance†						
5 percent Tide	Fair	Excellent	Excellent	Excellent	Excellent	Excellent
Alcohol	Poor	Fair	Good	Excellent	Excellent	Excellent
Nail-polish resistance	Poor	Fair	Good	Good	Excellent	Excellent
Mustard	Excellent	Excellent	Excellent	Excellent	Excellent	Excellent
Tea	Excellent	Excellent	Excellent	Excellent	Excellent	Excellent
Ink	Fair	Excellent	Excellent	Excellent	Excellent	Excellent
Physical resistance†						
Taber CS-17	Fair	Good	Good	Excellent‡	Excellent	Good plus
Taber sandpaper	Poor	Fair	Fair	Excellent	Good	Excellent
Scratch hardness (phonograph needle)	Poor	Poor	Fair	Excellent	Fair to good	Poor to fair
Gardner modified flail surface	Poor	Fair	Fair	Excellent	Good	Excellent
Edge flail impact pendulum	Fair	Fair	Fair to good	Excellent	Good	Poor to fair
Basic properties†						
Boiling water	Fair to good	Good to excellent	Excellent	Excellent	Good to excellent	Excellent
Hot wax, 342°F	Poor	Poor	Good	Fair to good	Good to excellent	Good to excellent
Natural quality appearance	Excellent	Excellent	Excellent	Excellent	Excellent	No
Interesting figure variation	Excellent	Excellent	Excellent	Excellent	Excellent	No
Repairability	Excellent	Excellent	Excellent	Good	Excellent	Poor
Design flexibility	Excellent	Excellent	Excellent	Good	Excellent	Poor
Bonus value						
Cigarette resistance	Poor	Poor	Fair§	Good	Excellent§	Good

* Under scientific testing, proportionate values from 1 to 20 are determined for each test. However, in this table the following terms are used: excellent, good, fair, poor.

† Under the Family Benefits Plan, a maximum total of 330 points is assigned to each of the first three categories. In all surfaces, resistance to one type of hazard must be modified to attain maximum resistance to another type.

‡ Twice plastics.

§ With foil.

- Sand with the grain; otherwise hard-to-remove scratches result.
- Use a sanding block on flat surfaces; cut sandpaper strips to fit curved surfaces.
- Wipe surface with a tack cloth to remove every trace of dust.

16.12.1 Filling   Close-grain woods (pine, birch, cherry, maple, fir) need no filler. Open-grain woods (oak, mahogany, walnut, chestnut, and imported woods such as primavera, samara, and Korina, which usually are in the form of plywood) should be filled for best results, but filling is not absolutely necessary. Most such fillers come in various colors or in a neutral shade which can be tinted with oil colors.

An application specification should generally include the following procedure. First, thin the filler to brushing consistency with turpentine or mineral spirits. Then apply it liberally with a brush. Fill only a small area at a time, then let the filler dry for a few minutes (usually about 10 or 20 min) until it begins to set and lose its gloss. Wipe briskly across the grain with burlap or another coarse cloth, pushing the filler into the open grain of the wood, then wiping lightly with the grain. Do not wipe the filler out of the grain. Allow the filler to dry for at least 24 h before proceeding. Then lightly sand with the grain to remove any excess filler that may have adhered to the top surface. Failure to remove all traces of filler on the top surface of the wood may result in streaking and a cloudy appearance of the finished job. The only function of the paste wood filler is to fill small cracks and the open grain of the wood.

16.12.2 Staining   Different woods accept stains differently because their natural color influences the stained result. For this reason, it is recommended that a test application be made on a sample of the wood to determine the exact color that will result on the particular type of wood being stained. To determine the true finish color, the finish coat should be applied over the stained sample and allowed to dry. If the stain is too dark, it may be lightened by thinning it with mineral spirits or a paint thinner.

When the wood is to be left in its natural color, only a clear varnish or sealer is applied. A clear varnish will protect the wood surface without changing its natural appearance. When the wood is to be finished in a shade different from its natural color, a stain is applied, followed by varnish or a sealer to protect the wood.

Open-grain woods can be stained and filled at the same time by intermixing the stain and paste wood filler in equal proportions, substituting the stain for thinner in the filler. When staining soft woods, some wood finishers prefer to apply a thin wash-coat sealer before applying the stain coat in order to obtain more uniform results. Fill nail holes with stain mixed with putty.

For plywood or other wall surfaces sand the bare wood surface, finishing the job with fine sandpaper. Remove sanding dust. Brush stain and allow it to set for 10 to 12 min, depending on the depth of stain desired. (The stain may also be applied with a pad of lint-free cloth.) Remove excess stain by wiping first across the grain, then lightly with the grain. Wiping stains are often chosen because the depth of the color can be controlled by the length of time that the stain is left on the surface before being wiped off and by the pressure applied during the wiping process. After the stain has dried (at least 6 h in most cases), apply a coat of sealer. After the sealer has dried, sand the surface lightly and remove sanding dust. Then apply the finish coat.

Be sure to specify the total coatings system from filler through topcoat in order to make sure that each material is compatible with, and will adhere to, the preceding coat. For example, some sanding sealers contain stearate, soaplike ingredients that impair the adhesion of hard topcoats such as urethanes. Also, use the same batch number (see coding on paint cans) to assure color uniformity, or intermix all the cans completely before starting the job. Brush the stain by working it into the surface to avoid an orange-peel effect in the finished job. Start at the top of the job and work down.

16.12.3 Flooring Sand the floor smooth and remove the sanding dust. Fill open-grain wood with wood filler. After the filler has flattened down, remove the excess with burlap or rough cloths, rubbing across the grain. After the filler has dried overnight, sand the floor smooth and remove sanding dust. Apply a sealer or a first coat of varnish. Sand lightly and remove sanding dust. Finish with one or two coats of varnish, sanding and dusting between coats. After at least 2 or 3 days, apply a coat of paste wax and then buff. If the floors are to be stained, they can be filled and stained in one operation.

Previously varnished floors require extensive surface preparation if the specification does not call for total resanding to expose the bare wood. If any wax is left on the floor, the new coat of varnish will neither dry nor adhere to the wood. It is exceedingly difficult to remove wax thoroughly enough to get the floor adequately clean to accept the new varnish. There are several materials that may be used, such as wax removers, paint thinner, or turpentine. These too must be removed entirely to avoid adhesion problems.

16.12.4 Bleaching If you prefer a very light natural wood finish or if you need to remove stains or other discolorations or to change a previously stained surface to one lighter in tone, you may choose to bleach the surface. Most bleaches are two-solution preparations. Some should be mixed at the time of application; others are applied in two successive steps. Almost all contain chemicals that are harmful to the skin as well as to clothing. This is why you should wear rubber gloves and an apron or other protective covering. Bleaches can also be harmful to the patina of fine mellowed wood.

Apply your bleaching agent with a brush or synthetic sponge, and allow it to dry. If you wish to bleach darkly stained wood, you may need to repeat this operation several times. The manufacturer's instructions will tell you whether or not you need a neutralizer. If not, simply sponge off the surface with water and allow it to dry for about 12 h. Finally, sand lightly to remove the slight grain that the bleach has raised.

When bleaching redwood, allow new decks (built with garden-grade redwood) to stand 1 month in good weather before applying bleach or stain. Garden grades are not kiln-dried and must air-dry to allow penetration of these types of finishes. For removing stubborn spots and stains on redwood, use 1 oz (0.028 kg) of oxalic acid crystals dissolved in 1 qt

Table 16.7 Interior Finishes for Redwood

Finish choices	Where to use	Application	Maintenance	Effects
Unfinished	Light-traffic areas, ceilings, walls, panels, and trim not cleaned often. Do not use in kitchens, bathrooms, or other areas exposed to moisture and grease	A danish oil can be used to give added richness to the wood; brush on two coats and wipe off excess	Smooth-surfaced redwood: lightly sand with fine-grade sandpaper. Resawn surfaced redwood: clean with wire brush or coarse sandpaper	Completely natural appearance; wood may darken with time
Wax	All interior uses except kitchens and bathrooms	Follow manufacturer's directions; on smooth-surfaced wood, use a soft cloth and rub with wood grain; on resawn surface, apply with a stiff brush	Wash with mild detergent; rinse with damp cloth. Remove grime with nonmetallic scouring pad. Restore appearance with new coat of wax; wipe off excess	Adds soft luster to redwood, and touch-up is smooth, easy, and even. Applying wax over two coats of clear lacquer makes it easier to remove or paint over
Alkyd resin and polyurethane sealers; danish oil	All interior uses except bathrooms; good for use on resawn surfaces	Two coats, brushed on	Clean with damp cloth	These clear, flat penetrating sealers will darken redwood appreciably
Clear lacquer	For ceilings, walls, dividers, trim, and any other places that need only dry cleaning; not for bathrooms, kitchens, or other areas requiring scrubbing	Apply by spraying (beware of open flames), or brush on; a coat or two of wax over lacquer gives rich luster; buff with a soft cloth	Clean with soft cloth dampened with turpentine, mineral spirits, or water. See maintenance suggestion under "Wax" for lacquer finishes topped with wax	Natural appearance with some protection from dirt. Although it forms a film, clear lacquer is not glossy unless many coats are applied. It will darken wood slightly
Alkyd resin polyurethane varnishes	All interior uses on surfaced redwood; multiple coats good for kitchens and bathrooms	Brush on two coats for most uses, up to six coats for kitchens and bathrooms; let dry and sand lightly between coats	Clean with soapy water and soft cloth, or use turpentine or mineral spirits. Touch up scratches or nicks with tinted wax	Available in flat, semigloss, or glossy textures; varnishes seal better than lacquers and withstand hard scrubbing. They darken and deepen wood tones and may show scratches

| Pigmented stains | Any interior use; protect from liquids, soiling, or frequent cleaning by covering the stain with a clear finish or stain sealer | Brush, roll on, or spray; more coats mean deeper color effect. To emphasize grain or texture, apply one coat of stain, wipe surface before dry. A coat of thinned paint, wiped off, gives a similar effect | Easy maintenance; avoid heavy scrubbing that may smudge wood | Recommended over paint. Stains only partially obscure redwood grain and texture and are available in many colors |
| Stain and clear combinations | Same as "Pigmented stains" | Same as "Pigmented stains." Protect from grease, dirt, and liquids with a coat of wax, lacquer, or varnish | Follow maintenance instructions for whatever finish is used as the overcoat | Stain provides color tone desired; overcoats provide protection |

SOURCE: California Redwood Association, San Francisco, Calif.

NOTES:

Interior finishing rules. The following rules will facilitate application and help assure best finish performance:

- Interior wood should be clean and dry.
- Follow finish manufacturer's directions for application and thinning.
- Do not mix incompatible finishes.
- Work in a well-ventilated room.
- If you are not familiar with a finish, try it out first on a separate piece of wood.
- Don't shake or stir varnishes vigorously because bubbles will then appear; bubbles are difficult to remove. (Straining through several layers of cheesecloth or burlap can remove bubbles.)
- Use good brushes for application; clean them well, and rinse with paint thinner, followed by a water rinse. Wrap brushes in newspaper to preserve their shape.

Removing a finish. Remove paint, varnish, or lacquer with a paint-and-varnish remover. Swab the wood with turpentine or mineral spirits to remove any residue deposited by the paint remover. Scrub waxes with mineral spirits and a soft brush. After removing any finish, sand lightly before refinishing (smooth-surfaced lumber only).

Preparation. For new redwood paneling (smooth-surfaced only), sand lightly before finishing to remove dirt, hammer marks, or other blemishes. Remove grease spots with a solvent such as benzene, mineral spirits, or turpentine prior to sanding. If nails are countersunk, nail holes can be filled with putty (if a paint or heavy-bodied stain is to be used) or with a nonoily filler (if a natural or light-stain finish is planned). Wood fillers are available in a variety of tones, and fillers can be tinted with dry pigments to match closely the color of the wood. A natural-looking filler can be made by mixing sanding dust from the wood with a clear household cement. Mix a small quantity at a time since the mixture dries quickly. After filling, sand flat and dust the surface before finishing. Nail holes can be left unfilled if the nails are carefully placed for an attractive pattern.

(0.946 1) of alcohol. Swab the wood with a cloth or a soft brush, covering the entire wall to prevent any contrast between treated and untreated areas. When the surface is dry, rinse the wood with alcohol to remove the acid. More than one application may be necessary. Use caution because oxalic acid is toxic.

The California Redwood Association, 1 Lombard Street San Francisco, California, has detailed recommendations for clear finishes on both interior and exterior redwood (see Table 16.7 and the accompanying box).

EXTERIOR FINISHES FOR REDWOOD

Do not specify oil treatments or varnishes for exterior redwood. Oil treatments are highly susceptible to mildew, especially in damp climates. Varnishes, polyurethanes, and synthetic clear finishes are difficult and expensive to apply and deteriorate within 1 year. With deterioration a varnish almost always turns yellow and begins to crack and peel. This will occur regardless of film thickness. Then expensive complete removal becomes necessary.

Use noncorrosive nails if the wood will be exposed to moisture. Top-quality, hot-dipped galvanized nails may be suitable if the galvanized coating is not damaged during nailing; recommended noncorrosive nails are stainless steel or aluminum alloy. Iron and copper nails will cause staining. Like ordinary corrosive nails, low-quality galvanized nails will react with redwood's natural extractives and cause stain streaks. If nails are countersunk, use a nonoily wood filler to fill nail holes when a natural finish is to follow. Use putty only if the wood is to be painted.

For redwood, use only a water-repellent treatment or a flat pigmented stain that is approved by the California Redwood Association. One such finish is Federal Specification TT-W-527B: *Composition A.*

NOTE: Some water repellents are toxic; do not specify spray application. Also, these repellents can damage nearby plants and cause skin irritation. Take appropriate precautionary measures.

16.12.5 Shellac   Gum shellac, called orange shellac or spirit varnish, is a solution of orange-colored resin dissolved in alcohol. (The resin is found in India and Tibet and is formed by the hardening of certain insect secretions.) It can be bleached with chlorine or other oxidizing agents. Then the material is called bone-dry–bleached shellac or white shellac. This should not be confused with pigmented shellac that contains ground titanium dioxide.

Bone-dry shellac is not as soluble as orange shellac, and for general-purpose work not requiring a water-white finish orange shellac is quite suitable. Normally it is purchased as 5 lb cut (5 lb of shellac in 1 gal of alcohol). In Europe the solution contains less shellac.

Shellac should be applied in relatively thin coats so that alcohol evaporates easily without becoming entrapped by resin that surface-dries. Thick coats often become cheesy and may not harden for long periods. Even when applying thin coats, using more than two coats can result in a tacky condition that destroys one of shellac's chief benefits—quick drying. One advantage of shellac is that it dries by evaporation rather than by oxidation, and once the surface is set, it can be handled almost immediately. Obviously it cannot be used in an area subject to alcoholic beverages or the film will redissolve. Also, shellac will soften in heat, so shellacked objects cannot be located near fireplaces or in direct hot sunlight.

There are other spirit varnishes (dammar, rosin, sandarac), but today these are not commonly used.

BIBLIOGRAPHY

"Alkyd Resins," *Encyclopedia of Chemical Technology,* vol. 1, Interscience Encyclopedia, Inc., New York, 1947.

Arno, J. W.: "Milk Paint," *Fine Wood,* March 1979, p. 15.

Battista, O. A.: *Fundamentals of High Polymers,* Reinhold Publishing Corporation, New York, 1958.

Billmeyer, F. W., Jr.: *Textbook of Polymer Chemistry,* Interscience Publishers, Inc., New York, 1957.

Bovey, F. A.: *Emulsion Polymerization,* Interscience Publishers, Inc., New York, 1958.

Buist, J. M., and H. Gudgeon: *Advances in Polyurethane Technology,* McLaren & Sons, Ltd., London, 1968.

D'Alelio, G. F.: *Fundamental Principles of Polymerization,* John Wiley & Sons, Inc., New York, 1952.

Dow Corning Literature on Silicone and Silicone Alkyd Coatings, Dow Corning Corp., Midland, Mich., 1979.

Dimer Acids in Urethane Coatings, Emery Industries, Cincinnati, 1979.

Hicks, Edward: *Shellac,* Chemical Publishing Company, Inc., New York, 1961.

Newell, Adnah C.: *Coloring, Finishing and Painting Wood,* rev. ed., C. A. Bennett & Co., Peoria, Ill., 1961.

Raw Materials Index, National Paint and Coatings Association, Washington, various dates.

Stalker, John, and George Parker: *A Treatise of Japanning and Varnishing,* Quadrangle Books, Inc., Chicago, 1960.

Williams, Alec.: *Paint and Varnish Removers,* Noyes Data Corp., Park Ridge, N.J., 1972.

Chapter **17**

Specialized Functional Coatings

WILLIAM LAWRENCE
Consultant

17.1 Introduction In addition to aesthetic qualities, the average coating is expected to have mildly protective properties. Under normal conditions coatings have token resistance to heat and chemicals and other environmental conditions. This chapter proceeds beyond the average coating to deal with finishes exposed to extreme conditions. Although

17-1

the architect and the specification writer normally may not rely on this information for the average specification, they should be aware of the tests that determine chemical resistance and the ratings of commonly used vehicles. These data will supplement information contained in Chapters 2 and 4.

17.2   Chemical-Resistant Coatings   Chemical groups that normally come in contact with coatings are acids, alkalies, corrosive salts, and water.

17.2.1   Reasons for chemical resistance   Vehicles that resist chemicals should have a high degree of impermeability to prevent chemicals from reaching the substrate. To assure protection against strong chemicals, the film must be completely inert in order to prevent permeation of the chemicals.

The specification writer should choose coatings to resist specific chemicals. This choice requires basic information. For example, nonpolar vehicles such as chlorinated rubber, polyethylene, and polytetrafluoroethylene are recommended for polar reactants such as strong acids, bases, and corrosive salts. On the other hand, ester groups or unsaturation in the backbone structure of the binder should be avoided because such vehicles are readily attacked by alkalies. Therefore coatings based on drying oil, varnishes, alkyds, or esterified epoxies are not preferred for chemical resistance.

Also, because some chemicals cause paint films to swell, the coating must not only be impermeable but also be thick enough to prevent penetration to the substrate. A thickness of 5 mils (0.127 mm) is considered adequate. The concept of applying several coats is deemed superior to one-coat application, since holidays can thus be avoided. At least three coats are considered satisfactory. A very important factor in obtaining optimum resistance is proper surface preparation.

17.2.2   Methods of testing   The following methods of testing are important to the paint specifier. These tests are documented in the testing procedures published by the American Society for Testing and Materials (ASTM) and in *Paint Testing Manual: Physical and Chemical Examination of Paints, Varnishes, Lacquers, and Colors.*[1]

ASTM tests
Evaluation of Painted Specimens to Corrosive Environments: D-1654. This method covers the preparation of previously painted specimens for accelerated and atmospheric exposure tests and their

[1] Henry A. Gardner and G. G. Sward, 12th ed., Gardner Laboratory, Inc., Bethesda, Md., 1962.

evaluation in respect to corrosion, blistering, or loss of adhesion. Each specimen is scribed vertically in the center to a distance of ½ in (12.7 mm) from the edges. It is then exposed to salt spray (ASTM B-117), acetic acid and salt spray (ASTM B-287), or exterior exposure tests (ASTM D-1014). Two methods of treatment are suggested:

Method A. After the exposed panel has been rinsed with water at 100°F (38°C), the entire surface is blown for 15 min with compressed air (at 40 to 90 psi, or 275,790 to 620,528 Pa), with short blasts on the scribed line and any other corroded spots.

Method B. After the exposed panel has been washed with water at 100°F, the entire panel is scraped vigorously with a dull knife or a piece of sheet metal for 15 min.

The coated panels are then rated for their ability to resist exposure by the following rating schedules, which show the average measure of failure from the scribed line:

Inches	*Rating*
0	10
$1/64$	9
$1/32$	8
$1/16$	7
$1/8$	6
$3/16$	5
$1/4$	4
$3/8$	3
$1/2$	2
$5/8$	1
1 or more	0

Method of Salt-Spray (Fog) Testing: B-117. In this test suitably conditioned compressed air is used to form a fog of salt water with the help of atomizing nozzles. The dried coated panels to be tested are properly scribed, and the edges are protected with wax. The panels are placed in a cabinet at 15 to 30° from the vertical. A 5% NaCl solution is then sprayed by means of compressed air at 10 to 25 psi (68,948 to 172,369 Pa) at 95°F (35°C). The fog collector should collect from 1.5 to 3 ml of water for every 80 cm² of horizontal collection area.

Method of Acetic Acid–Salt-Spray (Fog) Testing: B-287. Conditions of operation and apparatus are the same as in ASTM B-117. The coated dried panels, properly scribed and wax-sealed, are exposed to a fog spray of 5% NaCl solution brought to a pH of 3.0 by the addition of 0.2 percent of glacial acetic acid.

Exterior Exposure Tests of Paint on Steel: D-1014. Three kinds of steel panels may be used:

1. *Hot-rolled structural-steel angles (4 by 4 by ⅜ in, or 101.6 by 101.6 by 9.525 mm)*. Remove oil by solvents, and remove mill scale by preexposure and wire-brushing.

2. *Hot-rolled steel plate (12 by 12 by ¼ in, or 304.8 by 304.8 by 6.35mm)*. Clean as in Paragraph 1.

3. *Cold-rolled steel strip (6 by 12 in, or 152.4 by 304.8 mm)*. Clean by the method ASTM D-609, using a solvent to remove the oil film.

Paint may be applied by brush, manual spray, automatic spray, automatic dip coater, automatic doctor blade, roller coating, or curtain coating. The backs and edges of all panels are coated with the same system as the fronts.

After sufficient exposure, each coating is rated for degree of blistering by comparing it with photographs of ASTM D-714, degree of chalking as in ASTM D-659, and degree of rusting as in ASTM D-610.

Water and Alkali Resistance of Dried Varnish Films: D-1647

1. *Water resistance*. The coating is applied to tinplate panels. After the panels have been permitted to dry for 48 h, they are placed in distilled water for a predetermined time (18 h or more). They are removed from the water and rated according to the degree of blooming and rate of recovery by the following ASTM standards:

Grade	Observation
1	Not visibly affected.
2	Whitening disappears in 20 min.
3	Whitening disappears in 2 h.
4	Whitening disappears in 24 h.
5	Whitening disappears in more than 24 h.

2. *Alkali resistance*. Twenty 6-in (152.4-mm) test tubes are dipped in the varnish to be tested. Each tube is immediately inverted and allowed to dry for 72 h. The tubes are exposed to a 3% NaOH solution for periods of 1, 2, 3, 4, 5, 6, 7, 8, 16, and 24 h. Then they are rinsed, allowed to dry for 30 min, and examined for film whitening, blistering, or removal.

Water-Fog Testing of Organic Coatings: D-1735. The apparatus is the same as that for salt-spray testing (ASTM B-117), except that only demineralized water at a pH of 7 is used and the cabinet is maintained at 100°F (38°) instead of 95°F (35°C).

Effect of Household Chemicals on Clear and Pigmented Organic Finishes: D-1308

1. *Spot test, covered*. After the coating of panel after has been allowed to dry for 1 week, 1 ml of reagent is placed on it. The spot is covered with a watch glass.

2. *Spot test, uncovered.* The procedure is the same as in Paragraph 1 but without the watch glass.

3. *Immersion test.* Half of the coated panel is immersed in the reagent.

The following reagents are considered household chemicals: cold distilled water, hot distilled water, ethyl alcohol (50 percent by volume), vinegar (3% acetic acid), alkali solution, acid solution, soap solution, detergent solution, lighter fluid and other volatile reagents, fruit (open side of fruit in contact with coating on the panel), oils and fats (butter, margarine, lard, shortening, vegetables, etc.), condiments (mustard, catsup), beverages (coffee, tea, cocoa), lubricating oils, and greases. The effect on the film is noted after a specified time of exposure.

Perspiration Resistance of Organic Coatings: D-2204. Panels measuring 2 by 1½ by ¾ in (50.8 by 38.1 by 19.05 mm) and coated with the required films are held with the coating against the skin of human wearers for 1 week. Changes in gloss, color, smoothness, hardness, abrasion resistance, blistering, and adhesion are then noted.

A simulated perspiration mixture uses the following ingredients: urea, 1 ml; lactic acid, 5 ml; Na_2HPO_4 (10%), 1 ml; NaCl (1 normal), 100 ml.

Gardner tests

Battelle Chemical Resistance Cell Test. Coated panels are exposed to chemical reagents under standard test procedures.

Bratt Conductivity Cell for Determination of Chemical Resistance. Film conductance indicates lower chemical resistance.

Gearhart-Ball Solvent-Resistance Tests

1. *Cup test.* Film is fastened over the top of a beaker or dish, and a small amount of solvent is poured onto the film. The time required to puncture the film is noted.

2. *Distensibility.* Fasten one end of a tensile-strength film strip to a clamp. Immerse the strip in the required solvent until it has been stretched 1 in (25.4 mm). Note the time and percentage of elongation.

Spot Test of Chemical Resistance. Make a ring from inner-tube stock or the gasket for a jar. Cement the ring to the coating with wax. Then place a small amount of the reagent within the ring limits, and wipe off the excess. Examine the spot after a specified time.

17.3 Chemical-Resistant Vehicles   These materials are discussed in the following paragraphs. Resistance properties are summarized in Table 17.1.

17.3.1 Polyvinyl and polyvinylidene copolymers   Vinyl finishes are binders based on the combination of the copolymers of vinyl acetate and vinyl chloride. While they have excellent chemical resistance, their adhesion must be properly ensured by use of an appropriate metal con-

TABLE 17.1   Chemical-Resistance Properties of Vehicles

Properties	Polyvinyl chloride	Cellulose aceto-butyrate	Polyethylene	Nylon	Chlorinated polyether	Fluoro-carbons	Polyesters
Exterior durability	E*	E	F	G	E	E	E
Salt spray	E	VG	G	G	E	E	E
Water	E	VG	G	G	E	E	E
Solvents							
Alcohols	E	F	G	G	VG	E	VG
Gasoline	E	G	E	E	E	E	VG
Hydrocarbons	G	G	E	E	VG	E	VG
Esters, ketones	P	P	G	VG	G	E	F
Chlorinated	P	P	F	E	G	E	F
Salts	E	VG	E	VG	E	VG	G
Ammonia	E	P	E	G	E	E	G
Alkalies	E	P	VG	G	E	E	F
Mineral acids							
Dilute	E	G	E	F	E	E	VG
Concentrated	E	P	VG	P	G	E	G
Oxidizing acids							
Dilute	E	P	VG	P	F	E	G
Concentrated	G	P	P	P	P	E	G
Organic acids							
Acetic, formic	F	P	VG	P	G	E	G
Oleic, stearic	E	F	VG	VG	G	E	G

NOTE: E = excellent; VG = very good; G = good; F = fair; P = poor.
* With ultraviolet absorber.

ditioner followed by a wash primer which contains polyvinyl butyral. Although two coats are normally acceptable, three and four coats are preferred to assure proper film thickness.

Pure polyvinyl chloride, which is the basis for both organosols and plastisols, is much more resistant to such chemicals as strong acids and bases than the above-mentioned copolymers, but its adhesion is even poorer. In addition to use of a carefully selected primer, the baking schedule usually required for organosols and plastisols improves the adhesion to a great degree.

Polyvinylidene chloride has been used quite successfully in copolymeric form with acrylonitrile, which is known as Saran coating. Its films are resistant to strong acids (e.g., glacial and concentrated nitric and hydrochloric acids) but have poor resistance to alkalies. Moisture and gases cannot penetrate Saran.

17.3.2 Chlorinated-rubber and synthetic-rubber derivatives   Hypalon, or chlorosulfonyl polyethylene, is an unusual synthetic-rubber type of coating resin, in that it can be vulcanized by reaction with such metal oxides as lead oxide and magnesium oxide in the presence of catalysts, forming a highly elastic product with excellent resistance to strong acids such as concentrated hydrochloric and sulfuric acids and to oxidizing agents like ozone, hydrogen peroxide, and chromic acid. It is used to

Epoxy	Poly-propylene	Acrylic	Chlorinated rubber (Parlon)	Neoprene	Hypalon	Saran	Phenolic resins (Methylon)	Bituminous coatings
F	G	VG	E	E	E	E	E	P–E
VG	VG	VG	E	E	E	E	E	E
G	E	VG	E	E	E	E	E	E
E	G	F	E	E	E	P	G	P
E	E	F	P	P	P	P	G	P
E	E	P	G	P	P	P	G	P
F	G	P	P	P	P	P	G	P
E	F	P	P	P	P	P	G	P
E	E	G	...	...	...	...	E	E
P	E	VG	...	...	...	...	E	E
VG	VG	VG	E	E	E	P	E	E
E	E	G	G	E	E	E	G	E
G	VG	G	P	G	G	E	G	E
G	VG	G	P	G	G	E	...	E
P	P	P	P	G	G	E	...	E
F	VG	F	P	P	P	G	G	P
E	VG	F	P	P	P	G	G	P

great advantage in various chemical baths such as chrome-plating baths which contain chromic acid. Hypalon may be combined with chlorinated rubber, urea resins, and epoxy resins.

Neoprene is a polychloroprene with high acid and alkali resistance, although it has been susceptible to water. It has been used successfully on chemical-process equipment and chemical trailer trucks.

Parlon is a chlorinated-rubber resin which has good resistance to normal acid and alkaline conditions but is sensitive to strong oxidizing acids such as nitric and acetic acids as well as to fatty acids. In addition, it will react with sulfur dioxide, concentrated ammonia, and most organic solvents. Because it has poor adhesion, Parlon needs a special primer.

17.3.3 Epoxy resins and combinations   Epoxy resins have good chemical resistance, particularly to alkalies, because of the absence of ester groups in their backbone structure. They also have excellent adhesion because of the ether linkages and hydroxyl groups present. The large number of aromatic rings contributes to their hardness, while their excellent flexibility is attributed to the long distance between the hydroxyl groups.

A low–molecular-weight epoxy with the appropriate curing agent in a two-container system makes a useful, highly chemical- and solvent-resistant finish when baking is impractical. These epoxies have been used

TABLE 17.2   Typical Interior Chemical-Resistant Finishes

For a painted finish on	Finish	Type
Wood	High gloss	Two-component polyamide epoxy
	Gloss	Catalyzed alkyd
	Semigloss	Catalyzed alkyd
	Satin	Catalyzed alkyd
Plaster	High gloss	Two-component polyamide epoxy
Drywall	High gloss	Two-component polyamide epoxy
Metal	High gloss	Two-component polyamide epoxy
Concrete block and pumice block	High gloss	Two-component polyamide epoxy

to coat equipment in chemical plants (Table 17.2). Epoxies are used in large quantities with coal tar or asphalt to make exceptionally tough, chemical-resistant coatings. Phenolic-epoxy coatings are useful as baking finishes with excellent chemical and solvent resistance for the lining of food cans and steel and aluminum containers and for wire enamels, collapsible tubes, washing machines, and pipe linings. Further modification with silicones improves the water resistance of epoxies.

17.3.4   Bituminous materials and blends   Because bituminous coatings have good water resistance, excellent adhesion, and good chemical resistance, they are readily used to coat steel, wood, and concrete. Used alone, bituminous coatings are thermoplastic and tend to cold-flow at moderate temperatures. Sunlight tends to craze and alligator these exterior coatings.

The cold-flow tendencies of both coal-tar and asphalt coatings will be improved either by adding fillers or by combining them with another resin such as acrylonitrile, epoxy, or isocyanate. Coal tar has been preferred for combination with these resins, although asphalt has been used successfully.

Hot-applied bitumens are used extensively on the exterior of underground pipelines for oil and gas as well as on both the interior and the exterior of water pipelines.

Coal-tar emulsions are improved by the addition of acrylonitrile copolymers, which successfully reduce their water absorption and increase their oil resistance.

17.3.5   Polyfluoroethylene polymers and copolymers   Teflon (polytetrafluoroethylene) has a high degree of symmetry and also is very polar because of its four fluorine atoms per carbon atom. Consequently there are extremely strong attractive forces between its molecules. Because of its high degree of crystallinity, Teflon has a relatively high melting point (1382°F, or 750°C). Since the very strong secondary valence forces will not permit ordinary solvents to separate the molecules, Teflon is relatively insoluble in common solvents. It is actually the most highly chemical-resistant of all known linear organic polymers, being completely inert to halogens, fuming mineral acids, strong alkalies, and oxidizing agents. It can be attacked only by molten alkalies. Because it is incompatible with all known plasticizers and solvents, it cannot be applied as a typical coating in the solvent or organosol form. Flame or powder spraying is the only method available for proper application of this resin.

17.3.6 Polyethylene   The molecule of polyethylene is quite symmetrical but extremely nonpolar. It is sensitive to nonpolar solvents, especially above 122°F (50°C). When stretched, it develops crystallinity and increased chemical resistance.

   Both polyethylene and polypropylene, which are widely used as plastic material, have excellent water and chemical resistance as well as flexibility. However, they are difficult to apply as coatings since they have poor solubility and poor adhesion to the substrate. They can be applied directly by flame spraying.

17.3.7 Polysulfides   These are synthetic rubbers that contain pairs of sulfur atoms as part of their backbone structure of ether linkages. This unusual chain offers a tough, rubbery film which has excellent flexibility and adhesion as well as water, solvent, and chemical resistance. Polysulfides are frequently used alone or in combination with epoxies, Saran, and other binders.

17.3.8 Polyesters   These films, which are made by the catalysis of unsaturated esters, are becoming increasingly popular in the furniture and appliance industry. They are quite resistant to weak acids and bases but are attacked by strong alkalies and some strong acids. While they are resistant to most solvents, they are attacked by ketones and chlorinated solvents.

17.3.9 Polyurethanes   There are five types of polyurethanes (see Table 16.2 in Chapter 16):

▪ Two-container system with an alcohol component containing the curing agent of polyester, polyether, or epoxy. It has excellent chemical resistance.
▪ Two-container system with an alcohol component containing the amine curing agent, characterized by excellent chemical resistance.

▪ One-container baking polyurethane. This prepolymer has a phenol blocking system built into the molecule which splits off when baked and permits the polymer to continue its curing at the higher temperature. Chemical resistance is excellent.

▪ Moisture-curing single-container system. It has very good chemical resistance.

▪ Air-curing single-container system (these polyurethanes are also known as uralkyds). Chemical resistance is good.

17.3.10 Nylon In recent years, with the advent of electrostatic powder spraying, nylon has been used quite successfully as a coating in spite of its poor adhesion. It is resistant to all alkalies and to weak acids as well as to all common solvents. Strong acids will react with the film. Nylon will be dissolved by phenol and chlorinated solvents.

17.4 Chemical Resistance For general architectural purposes, commonly available epoxy or catalyzed-alkyd systems offer good chemical resistance (Table 17.2). You also must consider the chemical resistance of pigments (Table 17.3). For example, lead pigments tend to convert to plumbates, ruling out the use of white lead, chrome yellows, oranges, and greens, and molybdate orange for alkali-resistant pigments. Prussian blue, ultramarine blue, zinc oxide, and whiting (calcium carbonate) are sensitive to acids and should not be used in a chemical-resistant paint designed particularly to resist acids.

The chemical resistance of white pigments is as follows (B = borderline; E = excellent; F = fair; G = good; P = poor):

Pigment	Acid	Alkali
Titanium dioxide (TiO_2)	E	E
White lead ($2PbCO_3 \cdot Pb(OH)_2$)	G	P
Zinc oxide (ZnO)	P	E

The chemical resistance of extender pigments is as follows:

Pigment	Acid	Alkali
Barites ($BaSO_4$)	E	E
Silica (SiO_2)	E	E
Whiting (limestone; $CaCO_3$)	P	E
Mica ($K_2O \cdot 3Al_2O_3 \cdot 6SiO_2 \cdot 2H_2O$)	E	E
Clay	G	G
Talc	G	G

TABLE 17.3 Characteristics of Colored Pigments

Color	Organic	Inorganic	Alkali	Acid	Exterior durability	Heat resistance
					Resistant properties	
Red	Toluidine		P	P	G	P
	Monastral		G	G	G	G
		Iron oxide	G	G	G	G
		Molybdenum	G	B	G	B
		Cadmium	B	B	G	E
Orange		Red iron oxide	G	G	G	G
		Molybdenum orange	F	B	G	B
Yellow		Yellow iron oxide	G	G	G	G
		Chrome yellow	P	F	G	P
		Zinc chromate*	.	.	.	.
	Hansa		P	P	P	P
		Nickel TiO$_4$ titanate	G	G	G	G
Green		Chrome oxide	G	G	G	G
		Chrome green	P	F	G	P
	Phthalocyanine		E	E	E	E
Blue		Prussian blue	P	F	G	P
	Ultramarine blue		P	P	P	P
	Phthalocyanine blue		E	E	E	E
Black	Lampblack		G	G	G	G
	Carbon black		G	G	G	G
Metallic		Aluminum flake	P	P	E	E
		Zinc dust	P	P	E	E
		Stainless-steel powder	E	G	E	E

SOURCE: Reliance Universal, Inc., Houston, Tex. (edited).
NOTE: B = borderline; E = excellent; F = fair; G = good; P = poor.
* Used in primers.

17.5 Heat-Resistant Finishes   Some industrial and maintenance finishes are exposed to a wide range of temperatures either as part of their application operation, as with baking finishes, or after application, as in continuous or intermittent exposure to heat, exemplified in smokestack coatings or finishes covering exhaust or boiler doors (Tables 17.4, 17.5, and 17.6).

TABLE 17.4   Heat Resistance of Various Other Coatings

Binder	Maximum service temperature	Continuous heat resistance
Polyvinyl chloride	200°F (93°C)	140–175°F (60–79°C)
Cellulose acetobutyrate	180°F (82°C)	140–212°F (60–100°C)
Polyethylene	160°F (71°C)	180–210°F (82–99°C)
Nylon	180°F (82°C)	175–250°F (79–121°C)
Chlorinated polyether	290°F (143°C)	290°F (143°C)
Fluorocarbon	500°F (260°C)	400°F (204°C)
Polyester	250°F (121°C)	250°F (121°C)
Epoxy	350°F (177°C)	250–550°F (121–288°C)
Polypropylene	250°F (121°C)	250–320°F (121–160°C)

Most organic finishes tend to degrade or yellow when baked to 300°F (149°C) or over. They fail because of binder oxidation, as in air-drying alkyds, or because of decomposition, as in polyvinyl chloride. Decomposition is in large part due to the weakening of bonds between the aliphatic carbons of the binder. The temperature of breakdown is lowered as unsaturation and ester linkages are introduced in the backbone structure.

Fluorocarbons have been found to be quite heat-resistant. Crystals of Teflon melt at over 750°F (399°C). Likewise, organoinorganic compounds are quite effective in heat-resistant finishes exposed to temperatures from 600°F (316°C) up. These binders (silicones and silicon esters) consist of a range of compounds which are monomers, linear polymers, or cross-linked polymers (Fig. 17.1). Thus, an aluminum smokestack coating, consisting of powdered leafing aluminum dispersed in such a silicone vehicle, will resist heat to 1100°F (593°C) by splitting off the organic portions of the polymer and fusing the silicone backbone to both the aluminum and the metallic substrate of the smokestack. Heat resistance would be similar to that of quartz, the chemical formula of which is SiO_2.

TABLE 17.5   Typical Curing Processes

Temperature range	Process	Type of coating
140–200°F (60–93°C)	Forced dry	Alkyds, varnishes, epoxies, polyurethanes, polyesters
200–400°F (93–204°C)	Normal baking	Urea-formaldehyde or melamine with nonoxidizing alkyds, cross-linked acrylics
400–600°F (204–316°C)	High-temperature baking	Teflon, silicones

TABLE 17.6   Typical Heat Resistance of Specialized Coatings after Curing

Temperature range	End use	Type of coating
150–250°F (66–121°C)	Lighting fixtures, stoves, etc.	Short-oil phenolics, urea-formaldehyde–alkyds, melamine-alkyds, acrylics
500–1200°F (260–649°C)	Exhaust manifolds, boiler doors, smokestacks, electrical equipment	Silicones, ethyl silicates, organotitanates, with zinc dust or aluminum
1000°F (538°C)	Missiles	Butyl titanate

17.5.1   ASTM testing for heat-resistant finishes

Determining the Effect of Overbaking on Organic Coatings: D-2454-66T. To determine the time and temperature effect on the physical and chemical properties of lacquer and organic enamel films, four panels are suitably prepared. Two of the panels are baked at normal temperatures, using the recommended schedule of baking. The remaining two panels are immediately subjected to overbaking as agreed to by

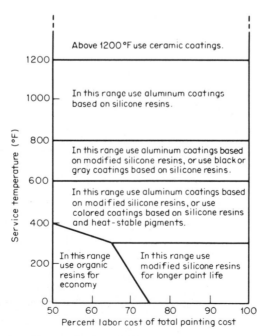

Fig. 17.1 Specifier's guide to selection of silicone coatings. [*Dow Corning Corp., Midland, Michigan*]

buyer and seller. After the panels have been suitably conditioned, they are evaluated for gloss, color, flexibility, adhesion, impact resistance, and resistance to reagents.

Resistance to Elevated Temperatures during Service Life: D-2485-66T. To evaluate the heat-resistant properties of coatings designed to protect steel surfaces exposed to elevated temperatures during their service life, panels suitably coated with the material under test are exposed by one of two methods, depending on whether they are intended for interior or exterior service.

Method 1: for interior service. After two suitably coated panels have been baked in a furnace for 24 h, one of the panels is plunged in water at 70°C (21°C). This panel is examined for evidence of film failure such as dulling, blistering, cracking, and loss of adhesion. The other heated panel is cooled to 75°F (24°C), and after 1 h it is rapidly bent double over a ½-inch (12.7-mm) mandrel. The panel is then examined for such film degradation as cracking and loss of adhesion.

Method 2: for exterior service. Panels are placed in a furnace for the following schedule:

Temperature	Hours
400°F (204°C)	8
500°F (260°C)	16
600°F (316°C)	8
700°F (371°C)	16
800°F (427°C)	8

Visual inspection at each temperature level gives evidence of failure such as peeling, cracking, blistering, abnormal discoloration, or loss of adhesion. At the end of the temperature exposure, the panels are removed from the furnace, allowed to cool at the ambient temperature for a minimum of 1 h, and again inspected for evidence of failure. If the panels successfully pass the heat test, proceed to test for salt-spray resistance according to the method of ASTM B-117 for 24 h. Also, test a separate pair of panels for weather exposure, using ASTM D-1014.

17.5.2 Gardner testing for heat-resistant finishes

Houston Heat-Exposure Test. Paint is applied to the outer of two concentric pipes that are held vertically over a Meker or Fischer burner. Temperatures reach 300°F (149°C) at the cool end and 1050°F (566°C) at the hot end. This test is an accelerated one and has fairly good reproducibility.

Simultaneous Heat and Exposure Test on Heat-Resistant Paint. This test uses a cylindrical electric heater, replaceable test surfaces (in the form of sheet-metal pipes), and racks for exposing test specimens when not under heat. Temperatures range from 300 to 900°F (from 149 to 482°C).

Melting-Point Bars for Testing Heat-Resistant Paint. The effect of heat on a metal finish may be determined on a melting-point bar, which is heated electrically at one end. The exact temperature is determined by a thermocouple or by the use of pure crystals of a known melting point.

17.6   Relatively Impermeable Coatings   A high degree of impermeability is mandatory for a successful chemical-resistant coating. Permeability is a complex, little-understood property of films. Water diffuses at a rate which is affected by many factors, such as temperature, vapor-pressure differential, osmosis, electric-potential gradient, and porosity. While normal coatings do not contain pores per se, a microporous or macroporous structure may result from weathering (or improper formulation), and cracks develop in the coating.

It is possible to compare the permeabilities of films exposed to polar and nonpolar solvents. Solvents such as water, ethyl alcohol, ethyl acetate, acetone, and benzene are allowed to permeate films of linseed oil, varnishes, alkyds, nitrocellulose lacquers, rubber, shellac, paraffin wax, glue, and gelatin. Permeability depends to some degree on the mutual solubility of solvent and film (this is not the case, however, with benzene and acetone in respect to gelatin and shellac).

The key point for specification writers is that they consider exposure conditions. Whenever a solvent or a chemical that can dissolve the vehicle of the paint film is present, there is serious reason to consider rewriting the coatings specification. Chemical absorption by the paint film is also a consideration, especially in the case of water (Table 17.7). Such absorption normally causes swelling and film deterioration.

TABLE 17.7   Water Absorption of Coatings

Binder	Water absorption (percent H_2O absorbed in 24 h)
Polyvinyl chloride	0.45
Cellulose acetobutyrate	1.2–2.1
Polyethylene	0.04
Nylon	2.9
Chlorinated polyether	0.01
Fluorocarbon	0.01
Polyester	0.5–2.50
Epoxy	0.08–0.15
Polypropylene	<0.01

For coatings systems which will be in immersion service, it is imperative that the inspector isolate all sags and runs and have the contractor remove them. Such excessively thick spots are capable of harboring paint solvents which can lead to film blistering on immersion. All sags and runs should be sanded smooth and the affected areas given one additional coat of the desired finish coat of material.

Coatings systems going into immersion service quite often require force drying to remove all residual solvent fumes. For lacquer-type coatings such as the chlorinated rubbers and vinyls, this can best be accomplished by circulating air warmed to 140°F (66°C) for approximately 48 h. If final drying is to be accomplished at ambient temperatures, it is a wise precaution to circulate outside air through the vessel for a minimum period of 72 h.

After ensuring that the surface is prepared properly and each coat applied in accordance with the applicable specifications, the next responsibility of the inspector is to ensure film continuity. For thin-film coatings, this is almost always accomplished by using low-voltage (67½-V) nondestructive holiday detectors.

BIBLIOGRAPHY

Book of ASTM Standards, parts 20 and 21, American Society for Testing and Materials, Philadelphia, 1967.

Application of Organic Coatings to the External Surface of Steel Pipe for Underground Service, National Association of Corrosion Engineers, Katy, Tex., 1975.

Briber, Alex A.: "Fire Tests for Surface Flammability," *Paint and Varnish Products,* vol. 61, no. 9, September 1971.

Cross, Thomas A.: "How to Evaluate Coatings," *Chemical Engineering,* Oct. 22, 1979, pp. 153–156.

Gardner, Henry A.: *Physical and Chemical Examination of Paints, Varnishes and Lacquers,* H. A. Gardner Laboratories, Inc., Bethesda, Md., 1962.

Hausner, Henry H.: *Coatings of High-Temperature Materials,* Plenum Publishing Corporation, New York, 1966.

Huminik, John: *High-Temperature Inorganic Coatings,* Reinhold Publishing Corporation, New York, 1963.

Meals, Robert N.: *Silicones,* Reinhold Publishing Corporation, New York, 1959.

Mellan, Ibert: *Corrosion Resistant Materials Handbook,* Noyes Data Corp., Park Ridge, N.J., 1971.

Miller, Irene R.: "New Maintenance Coatings," *Chemical Engineering,* Mar. 24, 1980, pp. 59–64.

Nylén, Paul, and Edward Sunderland: *Modern Surface Coatings,* Interscience Publishers, New York, 1965.

Parker, Dean H.: *Principles of Surface Coating Technology,* Interscience Publishers, New York, 1965.

Payne, Henry F.: *Organic Coating Technology,* 2 vols, John Wiley & Sons, Inc., New York, 1954–1961.

Protective Coatings in Petroleum Production, National Association of Corrosion Engineers, Katy, Tex., 1961.

Resolving Corrosion Problems in Air Pollution Control Equipment, National Association of Corrosion Engineers, Katy, Tex., 1976.

Specification and Plant Coating Guide, pocket ed., National Association of Pipe Coating Applicators, Shreveport, La., 1970.

"Storage of Volatile Organic Materials in White Tanks," Regulation NJAC 7:27–16, Sec. 16.2(a), New Jersey State Department of Environmental Protection, Trenton, N.J., 1978.

Chapter **18**

Application Techniques

GARY BOYD CHARLESWORTH
Colorite, Inc.

GUY E. WEISMANTEL
Chemical Engineering

18.1 Selecting the Contractor   There is an important step that precedes choosing the application method. It is choosing the contractor. The choice can be simple if an architect maintains a constantly updated bid list. Specifiers should develop a set of strict criteria for contractor selection (Table 18.1). The bid list should include the names of contractors who have performed well on previous contracts. Professional and experienced people are needed to apply properly high-performance maintenance coatings. Do not attempt to get by with the low bidder or the inexperienced applicator on a large-scale job. The problems that accompany inexperience can lead to severe setbacks through loss of time, material, and quality performance. Selection of a good contractor can help solve the innumerable problems that constantly arise. An experienced contractor can cope with these problems and make a job flow smoothly on schedule.

A competent painting contractor will use foremen and quality control inspectors who are familiar with the physical-property requirements of paint, such as viscosity, antisettling properties, drying time, adhesion, coverage (or solids) by volume, weight per gallon, and perhaps cost per square foot. They should also have knowledge of basic performance tests. Good reference works for these tests are *Federal Standard No. 141a* and ASTM Parts 27, 28, and 29.

It is not the responsibility of the painting contractor to choose the paint system, although the contractor should know exactly how application performance will be evaluated. Many companies use numerical-rating systems that quantify both the paint and its application. Normally, faults in application are brought out in field trials, but unfortunately the cho-

TABLE 18.1   Contractor Selection Questionnaire

Every architect or engineer should maintain a file on prequalified contractors that is updated at least every 18 months to 2 years. A contractor answering the following questionnaire would certainly more than substantiate that he or she is capable of handling a particular painting project. Any qualified contractor would be glad to submit this information as a prequalification requirement.

I. Name of business
   Street and mailing address_____
   City _____ State _____ Zip _____
   Telephone number with code (   ) _____ Teletype_____
   Applicant's name _____ Title _____
   Licensed in states of _____

   Other offices: Please attach list of sales offices, and representatives including names, addresses, and telephone numbers.

II. Organization
   A. Type of organization
      Sole ownership_____ Partnership_____ Corporation_____
   B. Date business founded _____ Under present
      management since_____
      1. Principals of company _____
   C. Net worth _____
   D. Annual dollar volume (each of last 3 years)
      _____ _____ _____
   E. Indicate maximum and minimum job cost range within which you prefer to bid:
      Maximum_____ Minimum_____
   F. References on following services with contact name and telephone number:
      Banking _____
      Bonding _____
   G. Attach a current financial statement, preferably by a certified public accountant.

III. Bidding interests
   A. Mark classes of work in which you are interested in bidding:
      1. Commercial painting
         Wall covering
         Decorating
         Others (specify)
      2. Industrial
         a. New construction:
            Heavy industrial; petrochemical, chemical, paper, etc.
            Sandblasting
            Light industrial; manufacturing facilities
            Other (specify)
         b. Maintenance
            Heavy industrial; petrochemical, chemical, paper, etc.
            Light industrial; manufacturing facilities
            Sandblasting
            Other (specify)
            Urethane foam

TABLE 18.1   Contractor Selection Questionnaire   (Continued)

    3. Residential
      a. New houses
         Tract work
         High rises
      b. Maintenance
         Houses
         High rises
         Other

IV. Labor relations: shop and field

List all trades with which you have contracts and/or working agreements. Include expiration dates and information on any recent labor difficulties.

V. Complete if applicable:

    A. Describe shop operations and type of work done.

    B. How and where do you use subcontractors?

VI. Using the attached format, please provide a brief résumé of important jobs completed by your firm in the past 5 years and attach a brochure, if available.

Résumé of work performed in last 3 years					
Owner and/or client and representative (name, title, address)	Magnitude of contract ($)	Prime or subcontract (list general contractor)	Type of work performed	Location of job	Year completed

VII. Equipment inventory list (use separate paper)

    A. Describe type and number of major pieces of equipment which are *owned* (not leased): air compressors, blasting units, power staging, trucks, and special equipment.

    B. Describe home plant facility and give its location.

VIII. Quality control

Describe briefly all forms which contractor normally uses to document quality control. Attach all inspection forms used, including material certification and inventory control, daily inspection sheets, acceptance and rejection sheets, etc.

IX. Face-to-face interview

Prepare a checklist of questions relating to contractor selection. These should verify what has been written in Secs. I–VIII and bring out qualifications (or the lack of them) that might not show up in writing. (See following checklist for sample questions.) Also conduct interviews with job foremen to verify qualifications and understanding of the type of job to be performed. Visit jobsites.

Checklist for contractor selection

Does the contractor

- ✓ Have proven knowledge of the surface-preparation method to be used on the job?
- ✓ Have necessary inspection tools, coupons for surface-profile comparison, etc., for monitoring the job (1) during surface preparation, (2) during application, and (3) after application?
- ✓ Have the specified application tools and qualified people to use them?
- ✓ Have experience in the painting system being used (especially important in two-component and high-solids coatings)?
- ✓ Strictly follow Occupational Safety and Health Administration and other applicable regulations?
- ✓ Have necessary bonding?
- ✓ Give guaranteed performance?
- ✓ Use an organized work system to assure that correct paint is used on the job, including mixing, thinning, color, mil thickness, etc.?
- ✓ Have a standard operating procedure for reconciling problems that arise.
- ✓ Consistently use high application standards as proved by past successful jobs?
- ✓ Have recommendations by past customers?
- ✓ Use experienced craftspeople schooled (holding certificates) in tools specified (e.g., water-blast equipment, airless spray, etc.)?
- ✓ Have a reputation for meeting schedules?
- ✓ Have good management (especially job foremen) and the necessary financial requirements for the job?
- ✓ Use a quality-assurance program to include (1) auditing, (2) quality control of the paint and application procedure, (3) inspection, (4) documentation and record keeping, and (5) reporting procedures?
- ✓ Use standard practices as required by the Painting and Decorating Contractors of America, the National Association of Corrosion Engineers, the Steel Structures Painting Council, the American Society for Testing and Materials, the American Institute of Architects, and other applicable organizations?
- ✓ Handle accelerated work? How is this invoiced?

SOURCE: Certified Painting Systems Inc., Louisville, Ky.

sen application contractor often has not been present at the trials. Consequently, the coating selected is often new to the contractor.

18.2 General Performance   The architect or general contractor expects the painter to do a good job, follow specifications, be courteous to clients, and avoid poor workmanship. These expectations imply that the painting contractor can meet deadlines and has knowledge of the paints being used on the job, color mixing, and paint equipment (particularly surface-preparation equipment) and how to repair it. The painting contractor must have and know how to use field testing equipment.

18.2.1 Scope of work   It is the intent of a specification to include all painting throughout the interior and exterior of a building, on wood, plaster, metal, or other surfaces requiring paint, as designated in the drawing and specifications, so that the job may be complete in every respect. A painting subcontractor usually understands that this work includes everything necessary for or incidental to executing and completing all painting work except that specifically excluded. The subcontractor also knows that all necessary scaffolding shall be furnished and installed by the painting subcontractor and that such scaffolding shall conform to regulations of the state industrial accident commission and local ordinances. Even so, a good paint specification will review these points in writing to prevent any misunderstanding.

18.2.2 Work in other sections   During preconstruction conferences it is wise to have the painting subcontractor examine the drawings and specifications for the section being painted and also the drawings and specifications for painting work in other sections. The subcontractor shall advise the architect of any conflict between his or her work and that of other trades and of any errors, omissions, or impractical details. Many design factors influence coating performance, among them welded joints, back-to-back angles, erector-set steel, rivets and bolts, edges and corners, concealed surfaces, and surfaces that are constantly cold, hot, or damp.

The painting subcontractor should also examine all the surfaces to be finished under the contract and see that the work of other trades has been left or installed in satisfactory condition to receive the paint, stain, or specified finish. Before starting work, the subcontractor should notify the architect in writing of any surfaces that are unsatisfactory for a proper paint finish. The application of the first coat of any finishing process often constitutes acceptance of the surface. Therefore, it is imperative that no exterior painting or interior finishing be carried out under conditions which jeopardize the quality or appearance of painting or finishing.

18.2.3 Protection of work   The painting subcontractor normally takes the necessary steps to protect his or her work and the work of other contractors while painting is in process. The paint specification should spell out these responsibilities, including the parties responsible for any and all damage to work or property caused by paint application. Other contractors are responsible for any damage that they cause to the paint job after painting has been completed. Specify responsibilities.

18.2.4 Workmanship   Paint must be applied by the method specified. Arbitrary substitution of one method for another (e.g., substituting spray painting when brushing is called for in the specification) can lead to serious adhesion and failure problems. The architect should include specific application instructions and testing procedures, with a preamble such as the following:

Each coat of paint shall be applied at the proper consistency and brushed evenly, free of brush marks, sags, and runs, with no evidence of poor workmanship. Color between coats of paint shall differ (color variations between coats should be large enough to distinguish color change but not large enough to impair hiding). Care shall be exercised to avoid lapping of paint on glass or hardware. Paint shall be sharply cut to lines. Finished paint surfaces shall be free from defects or blemishes.

Protective covers or drop cloths or masking paper and tape shall be used to protect floors, fixtures, and equipment. Care shall be exercised to prevent paint from being spattered onto surfaces which are not to be painted. Surfaces from which such paint cannot be satisfactorily removed shall be painted or repainted as required to produce a finish satisfactory to the architect.

Application of paint shall be by methods approved by the [local] joint labor committee. The architect's approval of spray applications shall be in accordance with and subject to union agreements for sprayable surfaces and types of finishes.

Each specification should also include complete lists of "work included" and "work not included" in the paint job. These lists will help to keep the price down if painters know what they are to paint and will also protect painters from other crafts that submit prefinished materials with only a primer.

18.2.5 Materials storage   Every specification should provide detailed storage information for every kind of paint on the job. For example, water-based paints can freeze (during winter shipment in noninsulated or nonheated trucks), and freezing can affect the finished quality of the paint on the wall (the paint may look all right, but the film may not coalesce properly). High temperatures can also be detrimental. The specification should include storage specifications consistent with manufacturers' recommendations (if you do not have this information, request it). The painting contractor should store all painting materials and equipment not in immediate use in a room or rooms assigned for the purpose. All painting materials should be received, and all mixing done, in this room to avoid pilferage and maintain quality control. All necessary precautions should be taken to prevent fire. Rags and waste soiled with paint should be removed from the premises at the end of each day's work or be stored in metal-covered metal containers half filled with water.

18.3 Job Preparation   In starting a job it is important to be completely prepared; otherwise the work will cost additional time and money. Here are important considerations:

▪ Make sure that the paint itself is approved before it reaches the jobsite. All cans must be coded; use only approved batch numbers. Thin material with the proper solvent according to the actual paint specification. Even today water is still mistakenly added to solvent-based coatings and vice versa.

▪ Request and follow a complete color schedule. The owner or the architect may request samples, which must be provided. Accurately figure the quantity of each color needed. If a job is large, paint is normally shipped from the factory. For special orders shipment can take from 2 to 3 weeks. Quantities of less than 50 gal (189.3 l) are usually obtained from local distributors, and if the quantity is very small, the paint can often be mixed at the jobsite.

18.3.1 Estimating paint requirements   Estimating the amount of paint needed is not always easy because some surfaces are much more porous than others. For example, block and concrete or even different block-company products are different, and paint may go 100 ft (30.5 m) further per gallon (3.785 l) on some block surfaces than on others. Use the following general estimating procedure:

▪ Determine the number of square feet to be finished (Figs. 18.1 and 18.2).
▪ Divide the total area of each surface to be finished by the estimated number of square feet that each gallon of finish material will cover.
▪ Refer to the spreading rate on the manufacturer's product data sheets for an estimated guide. Variables that affect the spreading rate include the type and condition of the surface, the tools used for application, thinning, and individual applicator techniques. Experience and good record keeping should be considered when using estimated spreading rates.
▪ In estimating paint requirements when colors are involved, all paint should be from the same batch, whether it consists of factory-mixed colors or colors made by tinting white bases, in order to avoid slight color variations. When there is doubt about the material estimate, reserve additional paint from the same batch until the job is complete. If insufficient material from the same batch is on hand, boxing (mixing) all batches together is necessary to avoid slight color variations.

18.3.2 Scaffolding techniques   There are many different types of scaffolding (Table 18.2). As time progresses, scaffolding improves. The best scaffolding for a job will have the required strength and will be easy to erect, disassemble, and move. New government safety regulations closely control what you can and cannot do. In most cases, handrails or safety belts are required. New rules require 20-in (508-mm) planks where 12- or 16-in (304.8- or 406.4-mm) floats have been used. High areas create significant problems.

HOW TO ESTIMATE SQUARE FOOT COVERAGE FOR DIFFERENT SHAPES

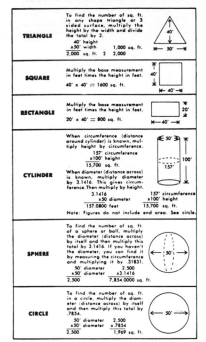

TRIANGLE	To find the number of sq. ft. in any shape triangle or 3 sided surface, multiply the height by the width and divide the total by 2. 40' height ×50' width 1,000 sq. ft. 2,000 sq. ft. 2 2,000
SQUARE	Multiply the base measurement in feet times the height in feet. 40' × 40' = 1600 sq. ft.
RECTANGLE	Multiply the base measurement in feet times the height in feet. 20' × 40' = 800 sq. ft.
CYLINDER	When circumference (distance around cylinder) is known, multiply height by circumference. 157' circumference ×100' height 15,700 sq. ft. When diameter (distance across) is known, multiply diameter by 3.1416. This gives circumference. Then multiply by height. 3.1416 ×50 diameter 157.0800 feet 157' circumference ×100' height 15,700 sq. ft. Note: Figures do not include end area. See circle.
SPHERE	To find the number of sq. ft. of a sphere or ball, multiply the diameter (distance across) by itself and then multiply this total by 3.1416. If you haven't the diameter, you can find it by measuring the circumference and multiplying it by .31831. 50' diameter 2.500 ×50' diameter ×3.1416 2,500 7,854.0000 sq. ft.
CIRCLE	To find the number of sq. ft. in a circle, multiply the diameter (distance across) by itself and then multiply this total by .7854. 50' diameter 2.500 ×50' diameter ×.7854 2,500 1,969 sq. ft.

WALLS—When figuring the square foot area, openings of less than 100 square feet should not be deducted.

OPEN WEB STEEL JOISTS

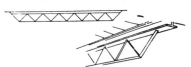

Original equipment manufacturers and fabricators generally dip these joists, as a first or shop coat. On all repaint work, by spray, these manufacturers recommend the paint be estimated by thinking of the joist as a solid rather than open web. Double for both sides.

STACKS

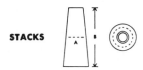

To compute the square foot area of a stack multiply height (B) by the average diameter (A) and multiply that total by 3.

EXAMPLE: Diameter of stack at the top — 5 feet. Diameter of stack at the bottom — 15 feet. Average diameter — 10 feet (2 | 5 + 15). Height 60 feet. 60 × 10 = 600. 600 × 3 = 1800 square feet of surface area.

CHAIN LINK FENCES

Figure the square foot area as a solid and double it to allow for both sides. For spray, this will be fairly accurate since overspray must be considered. For roller, the coverage indicated in the product data sheets should be increased.

DOWNSPOUTS & GUTTERS—Measure the circumference in inches. To figure square feet, divide the circumference by 12 and multiply by total length in feet. Double this figure if the inside is to be painted.

PICKET FENCE—Multiply the height by the length and multiply the result by four.

Fig. 18.1 *[Fuller-O'Brien Corp., South San Francisco, California]*

Steel scaffolding is fireproof and has the advantage of low wind resistance. It also has wheels on which to move it and a jack with which to balance it if the ground is uneven.

Aluminum scaffolding is also available. It is much lighter, but its load-carrying capacity is considerably lower than that of steel, and it cannot be

SURFACE AREA OF VARIOUS SIZE ELEVATED WATER TANKS*

CAPACITY (Thousand Gallons)	RISER (Diameter)	INSIDE AREA (Square Feet)	OUTSIDE AREA + (Square Feet)
50	4'	3,150	6,500
100	4'	4,300	8,000
150	4'	5,100	9,900
200	4'	5,900	11,100
250	4'	6,700	12,700
500	5'	10,000	19,600
750	Dry 8'	13,600	29,100
1000	Dry 8'	17,000	36,900

* Low Water Level 100' above grade.
+ Includes supporting columns.

The above chart is applicable to the tanks shown below.

In estimating the square foot area of a tank different than those shown on the preceding page, do the following —

1. To find the end areas of a tank: Multiply the square of the diameter by .7854.

2. To find the circumference of the tank: Multiply the diameter by 3.1416.

3. To find the area of the walls of the tank: Multiply the height by the circumference.

EXAMPLE:

Suppose the tank is 30 feet across and 50 feet high. The square of the diameter then is 900 feet (30 x 30). Which when multiplied by .7854 shows 706.9 square feet at the top of the tank. The diameter of 30 feet multiplied by 3.1416 shows that the tank is 94.3 feet around. The circumference of 94.3 multiplied by the height of 50 feet equals 4,715 square feet — area of the wall. Total area of approximately 5,425 square feet.

Any accessories such as piping, valves, rails, structural work, etc., would have to be estimated separately.

HOW TO FIND SURFACE AREAS OF:

CORRUGATED METALS

2½" Corrugated Sheet — to find width before corrugation multiply the width after corrugation by 1.08. Assume depth to be ⅝".

1¼" Corrugated Sheet — To find width before corrugation multiply the width after corrugation by 1.11. Assume depth to be ⅜".

ROOF DECK

If the roof deck has a cross-section view similar to that shown, first figure the square foot area then multiply by 2.42 to obtain the actual surface area.

If the roof deck has a cross-section view similar to that shown, figure the top side as just the square foot area of surface. Figure the underside as follows —

A For each square foot area multiply by 1.63 for actual surface area.

B Multiply by 1.75.

C Multiply by 1.92.

(A) 4½"   (B) 6"   (C) 8"

SIDING

If the siding has a cross-section view similar to that shown multiply each square foot of area by 1.5 for actual surface area. Double for both sides.

If the siding has a cross section view similar to that shown multiply each square foot of area by 1.42 for actual surface area. Double for both sides.

If the siding has a cross section view similar to that shown multiply each square foot of area by 1.75 for actual surface area. Double for both sides.

If the depth is 3" multiply by 1.5. Double for both sides.

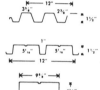

SURFACE OF SPHERES

Diameter in Feet	*Surface of Sphere in Square Feet
20	1,257
25	1,963
30	2,827
35	3,848
40	5,027
45	6,362
50	7,854
55	9,503
60	11,310
65	13,273
70	15,394

* Outside surface area only — double surface area for inside and outside.

Fig. 18.2 [*Fuller-O'Brien Corp., South San Francisco, California*]

built as high. Because of its weight it is very easy to move. Since most aluminum scaffolding is designed to be snapped together, it is easier to erect than steel scaffolding.

You may also have access to a ladder truck or a bucket truck. These are fast and enable the painter to reach high areas and areas behind objects.

TABLE 18.2   Scaffolding Guidelines (July 1980)

Types	Economic data	Advantages	Disadvantages
Extension ladder	$21 per week, up to 24 ft	Easy to put up	Very limited on height and safety
Ladder and plank	$23 per week for one plank, up to 24 ft	Easy to erect	Very limited on height and safety
Single-pole wood scaffold	Not rented; build your own	Cheap construction	Cannot be moved; it is torn down each time
Independent wood scaffold	Not rented; build your own	Cheap construction	Cannot be moved; it is torn down each time
Outrigger scaffold	$20 per week	Good for extending working surface	Care needed to avoid tipping scaffold over
Hand-operated suspension scaffold	$75 per week	Good for exterior of building	Hand-operated
Boatswain's chair	$20 per week	Good for high areas	Covers small area; hand-operated
Freestanding aluminum	$20 per section per week	Easy to get into standard doorways	Limited in height because of light metal
Freestanding steel	$15.75 per 6-ft section per week	Easy to get into small openings	Heavy weight
Spider staging	Cost, $800; rental per week, $250	Extends to great heights	Another scaffold may be needed to set it up
Boom bucket (ladder trunk)	Rental per week, up to $1075, depending on height	Good for small, high jobs	Exterior use only on a large area
Scissor lifts	Rental per week, $165 to $420, depending on height	Easy to change height rapidly	Cannot be moved into a room with small openings

NOTE: The price of rental equipment varies with the length of the rental.

Additional supports are required for trucks, which otherwise might easily be tipped over. Moreover, the area that can be reached in one move is very limited. Another alternative is an air-operated or electric spider; the electric spider moves more rapidly up and down. Used in conjunction with scaffolding, spiders can increase productivity.

For lower areas use a trestle ladder or two trestle ladders with an extension trestle and plank. Use an extension ladder consisting of two or more sections, arranged to permit adjustment of the ladder length. You can also choose ladder jacks and planks to be used with extension ladders. Wood scaffolding is now used sparingly because it will not meet government safety codes. Scissor lifts are fast becoming the most widely used type. They can be controlled from the working platform to go up or down and forward or backward, and they reach heights of 50 ft (15.2 m). For extremely high areas, however, the service of a crane must sometimes be obtained.

18.4   Prepaint Techniques   The surface is usually prepared by the painting contractor. This part of the paint job can be expensive.

18.4.1   Acid-etching   This technique is most commonly used on concrete to assure a complete bond and to remove any effervescence, grease, etc., from the concrete. Handle the acid with great care, and do not make the solution too strong. Use rubber gloves to protect the skin. After the etching has been applied and allowed enough time to work, wash it off completely with clean water before applying any paint.

18.4.2   Etching   Wash solutions are used to etch metal. They are easy to apply either by spraying for large surfaces or by using a brush or a rag for small or narrow, elongated surfaces such as gutters, flashings, and downspouts. If a commercial etch is not available, vinegar may serve as a substitute.

18.4.3   Water-and-soap washing   This method is generally applied with a high-pressure portable unit similar to that used in a car wash. It is designed to remove dirt and grease from large areas. In a smaller, constricted area or in situations in which products could be damaged, a bucket, sponge, and water are more commonly used. Make sure that all surfaces are rinsed properly and given adequate time to dry before applying paint.

18.4.4   Solvent cleaning   This method is used in conjunction with water-and-soap washing. It is employed in areas where grease or stains are concentrated and are difficult to remove. Solvent cleaning can become costly, and it entails a disposal problem.

18.4.5   Sandblasting and water blasting   Sandblasting is expensive, but it is the most effective preparation technique. Basically it is used on metal to remove rust, and it is also used on concrete, blast glass, brick, and, in some cases, wood. It is employed to expose the aggregate of the concrete and give it texture. There are two basic types of sandblasting units. One is the Vacu-Blast, which uses a steel grit that is recycled through the machine. The other is the regular sandblaster, in which the sand or grit is not recycled but falls to the ground and in most cases is not reusable.

In the water-blasting method, water is applied at a very high pressure. This method is best used on concrete, as it may cause steel to rust.

18.4.6 Sanding   Surfaces should be properly prepared by sanding. This method is used not only prior to painting but also between coats. If surfaces that require sanding are not properly prepared, the finish job will be rough, especially if enamels are being used. The finish coat can be no smoother than the surface to which it is applied. On most finishing jobs sanding is necessary to remove unevenness and fuzziness from the surface. Power sanders can be used for large jobs.

18.4.7 Applying spackle, putty, and caulking   These products must be applied properly to assure a good job. Spackle is basically designed to assure an even surface for Sheetrock. Putty is designed to fill nail holes and cracks in wood, and caulking to fill cracks between different surfaces. These goods can be purchased in a stick, a squeeze tube, or a caulking tube or in bulk containers holding from ½ pt to 5 gal (from 0.237 to 18.93 l).

18.4.8 Bleaching wood   Bleaching is employed to make the color of wood uniform and also to lighten dark wood. There are two types of commercial bleaches. One consists of two solutions that are applied in separate coats, and the other of two solutions that are mixed together just prior to using. Use gloves to protect your skin. Also, regular household bleach may be used. In most cases bleach is needed after old wood has been stripped in order to remove dark spots caused by stain and age.

18.4.9 Paint stripping   Stripping is a method of removing old material from a surface. It is most commonly used on wood to restore the wood or to change its color. Strippers are used on many surfaces such as metal and concrete when sandblasting is not practicable because of the problem of waste removal. They are applied to the surface and allowed to set for a few minutes so that the chemical process may work on the paint and loosen it from the surface. It may be necessary to strip the surface two or three times to remove all the paint.

18.5 Application Equipment   For some coatings applications many different methods can be used, while for others only one type may be entirely suitable. The following is a list of the various ways in which a coating can be applied:

- Brush
- Roller (hand)
- Applicator pad
- Aerosol
- Dipping
- Flow coating

- Gloves or mittens
- Hot-air spray
- Cold-air spray (automatic or hand)
- Electrostatic air spray (powder or fluid)
- Fluidized bed
- Airless spray
- Airless electrostatic spray
- Machine roller coating
- Curtain coating
- Electrodeposition
- Bonding
- Transfer coating
- Barreling
- Silk-screen application

This *Handbook* serves architects and engineers who are concerned mainly with writing specifications covering brush, roller, and spray methods.

The paint specification must take into account the possible method of application, relating the method to actual conditions on the job. Faster drying may be required if there is an overspray problem or poor weather conditions. Brush or roller application may be necessary in small or hard-to-reach areas or when overspray cannot be tolerated.

18.5.1 Brushes   The brush is the oldest and best-known way of applying paint, but it is also the slowest. A good bristle brush will hold more paint than a nylon one, with less dripping and spattering, and will also put on a smoother coating and cut a clean, even edge. Nylon-fiber brushes are normally used to apply water-based paints. Nylon will last 3 to 4 times longer and will not go soft when water is applied to it, but it does not hold as much material as a bristle brush. In industrial painting the brush is used as an adjunct to spraying. Brushing is often employed to apply paint to a surface next to an area that does not require paint or is painted in a different color. It is also suitable for small items or in situations in which overspray cannot be chanced.

For commercial work as well, the brush is the backup for the roller and the sprayer. It must be used on many surfaces, such as window moldings, doors, and door casings, and for cutting in alongside different surfaces.

Residential painting requires much more technical work with brushes. There are a variety of jobs to be handled with the brush, from painting trim to graining or antiquing cabinets.

18.5.2 Paint rollers   Rollers have many advantages over brushes and some advantages over spraying. They are ideal for large areas and adapt well to hard-to-reach places like high stairways or high ceilings. They are also used to stipple finish paint. When there is a great amount of masking to be completed, rollers are often faster than spray equipment.

Rollers are not often used on industrial jobs unless spraying is impossible. Then they are used with brushes and mitts as a substitute for spraying. There are several types of rollers. Pipe rollers are rollers that when pressed on a pipe take the form of the pipe. There are also bucket or pan rollers. Texture rollers fall into this class. A texture roller leaves a design on the surface and part of the undercoat exposed. Carpet rollers impart a stippled effect. Mohair rollers give a very fine, smooth sprayed look. There is also a magazine type of roller that stores paint inside a hollow cylinder. A pressure roller feeds paint through a hose to the inside of the roller.

Rollers are often used in commercial and residential work, especially to apply stipple paints and to cover areas where the finish floor and trim have already been installed, and in applying block filler. They are frequently used because they are much faster than brushes and fill voids and pores just as well.

18.5.3 Spray painting   This is one of the fastest methods of applying paint, and it also is one of the quickest to lead to some kind of paint failure (lack of adhesion, holidays, etc.). Most manufacturers of paint-spraying equipment conduct courses on proper spraying techniques. It is not unreasonable, particularly on important jobs, to require that applicators have a certificate from such a course. Graduates will normally work more rapidly and will use proper spraying techniques to apply the right amount of paint to produce a properly painted surface of uniform film thickness.

Economically, the method selected to apply most coatings is either an air spray or an airless spray; generally, the choice is left to the contractor. Some coatings, however, have certain limitations. Do not apply inorganic zinc-rich paints, for example, by brush. These coatings are sensitive to mud cracking, particularly when brushed or when applied in too heavy a film thickness. While it is the job of the painting contractor to follow specific instructions for the application of paint, the specification writer should be familiar with air pressure, paint viscosity, temperature, humidity, type of spray equipment, and drying time between coats, because these factors are essential parts of good application practice. Certain solvents may be required to thin the paint; amounts and mixing instructions should be included as part of the coating specification. Most oil-based coatings require 24 h to dry between coats. When this drying time is necessary, scheduling the job in good summer weather is often important. Do not expect proper application of such paints in cold, rainy weather.

Spray painting was designed for industrial surfaces with many irregular shapes, sizes, and textures and with hard-to-reach areas. Industrial painting also involves equipment, structural steel, and piping. Spray painting is by far the most economical method of paint application. The big problem here is overspray. Again, spraying requires good equipment and trained operators. It is most economical on large areas that are not broken up by unpainted items. Sometimes much masking is required to protect unpainted surfaces or surfaces of another color. This can add to the cost and make spraying unprofitable.

Residential spray painting varies with the price of the house. Cheaper housing is completely sprayed, whereas in the most expensive houses about the only items to be sprayed might be cabinets, doors, exterior wood, grills, etc.

Airless-spray equipment has made big advances in the painting industry. It saves labor and applies 25 percent more material than regular spray equipment. It is possible to paint near edges without shields because the straight up-and-down fan shape can be closely controlled by the operator.

18.5.4   Miscellaneous application methods

- *Mitt painting.* This method is used for pipes and handrails in industrial work, mostly when spraying cannot be employed. A painter who is well qualified in the use of a mitt can work fast.
- *Staining with rags.* Doors and cabinetwork require wiping a stain with a soft rag so that the grain will show and the stain will be even.
- *Corner rollers.* These are used on 90° angles.
- *Pads.* While not used regularly, pads do a good job on wood surfaces.
- *Filling with rags.* After wood filler has been applied, it must be wiped with burlap to remove any excess.
- *Texturing tools.* Sponges, whisk brooms, and brush tips are used with other tools to texture and create designs in paint.

18.6   Guide to Applying Specific Coatings

Quite often the vehicle type (Table 18.3) determines the special application instructions that should become part of the paint specification (see box). From another viewpoint, certain special effects such as graining or marbleizing require application techniques that are almost lost arts.

COATING GUIDELINES FOR APPLYING ZINC-RICH PAINTS

Apply zinc-rich coatings only after all procedures set forth in the surface-preparation specification (sanding, abrasive, cleaning profile, inspection) are complete. Then apply the coatings system in accordance with the number of applications and the mil thickness required.

TABLE 18.3   **Application Characteristics of Common Paints**

Coating type	Possible problem areas	Application guidelines			
		Brush	Spray	Roll	Other
Coal tars	Necessity to recoat in a certain number of hours; odors	Only if needed	X		
Epoxies	Odors; short pot life; possibility of wasting material	X	X	X	
Latex-based acrylics	Possibility of roping up. If it is too warm, paint may bubble. Paint may dry too quickly for good control	X	X	X	Mitt pad
Alkyd-based paints	Paint skims over if left in the air in the pot for a long time	X	X	X	Mitt pad
Oils	Mixing with other products	X	X		
Stains	If wood has had water on it, spots may show up	X	X	X	Rag pad
Varnishes	Obtaining a smooth job without material in the air getting into the varnish	X	X	X	
Lacquers	Odors	Brushing type only	X		
Wood fillers	Getting the filler off the surface after application	X		X	Burlap
Block fillers: oil and latex	Getting voids filled	Sometimes required	X	X	
Texture coatings	Getting coating uniform		X		

When using automatic blasting equipment, the steel plate must be run through the wheel at speeds to ensure SSPC SP-10 near-white conditions.

Immediately after the steel has left the blasting unit, blow the surface with clean air to remove all dust and abrasive.

Apply enough zinc-rich paint to assure the dry mil thickness desired. Check the dry mil thickness between coats to assure the proper thickness of the primer and the topcoat respectively.

For all equipment in which assembly will create inaccessible areas for coating, such as gantry trucks, apply the full system of zinc-rich coatings prior to assembly.

When painting galvanized surfaces such as conduit, junction boxes, etc., make sure that the surface has been cleaned and degreased. Then apply primers and topcoats. If blasting has damaged the galvanized surface, use an appropriate primer to seal the scratches and damage.

When factory-finished equipment is used on exterior portions of equipment, including electric motors, switches, hydraulic equipment, etc., or when the paint or coating on the equipment is considered insufficient for exposure, sand the painted surface by hand or with power equipment to ensure the proper profile, and apply topcoats compatible with the existing coating. Some patch-test work may be necessary to make sure that additional coats will not lift the existing coating.

Take mil readings on each coating as applied, both by wet-film gauges and by Mikrotest gauges on dry film. Make sure that each coating is inspected and approved by the owner's representative before proceeding to the succeeding coating.

In *graining,* use the following steps. Sand the surface and then prime. Putty all voids. Then apply a second coat of enamel underbody and resand. Next apply a split coat of eggshell enamel and enamel underbody to match the ground color. Mix the glaze to the color of the grain, and apply it to the surface with a small brush or use graining combs to get the desired affect. Use a blending brush to pull out the grain; brush only with an upward motion. Use the same glaze, making it a little darker as an overgrain, and brush it on the surface. Then take a flogging brush and flog the entire area. After overnight drying, brush on one coat of sanding sealer. Sand with No. 220 sandpaper. Apply one or more coats of varnish in flat, satin, or high gloss. If desired, rub out with pumice and oil.

In *marbleizing,* apply a light cream base color in most cases. Use a piece of marble as a guide. When the surface is dry, paint the area with a thin white paint or apply the paint with a sponge. While the paint is still wet, put in patches of the desired color. Shade the paint outside the patches with a thinner mixture, using a clean rag. Dab lightly to soften the outline. Put in dark veins by using an artist's brush and then blend. Follow the dark veins and blend, using a thin white paint.

In *gilding* with metal leaf, first apply a clear sizing material. Then place the metal leaf on it and smooth it down. When metal leaf is used inside, it is generally varnished to prevent it from coming off. The metal leaf most commonly used in the painting industry is gold, imitation gold, palladium, or aluminum.

18.7 Application Problems   The problems discussed below are not always caused by poor application practice, but to make sure that application is not the cause of failure the specification writer should include specific information on correct application in his or her specification. To correct these problems see Chapter 19, "Troubleshooting and Inspection."

18.7.1 Blistering

▪ *Painting in hot sunlight.* The surface of the film dries too rapidly and traps solvent in the underlying film. Heat from the sun vaporizes the solvent, which expands and forces paint into a blister.

▪ *Application of an excessively thick coat of paint.* The surface film dries before the underlying layer, trapping the solvent.

18.7.2 Brush or roller marks

▪ *Similar marks in the previous finish.* Since the new coating will follow the contour of the previous finish, the surface must first be sanded smooth to eliminate marks.

▪ *Too porous a surface.* A poorly sealed surface absorbs too much vehicle, thus reducing flow and leveling. A primer, sealer, or undercoater may be necessary.

▪ *Excessive working of the wet film.* Excessive brushing or rerolling reduces flow, leaving applicator marks and causing an uneven film thickness.

▪ *Working in a draft and/or a high temperature.* Evaporation of solvents is accelerated before the film can level out.

▪ *Application of too little material.* Coatings are designed to flow and level at recommended spreading rates. When the application is too thin, the resulting insufficient body will not permit flow.

- *Insufficient drying time between coats.* Application of the second coat resoftens the original film and sets up a tacky condition that prevents normal flow.
- *Use of a poor-quality or incorrect paint applicator.* Incorrect or cheap brushes and roller covers fail to deposit a uniform and correct amount of coating, usually causing excessive working that inhibits flow.
- *Insufficient or wrong thinner.* The recommended type and amount of thinner for a particular material and application method maintain the desired flow and leveling.
- *Low temperature of coating, surface, or air.* The coating thickens at lower temperatures, with a resultant loss of flow.

18.7.3   Cracking, checking, scaling, and flaking

- *Thickness of the paint film.* The fact that the paint film usually is too thick may be due to the heavy application of a single coat. However, this problem is characteristic of older houses where many coats of paint have been applied, frequently without sufficient weathering time between coats to reduce film thickness, and the layers have become brittle with age.
- *Failure adequately to prime and protect pressed composition board and siding.* Composition siding comes in many varieties, often factory-primed, and should be reprimed immediately after installation and before applying finish coats. This type of board is especially sensitive to moisture, and the softer varieties will swell and crack. Cracks in the board will eventually become cracks in the paint.

18.7.4   Bubbling and cratering

- *A new roller cover.* A new cover should be saturated with solvent or water and rolled out to expel air before use; otherwise bubbles may be introduced into the wet paint.
- *Overshaking.* Overshaking causes excessive foaming and bubbling. This is more likely to occur in water-thinned coatings than in solvent-thinned types. Let the material set until bubbles and foam disappear.
- *High temperatures during application.* Material dries too quickly, preventing bubbles from breaking and flowing out. Add a small amount of thinner.

18.7.5   Crawling

- *A surface that has not been thoroughly cleaned.* Wax, oil, grease, silicone, residue from detergent or soap solutions, and other contaminants that are left on the surface will cause crawling.
- *Use of a liquid preparation to dull a glossy, solvent-thinned finish before applying a latex coating.* Oils are brought to the surface of the previous coating, causing the latex to crawl.

• *Moisture on or in the surface.* Dew on exterior surfaces is a frequent cause of crawling when painting early in the day. Moisture may be present on interior surfaces because of condensation.

18.7.6 Uneven gloss

• *Inadequate sealing of a surface that varies widely in porosity.* Too much vehicle from the finish coat sinks into the surface, leaving an excessive amount of pigment, too little binder, and a corresponding loss of gloss. This may occur on repainting as well as on new work.

• *Varying film thickness.* Gloss or sheen is raised with increased film thickness. Two coats will have a higher gloss or sheen than one coat.

• *Moisture on the paint film during the drying cycle.* Moisture is common on exterior surfaces painted with high-gloss finishes late in the day. Nighttime dew or frost touches areas of the paint film, and flatting results. Moisture may also occur on interior work because of condensation when warm air comes in contact with a cold outer wall, condensation drops from water pipes, etc.

• *Uneven or rough surface.* Light rays are reflected at different angles from rough surfaces such as building blocks, resulting in areas that differ widely in gloss or sheen.

• *Temperature variation during the drying cycle.* Localized heat caused by radiators, electric wall heat, etc., will cause variations in gloss or sheen.

• *Overlap areas and streaks of heavy paint film caused by uneven brushing.* Increased film thickness increases gloss or sheen. The effect is exaggerated in deep colors.

18.7.7 Holidays and poor hiding

• *Too much surface and too little paint.* Measure the number of square feet covered per gallon. Then refer to the recommended spreading rate listed on the paint label, making allowances for the type of surface, whether smooth or porous.

• *Radical color change.* A radical change in color may require a prime coat tinted toward the finish color. A white or off-white background may be more difficult to hide than a light-colored surface.

• *Porous underlying surface.* If the surface is sufficiently porous, the paint will "strike in" (penetrate), carrying hiding power and color with it. Priming may be necessary.

• *Incorrect or poor-quality brush or roller.* The applicator manufacturer's recommendations for type, nap size, etc., should be followed. An applicator of poor quality or of the wrong type will not provide the uniform and proper film thickness necessary for good hiding.

• *Excessive thinning of paint.* Excessive thinning will result in spreading paint over too large a surface and will reduce the hiding power available. Refer to the recommended thinning listed on the paint label.

▪ *Inadequate mixing of paint.* Pigment that settles on the bottom of the container is a loss in available hiding power.

18.7.8 Lap marks

▪ *Working for too long a time in one area.* Coating begins to set and produces a thicker film wherever overlapping occurs, altering color and sheen.

▪ *Too much heat and/or draft during application.* Rapid loss of solvent speeds the set time and results in a thicker film wherever overlapping occurs, altering color and sheen. This problem frequently occurs on exterior applications.

▪ *Sealing or priming required by surface.* A porous or open surface will absorb too much vehicle from the coating, resulting in spots that have a different sheen if the porosity is uneven.

▪ *Spreading material too far.* Check the paint label for the recommended spreading rate. An overly thin film sets up at a rapid rate, produces a thicker film where overlapping occurs, and thus alters color and sheen.

▪ *Use of improper thinner.* A solvent that evaporates too rapidly shortens the working time, producing a thicker film where overlapping occurs, thus altering color and sheen.

18.7.9 Lifting

▪ *Application over a film that appears sound but has spots of poor adhesion* (*blisters sometimes shrink and flatten with age*). Shrinkage of the new coating as it dries applies pressure to the previous coating and dislodges areas of poor adhesion.

▪ *Application of a coating containing solvents stronger than the solvents in the previous coating.* Strong solvents resoften the previous coating and loosen the bond. As the new coating dries, the shrinkage pulls the original coating away from the surface.

▪ *Application of a finish coat before the previous coating is thoroughly dry.* Drying of the final coat will trap solvents in the first coat; lifting and/or wrinkling usually results.

▪ *Application of a hard coating over a softer coating.* The contraction of the finish coat as the film dries pulls the softer underlying film away from the surface.

18.7.10 Peeling

▪ *Inadequate cleaning and preparation of the surface before painting.* Check the surface behind the paint film for evidence of dirt, mildew, rust, wax, oil, residue from detergent solutions, or other contaminants.

▪ *Failure to dull glossy surfaces by sanding.* Check the area behind the peeled surface to determine whether gloss is present.

• *Exterior painting over a chalky surface.* Examine the back side of the flakes and the surface from which the flakes have been removed. White powder indicates chalk, which does not permit sufficient penetration to establish firm contact with the solid surface underneath.

• *Failure to apply a finish coat for several months after primer has been applied over an exterior surface.* Exterior primers are not formulated to weather as a finish coat does. Premature chalking may occur, preventing penetration of the finish coat; or dirt, grease, oil, etc., may contaminate the surface and prevent a good bond. When applying catalyzed coating materials, it is imperative that the inspector guarantee that not too much curing time elapses between coats. Manufacturers commonly report that this time lapse should not exceed 7 days. In southern latitudes even this period is too long, and a maximum curing period of 48 h should be observed. Excessive curing time between coats induces film lamination and early failure.

• *Resin streaks or knots in wood.* These areas require special preparation and priming. However, extremely resinous areas may cause continuing problems.

• *Undereave painting over salts.* Water-soluble chemicals are leached from paint films by constant wetting and drying. As moisture evaporates, these salts are deposited on the surface. In exposed areas, weathering removes the deposit, but it accumulates in protected areas, resembling a white powder on dark colors but being hardly noticeable on light tints or white. Removal requires detergent washing and, particularly, flush rinsing under a high-pressure hose.

• *Failure to remove rust from metal or to paint immediately following cleaning.* Examine the back of paint chips for rust. Paint may bond to rust, but rust is an unstable surface that may slough off, bringing with it the paint film. Since air and humidity cause metal to rust quickly, immediate painting following cleaning is necessary.

• *Failure to etch or properly prime galvanized metal.* Galvanized surfaces require either etching or special primers to assure the bonding of paint.

• *Failure to prepare properly and seal a masonry surface with an alkali-resistant primer.*

18.7.11   Sagging

• *Application over a hard and/or glossy finish that should have been sanded dull to provide tooth to which the new coating will adhere.* Sanding is always the safest procedure. Liquid-type preparations that are designed to soften and dull an old hard coating do not always work. Many types are not suitable for latex finish coats.

• *Application of too much paint.* Measure the number of square feet covered per gallon to determine if the spreading rate is excessive.

Amateur brushing, especially on cabinets and panel doors, is a frequent cause. Also, there may have been an attempt to force one-coat hiding when it was not practicable.

■ *Excessive thinning.* Determine if the paint was thinned, with what, and by how much. Paint stored in a cold area will increase in viscosity and sometimes is thinned before it is brought back to room temperature.

■ *Residual film left from a washing compound (detergent type or other).* This occurs most frequently on walls and woodwork but also is found on machinery, fixtures, etc. Inadequate rinsing may leave a film that interferes with a bond directly to the substrate or to the previous coating.

18.7.12 Staining

■ *Contaminants on or in the surface being painted.* Unless they are removed or sealed off, many types of surface contaminants will dissolve and become part of the fresh paint coating or will bleed into the dry film. Among typical sources of stain are oil, grease, spatters, pen and pencil marks, coloring matter left in the substrate from water leaks such as ceilings, etc.

■ *Painting over wood that contains large amounts of water-soluble coloring matter such as redwood and red cedar.* This problem occurs most frequently when painting with latex, but it may occur in a solvent-thinned paint as well.

■ *Staining problems caused by building materials such as asphalt shingles, creosote-treated wood and wood shingles, composition siding made from pressed materials, etc.* The almost endless variety of building materials requires careful analysis and experience to determine whether the surface can be properly primed and coated to prevent staining.

■ *Migration of plasticizers from synthetic-rubber and plastic surfaces.* Examples are plasticizers that migrate into a paint film from vinyl molding, synthetic-rubber tires, masking tape, etc.

18.7.13 Wrinkling

■ *Application of too much paint.* The surface of the film dries, but the bottom layer remains soft.

■ *Application of a second coat before the first coat is thoroughly dry.* The finish coat shrinks as it dries and wrinkles the softer, partially dried coating underneath.

■ *Painting in the hot sun.* Even though a proper amount of paint has been applied, forced rapid drying of the surface traps solvents in the underlying layer, which remains soft.

■ *Application of a hard finish over a softer coat of paint (for example, a trim enamel over a house-paint primer).* Improper choice of a primer-finish system is a frequent cause of wrinkling.

■ *Wrinkling when paint is applied over a cold surface.* In this case the surface of the film skin dries, while the colder bottom area remains soft.

■ *Painting over a glossy surface.* Failure to bond to the previous coat will result in wrinkling as the new film shrinks upon drying.

18.7.14   Weather problems   The best time to paint is when the weather is warm and dry with little wind. Obviously, many coating projects cannot be delayed until these ideal conditions prevail. The following are common problems:

Wet or damp weather   When humidity is high, condensation often occurs. Condensation on the substrate interferes with the bonding of the coating. Condensation on the surface of a freshly applied coating may alter its curing process or change the gloss of the finished paint job.

Dry weather   Very low humidity can also be a problem, especially with water-based products. Rapid flash-off of the water may result in film cracking. Low humidity can also cause poor curing rates for solvent-based and ammonium types of inorganic coatings. In addition, lap marks may appear on siding because the paint does not maintain a wet edge during application.

Low temperature   At low temperatures the film thickness of high-build, or thixotropic, coatings becomes more difficult to achieve. Curing reactions slow down or stop for many materials. Water-based products may freeze. Solvents evaporate more slowly. Furthermore, when relative humidity exceeds 70 percent, condensation is almost sure to develop.

High temperatures   Although heat has many beneficial effects in the application of coatings, it often increases overspray (dry fallout), trapped air or solvent bubbles, and, in the case of zinc inorganics, the incidence of film cracking. It also reduces the pot life of catalyzed materials.

Strong winds   Wind is a nuisance, particularly in spray painting. As the material leaves the spray gun, it can be deflected from the target. Solvent tends to flash off, creating excessive dry spray at the edges of the spray pattern. Lap marks become more evident, and dirt and other debris may become embedded in the wet film.

Dew and condensation   Moisture problems can occur when humidity is high. If possible, avoid painting surfaces whose temperatures are below that of the ambient air. Unfortunately, on large-scale projects primers are often applied late in the workday and sometimes at night. Abrasive blasting is a slow process, while applying a primer by spray is very rapid. Because of this wide difference in work rates, the contractor may take 6 h of an 8-h workday to prepare steel for 1 to 1½ h of primer application. A contractor who follows this practice should assure that evening or overnight condensation does not create a painting problem. This is especially true in tank-lining work.

18.7.15 Overspray   Serious problems result whenever painters do not consider the consequences of cleaning or spray-painting a surface, especially in high places. Chemical cleaning agents, blasting material, or paint can drop onto cars, other equipment, or personnel, causing injury, damage, and legal problems. In addition, some coating materials, particularly lacquer-type products, have a propensity for overspraying when applied by a painter of limited ability who tends to arc the spray gun at the end of each pass. The inspector should look for dry spray and insist that it be removed by sanding before application of the next coat of the coating system.

18.8 Inspecting Paint Application   Key factors to assure a quality job that meets specifications include the following:

- Examine equipment, rigging, and air supply (for painting and sandblasting).
- Check quality of metal-surface preparation.
- Check quality of abrasive.
- Determine anchor-pattern profile (see Chapter 6, "Surface Preparation: Part II").
- Measure paint film thickness (average five or six readings per square foot) for each coat on the areas being measured.
- Examine total paint film thickness.
- Inspect coverage around sharp edges, crevices, welds, and fillets.
- Test the paint itself. Check total solids, viscosity, and total pigment-volume concentration (PVC). Check vehicle by using infrared techniques.
- Do not allow poor workmanship or unclean, unsafe, or hazardous conditions.

Simple field tools are readily available to complement laboratory test procedures. Recent (1979) rule-of-thumb surface-preparation and application costs are shown in Figs. 18.3, 18.4, and 18.5.

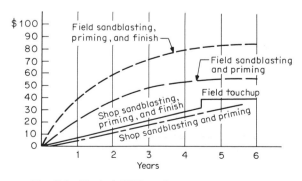

Fig. 18.3   Typical 1979 surface-preparation costs.

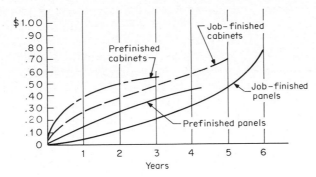

Fig. 18.4 Economics of using prefinished wood compared with job-finished wood.

In addition to job inspection, not less than 9 months or more than 12 months after job completion all coated surfaces should be reinspected. All areas that are found not to meet these specifications shall be repaired. The manufacturer of the specific coating materials used will recommend suitable procedures.

FIELD TEST TOOLS

- *Knife and/or spatula.* This tool is used to determine the adhesion of the film to the substrate and the flexibility or brittleness of the film (it indicates the condition of the coating).
- *Surface thermometer; temperature sticks.* These indicate the temperature that the coating must resist or tolerate during application.
- *Continuity tester.* This instrument is important in checking the porosity or continuity of the film.
- *Hardness pencils.* These are used to determine the hardness or softness of the film, which is a factor in recoatability.

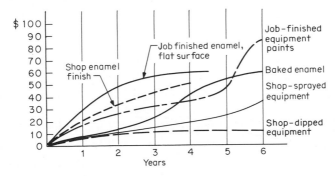

Fig. 18.5 Cost of applying enamel finishes.

> ▪ *Magnifying glass.* This is used to inspect the condition of the film and the substrate so as to determine the quality of surface preparation, the number of coats on the surface, and the presence of pinholes in the film.
>
> ▪ *Dry film thickness gauge.* This is used to determine the total film thickness and the thickness of the primer.
>
> ▪ *Spot tests.* Solvents are used to determine surface conditions or the type of coating present. Methyl ethyl ketone (MEK) will not attack cured epoxies and urethanes. Copper sulfate is an indicator for bare steel. An inspector should carry varnish remover to remove the coating in order to inspect the substrate for underfilm corrosion.
>
> ▪ *pH paper.* This is used to check the acidity or alkalinity of the surface.

18.9 Safety Precautions   When coatings are applied inside tanks or confined spaces where air circulation is limited, a safety engineer must verify the operational procedures. These may become part of the specification, especially if it is necessary to protect personnel against toxic hazards or to prevent fire or explosion.

Sufficient ventilation must be provided to keep the concentration of solvent vapor below the lower explosive limit. The term "lower explosive limit" refers to the volume of solvent vapor in air that is just below the point at which an explosion will occur if the mixture is ignited with a flame or spark. If the concentration of solvent vapor does not reach the lower explosive limit, no explosion can occur. Adequate ventilation is therefore the key to safe application, and it follows that if the applicator provides adequate ventilation and takes recommended precautions against open flames and sparks, hazards from fire or explosion will be negligible.

Vapors from all organic solvents are unsuitable for breathing in high concentrations. It may or may not be possible to detect the presence of harmful concentrations of vapors by sharp or pungent odors for any given solvent. Instruct employees to watch for warning signs such as a definite stinging sensation in the eyes and nose. This will occur at concentrations well below a toxic level, thereby warning that more ventilation is needed. Although many solvents are not extremely toxic, all workers should wear compressed-air masks, and adequate ventilation should be provided to keep fumes below explosion levels.

The following rules summarize the precautions to be observed when applying or using solvent-based products in tanks or confined areas:

1. *Provide sufficient ventilation to keep solvent concentrations in all parts of the tank below the lower (or upper) explosive limit.* Figure 18.6 shows the recom-

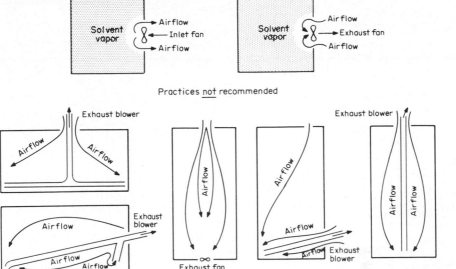

Fig. 18.6 Typical venting practices. Consult a safety engineer before starting any hazardous work. [*Reliance-Universal, Inc., Houston, Texas*]

Blower Capacities Required to Maintain Vapor Concentration Well Below Lower Explosive Limit

Volume of tank (gallons)	Required blower size* (cubic feet per minute)
500– 5,000	1,000
5,000– 20,000	2,000
20,000– 100,000	5,000
100,000– 250,000	10,000
500,000	15,000
1,000,000–2,000,000	20,000

* All blowers to be suction type.

mended blower sizes for a variety of tank sizes. This ventilation must be maintained during the entire application period and for at least 3 h thereafter.

2. *Require the use of an approved compressed-air mask by all workers in tanks and closely confined areas.* A compressed-air mask is the only type that should be used.

3. *Remove the solvent vapors from the tank by suction.* Do not force air from the outside into the tank. Most paint solvents are heavier than air and therefore tend to settle in the lowest part of the tank. In setting up the ventilation system, the most remote and the lowest areas should receive

special attention. Figure 18.6 shows the practices that are recommended and those that are not recommended.

4. *Use explosion-proof and spark-proof equipment.* All electric cable, motors, and lighting equipment must be of an approved explosion-proof type. No electric junction boxes should be permitted inside the tank. All droplights used by workers must be explosion-proof. Workers should be cautioned not to cut or stretch electric cables since sparks will result if a cable parts. Within the hazardous area, all metal equipment and hand tools must be of a nonsparking type, and workers' shoes must have rubber soles and heels.

5. *Prohibit smoking, matches, flames, or sparks of any kind.*

6. *Provide properly treated respiratory air.*

BIBLIOGRAPHY

Application Techniques, Physical Properties, and Chemical Resistance of Chlorinated Rubber Coatings, National Association of Corrosion Engineers, Katy, Tex., 1975.

Barker, J. L.: *Inspection,* Eighth Liberty Bell Corrosion Course No. 4, Drexel University, Philadelphia, September 1970.

———: "Painting Economics," *Materials Protection and Performance,* August 1972.

Glossary of Terms Used in Maintenance Painting, National Association of Corrosion Engineers, Katy, Tex., 1965.

Gross, William F.: *Application Manual for Paint and Protective Coatings,* McGraw-Hill Book Company, New York, 1970.

Industrial Maintenance Painting, 3d ed., National Association of Corrosion Engineers, Katy, Tex., 1970.

A Manual for Painter Safety, National Association of Corrosion Engineers, Katy, Tex., 1963.

Painting, Inside and Out, U.S. Department of Agriculture Home and Garden Bulletin 222, October 1978.

Tatton, W. H.: *Industrial Paint Application,* D. Van Nostrand Company, Inc., Princeton, N.J., 1964.

Vanderwalker, Fred N.: *House Painting Methods with the Brush and Spray Gun,* Frederick J. Drake & Co., Inc., Chicago, 1945.

Union Carbide Uniform Paint Evaluation Manual, New York, various dates to 1979.

U.S. Army Corps of Engineers, *Painting Hydraulic Structures and Appurtenant Works,* CE-14900.

Ziegeweid, J. E.: "Applying Organic Coatings," *Metal Finishing,* July 1979, pp. 37–41.

Chapter 19

Troubleshooting and Inspection

NEIL B. GARLOCK
Consultant

19.1 Introduction   Most paint problems are avoidable. This chapter suggests practices for remedial action once problems do develop.

19.2 Inspection during Construction   Many difficulties encountered in older buildings are the result of poor construction practices. Competent inspection during construction is very important. Problems also appear after years of weathering; these are not visible early in the life of a coatings system.

A valuable tool in solving paint problems is the *Exposure Standards Manual*. It defines, describes, and gives photographic standards for many of the common paint film defects, including blistering, chalking, cracking, erosion, corrosion, flaking, mildew, rust, abrasion, and chipping. Prepared in cooperation with ASTM, it is available from the Federation of Coatings Technology, 1315 Walnut Street, Philadelphia, Pennsylvania 19107.

Water, either rain or high humidity, can cause problems which may not be apparent until after a building has been occupied. Groundwater should be stopped outside the building either by waterproofing the outside of the foundation or by providing adequate outside drainage. Caulking all joints and constructing a tight roof also help to avoid problems. Do not apply paints to masonry until it is dry.

Wood, especially plywood, requires weather protection. Inspect wood surfaces properly.

Apply paint to plaster only after it has dried properly; otherwise paint problems will develop. Likewise, if chalk appears on a surface before the finish coat is applied, it must be removed with a detergent solution and a stiff brush. Then wash the surface with clean water, and apply the topcoat by brush to wet any remaining chalk.

Paint does not stick to galvanized steel without proper pretreatment or priming. Improperly cleaned brick causes paint problems, and problems also occur when concrete is not properly treated. When a muriatic acid wash is used, the surface must then be neutralized. A common error is to assume that all surfaces are clean when they are turned over to the painter.

Steel structures and appurtenances need especially good inspection. Paint is an effective barrier to the oxidation of steel and provides longtime protection, but rust must be removed prior to painting.

Fig. 19.1   Keane-Tator comparator. [*KTA-Tator Associates, Inc., Coraopolis, Pennsylvania*]

Fig. 19.2 Nordson wet-thickness gauge. [*KTA-Tator Associates, Inc., Coraopolis, Pennsylvania*]

Steel comes from the mill with an oxide coating called mill scale. While tight mill scale forms a good seal for steel, mill scale rarely is completely tight and free from holes. Corrosion will be aggravated at holes or breaks. Clean steel surfaces to remove scale and any incipent rust. Do not rely on weathering as a method to clean steel.

If abrasive blasting is used, it is important to determine how deeply the blasting has indented the metal. It is not unusual to have blast patterns 3

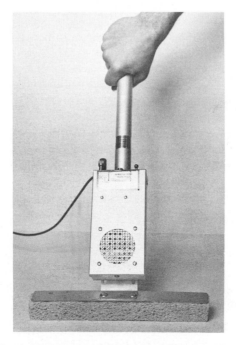

Fig. 19.3 K-D bird-dog hole detector [*KTA-Tator Associates, Inc., Coraopolis, Pennsylvania*]

Fig. 19.4   Elcometer thickness gauge. [*KTA-Tator Associates, Inc., Coraopolis, Pennsylvania*]

mils (0.0762 mm) thick, and if only 3 mils of paint is applied, the tops of the indentations will not get as much paint as the valleys. By using a profile comparator to determine blast depth, the specifier can determine the number of coats required for good coverage (Fig. 19.1).

After paint has been applied, determine film thickness by means of a wet-thickness gauge. Then if additional paint is needed, it can be applied while the painter is still available (Fig. 19.2).

A single coat of paint contains pores that can allow moisture to penetrate the film, causing corrosion or peeling. This is largely prevented by applying two or more coats of paint. While the pores are not likely to line up, holidays (spots where paint is thin or missing because of skimpy application) can still occur. The bird-dog hole detector (Fig. 19.3) is an excellent inspection tool to reveal holidays so that they can be covered with paint. Usually it is the primer which should be pore-free. It may be necessary to apply several coats to attain this state.

After a coat of paint has dried, dry film thickness is important, for it largely determines how long the structure will be protected. There are several dry-thickness gauges. Often they depend on the fact that a film between a magnetic metal and a magnet reduces the attraction of the

Fig. 19.5   Inspector thickness gauge. [*KTA-Tator Associates, Inc., Coraopolis, Pennsylvania*]

TABLE 19.1   Inspection Instruments

Instrument	Description	Use
Elcometer	Pocket size. Magnetic attraction of the substrate is controlled by the thickness of the coating and measured on a dial	Measure dry coating thickness by holding two prongs against the coated surface. Use instrument on any coated magnetic surface
Inspector gauge	Tubular; 9 in (228.6 mm) long. Magnetic attraction of the substrate is measured by adjusting a knurled knob until contact with the surface is broken	Measure dry coating thickness by setting the dial so that contact is made. Back the dial until contact breaks. Use instrument in any position
Pinhole detector	Low voltage demonstrates contact with the conductive surface when a break in the coating is encountered	After paint is dry, the instrument is passed over the coated surface. A spark indicates a break in the coating
Wet-thickness gauge	Plate with notches of different depths in its edge	A notched edge is pushed through a fresh coating until contact is made. The boundary between wet and non-wet prongs shows the film thickness
Profile comparator	Calibrated graduations are compared microscopically with the anchor pattern of the blasted surface	With the aid of a portable magnifier the depth of the anchor pattern is determined by comparison

magnet for the metal. The reduction is related to the thickness of the paint film (Figs. 19.4 and 19.5 and Tables 19.1 and 19.2).

PRECAUTIONS IN MEASURING SURFACE PROFILE, FILM THICKNESS, AND POROSITY

Recognize that preparing a surface to an incorrect profile has certain adverse effects. Excessive peaks and valleys in a surface are a particular problem with thin coats because the peaks will not receive enough paint.

TABLE 19.2   Suppliers of Test Instruments*

Elcometer	
Caltaro Scientific Control Co. Route 1, Box 236 Chantilly, Virginia 22021	Gardner Laboratory, Inc. P.O. Box 5728 Bethesda, Maryland 20014

Inspector gauge	
Kenneth Tator Associates 2020 Montour Street Coraopolis, Pennsylvania 15108	Nordson Corporation Amherst, Ohio 44001

Pinhole detector	
Kenneth Tator Associates 2020 Montour Street Coraopolis, Pennsylvania 15108	Gardner Laboratory, Inc. P.O. Box 5728 Bethesda, Maryland 20014

Wet-thickness gauge	
Nordson Corporation Amherst, Ohio 44001	Kenneth Tator Associates 2020 Montour Street Coraopolis, Pennsylvania 15108

Profile comparator	
Kenneth Tator Associates 2020 Montour Street Coraopolis, Pennsylvania 15108	Gardner Laboratory, Inc. P.O. Box 5728 Bethesda, Maryland 20014

Clemtex, Ltd.
248 McCarty Drive
Houston, Texas 77020

* These are not the only instruments available for paint testing.

Problems also occur when instruments are used incorrectly, beyond their limitations, or when calibrated incorrectly. For example, if an Elcometer is set for a given piece of steel, the shim rests essentially on top of the profile. Thus, in subsequent tests the thickness of the coatings is the film thickness above the profile tips. Mikrotest units calibrated on smooth steel, on the other hand, may read film thickness as much as 1 mil (0.0254 mm) below the profile tips on steel blasted to a profile of 4 mils (0.1016 mm). While normal calibration gives readings of about 0.5 mil (0.0127 mm) below the blast-profile tips, the specification writer must take instrument tolerance into consideration when calling for minimum film thickness. Furthermore, this tolerance can affect the amount of paint needed for a particular job.

Moreover, while specification writers do not normally concern themselves with the details of blasting, they should recognize that unless a compressor of sufficient size is used, there will not be enough pressure at the nozzle to perform the work efficiently. This causes slower production and increases the cost per square foot of a sandblasting job, and it also results in an insufficient (shallow) anchor profile that can lead to loss of adhesion and thus shorten coating life. In contrast, deep blast profiles can create peaks that penetrate the coating system, creating pinholes that open the surface to attack from moisture or chemicals. Thus, coating thickness alone, important as it may be, cannot guard entirely against film degradation. Inspection procedures should couple film-thickness measurements with porosity-detection instrumentation.

19.3 Maintenance Costs

19.3.1 Ferrous surfaces
There are sound economic reasons for preparing a surface properly before a paint coating is applied. Further, it is good business not to let a ferrous surface become rusty and pitted. Such corrosion will happen if the surface is allowed to remain unprotected.

Industrial Maintenance Painting[1] gives costs for preparing surfaces that have different degrees of corrosion. In one case a surface has been primed, presumably at the factory, and has just begun to rust. In a

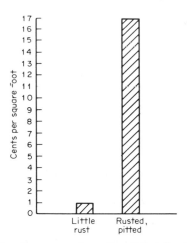

Fig. 19.6 Costs for cleaning ferrous surfaces (in 1975 dollars): surface with little rust, 1 cent per square foot; rusted and pitted surface, 17 cents per square foot.

[1] Paul E. Weaver, *Industrial Maintenance Painting*, 3d ed., National Association of Corrosion Engineers, Houston, Tex., 1967.

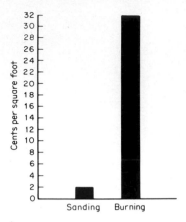

Fig. 19.7   Costs for cleaning wood surfaces (in 1975 dollars).

second case the surface has severely rusted and pitted. Values are based on cleaning 4000 ft² (371.6 m²) in a tank. The slightly rusted tank is cleaned by hand, and the worker is paid $5 per hour. Sandblasting is required for the other tank, at 17 cents per square foot. If there were slightly more rust than in the first case, abrasive blasting would suffice, at a cost somewhere between the costs of little rust and pitted rust (Fig. 19.6).

On ferrous surfaces, the specification writer should include curing instructions when zinc-rich primers are applied. These coatings cure from the top and require moisture. When they are inadequately cured, a hard top such as an alkyd or an epoxy ester will pull away from the soft interior. Desertlike climates may actually require a daily water spray for as long as 3 months.

19.3.2   Wood surfaces   It is well established that in the long run lack of maintenance costs more than good maintenance of a painted surface. Unfortunately there is not a straight-line relationship between the costs of good and bad maintenance. A contractor can use cheap labor to sand and putty a house in good condition, but an experienced worker is needed to burn off the paint. The *Estimating Guide* of the PDCA[2] presents factors for figuring costs per square foot for sanding and puttying versus burning (Fig. 19.7), as does the McGraw-Hill Cost Information Systems 1980 *Dodge Manual for Building Construction Pricing and Scheduling.*

[2] *Estimating Guide,* 2d ed., Painting and Decorating Contractors of America, Falls Church, Va., 1969.

19.4   Specific Paint Problems

19.4.1   Wood surfaces   Wood surfaces experience a variety of problems including some that are strictly cosmetic and others that have severe economic consequences.

Plywood   Plywood is frequently used for siding in which either board-and-batten construction or unrelieved sheets of plywood are used. The boards are sensitive to moisture, and if they are not protected until they are painted, serious problems can result. In some developments house sides are constructed as units. If plywood is the siding material and the units lie in a builder's yard for a time, paint applied to the wood after erection will usually peel in a short time. When the plywood is exposed to the weather, small wood checks form, and when paint is applied, these checks usually are not filled. Moisture enters the checks through the cracks at their edges, and paint will not stick to the wet wood. If checked plywood is brush-painted and care is taken to wet the cracks, less peeling will develop, but it is far better to prevent checks if possible (Fig. 19.8).

Some lower-grade plywoods contain inserts. These are wooden plugs, known as boats or cat eyes, that are used to replace imperfections in the wood and are very hard to paint so that they are not noticeable.

Protection from weather helps prevent plywood checks, but humidity may still cause checks to form. Stresses created in plywood during its manufacture also cause checking. Treatment of plywood with a water-repellent solution will greatly reduce checks. Two common types of repellent solutions are available: pentachlorophenol and copper naphthe-

Fig. 19.8   Plywood cracking.

Fig. 19.9   Untreated plywood exposed for 2¾ years.

nate. Sometimes contractors use plywood which has been patched. The oval-shaped patches show through paint if moisture gets to them. Specify a paint system or a repellent solution to avoid plywood-patch problems (Figs. 19.9 and 19.10).

Some plywood is factory-primed. If such sheets are exposed to weather before topcoating, chalk may develop and cause adhesion problems. If the chalk is more than a slight dust, take remedial action.

Fig. 19.10   Treated plywood exposed for 2¾ years.

The American Plywood Association, 1119 A Street, Tacoma, Washington 98401, is conducting an exposure program to determine the best paints for plywood. Its experiences are available on request.

Knotty siding   For obscure reasons builders sometimes use wood that contains numerous knots for siding. Frequently the knots are ignored during painting, with the result that the rosin in the knots makes a definite pattern on the paint (Fig. 19.11).

It is possible to seal knots so that they will not bleed. One of the best sealers is common shellac or pigmented white shellac, which is applied to the bare knots before any paint is applied. Some authorities have recommended aluminum paint to seal knots, but its use can be risky. The vehicle in some aluminum paints will dissolve rosin and aggravate bleeding. Aluminum-pigmented shellac has been used successfully. It is also possible to heat the knots with a propane torch or a blowtorch and thus boil out enough rosin so that bleeding is not likely. A pigmented primer-sealer conforming to the Western Wood Products Association's formula WP-578-P is another suitable material. Whatever method is used, bleeding should be prevented; such blemishes on a paint job are unnecessary.

Southern pine   Southern pine, which is usually lower-priced than other siding, is widely used in some parts of the United States. Unless care is taken long before the wood is used, painting southern pine may be unsatisfactory. The tree is a fast-growing pine with wide growth bands and considerable rosin. Both of these factors contribute to adhesion difficulties. Southern pine is cut as drop siding, in contrast to redwood and red cedar, which are cut as bevel siding.

If southern pine is sealed at the mill in plastic and the plastic is not broken until the wood is ready to be painted, adhesion will be better. This

Fig. 19.11   Unsealed knots.

adds to the cost of the siding, and the chief advantage of southern pine is thus lost. However, a homeowner with peeling problems would have been willing to pay a premium had he known of the advantages of a plastic seal (Fig. 19.12).

When a house sided with southern pine has peeling problems (as most such houses do), the best way to keep it looking good is to maintain paint on the wood at all times. The wood is affected by weather. However, keeping paint on the siding may be discouraging, for when the house begins to peel, it is very difficult to discover all the loose paint. This is true because there is nearly always some paint which has lost adhesion but is lying flat on the wood. If this paint is not removed by scraping, it will be lifted when a fresh coat of paint is applied over it. The new paint will shrink slightly, causing the old paint to peel and pushing the new topcoat

Fig. 19.12   Peeling on southern yellow pine.

off as well. Peeling on a newly painted house is frustrating, but bare areas should be painted to avoid costly water damage. If such careful maintenance is carried out, a house sided with southern pine will at least be nice-looking.

Hardboard   Problems with painted hardboard are due to several causes. Hardboard is reconstituted wood which is reduced to small particles and then pressed and rolled into boards. By their nature, the boards are quite susceptible to water absorption, and one cause of paint problems is inadequate sealing of the boards. If the paint is thin because too few coats have been applied or because it has weathered, water getting to the boards will probably cause peeling.

There are two types of hardboard, tempered and standard. Tempered hardboard is much superior for exterior use, but since the fact that it is tempered is not obvious to an owner, it is very important to keep surfaces adequately painted. Specification writers should call for tempered hardboard for exteriors.

Sometimes hardboard is factory-primed. The primer is probably a good sealer, but it can be damaged and it will be sawed. Exposed areas (especially edges) must be painted to avoid problems.

For interior construction, moisture is not so great a problem, but in the presence of high humidity particles may lift from the boards and be unsightly. Sufficient paint to seal the boards will prevent particle lifting.

Windowsills   Paint peeling and wood cracking are common on wooden windowsills. New sash lumber is treated with water-repellent chemicals, but this treatment is not permanent. It is gradually expended, and then water gets to the wood, causing both wood and paint checks. Water

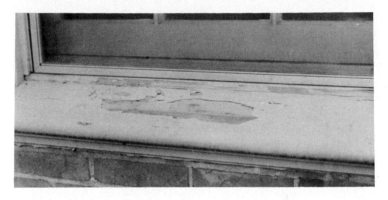

Fig. 19.13 Windowsill peeling.

stands on sills much longer than on vertical surfaces, and if the paint film is thin or cracked, water gets through (Fig. 19.13). Since paint receives more sunlight on the horizontal sills, it weathers more rapidly and requires more frequent maintenance.

It is best to paint a windowsill before peeling starts. If peeling occurs, the wood is likely to be cracked, and wood cracks are hard to seal with paint. If the peeling extends over most of the sill, it is wise to remove all the paint possible. Then all bare wood should be flooded with a water-repellent solution, which should be allowed to soak into the wood to the greatest extent possible. When the repellent is dry, two or, preferably, three coats of paint should be applied. After that, careful maintenance will reduce and eventually stop the peeling.

Interior high humidity   Probably the most common cause of paint peeling on the exterior of a house or other building is excess humidity in the interior. Water from the moist inside air condenses on the inside, unexposed surface of the siding, wets the siding, and destroys the adhesion of

Fig. 19.14 Peeling outside a bathroom.

the paint. The paint then peels; in essence the paint is pushed off by moisture trying to escape to the outside.

Condensation on siding, caused by moisture buildup in a tightly closed house, occurs in cold weather. Vapor barriers specified in new construction help to prevent moisture buildup. These barriers definitely restrict the flow of moisture if they are constructed properly, but often the pressure in a house is still great enough so that the barrier is penetrated. When the outside humidity is from 10 to 15 percent and the inside humidity is from 60 to 65 percent, the difference is great enough for moisture to reach the siding (Fig. 19.14).

If the humidity in a house is kept at 40 percent or lower, there is little penetration of the walls, and 40 percent humidity is a healthy humidity. If high humidity persists, a specifier should provide for a satisfactory way to lower humidity by exchanging some of the wet inside air with drier outside air. This can be done by using an exhaust fan or by opening windows.

Examination of paint chips will sometimes reveal a thin layer of wood on the backs of the chips. This occurs because the wood is wet for an extended time and softens. Such peeling is caused by interior moisture or leaks.

Wet siding is sometimes localized in a house. The siding outside a bathroom, for example, can exhibit peeling caused by high humidity in the bath. The moisture moves through the nearest wall (Fig. 19.15). Similar conditions can be found less frequently outside a kitchen.

Unsealed joints Whenever there is a corner (either interior or exterior) in the construction of a building, water can get to the wood

Fig. 19.15 Peeling from interior moisture.

through cracks at the edges of siding or the corner edges. When it does, peeling follows. Problems normally occur along window and door frames, along roof lines, and under eaves as well as in other places (Fig. 19.16).

When caulking can be used, it prevents the penetration of water and consequent peeling. Peeling caused by unsealed joints is unsightly, but it usually extends over a relatively small area. The gaps between siding ends can let water into the wood if they are not closely joined, but they can be caulked.

Architects of high-rise and curtain-wall projects should pay particular attention to proper caulking. Inspection after a failure often shows fracture of the original caulking material, exposed joints, joints that are

Fig. 19.16 Peeling from an unsealed joint.

Fig. 19.17 Peeling below a clogged gutter.

barely bridged with caulk, and shifting of the elastomerics used to cushion the locking area between panels during building erection. These conditions lead to paint failure.

Overflowing gutters   If gutters are not kept clean, they can overflow. The cascade of water will wet siding and other joints and cause peeling. Gutters filled with leaves can overflow, and so can gutters filled with ice. There is another hazard from winter ice: it can block the flow of water and divert it to inside walls. This causes problems which will be discussed later (Figs. 19.17 and 19.18).

Siding in contact with the ground   Sometimes houses are so constructed that siding runs low enough for edges to touch the ground. If the ground is wet, the siding will get wet and paint on the siding will peel

Fig. 19.18 Peeling from a gutter overflow.

Fig. 19.19 Peeling from contact with the ground.

(Fig. 19.19). A similar condition can develop above steps where snow accumulates. These problems are difficult to correct structurally. If a single coat of latex paint is applied to the cleaned siding, the paint will breathe and not peel. However, if more than two coats are applied, breathing stops. It may be necessary to strip the wood every few years after too much paint has been applied (Fig. 19.20).

Wooden decks and porches   Since these surfaces are horizontal, they receive the full brunt of sun and rain unless they are covered. They must be repainted more often than vertical surfaces, and on older porches that have a heavy buildup of paint the paint may have to be removed before repainting. Sometimes porches peel shortly after they have been painted. This probably is due to inadequate cleaning or, if the porches are cleaned with soap, to inadequate rinsing. If water gets between the flooring boards and does not drain away, peeling can take place. If the flooring is porous from the weather, it may be hard to prevent the entry of

Fig. 19.20 Peeling from contact with a step. Caulking is missing.

moisture with a single coat of paint. If the floor is sealed and properly maintained, however, these problems will not appear.

Cedar stains Both red-cedar and redwood siding are quite common, and they provide an excellent surface for paint. If they are not properly painted, however, there is a strong possibility that the wood will cause stains to appear (Fig. 19.21). The color is due to a pigmented material in the wood that leaches out if water gets to the wood. A paint film that is thin (because of weathering or poor workmanship) will allow water to penetrate to the wood. Hail damage or other breaks in the film also can cause stains, as can moisture from the house interior. Stains from exterior moisture should be covered with paint. If stains are caused by interior moisture, follow the recommendations in the section "Interior High Humidity."

Fig. 19.21 Cedar stains.

Millwork stains Occasionally rather mystifying stains appear on millwork, windows, and doors. The wood, which usually is pine, is treated for water repellency and rot resistance at the mill. The stains usually are buff-colored and are long and narrow. Considerable investigation as to the cause has failed to yield a definite answer. The stains are thought to be related to the quantity of treatment solution applied and possibly to the grain pattern. Even though the cause is not definitely known, the stains can be sealed in the same manner as knots.

Hail Hail can do much damage to the paint on a house. Hailstones break the paint film and expose the wood to moisture. If rain accompanies the hail, as it usually does, water penetration is immediate. This water can cause stains or peeling. When hail strikes, it is good practice to touch up the broken paint promptly to seal the wood. Insurance may cover a new paint job. If so, the damaged spots should be primed before the full paint coat is applied.

A variation of this condition can develop in an area that has frequent hailstorms. If a full coat of paint is applied whenever it hails, the paint will soon get too thick. A touch-up should suffice in such conditions.

Uneven gloss   Sometimes a rather unusual problem develops on a freshly painted house. If the paint is applied late on a day when the humidity is high, an uneven pattern develops (Fig. 19.22). The paint has had dew form on the insulated portions of the wall, which are slightly cooler than the studs, and the dew has marred the gloss enough for the effect to be visible at an acute angle. This pattern will soon disappear as the paint chalks. The problem can be avoided by not painting late on a very humid day.

A similar condition can develop in an area which has evening fog. The fog will act as did the dew and cause an uneven or blotchy gloss. The paint is said to be "fog-struck"; sometimes the condition is so severe that an enamelized house paint looks flat.

Fig. 19.22   Uneven gloss.

Painting over chalk   Some latex exterior paints do not easily wet chalk, and painting over chalk thus presents problems. Formulators are improving latex paints so that they do adhere to moderate chalk, but it is still wise to remove heavy chalk. When suitable, an oil primer brushed on the chalk will go far toward wetting the remaining chalk. Chalk is removed with a stiff brush and a detergent solution followed by a good rinse. Do not allow the surface to rechalk before painting.

Cracking under eaves   This type of cracking has several causes. Because there is very little weathering under an overhanging eave, the paint can get thick enough to be too rigid to move with the wood and so may crack. Cracking and peeling may also be the result of wood wet because of interior moisture.

North side of a building   Here again, lack of weathering will cause paint to become excessively thick. Whenever a house or other structure is painted, the north side is included. Usually this side is dirty but is not chalking. The dirt retention and the lack of chalk mean that no erosion takes place between paint jobs. On other sides of the house sun and rain cause steady erosion, so if paint jobs are spaced several years apart, the paint film thickness will be considerably reduced. Meanwhile, the film thickness on the north side increases. After several paint jobs the paint on the north side becomes inflexible. It does not move as the wood moves, and cracking starts. To correct this condition use only a very thin coat of paint on the north side, and if washing will clean the surface, do not paint it.

Nailhead rusting If common nails are used to fasten siding, rust stains usually appear. If it is not practical to countersink the nails, they should be cleaned to remove paint and rust. Then a metal primer should be applied with a small brush. A zinc chromate primer will be easier to hide than a red-lead or iron oxide primer. When the primer is dry, it can be touched up and painting completed.

Fig. 19.23 Peeling from a gutter.

19.4.2 Metal surfaces

Rusty metal In spite of good surface preparation many metal paint problems are traceable to painting rusty metal. This condition is usually rather easy to determine since in most instances rust stains are found on the paint. In advanced cases in which paint has peeled, there is uncertainty as to whether the metal rusted before painting or after peeling. A careful examination of the surface, especially under the remaining paint, will usually determine whether or not the paint inhibited rusting. There are paints which can be applied to rusty metal if the rust is tight. They do not protect metal as well as it is protected by proper surface preparation prior to painting, but their use is indicated in some cases.

Galvanized steel One problem is paint peeling from gutters. Most gutters are made of galvanized steel, and the problem is usually traced to using the wrong paint. Several remedies are widely used prior to the original paint application, but they are of questionable value. They include vinegar wash, acetic acid wash, and short-time weathering. Often there is no treatment, and the painter uses the same paint on the gutter as on the rest of the job, and the painter is gone before peeling starts, which sometimes does not take a long time (Fig. 19.23). Painters do not like to apply a treatment coat to gutters, especially when two additional coats are required to hide the treatment.

The time-proven primer for galvanized steel contains zinc dust. It gives good adhesion to the metal but is hard to find in paint stores. This primer has a poor shelf life, and there is not a large-volume demand for it. Another good primer is the wash primer, which is marketed under several brand names and is covered by military specification MIL-P-15328. It has little color and gives improved adhesion to most metals including galvanized steel. There are also latex paints which are good primers for galvanized steel. The important requirement is that a primer be used to offset the thin oil film left during the galvanization process.

Fig. 19.24   Peeling on a galvanized-sided building.

When a peeling problem demands correction, the best first step is to do nothing until most if not all of the paint has peeled. Since this step is not likely to be accepted by a client, the next-best first step is to remove as much of the old paint as possible. Then a good primer should be applied to all the bare metal. At this point, if the client approves, the gutters should be painted, but the owner should be advised that more peeling is likely to occur until all the old paint is gone.

Poor adhesion to galvanized steel is not limited to gutters (Fig. 19.24).

Weathering has been cited as a method of improving adhesion. While it will work, the problem is knowing how much weathering is necessary. The only way to tell if a surface is ready for regular house-trim paint is to paint it. This step is self-defeating, and it is much wiser to use a proven primer.

Copper staining   The use of copper screen wire and copper fixtures is quite common. In time light-colored paint under the copper will be stained by oxides washing down onto the paint. While it is easy to paint over the stains, they will reappear. If the copper color is desired, an acrylic lacquer applied to clean copper will seal the surface and prevent stains. Clear coatings have a limited life and should have periodic inspection. When the gloss of the lacquer is gone, the copper should be recoated. If painted screens are acceptable, less frequent recoating will be required. Since the copper is usually selected for its color and nonrusting property, the use of paint is generally unacceptable (Fig. 19.25).

Abrupt temperature drops   In areas where winter temperatures drop by 20°F (11°C) or more in a short time, paint may pop off. This has happened with baked enamel on highway signs. It can also happen on galvanized steel. To prevent a recurrence a wash primer should be applied before the new paint. Also, specify a more flexible paint that will withstand abrupt temperature drops.

Fig. 19.25 Copper stains.

An abrupt drop of lesser magnitude can prevent a latex paint film from coalescing properly if the temperature changes before the paint is dry. An imperfect paint film is left on the surface. Outside, the only way to avoid this problem is to heed weather predictions. Inside, it is possible to heat the rooms until the coating cures.

Marine-paint problems   Water is a very aggressive environment for paint. Water problems can usually be traced to an improper choice of paints or improper surface preparation. Surface preparation is of great importance for immersed articles. If rust has developed, the only satisfactory way to correct the problem is to clean the metal by abrasive blasting and to recoat it with proper paint. It is not practicable to touch up immersed articles.

For structures in a marine atmosphere but not immersed, the environment is still aggressive. Saltwater spray aggravates corrosion. Here good painting practice is essential. Marine paints frequently blister and develop underfilm corrosion.

Aluminum   It is virtually impossible to remove the oxide from aluminum long enough to paint the metal. Since oxides are not good bases for paint, an aluminum pretreatment is required prior to priming. Painted aluminum usually presents few problems provided dirt and excessive chalk are removed.

19.4.3 Cementitious surfaces

Plaster   Wet plaster, as distinguished from drywall, is becoming increasingly scarce. In nearly all new buildings drywall plaster is used. However, there are thousands of buildings that have wet plaster from 5 to 50 years old. Most do not present problems, but there are enough that do to make many owners unhappy. The trouble lies in the white or finish coat. After a time chalking may appear on this coat of plaster. It destroys

Fig. 19.26   Peeling from chalky plaster.

the bond of paint on the plaster, and the paint peels off. If the peeling is slight, it is usually possible to apply an oil primer-sealer by brush so that the chalk is wet. This provides a good surface for paint, and trouble is avoided at least for a while, until the plaster forms more chalk. Do not use latex primer-sealers, because latex does not wet the chalk, and good adhesion is not achieved (Fig. 19.26).

Often there is more than slight chalk on the plaster. The chalk can be heavy enough to prevent wetting by any paint. If a scrub brush and a detergent solution will remove most of the chalk, an oil primer-sealer will give good adhesion. Heavily chalking plaster walls will not accept paint, and the best solution to the problem is relining walls and ceilings with lightweight hardboard.

In some older houses, hotels, and apartments, poor paints with little vehicle were used. These paints do not adhere well to walls or, especially, to ceilings. Repainting, particularly with water-based paints, will cause peeling. The drying forces in the topcoat simply pull the original coat off the wall. Sometimes the problem is solved by using cheap high-PVC paints with little binder, but the best solution is to remove all chalk and paint down to the plaster prior to repainting (unless hardboard relining is used).

Plaster becomes crumbly when it is wet for an extended time (Fig. 19.27). This sort of problem usually is caused by a leak in the roof or from plumbing, and the paint peels. If the area involved is large, the paint should be removed and the surface replastered. When a leaky roof is encountered, the damaged plaster may not be close to the leak, since water can run for some distance before penetrating to the plaster.

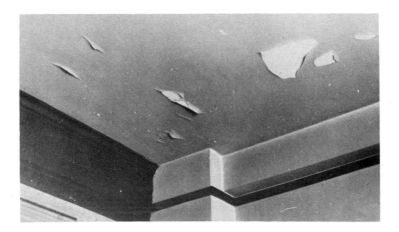

Fig. 19.27 Peeling from crumbly plaster.

Plaster cracks sometimes develop under paint (Fig. 19.28). Then the paint cracks as well, and at times it is difficult to tell whether plaster or paint cracked first. The difference is important. Paint which has cracked can be removed, but plaster cracks need repair. A texture paint can be used to hide plaster cracks if such a finish is acceptable; otherwise the paint must be removed and the crack spackled.

Do not use water-based sealers on the newer keen, hard plasters. Penetration is nearly impossible, and the emulsion-based products simply won't hold onto the surface.

Stucco   Painting is nearly always done during the lifetime of a stucco structure, and many times the paint does not stick. When this happens, it is often a result of the stucco surface's not adhering to the rest of the stucco. In other words, the stucco surface has crumbled and sloughed off, bringing the paint with it. This crumbling can happen because of continuous water exposure, but most times it is simply disintegration of the stucco, probably related to the gauging and possibly to the troweling. Treatment of bad stucco is beyond the scope of this *Handbook,* but specifiers should choose stucco carefully, provide for adequate inspection, and be aware that painting problems will result from poor stucco or stucco application techniques.

Drywall plaster   Drywall plaster does not present the variety of problems that wet plaster does, but problems can develop. If water leaks or water from other sources wets the face of drywall, paint on the wall will peel. Water behind the wall will migrate through the wall and cause peeling. Persistent water will cause the plaster to crumble as does wet

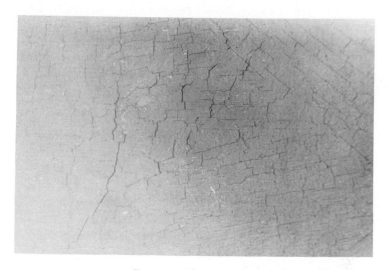

Fig. 19.28 Plaster cracks.

plaster. If the plaster is very dry when it is painted, the drywall may absorb enough primer vehicle to prevent good bonding and cause peeling.

Asbestos fire retardants   The U.S. Environmental Protection Agency (EPA) wants sweeping controls imposed on asbestos, including minimizing personal exposure to the substance. The EPA is particularly concerned with how to control the large amount of asbestos used as a fire-retardant in public buildings, especially schools. At this time, Battelle Columbus Laboratories of Columbus, Ohio, is evaluating various recommendations for handling the asbestos problem. These include removal, encapsulation, enclosure, and deferred action.

A guidance document, No. C00090, has been prepared, and the EPA has established an asbestos hot line that anyone can call for free advice: 800-424-9065, Asbestos Containing Material in School Buildings Consulting Service. According to the EPA, the details can be applied to any building.

In addition, Battelle is preparing a detailed technical report and application manual for encapsulation procedures. It has evaluated 74 sealants (coatings, etc), and the report should be available by the end of 1980.

Asbestos siding   Asbestos siding or shingles are not painted normally until they become weathered and water-absorbent, usually after many years. If a latex paint is used, it is wise to inspect the surface for chalk. If there is a significant quantity of chalk, it should be removed. Chalk is not normally expected on asbestos-cement surfaces, but it can develop. By the time that the shingles have weathered long enough to chalk, there is no alkali on the surface and an oil primer can be used. It is more likely to wet the chalk and give a good bond. Peeling can be caused by the loosening of the surface of the shingle or sheet. When this happens, the shingles should be replaced if possible. Patching is not conducive to a smooth paint job. If the surface is very absorbent, specify a seal coat to prevent uneven gloss. Also, remove any mildew.

Glazed asbestos-cement shingles do not weather as rapidly as do nonglazed shingles. If gloss remains, the shingles require etching with muriatic acid and should be thoroughly rinsed before painting. Use caution because painting glazed shingles can lead to severe peeling problems.

Foundations and basement walls   Under certain conditions mortar between foundation blocks will disintegrate and slough off, bringing paint with it. This condition is corrected by repointing the joints and painting with an alkali-resistant paint. When mortar joints become soft, they should be removed whether or not paint is involved. Water will penetrate the foundation if joints are not tight.

If paint on a foundation fades unevenly, this condition is sometimes caused by alkali in the mortar that affects the pigments in the paint, in a process called "burning out." This is common with lead chromate pigments, but oxides usually do not burn out. Burning generally does not happen on foundations which have weathered for several months. Sometimes stucco surfaces are "hot" with alkali and can burn out the color of pigments.

Painting inside basement walls also presents problems. Often water is not stopped outside the foundation, and as it moves through the wall, it destroys paint adhesion on the inside. Another problem is efflorescence, a deposit of crystals of salts which have been picked up by the water moving through the foundation. They are dissolved until they reach the air, when they form on the wall. The salt crystals should be brushed off. If an attempt is made to wash them, new crystals will form.

It is important that the foundation be made waterproof before backfilling. If this is not done, problems are more difficult and expensive to correct. The best way to stop water is to dig around the foundation and waterproof the outside walls with heavy asphalt or plastic sheets. Using drain tiles and sloping the ground are other possibilities, but these are not as positive as waterproofing.

If the water is only seepage, it is possible to chisel out some of the mortar from the inside and fill the space with mortar that includes iron filings. The filings will rust and swell sufficiently to stop the seepage. If the water flow is continuous, one alternative is to channel to the basement drain. This may not permit complete painting of the basement wall, but the blocks can be painted and the mortar left unpainted. Sometimes a latex paint will adhere because the film breathes. Paints which are claimed to prevent water seepage are on the market. Such claims should be checked thoroughly before specification because it is hard to stop water under pressure by using a coat of paint.

Concrete Good construction practice requires that concrete be completely dry before it is painted. This requirement is frequently overlooked or underestimated. As an example, a department store in a shopping center had several large concrete columns erected at a downhill entrance. The columns, which were 3 ft (0.9 m) in diameter and 20 ft (6.1 m) high, were poured at a remote site and delivered wrapped in heavy paper to prevent scarring. After erection and an interval of time, they were painted white. After the store had its opening, customers began to complain that they found white powder on their coats and dresses when they used that entrance. This concerned the store manager, and the white powder was traced to the columns. The paint on them had reacted with the alkali in the nondry concrete, and its vehicle was largely gone. The store paid for cleaning several garments and replaced a few. The powdery paint was removed, and the concrete was allowed to dry unpainted for several months.

Structures like drive-in–theater screens can be very slow-drying. Usually the builder is in a hurry to get them painted and complete the job, but without patience and good painting practice there will probably be unsightly paint peeling (Fig. 19.29).

Concrete slabs such as porches laid directly on the ground are hard to keep painted. Water from the ground migrates up through the concrete and destroys the paint bond. A latex paint will allow some breathing, but breathing usually ceases when more than two coats of paint are applied. Concrete porches above the ground soak up water, and good drainage is essential.

Brick Many brick structures are eventually painted. In some of them the paint stays on and looks good for a long time, but when others are painted, the paint erodes from the bricks (Fig. 19.30). Part of the difference in paint-holding power lies in the density of the bricks. Dense bricks do not absorb water as easily as do porous bricks, and water getting into

Fig. 19.29   Peeling from nondry concrete.

porous bricks causes the paint to erode or peel. Porous bricks need a sealant and two coats of paint to produce a job that should look good for many years. Cheap bricks, such as those imported from Mexico, crumble and fail. You cannot expect a coating to solve this problem.

Sometimes brick houses have white shutters. If chalking occurs, the chalk washes down onto the brick, creating an unsightly area. Painted siding over brick can cause similar stains (Fig. 19.31). The chalk embeds itself in the pores of the brick and is difficult to remove. Sometimes a stiff

Fig. 19.30   Peeling from bricks.

Fig. 19.31   Paint chalk washed down onto brick.

brush and a detergent solution will flush the chalk away, but often the only way to get clean bricks is to paint the wall with brick-colored paint, using a lighter-colored paint for mortar lines to maintain the brick appearance.

With brick chimneys it is desirable to apply paint to seal the bricks. Several coats may be required to prevent soot stains.

Brick walks may not hold paint, just as concrete slabs on the ground do not. Here again latex paint is sometimes used until the film stops breathing because of recoating.

A brick wall can develop efflorescence (Fig. 19.32) if water gets behind it or if there is a leak. Overflowing gutters also cause efflorescence. Water

Fig. 19.32   Efflorescence on brick.

will not remove efflorescence because it brings out more salts. A wire brush is a good tool for removal. Cinder-block walls also develop efflorescence when water gets to their salts.

Swimming pools   Paint on swimming pools is subjected to a rather severe environment, and great care must be taken in surface preparation. Most pools are constructed of masonry. Some are metal, but either type should be abrasive-blasted to get a clean surface. With metal pools an inhibitive primer is essential, but this is not true of masonry pools unless they need sealing. Epoxy coatings and chlorinated-rubber coatings are probably the best available, and one or the other type should be used, but both require very clean, tight surfaces.

When repainting a previously painted pool that is in good condition, washing will suffice for surface preparation, either by itself or with light sanding to dull the gloss. It is important that the surface be rinsed very thoroughly, as any soap residue will prevent paint adhesion.

If flat surfaces around the pool are painted, it is good practice to embed some abrasive in the paint to reduce the danger of slipping and falling.

19.4.4   Roofs

Various types of roof are painted for different reasons. At best roof paints experience a severe environment and have relatively short lives.

Metal roofs   Roofs, especially on industrial buildings, are frequently constructed of galvanized steel. Usually they are not painted for several years after construction, so problems of poor adhesion to the galvanized steel are removed. If rusting has started, cleaning is necessary (see section "Rusty Metal"). If the roofs are painted shortly after erection, priming is necessary. If peeling starts from a relatively new galvanized roof (Fig. 19.33), the primer was probably omitted. On the other hand, if erosion or washing away develops after 2 or more years, the condition is normal. Paint vehicles break down from sun and moisture, and the pigment is left to wash away.

Asphalt or tar roofs   Dark-colored roofs are usually painted a lighter color to increase heat reflection. New galvanized metal reflects heat, and light-colored paint increases the reflectance of dark roofs quite dramatically. On asphalt or tar roofs many paints will experience bleeding because their solvents dissolve the bituminous roof material. As a result, exterior latex paints and cement paints are commonly used for these roofs.

In warm climates it is the custom to paint roofs white to reduce the sun's heat. A light-colored roof will reduce attic temperatures by as much as 15°F (8°C) from those under a black roof. Many of these paints are intended to last only one season, but where the sun is less intense, more durable paints are used with the expectation that they will last from 2 to 3 years.

Asphalt shingles   Asphalt shingles are painted to change their color, but some paints also extend the life of worn shingles. To avoid bleeding, the paint used on these shingles should not contain mineral spirits or any other petroleum solvent. If black is satisfactory, an asphalt paint is good to renew worn shingles, but if a reduction in heat input is desired, a lighter-colored latex or cement paint should be selected.

Slate roofs   Slate roofs are rarely painted, and the only problem with painting them to reduce heat input would be dirt on the shingles that would interfere with adhesion. Most exterior paints are suitable on slate; latex paints give long service.

19.4.5 Glass   Inflexible coatings like lacquers or enamels should not be used on glass to opacify a window or other large area. These coatings

Fig. 19.33 Peeling on a roof.

are not flexible enough to move with the movement of the glass, and the difference in expansion rates will cause alligatoring. For large areas a flat or low-gloss coating should be used. For lettering, an enamel or a lacquer can be used. If alligatoring has started, all the coating must be removed to get a satisfactory job.

19.4.6 All surfaces

Mildew or mold   Next to water, mildew is probably the greatest enemy of good paint service. It discolors paint, and it affects paint integrity. Mildew is not limited to any particular area, although it grows better in warm, moist climates on both interior or exterior surfaces (Figs. 19.34, 19.35, and 19.36).

Mildew spores are in the air continuously, and they need only food

Fig. 19.34 Mildew in a shaded area.

and moisture to start growing. Paint vehicles provide the food unless they are inorganic like cement. Spores prefer a warm temperature, but once started they grow in a wide range of temperatures. Mildew comes in a variety of colors, and it is easily confused with dirt. However, it can be identified rather simply in most instances. If what is believed to be dirt will not wash off, mildew should be suspected. Using a fresh bottle of 5% hypochlorite bleach, apply a few drops to the suspect area. If the color bleaches out in a few minutes, the problem is mildew. If it does not, it is very likely to be dirt.

If further evidence is desired, a low-power magnifying glass will probably reveal a fernlike growth (Fig. 19.37). However, this is not always

Fig. 19.35 Mildew on an unshaded house.

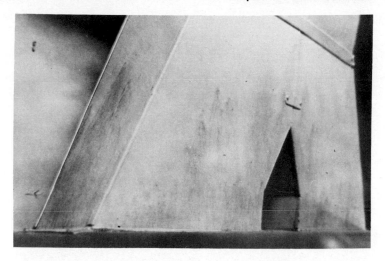

Fig. 19.36   Mildew on an interior duct.

true, and mildew can appear as tiny rods or dots. If it cannot be identified and identity is important, a biological laboratory can definitely identify mildew.

When mildew has been identified or suspected, the following solution will kill it:

⅔ cup trisodium phosphate scrubbing powder
⅓ cup strong detergent

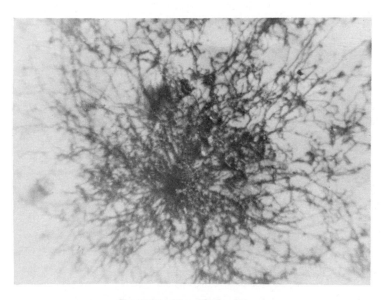

Fig. 19.37   Magnified mildew.

1 qt 5% fresh household bleach
3 qt (2.84 l) warm water, or enough to make 1 gal (3.785 l)

Use this solution without diluting it, applying it to the stain with a medium-soft brush. When the stain is bleached and the surface is clean, rinse thoroughly with fresh water from a hose. If only a small area is involved, household hypochlorite bleach can be used. Household chlorine-containing scrubbing powder is also suitable, but when large areas are involved, use the mildew solution given above. When a powder is used, it must be thoroughly removed before painting.

When repainting a surface that has been cleaned of mildew, use a paint containing mildewcide to prevent further mildew growth. While the wash solution kills the mildew, it has no residual effect.

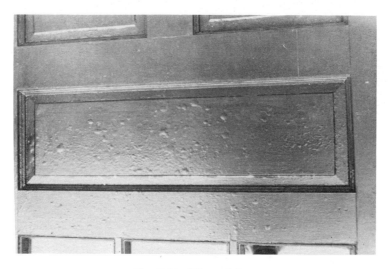

Fig. 19.38 Blistered door.

Mildew can cause mysterious peeling. If a house is primed in the fall and then not given a topcoat until the next spring, mildew can form on the primer during the winter, and when warm weather arrives, it begins to grow. If a fresh coat of paint has recently covered the mildew, there is ample food and moisture to provide good growing conditions. As mildew grows, it expands and in so doing pushes the new paint from the primer. If mildew is identified on the back of the peeled paint or on the primer, the peeling was caused by the mildew. It is likely that all the new paint must be removed if it has not peeled. Then the old primer surface requires treatment to kill the mildew prior to repainting.

Blistering Blisters can develop under paint on almost any surface in many environments. They are caused by gas or liquid being trapped under the paint and expanding while the paint is still flexible. If the paint were not flexible, the pressure would cause cracking and peeling. The blistered door in Fig. 19.38 was painted in very hot weather, and the surface skin of the paint film dried, or set, before all the solvent was removed. Blisters can also form from outside water, for instance, from steam. In addition, standing water causes blisters. Corrective measures vary with the type of surface and the cause, which must be determined.

Alligatoring   If a hard-drying topcoat, an enamel, for instance, is applied over a nondry coating, alligatoring can develop. This condition is so named because it resembles an alligator hide (Fig. 19.39). If a topcoat dries too rapidly from being in the sun or if the topcoat is a fast-drying coating and the undercoat is not hard, alligatoring can develop. Once the paints have reached full hardness, the cracking will stop, but refinishing such a surface is not easy. Cracks will still be evident under a thin coat of

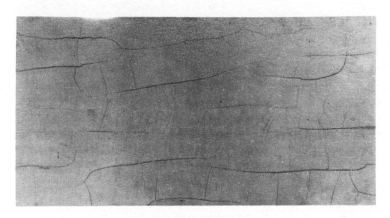

Fig. 19.39   Alligatoring.

paint. Sanding and the application of several thin coats will build a heavier coat, and the cracks will then be less conspicuous and may eventually disappear. A texture coating will obscure the cracks.

Smoke stains and odor   After a building fire, unburned areas frequently have smoke odor, blisters, or stains or all three. Blisters are not a problem, but odor and stains can be. A good sealer is shellac; it will stop stains from bleeding and seal smoke odor. The shellac can normally be overcoated.

Craters and fisheyes   This problem is usually found on concrete treated with a wax or a silicone compound whose purpose was to achieve water repellency. The treatment prevents the paint from wetting the surface, and the resulting film is discontinuous.

When a fisheye problem is suspected or encountered, use a mineral-spirits wash before painting to remove any wax. A power brush may be necessary to cut the wax easily. A patch test should then be made. If the problem still exists, additional weathering may remove sufficient repellent, making the surface suitable for painting.

Some cases require light sandblasting to assure a clean surface. If this and the other suggested steps do not accomplish removal, the surface should be left unpainted. The surface will be more attractive than if it is covered with mottled paint.

Fisheyes and craters can occur when repainting over certain hard enamels that contained silicone as part of the original formulation. Some painters are successful in overcoming this problem by sanding the surface and/or mixing in one of the commercially available additives used to control fisheyes. Since these additives often contain silicone, however, the repaint problem could recur.

Acid rain Architects and engineers specifying paint that will be exposed to acid rain should take this condition into consideration in specifications for coatings systems.

BIBLIOGRAPHY

Berger, Dean M.: "Detecting Film Flaws in Coatings," *Chemical Engineering*, Mar. 17, 1975, pp. 79–83.

———: "Inspecting Coatings for Film Thickness," *Chemical Engineering*, Feb. 17, 1975, pp. 106–110.

"Causes of Discoloration in Paint Films," cassette, Federation of Societies for Coatings Technology, Philadelphia, 1975.

"Efflorescence," *Concrete Construction*, March 1970, pp. 81–82.

Evaluation of Coatings on Coastal Steel Bridges, Report M&R 64513, State of California Business and Transportation Agency, Materials and Research Department, 1969.

David J. Engler, Finnaren & Haley, Inc.: "Color Difference Acceptability," Forty-Eighth Annual Meeting, Federation of Societies for Paint Technology, Boston, 1970.

———: "The Influence of Paint Storage Temperature on Drying Characteristics," Forty-Eighth Annual Meeting, Federation of Societies for Paint Technology, Boston, 1970.

Garlock, Neil B.: "Diagnosing Paint Problems and Correcting Them," *American Paint Journal*, reprint, 1967.

Hess, Manford: *Paint Film Defects*, Reinhold Publishing Corporation, New York, 1965.

"New Method Probes Coated Metal Corrosion," *Chemical & Engineering News*, May 19, 1980, pp. 7–8.

Peterson, Carl O.: "Concrete Surface Blistering," *Concrete Construction*, September 1970, p. 317.

"Rx for Winter Storm Damaged Homes," *Journal of Coatings Technology*, March 1980, pp. 21–24.

Trimber, Kenneth A.: "Conflicting Holiday Tests," *American Painting Contractor*, February 1979, pp. 14–17.

Weaver, Paul E.: *Industrial Maintenance Painting*, 3d ed., National Association of Corrosion Engineers, Katy, Tex., 1967.

Chapter 20

Economics

ALBERT H. ROEBUCK
Fluor Engineers and Constructors

GUY E. WEISMANTEL
Chemical Engineering

20.1 Introduction   Architects and specification and coatings engineers are concerned about total painting cost. Therefore an economic analysis of coatings systems and related costs involves questions of life expectancy as well as product application costs. An architect may address the question of life expectancy by emphasizing standards of acceptable appearance, while an engineer may place the greatest importance on corrosion protection. In either case the coatings specialist and the specification writer must be familiar with economic and accounting procedures such as discounted cash flow (DCF).

The reason is obvious: DCF is used by management to evaluate the economics of making an investment decision. The advantage of DCF is that projects (or parts of projects such as painting requirements) are evaluated in terms of the present and future value of money, cash flow, interest and discount rates, and profitability. DCF, then, is an important tool in economic justification. If coatings specialists are not familiar with DCF and other accounting procedures, they will be unable to communicate with management in terms and procedures that management uses in making business decisions.

This chapter of the *Handbook* offers architects and specifying engineers a guide for estimating the costs of a coating system for a given environment and will assist them in making correct economic decisions based on specific current data coupled with forecasts.

20.2 System Costs   A comprehensive economic evaluation includes the costs of:

- Surface preparation
- The paint itself (primer and topcoats)
- Paint application
- Maintenance painting
- Overhead and other expenses

20.2.1 Surface preparation   Abrasive blasting is the preferred and most widely used means of surface preparation for industrial painting. Other surface-preparation methods include hand- or machine-tool cleaning and chemical cleaning and degreasing. These procedures are all part of the process of making a surface paintable (ready to receive and hold paint). The total cost of all such operations, including expendable materials, should be estimated and charged as surface-preparation costs. As a rule of thumb, costs and prices are presented on a square-foot basis. To convert to a per-ton basis, simply multiply by 250 for a typical ton size.

20.2.2 The paint system   This cost includes the cost of all coats of paint to be applied over the completely prepared surface. It must include the total cost of all paints for the job. If there is leftover material and it is put in stock for use on a subsequent job, complete record keeping requires that a bookkeeping entry be made for the transfer. The same procedure should be applied to excess materials for surface preparation.

20.2.3 Paint application   The cost of paint application should be kept separate from the cost of surface preparation because these are distinct operations with different objectives. Also, the cost of extra operations that form part of the paint application process should not be overlooked. Among these operations are putting up and removing protective masking, mixing and thinning paint before and during application, downtime because of trouble with paint or equipment, equipment rental, scaffolding, cleanup during and after paint application, and inspection after each operation before proceeding to the following one. Some of these operations are more important to the painting contractor, but the architect or engineer writing the paint specification should have them in mind to assure proper planning by the contractor in order to avoid job delays.

20.2.4 Maintenance and repainting   Painting and paint maintenance constitute an expense item for accounting purposes. As the rate of corporate income taxes is roughly 50 percent, the actual cost to the corporation of expense items is 50 percent, if we assume that a profit is being made. This cost is often referred to as the 50-cent maintenance dollar. Accounting considerations have led some companies to delay topcoating until construction is complete or until a plant starts up. While the practice is less prevalent in buildings or commercial work than in plants, the architect must realize that painting delays for accounting purposes could cause a financial disaster if the prime coat were to fail and reblasting and repriming were to be required.

The thinking behind delayed topcoating is economics. As noted above, maintenance costs and their deduction as an expense item in the current tax structure produce an actual cost of 50 cents on $1. If a decision to delay applying the topcoat has been made and the specified primer does not protect the surface (e.g., steel) in the interim, degradation can occur. To avoid such degradation, it is imperative to include adequate preparation specification procedures. Incipient failure areas, scars, holidays, construction damage, etc., must be properly prepared so that topcoats will adhere properly.

Inorganic zinc primers are frequently used when a delay between priming and topcoating is planned. They offer excellent protection from corrosion and from handling damage. Also, aging an inorganic zinc primer before topcoating offers the advantage of minimizing blistering or topcoat bubbling problems. When freshly applied inorganic zinc is topcoated, it often suffers from blistering. Thus the delay in topcoating has certain technical as well as economic advantages in this case. Again, however, care must be exercised to topcoat before major primer degradation and incipient corrosion begin.

20.2.5 Overhead and other costs   The painting contractor should carry liability insurance or set up a contingency fund to provide for possible repainting costs due to premature paint failure. The cost of the liability insurance or the contingency fund is a cost of doing business to be included in each paint job. Architects and specification engineers should assure that the contractor has this kind of coverage. Without it, the contractor's warranty may be worthless when it is needed because of premature paint failure.

Large companies with their own painting crews have a somewhat different set of considerations. For them the questions to be answered are generally as follows:

- How much of our own painting should we do? Only the small, routine jobs or the big ones too? Or should we put out the big jobs for bid?
- If we do the big jobs ourselves, should we rent some of the bigger pieces of equipment, or should we buy them for these and future jobs?
- If we farm out the larger jobs, who will be responsible for paint inspection and maintenance?

20.3 Figuring Costs   Projects vary in size, complexity, mix of sizes and shapes, site accessibility, location, and climatic conditions. These and other factors offset job costs.

As in any business endeavor, return on investment is a key economic consideration. In this light, there is need to weigh the economics of coatings with a high initial cost versus those with a low initial cost. An architect is often asked to justify the choice of a high-cost, long-term performance system. In some cases, especially with new formulations of high-build coatings, justification based on experience is difficult. Thus, some owners may opt to return to a low-cost, short-term paint system, often with the result of high annual maintenance costs. In the case of tract homes, using cheap, low-quality paint has become almost a standard procedure because repainting costs become the problem of the new owners. It is common for flat wall paints to be of such high pigment-volume concentration (PVC) that they easily collect dirt and have little scrubbability. Likewise, exterior coatings are applied on unprimed surfaces and with a limited film thickness, leaving surfaces ripe for delamination, blistering, mildew attack, or erosion. This chapter does not address this problem of false economies. Suffice it to say that should *one* suit result from this practice, the legal costs involved would have more than paid for the use of a higher-quality paint (if we assume that failure is a product problem of the coating and not of poor workmanship).

20.3.1 General costs in surface preparation   One of the most important cost considerations in coatings work is surface preparation. Not only is proper surface preparation important as the foundation step in obtaining a good coatings job, it is also economically important because it is one of the larger expenses, often accounting for more than 50 percent of the total job cost. For example, sandblasting in the field to white metal (SSPC SP-5) can run 50 cents per square foot for surface preparation, while the total cost for the coatings job is $1 per square foot.

Although surface preparation by other procedures, such as hand-tool or power-tool cleaning, can be considerably less expensive, modern high-performance coatings systems require abrasive blasting. Architects or specification writers concerned with standard architectural coatings and surfaces do not normally encounter surface-preparation costs of more than 15 cents per square foot. Wallboard, for example, requires virtually no preparation prior to sealer application except for spot sanding at the joints or at patches. Similarly, a proper choice of coating can camouflage surface imperfections at wooden doorjambs and moldings, and some enamel undercoats require virtually no sanding prior to the application of the topcoat. Naturally, the greater the attention given to paint specification, including acquisition of an intimate knowledge of the product and its proper application, the lower the preparation costs. If an absolutely glasslike film is expected in an enamel topcoat, sanding is an expensive requirement to remove the brush marks of the undercoat. The specification writer must realize that considerable cost is involved whenever a painter must spend 15 min in sanding a doorjamb to meet a smoothness specification.

As a rule of thumb, the actual cost of the paint for most simple paint jobs (those that do not require exotic coatings or preparations) is between 20 and 30 percent of the total coating job. For certain internal lining jobs the coating cost will run higher, but a range of 20 to 30 percent is considered reasonable for most coating work. Obviously, a large proportion of the coating job consists of surface-preparation and application costs. When surface preparation is minimized, the contribution of paint costs to the total job could exceed 30 percent. Paying attention to other job specifications, such as the quality of wood molding, can reduce surface-preparation costs.

Surface-preparation costs are alarmingly expensive when a job involves blasting to high cleanliness (for example, SSPC SP-5 for white metal). Nonetheless, a high initial expenditure can prove to be less expensive in the long run. When estimating a job, many of the more progressive contractors will blast a small area to determine just how difficult it is to prepare the surface properly by removing rust, scale, old paint, etc. This procedure can prove advantageous on large or difficult jobs for which accurate job costs are required. It also provides the client with a sample area which can be used on the job as a reference to assure quality control.

While blasting can be a requirement for masonry surfaces (see box), hand or power cleaning does not effectively remove mill scale from steel.

If mill scale is left in place, corrosion continues to take place underneath it, and eventually it will delaminate. The time to delamination will vary with the coating and the exposure environment. Delamination usually occurs in a year or two, leaving the surface mottled in appearance and subject to corrosive attack at the areas of delamination. Repair costs are high, one reason being that mill-scale delamination may continue slowly over several years, making continuous maintenance necessary. For plants with life expectancies of more than 5 years, economics mandates the use of abrasive blasting to remove mill scale.

SURFACE PREPARATION OF MASONRY

Asking a contractor or an architect to sandblast a brand-new wall sounds ridiculous. Nevertheless, following this suggestion can avoid problems on many occasions.

Poured-concrete walls   Often, etching with acid is not an appropriate alternative. In addition to loose cement, dirt and dust collect on a poured-concrete wall. The contractor may be confronted with unexpected contamination such as form-release materials detrimental to adhesion. If oil is used as a form-release agent, the walls can be wiped with a solvent-soaked rag, but quite often the only thing that is accomplished is the movement of the contaminant from one area to another. The most practical recommendation that can be made for preparing a poured-concrete wall is to clean it by brushing the surface with a stiff brush and then to reduce the first coat of paint to low viscosity and solids. This procedure results in maximum penetration of the coating material into the surface. It improves adhesion and may mean the difference between success and failure.

Block walls   Generally, the greater the porosity of a masonry surface, the less surface preparation is required. Usually, the only requirement for a block wall is to let mortar joints cure out as long as possible and to remove any loose, protruding material with a stiff brush. If a smoother appearance is desired, a block filler is usually applied. A latex-type block filler is very satisfactory above grade, and a water-resistant block filler should be used below grade.

Brick walls   By the time that a decision is made to paint a brick wall, the brick is quite often old, dirty, soft, and in poor condition. Often an attempt to sandblast such a surface further pulverizes the brick. If such a condition exists, the only practical means of surface preparation is pressurized chemical or steam cleaning. After cleaning, the next step is to coat the brick with a good surface conditioner or sealer.

The sealer will penetrate the brick, help bind any loose material, and help prevent efflorescence. The economics of each step must be considered carefully in job costs.

Concrete floors   It pays to give great consideration to preparing these surfaces properly; otherwise there may be grave costs in the future. This type of surface probably holds the record for frequency of failure. If the floor has been laid on bare ground without a barrier, a floor paint is vulnerable to hydrostatic pressure from below. Much too often, however, failure occurs not as a result of these expected causes but because of insufficient or careless surface preparation.

Surface-preparation economics for concrete floors can be quite different from that for other masonry surfaces. Masonry contractors use the best method prescribed by their state of the art without much regard for what may be applied to the surface after their job is finished. As a result, they incorporate concrete hardeners and sealers and hand-trowel the surface until it is as smooth as glass. If sealers or concrete hardeners have been added to a floor, only two things can be done for surface preparation: painting the floor as it is and hoping for the best or sandblasting it. If there is such a thing as a cure-all method for the surface preparation of a concrete floor, sandblasting may be the one method for any condition.

If the use of hardeners and sealers is excluded, it has been proved that when a floor is smoothly troweled, as much as 32 mils (0.8128 mm) of laitance can be expected. In fact, there will be some laitance even if a different method is used. (Laitance is a thin layer of fine cement powder that floats to the surface as the mortar is being tooled and handled.) Laitance can be removed quite economically and successfully by a hydrochloric (muriatic) acid etch. A rule-of-thumb cost for chemical-cleaning economics is $1 per square foot of concrete (the cost will vary with the size of the job.)

Blast cleaning is typically considered a nuisance and is not allowed in certain cities and states. It is the most practical and effective means of properly cleaning surfaces. Blast cleaning removes mill scale and creates an anchor pattern which is basic to good adhesion, the foundation of a coatings system. While it is commonly considered more expensive than hand- or power-tool cleaning, service life for the same coatings system is appreciably lengthened by blasting, thus reducing the cost per year over the life of the system.

Tables 20.1 through 20.6 present cost data that reflect service life in

TABLE 20.1   1978 Field Painting Costs: Manual Blasting of Structural Steel

Field painting contractors-applicators (union shop); labor, equipment, and related costs*

Cleaning grade	Region of United States			
	East	Central	Gulf	West
SP-2 hand–wire brush	$.25	$ .30	$.30	$.25
SP-3 power tool	.35	.40	.35	.35
SP-7 brush-off blast	.25	.30	.25	.25
SP-6 commercial	.45	.60	.45	.50
SP-10 near white	.60	.80	.60	.65
SP-5 white	.80	1.00	.90	.90
Prime coats				
One-package	.10	.12	.10	.10
Epoxy	.15	.17	.15	.15
Zinc-rich	.20	.22	.20	.20
Touch-up on ground†	.10	.12	.10	.10
Additional coats				
One-package	.10	.12	.10	.10
Epoxy; urethane	.15	.17	.15	.15

SOURCE: Presented at the NACE Annual Conference, Atlanta, Ga., March 1979.

NOTE:

1. Prices shown are for structural steel on the ground.

2. For prices in the air, add 25 percent for simple structures and 50 percent for intricate structures. For tank work, deduct 25 percent. For piping 4 to 6 in in diameter, use above cost. For piping 1 to 2 in in diameter, add 25 percent. For piping 6 to 48 in in diameter, deduct on a sliding scale up to 25 percent for 48-in-diameter pipe.

3. To convert to typical ton-size cost, multiply by 250.

4. Prices are approximate, based on 1978 data.

* Blast clean; no paint cost included.

† 10 percent damage assumed. Calculate touch-up rate times total square footage of exposed steel.

respect to surface cleaning. Add 10 to 25 percent to these numbers to reflect 1979–1980 costs. Table 20.1 lists approximate 1978 labor, equipment, and related costs for field painting contractors and applicators by regions of the United States. Table 20.2 gives the same type of data for steel fabricators with automatic wheel blasting equipment. The 1978 material costs by generic types of coatings are included in Table 20.3. Table 20.4 presents the expected service life of major generic systems in nine environments.

TABLE 20.2   1978 Shop Priming Costs of Structural Steel
Steel fabricators with wheel blasting equipment; labor, equipment, and related costs*

Cleaning grade	Region of United States			
	East	Central	Gulf	West
SP-2 hand–wire brush	$.20	$.20	$.20	$.25
SP-3 power tool	.25	.25	.25	.30
SP-6 commercial	.20	.20	.20	.25
SP-10 near white	.25	.25	.25	.30
SP-5 white metal	.35	.35	.35	.40
SP-8 pickling†	.075	.075	.075	.075
Prime coats				
One-package	.05	.05	.05	.06
Epoxy	.10	.10	.10	.11
Zinc-rich	.15	.15	.15	.16
Additional coats				
One-package	.10	.10	.10	.12
Epoxy; urethane	.15	.15	.15	.17

SOURCE: Presented at the NACE Annual Conference, Atlanta, Ga., March 1979.
NOTE: Size of job for competitive pricing: 250 tons.
1. Prices shown are for commercial-size jobs by steel fabricators having wheel blasting equipment.
2. For steel fabricators without wheel blasting equipment or those using conventional blasting, prices will approximate contractor-applicator prices in the field (Table 20.1).
3. To identify qualified fabricators with wheel blasting equipment, get in touch with local or national centrifugal blasting manufacturers.
4. Prices are approximate, based on 1978 data.
* No paint cost included.
† When pickling steel plate, the cost per square foot is 10 cents. If both sides of the plate are to be coated, the effective cost per square foot of the area to be painted is 5 cents. If only one side is to be coated, the cost is 10 cents per square foot.

Table 20.19 shows the cost per year, on a square-foot basis, of the systems listed in Table 20.5 (East Coast, field-applied) for mild, moderate, and severe environments. The service life in years for each system and category shown in Table 20.4 was divided into the system cost (structural steel on the ground) to determine the cost per year. For illustration purposes, Tables 20.5 and 20.6 take these same generic systems and, for an East Coast location, show calculated total system costs for each, both with field topcoating.

Tables 20.1 through 20.6 present specific cost data. Table 20.4 shows

TABLE 20.3 1978 Typical Material Costs for Paint and Protective Coatings

Type	Typical dry film thickness per coat (mils)	Approximate cost per square foot at typical dry film thickness (cents)		
		Theoretical	Spray (practical)*	Roller or brush (practical)†
Alkyds				
Primer	1.5	1.7	2.5	1.9
Gloss topcoats	1.5	1.6	2.2	1.7
Silicone alkyd	2.0	3.2	4.5	3.5
Latex				
Primer	2.0	2.5	3.5	2.7
Topcoats	1.5	2.3	3.3	2.6
Epoxy ester				
Primer	1.5	1.9	2.7	2.1
Topcoats	1.5	2.3	3.3	2.6
Zinc-rich primers				
Inorganic; self-cured	3.0	6.1	8.7	‡
Organic; epoxy	3.0	7.4	10.5	8.2
Epoxy				
Primer	2.0	2.9	4.1	3.2
High-build intermediate coat or topcoat	4.0	5.5	7.9	‡
Finish (enamel) coat	2.0	2.6	3.7	2.9
Chlorinated rubber				
Primer	2.0	3.8	5.4	4.2
High-build intermediate coat or topcoat	4.0	6.0	9.2	‡
Finish (enamel) coat	1.5	2.7	3.8	3.0
Solution vinyls				
Primer	1.5	3.9	5.6	‡
High-build intermediate coat or topcoat	4.0	9.5	13.6	‡
Finish (enamel) coat	1.5	3.7	5.3	‡
Urethane				
Aliphatic urethane	1.5	4.1	5.9	4.6
Coal-tar epoxy				
Standard	8.0	5.8	8.2	‡
C200 version	8.0	6.2	8.9	‡

SOURCE: Presented at the NACE Annual Conference, Atlanta, Ga., March 1979.
NOTE: Costs are approximate, based on 1978 data.
* 30 percent spray loss.
† 10 percent loss by brush or roller.
‡ Spray only.

TABLE 20.4 Expected Service Life of Various Generic Systems in Typical Industrial Environments

System	Cleaning	Dry film thickness (mils)	Life in years									Dry heat resistance
			Mild	Moderate	Severe	Caustic	Acid	Fresh water	Salt water and brine	Halogens	Solvents and gasoline	
Two-coat alkyd	SP-2; SP-3	3.5	2	1	½	N	N	N	N	N	N	250°F*
Three-coat alkyd	SP-2; SP-3	4.5	4	2	1	N	N	N	N	N	N	250°F*
Three-coat alkyd	SP-6	4.5	7	5	3	N	N	N	N	N	N	250°F*
Two-coat latex	SP-2; SP-3	3.5	2	1	½	N	N	N	N	N	N	250–300°F†
Three-coat latex	SP-2; SP-3	5.0	4	2	1	N	N	N	N	N	N	250–300°F†
Three-coat latex	SP-6	5.0	7	5	3	4	4	N	N	N	4	250–300°F†
Two-coat epoxy	SP-6	6.0	7	6	5	4	4	N	N	2	4	250–300°F†
Three-coat epoxy	SP-6	10.0	10	8	6	7	6	6‡	5‡	5	6	250–300°F†
Three-coat inorganic zinc, high-build epoxy	SP-10	11.0	12	10	7	7§	6§	6	N	5	7	250–300°F†
Three-coat inorganic zinc, high-build epoxy, aliphatic urethane	SP-10	8.5	12	10	7	7§	5§	N	N	4	8	300°F†
Three-coat vinyl	SP-10	9.5	10	9	8	6	9	5	5	N	N	140–160°F
Three-coat chlorinated rubber	SP-6	10.0	8	7	6	4	7	6‡	6‡	N	N	140–160°F
Two-coat coal-tar epoxy	SP-6	16.0	8	7	6	6	4	8‡	7‡	N	N	160°F

SOURCE: Presented at the NACE Annual Conference, Atlanta, Ga., March 1979.

NOTE: Decoration and cosmetic aspects are not considered. N = not recommended.

Mild. Rural or residential atmosphere; no appreciable industrial fumes or fallout.

Moderate. Industrial plants present; no heavy contamination of industrial fumes or fallout.

Severe. Heavy industrial and chemical-plant area with high-level contamination of industrial fumes and fallout; can include proximity to salt water.

Caustic. Caustic soda up to 50 percent concentration with splash spills and fumes; dry heat to 160°F.

Acid. Mineral acids at approximately 10 percent concentrations, with splash spills and fumes; dry heat to 160°F.

Fresh water. Nondesalinated immersion at ambient temperature.

Salt water and brine. Immersion at ambient temperature.

Halogens. Chlorine and bromine splash, spills, and fumes; dry heat to 160°F.

Solvents and gasoline. Aromatic hydrocarbons, selected esters, gasoline, and alcohol; splash, spills, and fumes to 160°F; dry heat to 160°F.

* Thermoplastic; softens at 120°F, but protection remains; will pick up dirt.

† Colors will darken or yellow, with loss of gloss; protection remains.

‡ Near-white blast No. SP-10 required for immersion service.

§ It is assumed that topcoats are intact and zinc is not exposed.

TABLE 20.5 1978 Coatings Cost Data for Structural Steel on the Ground*

East Coast; all work at jobsite

System	Cleaning	Dry film thickness (mils)	Total material cost	Field labor, equipment, and related costs				Total cost
				Surface preparation	Prime coat	Intermediate coat	Topcoat	
Two-coat alkyd	SP-2; SP-3	3.5	$ .047	$ .35	$ .10	$ ...	$ .10	$ .597
Three-coat alkyd	SP-2; SP-3	5.0	.069	.35	.10	.10	.10	.719
Three-coat alkyd	SP-6	5.0	.069	.45	.10	.10	.10	.819
Two-coat latex	SP-2; SP-3	3.5	.068	.35	.10	...	.10	.618
Three-coat latex	SP-2; SP-3	5.0	.101	.35	.10	.10	.10	.751
Three-coat latex	SP-6	5.0	.101	.45	.10	.10	.10	.851
Two-coat epoxy	SP-6	6.0	.12	.45	.15	...	.15	.897
Three-coat epoxy	SP-6	10.0	.199	.45	.15	.15	.15	1.099
Three-coat inorganic zinc, high-build epoxy	SP-10	11.0	.245	.55	.20	.15	.15	1.295
Three-coat inorganic zinc, high-build epoxy, aliphatic urethane	SP-10	8.5	.225	.55	.20	.15	.15	1.275
Three-coat vinyl	SP-10	9.5	.328	.55	.10	.10	.10	1.178
Three-coat chlorinated rubber	SP-6	10.0	.238	.45	.10	.10	.10	.988
Two-coat coal-tar epoxy	SP-6	16.0	.164	.45	.15	...	.15	.914

SOURCE: Presented at the NACE Annual Conference, Atlanta, Ga., March 1979.

NOTE:

1. All prices on square-foot basis. To convert to typical ton size, multiply by 250.
2. 30 percent spray loss.
3. A 10 percent area for touch-up is assumed.
4. Material based on 1978 prices.
5. Labor based on $20 per person-hour, loaded rate.

* For prices in the air after erection, add 25 percent for simple structures and 50 percent for intricate and complex structures. See Tables 20.1, 20.11, or 20.12 for factors on other structures and surfaces.

TABLE 20.6 1978 Coatings Cost Data for Structural Steel on the Ground*

East Coast; shop-primed; touch-up and topcoats at jobsite

| System | Cleaning | Dry film thickness (mils) | Total material cost | Labor, equipment, and related costs | | | | | Total cost |
| | | | | Shop | | Field | | | |
				Surface preparation	Prime coat	Touch-up	Intermediate coat	Top-coat	
Two-coat alkyd	SP-2; SP-3	3.5	$ .050	$ .25	$ .05	$ .10	$ ...	$ .10	$ .550
Three-coat alkyd	SP-2; SP-3	5.0	.072	.25	.05	.10	.10	.10	.672
Three-coat alkyd	SP-6	5.0	.072	.20	.05	.10	.10	.10	.622
Two-coat latex	SP-2; SP-3	3.5	.072	.25	.05	.10	...	.10	.572
Three-coat latex	SP-2; SP-3	5.0	.105	.25	.05	.10	.10	.10	.705
Three-coat latex	SP-6	5.0	.105	.20	.05	.10	.10	.10	.655
Two-coat epoxy	SP-6	6.0	.124	.20	.10	.10	...	.15	.674
Three-coat epoxy	SP-6	10.0	.203	.20	.10	.10	.15	.15	.903
Three-coat inorganic zinc, high-build epoxy	SP-10	11.0	.254	.25	.15	.10	.15	.15	1.054
Three-coat inorganic zinc, high-build epoxy, aliphatic urethane	SP-10	8.5	.234	.25	.15	.10	.15	.15	1.034
Three-coat vinyl	SP-10	9.5	.334	.25	.05	.10	.10	.10	.934
Three-coat chlorinated rubber	SP-6	10.0	.244	.20	.05	.10	.10	.10	.794
Two-coat coal-tar epoxy	SP-6	16.0	.172	.20	.10	.10	...	.15	.722

SOURCE: Presented at the NACE Annual Conference, Atlanta, Ga., March 1979.

NOTE:
1. All prices on square-foot basis. To convert to typical ton size, multiply by 250.
2. 30 percent spray loss.
3. 10 percent area for touch-up is assumed.
4. Material based on 1978 prices.
5. Field labor based on $20 per person-hour, loaded rate.

* For prices in the air after erection, add 25 percent for simple structures and 50 percent for intricate and complex structures. See Tables 20.1, 20.11, or 20.12 for factors on other structures and surfaces.

the life of a three-coat alkyd system with an SP-6 blast versus an SP-2 or SP-3 cleaning. In a mild environment, the difference is 7 versus 4 years; in a moderate environment, 5 versus 2 years; and in a severe environment, 3 versus 1 year. The epoxy coating system simply lasts longer and performs better over bare steel free of mill scale. Even if the 50-cent maintenance dollar seems to favor a "cheaper" initial system, repainting and surface-preparation costs are actually higher because of pitting, etc., and under the best of conditions will greatly inflate total painting expenditures.

From the above figures, it is obvious that the economics of doing coatings work is becoming increasingly important as financial pressures are felt. For this reason, it is extremely important that initial construction coating systems provide long trouble-free service with a minimum of expensive maintenance requirements.

20.3.2 Blasting costs   Abrasive-blasting costs vary with the type of cleaning required (see box), the location of the blasting, and the way in which the blasting is performed. Tables 20.7 through 20.9 present data comparing manual and automated field blasting costs. Automatic blasting is less costly for jobs with large square footages of steel, while manual sandblasting is cheaper for small jobs.

TYPES OF SURFACE PREPARATION

Surface preparation shall be in accordance with the Steel Structures Painting Council's *Steel Structures Painting Manual,* Vol. 2, using the following requirements:

SSPC SP-1. Solvent cleaning to remove all oil, grease, dirt, soil, salts, and contaminants by cleaning with solvent, vapor, alkali, emulsion, or steam.

SSPC SP-2. Hand-tool cleaning to remove loose rust and mill scale by hand chipping, scraping, sanding, or wire brushing. Surface cleanliness shall be equivalent to Swedish pictorial standards.

SSPC SP-3. Power-tool cleaning to remove loose rust and mill scale

---

TABLE 20.7   1980 Blasting Costs for Tank Plate

Surface preparation	Manual blast with sand at jobsite	Automated blasting in shop	
		50,000 ft²	100,000 ft²
SSPC SP-6	$.31	$.19	$.16
SSPC SP-10	.38	.25	.21
SSPC SP-5	.44	.31	.26

TABLE 20.8   1980 Blasting Costs for Ladders, Handrails, Guardrails, Cages, Etc.

Surface preparation	Manual blast with sand at jobsite	Automated blasting in shop	
		50,000 ft²	100,000 ft²
SSPC SP-6	$.63	$.25	$.21
SSPC SP-10	.75	.35	.30
SSPC SP-5	.88	.40	.34

by power-tool chipping, descaling, sanding, wire brushing, and grinding. Surface cleanliness shall be equivalent to Swedish pictorial standards.

SSPC SP-5. White-metal blast cleaning to remove all visible rust, mill scale, and foreign matter by blast cleaning, using a wheel or nozzle with sand, grit, or shot. Surface cleanliness shall be equivalent to Swedish pictorial standards.

SSPC SP-6. Commercial blast cleaning until at least two-thirds of each element of the surface area is free of all visible residues. Surface cleanliness shall be equivalent to Swedish pictorial standards.

SSPC SP-7. Brush-off blast cleaning which will remove all except tightly adhering residues of mill scale and rust and will expose numerous evenly distributed flecks of underlying metal. Surface cleanliness shall be equivalent to Swedish pictorial standards.

SSPC SP-10. Blast cleaning to white-metal cleanliness except for 5 percent of the surface, which shall be tightly adhering oxides or paint residues or light discoloration. Surface cleanliness shall be equivalent to Swedish pictorial standards.

Most corrosion engineers agree that the cleaner the surface and the sharper the profile, the better and longer a coating system will perform. The development of blasting media with new and better techniques has started. For example, an alumina-containing by-product is ecologically

TABLE 20.9   1980 Blasting Costs for Heavy Structural Steel

Surface preparation	Manual blast with sand at jobsite	Automated blasting in shop	
		50,000 ft²	100,000 ft²
SSPC SP-6	$.38	$.24	$.19
SSPC SP-10	.44	.39	.31
SSPC SP-5	.50	.48	.38

acceptable and results in faster cleaning. It has a lower breakdown rate, provides excellent profile qualities (see box), and is recyclable to a limited extent. Zinc by-products, fly-ash by-products, and slag by-products also are being developed. In addition, there is ongoing research and testing on the use of carbon dioxide pellets for blasting. Until the new units become commercially available and environmental regulations are resolved, one acceptable approach is field blasting with enclosed automated and semiautomated blasting units.

RELATION OF ECONOMICS TO SURFACE PREPARATION

It is possible to achieve capital cost savings by deferring both surface preparation and painting, but to do this the specification writer and maintenance engineers must have intimate knowledge of the products used and employ a carefully monitored inspection program. For example, a new tank or vessel with a very tight mill scale can be allowed to rust for about 1 year. This procedure delays surface-preparation costs, and weathering makes it easier to remove the mill scale. The danger is that you must be sure to blast before pitting occurs. Also, with field blasting you run the risk of the abrasive's entering critical rotating equipment such as pump seals.

Choosing the proper blast abrasive is critical to economics. Using a very coarse abrasive produces higher surface profiles, creating deep valleys that consume more paint and high peaks that poke through primers (and even topcoats) to create pinholes which lead to poor

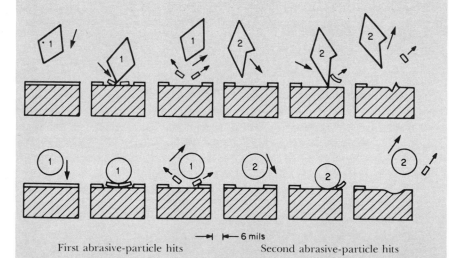

First abrasive-particle hits   Second abrasive-particle hits

Removal of mill scale with very coarse grit (above) and removal with round steel shot (below) leave different surface profiles, but both methods can lead to paint adhesion problems.

coating performance. On the other hand, round shot alone peens the surface, producing no roughness to enhance paint adhesion.

Ideally, selected medium or fine abrasives specifically designed for blasting are worth the additional abrasive cost. Experienced paint engineers recognize that coating performance and overall economics are greatly influenced by the surface profile. Paint manufacturers specify optimum surface profile, and it is important to adhere to their recommendations, especially for alkyd, chlorinated-rubber, vinyl, and epoxy paints that do not protect the surface cathodically (as do zinc-rich paints).

In general, the Steel Structures Painting Council does not recommend abrasives coarser than 16-mesh. But even if the correct abrasive is chosen, you must be concerned with the thoroughness of the blast in order to avoid hackles. These are very high peaks (up to 6 mils, or 0.1524 mm) that are more common with steel shot and grit than with nonmetallic abrasives.

After blasting, a postcured inorganic zinc primer can be applied and allowed to set for an additional period, say, up to 4 years, before applying a topcoat. Here again, careful routine inspection is an absolute requirement. Primer failure can lead to expensive recoating. Painting over cracked mill scale (poor surface preparation) will lead to coating failure and to corresponding expenses that will drastically affect the economics of the system.

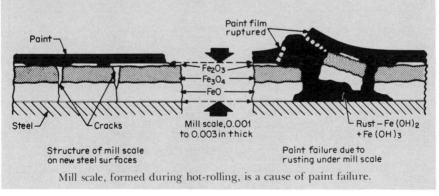

Mill scale, formed during hot-rolling, is a cause of paint failure.

Automated blasting produces quality coating surfaces and meets current industry regulations. The operation utilizes an automated or semiautomated blasting unit near or at the jobsite. With such a unit it is possible to blast random-length pipe, structural steel, and tank plate prior to fabrication. Immediately after blasting, the prime coat is applied.

One of the keys to automated blasting is the size of the job. If the job is not large enough, conventional sandblasting provides lower-cost surface preparation. However, when the job size for a given type of steel surface is in the range of 300,000 to 700,000 ft² (27,870 to 65,030 m²) or larger, automated blasting is lower in cost.

The usual form of automated blasting is a centrifugal wheel that throws recoverable steel abrasive. Other semiautomated blasting equipment employs compressed air and recoverable metallic abrasive. The exact cost of the blasting unit and its attendant facilities can vary greatly. The larger, more costly units have generally higher production rates. Figure 20.1 presents cost data for different-size jobs, comparing automated and manual blasting.

Because automatic centrifugal-wheel shop blasting costs about half as much as field blasting, it is not always true that blasting is more expensive than hand- or power-tool cleaning. Priming costs are significantly lower if priming is done in the shop rather than in the field. This is particularly true for piping (see box). However, the blasting prices of steel fabricators without automatic equipment frequently approximate field blasting prices. It is important, therefore, that qualified fabricators be included on your bid list if you want to get the price advantage of wheel blasting. You cannot overlook the quality of shopwork, which usually produces better jobs because conditions are controlled. Because no rigging or staging is required, application is easier on the ground; spray loss is reduced, and personal safety is increased. Compromises are often forced on field paint crews to accommodate construction schedules, field problems, and

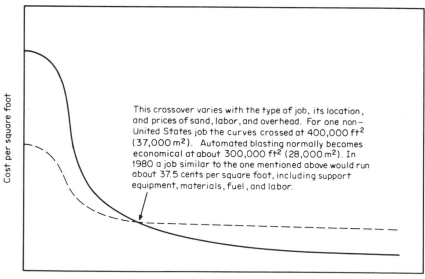

This crossover varies with the type of job, its location, and prices of sand, labor, and overhead. For one non-United States job the curves crossed at 400,000 ft² (37,000 m²). Automated blasting normally becomes economical at about 300,000 ft² (28,000 m²). In 1980 a job similar to the one mentioned above would run about 37.5 cents per square foot, including support equipment, materials, fuel, and labor.

Cost per square foot

Square feet to blast

Fig. 20.1   Cost comparison of manual blasting (broken line) and automatic blasting (solid line) at the jobsite.

the weather. These factors greatly influence the economics of a job. Shop painting gives greater control over such variables.

COST ESTIMATES: FIELD VERSUS SHOP FOR BLAST AND PRIME STRAIGHT-RUN PIPE

Item	Shop*	Field Nonunion	Field Union
1. Blast and prime	$.35†	$ .40	$ .45
2. Unload railcars	.018	.095	.095
3. Freight from shop to jobsite	.068		
4. Unload truck at jobsite	.069	.069	.069
5. Movements of pipe at field			
Railcars to laydown		.069	.069
Laydown to blast area		.069	.069
Blast area to laydown		.069	.069
Laydown to erection	.069	.069	.069
6. Clean up sand		.003	.003
7. Dunnage and yard preparation		.028	.028
8. Rain delay at 13 percent		.0776	.1076
Total	.574	.9486	1.0286
Difference		.3746	.4546
Savings possible at 500,000 ft²		$187,300	$227,300

* Color coding and heat number retention carry an extra price. For color coding add $0.015 per square foot; for heat number retention, $0.003 per square foot.
† Unloading and loading trucks included in paint price.

Assumptions

1. 200,000 ft of pipe = 500,000 ft² of blasting and painting (standard-weight pipe; average size, 10 in).
2. Productivity of field blasters and painters is estimated at 1000 to 1200 ft² per worker per shift.
3. Productivity of cherry picker = 2000 ft of pipe per day per unit.
4. Estimated rates for equipment and personnel per hour:

   a. Cherry picker $17.00
   b. Front-end loader   17.00
   c. Truck and driver   14.00
   d. Operator, cherry picker 10.00
   e. Operator, loader   13.00
   f. Swampers   8.00

5. Blast yard size of 5 acres proposed at $40,000 per acre; residual values of blast yard after project completion, $190,000.

Preparation	$200,000
Residual value	190,000
Cost	$ 10,000

6. Dunnage and rails estimated at $4,000.
7. Rain delays estimated at 4 days per month.
8. Sand required for blasting straight-run pipe, 300 tons.
9. Specifications for quality of blast and painting materials:

Surface preparation: SSPC SP-10 (near white)
Primer: 2.5 to 3 mils dry film thickness, inorganic zinc

Calculations

1. Field unloading of railcars

Item	Dollars per hour
1 cherry picker	$17.00
1 operator	10.00
4 swampers at $8	32.00
	$59.00

Unloading time: 800 h

Cost, 800 h × $59	$47,200
Cost per square foot, $47,200/500,000	$ .095

2. Field unloading trucks and movements in field

Item	Dollars per hour
1 cherry picker	$17.00
1 operator	10.00
2 swampers at $8	16.00
	$43.00

Unloading time: 800 h

Cost, 800 h × $43	$34,400
Cost per square foot, $34,400/500,000	$ .069

3. Railcar unloading at shop*

$$\text{Cost} = \frac{(\text{length})(\text{weight per foot})(\text{cost})}{(\text{length})(\text{square feet per linear foot})(100)}$$
$$= \frac{(40)(40.48)(0.125)}{(40)(2.81)(100)}$$
$$= \frac{202.4}{11240}$$
$$+ \$0.018 \text{ per square foot}$$

4. Freight to jobsite from shop at 44 cents per hundredweight†

$$\text{Cost} = \frac{(\text{length})(\text{rate})(\text{handling charge})(\text{weight per foot})}{(\text{length})(\text{square feet})(100)}$$

$$= \frac{(40)(0.44)(1.10)(40.48)}{(40)(2.81)(100)}$$

$$= \frac{783.69}{11240}$$

$$= \$0.069 \text{ per square foot}$$

5. Sand removal: 300 tons, 4 days' work

Item	Rate
1 front-end loader	$17.00
1 operator	13.00
1 truck and driver	14.00
	$44.00

$$\text{Cost} = \frac{32 \text{ h} \times \$44}{500,000} = \$0.003 \text{ per square foot}$$

SOURCE: Custom Pipe Coatings, Houston, Tex.
 * Similar calculations for 16-in, 20-in, and 30-in pipe sizes all equal 18 cents per square foot.
 † Calculations for freight are also the same for 16-in, 20-in, and 30-in pipe.

An actual case history shows that automated blasting costs are less than conventional manual blasting. On a project outside the United States, the engineer was faced with cleaning 1,500,000 ft² (139,350 m²) of tank plate and 1,400,000 ft² (130,060 m²) of aboveground pipe over an 18-month period. In addition, approximately 900,000 ft² (83,610 m²) of pipe for underground service required blasting, priming, and tape wrapping. An initial inquiry disclosed that good blasting sand was not available near the jobsite, and sand would have to be imported from outside sources. The closest available sand for blasting that was suitable for the project cost $45 per ton, exclusive of shipping charges. It was estimated that approximately 15,200 tons of sand would be required for manual blasting. If used, sand delivered to the jobsite would have cost more than $1 million. In addition to this sizable cost, the project was faced with weather that was not conducive to manual operation. The site was located in an area with high annual rainfall. To meet the schedule, alternative methods would be required.

Investigations were made of the feasibility of using facilities located elsewhere. Each alternative studied proved to be uneconomical. It was then decided that shipping and erecting a semiautomatic blasting unit at the jobsite possibly would be economical. After intensive cost studies, it was decided to utilize a system which fit the needs of the project, employing recoverable metallic abrasive. The blasting unit, blasting equipment, and transportation to the point of erection were negotiated for use during an 18-month schedule. In addition, labor rates and the costs of personnel accommodation were determined. The data in Table 20.10 represent a cost comparison of the use of the semiautomated unit versus manual blasting.

Again, Fig. 20.1 represents, in general, estimated costs to perform the work by using the two methods shown in the table. For this job, the automated-blasting cost curve crossed the manual curve at approximately 100,000 ft² (37,000 m²). These coats would not apply to other projects. Specific costs, considerations, and comparisons must be computed for each project, for they depend on factors relevant to the specific job.

TABLE 20.10   1978 Estimated Cost per Square Foot for a Jobsite Blasting Plant*

	Manual	Semiautomatic
Material (inorganic zinc)	$ .090	$.090
Sandblasting sand at $80 per ton	.320	. . .
Abrasive at $400 per ton	. . .	.012
Labor	.281	.092
Fixed costs (equipment, etc.)	.365	.185
Total	$1.056	$.379

* The difference of $.67 per square foot represents 64 percent savings achieved.

On a similar overseas project on which an automatic system was to be used, surface preparation including support equipment, materials, fuel, and labor ran approximately 29 cents per square foot. If this had been a manual blasting operation, the cost per square foot would have been approximately 46 cents. In addition, it was estimated that 111,000 person-hours would have been consumed to blast and prime the 2,300,000 ft² (213,670 m²) required.

The semiautomatic field blasting system has also been used on projects in the United States. No history or cost data are presently available, but there is little doubt that this system will prove more economical than manual blasting for large projects.

20.3.3 Field painting versus factory or shop priming plus field topcoating

Many houses are being built with factory-primed wood siding or factory-primed Masonite siding. In fact, the trend toward factory priming is spreading into several kinds of construction. Architects should realize the limitations of these primers, especially when lengthy

exposure to weather is anticipated. Chalking can occur, resulting in reduced intercoat adhesion and causing other paint problems.

Nonetheless, the economics of construction favor factory precoating
and shop priming in many cases, and paint specification writers should
weigh the dollar advantages carefully. The following examples illustrate
the principles to be used in making these cost comparisons. While they
cover epoxy and alkyd systems, the same exercise will provide comparative values for any factory-primed, field-topcoated system.

Example 1 shows a cost comparison of field painting and shop priming
with touch-up and topcoating in the field on the ground. The system is a
three-coat epoxy, using an SSPC SP-6 commercial-grade blast; an East
Coast location is assumed. See the eighth system in Tables 20.5 and 20.6.

Example 1

Three-coat epoxy, field-applied: Material cost practical† (Table 20.3)			Three-coat epoxy, shop-primed*; field touch-up and topcoated: Material cost practical† (Table 20.3)		
	Dry film thickness (mils)	Cost per square foot		Dry film thickness (mils)	Cost per square foot
Epoxy primer	2.0	$ .041	Epoxy primer	2.0	$ .041
High-build epoxy	4.0	.079	High-build epoxy	4.0	.079
High-build epoxy	4.0	.079	High-build epoxy	4.0	.079
Total system	10.0	$ .199	Touch-up (10 percent of primer)		.004
Labor, equipment, and related costs			Total system	10.0	$ .203
SP-6 blast and prime (Table 20.1)		$ .45			
Apply prime coat (Table 20.1)		.15	Labor, equipment, and related costs		
Apply two coats (Table 20.1)		.30	Shop blast (Table 20.2)		$ .20
Total system, labor, etc., cost		$ .90	Shop prime (Table 20.2)		.10
Material		$ .199	Touch-up (Table 20.1)		.10
Labor, etc.		.90	Apply two coats (Table 20.1)		.30
Total system cost		$1.099	Total system, labor, etc., cost		.70
			Material		$ .203
			Labor, etc.		.70
			Total system cost		$ .903

* Using wheel blasting equipment.
† 30 percent spray loss.

Example 2 involves two different systems being considered for the
same job, including estimated service life and cost per year. This project
is assumed to be located on the East Coast with a moderate environment.

The two systems being considered are a three-coat epoxy with an SSPC SP-6 commercial blast and a three-coat alkyd using SSPC SP-2 or SSPC SP-3 hand- or power-tool cleaning. All work is assumed to be done in the field. See the eighth system in Table 20.5 and the second system in Table 20.6.

Example 2

Three-coat epoxy, field-applied Material cost practical* (Table 20.3)			Three-coat alkyd, field-applied Material cost practical* (Table 20.3)		
	Dry film thickness (mils)	Cost per square foot		Dry film thickness (mils)	Cost per square foot
Epoxy primer	2.0	$ .041	Alkyd primer	1.5	$ .025
High-build epoxy	4.0	.079	Gloss alkyd	1.5	.022
High-build epoxy	4.0	.079	Gloss alkyd	1.5	.022
Total system	10.0	.199	Total system	4.5	.069
Labor, equipment, and related costs			Labor, equipment, and related costs		
SP-6 blast (Table 20.1)		$ .45	SP-2 or SP-3 clean (Table 20.1)		$ .35
Apply prime coat (Table 20.1)		.15	Apply prime coat (Table 20.1)		.10
Apply two coats (Table 20.1)		.30	Apply two coats (Table 20.1)		.20
Total system, labor, etc., cost		$ .90	Total system, labor, etc., cost		$ .65
Material		$ .199	Material		$ .069
Labor, etc.		.90	Labor, etc.		.65
Total system cost		$1.099	Total system cost		$ .719
Expected life in moderate environment		8 years	Expected life in moderate environment		2 years
Cost per year		$ .137			$ .36

* 30 percent spray loss.

The basic principles covered to this point center in knowing and defining the specific objectives, factors, and conditions of the plant to be painted. Work Sheets A and B (Tables 20.11 and 20.12) are designed to record this information. When used with the cost and service-life tables, they will guide the specifying engineer through the basic cost calculations, direct comparisons of systems being considered, and, ultimately, system selection.

Table 20.11 covers field painting, and Table 20.12 covers shop priming, field touch-up, and topcoating. Various factor percentages are included in the work sheets to aid in calculating costs for application to structural steel in place, piping, and tank work, estimated service life, and cost per year.

TABLE 20.11 Cost Work Sheet A
All work at jobsite

Location _____

Material cost practical* (Table 20.3)

	Type	Dry film thickness	Cost per square foot†
Prime coat	_____	_____	$_____
Intermediate coat	_____	_____	_____
Topcoat	_____	_____	_____

Total dry film thickness $_____ (1) Total material
                                         cost

Labor, equipment, and related costs
SP-_____blast clean (Table 20.1) $_____
Prime coat (Table 20.1)   _____
Intermediate coat (Table 20.1)   _____
Topcoat   _____

                    $_____ (2) Total labor,
                               equipment, and
                               related costs
                    $_____ (1) Material
                    $_____ (2) Labor, etc.

Total system cost for structural steel on the ground‡ $_____ (3)
If calculating in the air or on other surfaces, use
 percentages below to factor total cost   _____ percent
Adjusted total system cost per square feet $_____ (4)
Environment (Table 20.4) _____
Expected service life (Table 20.4)   _____ (5) Years
Cost per year (dollars per square foot per year);
 (3) or (4) ÷ 5   _____

SOURCE: Presented at the NACE Annual Conference, Atlanta, Ga., March 1979.
* 30 percent loss by spray; 10 percent loss by brush or roller.
† To convert to typical ton-size cost, multiply by 250.
‡ For prices in the air, add 25 percent for simple structures and 50 percent for intricate structures. For tank work, deduct 25 percent. For piping 4 to 6 in in diameter, use above cost (3). For piping 1 to 2 in in diameter, add 25 percent. For piping 6 to 48 in in diameter, deduct on a sliding scale up to 25 percent for 48-in-diameter pipe.

It becomes apparent that economic justification is as important to the selection process as the choice of coating. The analysis takes two forms:

▪ *Cost per square foot per year.* This is the sum of all costs required to apply and maintain a coating system on a surface divided by the life of that system.

▪ *Discounted-cash-flow analysis, or capitalized cost.* This analysis, somewhat more complicated, considers the value of the money used in the project. There are many variations of this analysis, but essentially it says that a dollar not used can be invested for return and spent at a later date.

In Example 2, it is true that the epoxy cost per year of 13.7 cents is substantially lower than the cost of 37 cents per year for the alkyd system. Is the saving really 22 cents per year per square foot? No! To get this

TABLE 20.12   Cost Work Sheet B
Shop-primed, touch-up and topcoats at jobsite

Location _____

Material cost practical* (Table 20.3)

	Type	Dry film thickness	Cost per square foot†
Prime coat	_____	_____	$_____
Intermediate coat	_____	_____	_____
Topcoat	_____	_____	_____
Touch-up (10 percent of primer cost)		Total dry film thickness	_____
		$_____	(1) Total material cost

Labor, equipment, and related costs

SP-_____blast clean (Table 20.2) $_____
Prime coat (Table 20.2)   _____
Touch-up (Table 20.1)   _____
Intermediate coat (Table 20.1)   _____
Topcoat (Table 20.1)   _____

                                                              $_____ (2) Total labor,
                                                                                equipment, and
                                                                                related costs
                                                              $_____ (1) Material
                                                              $_____ (2) Labor, etc.
Total system cost for structural steel on the ground‡ $_____ (3)
If calculating in the air or on other surfaces, use
  percentages below to factor total cost   _____ percent
Adjusted total system cost per square feet $_____ (4)
Environment (Table 20.4) _____
Expected service life (Table 20.4)   _____ (5) Years
Cost per year (dollars per square foot per year);
  (3) or (4) ÷ 5   _____

SOURCE: Presented at the NACE Annual Conference, Atlanta, Ga., March 1979.
* 30 percent loss by spray; 10 percent loss by brush or roller.
† To convert to typical ton-size cost, multiply by 250.
‡ For prices in the air, add 25 percent for simple structures and 50 percent for intricate structures. For tank work, deduct 25 percent. For piping 4 to 6 in in diameter, use above cost (3). For piping 1 to 2 in in diameter, add 25 percent. For piping 6 to 48 in in diameter, deduct on a sliding scale up to 25 percent for 48-in-diameter pipe.

cost-per-year "saving," it is necessary to spend 38 cents per square foot, or 53 percent more in construction dollars. When you consider the 50-cent maintenance dollar, the financial picture changes again.

20.4   Discounted Cash Flow   In DCF the annual cash flow is considered a periodic payment to an annuity whose length is equal to the life of the facility. The value of the annuity is equal to the sum of periodic payments (annual cash flows) plus accumulated compounded interest. While the discussion is beyond the scope of this *Handbook,* specification writers should understand that more sophisticated approaches to economic analysis would include inflation, depreciation, and other factors. The paint job is just one portion of the total investment, and the architect should be able to evaluate that part of the investment in terms of the value of money, cash flow, discount rate, and profitability. Previous sections of this presentation dealt with costs. Costs are basic as the starting point for all economic-justification calculations. In recent years coating costs have often been expressed as dollars per square foot per year of coating life.

As previously noted, when an epoxy system is compared with an alkyd system, it is seen that the epoxy system has a higher initial cost and requires more sophisticated and expensive surface preparation. It is less expensive than the alkyd system on the basis of cost, or dollars per square foot per year of coatings life. The epoxy coating system lasts longer.

By using these comparisons it is possible to evaluate more expensive coatings on the basis of their possible longer life. Additionally, coatings of higher corrosion resistance and greater initial cost usually require less maintenance. However, the benefits achieved in lower maintenance are not reflected in the usual expression of dollars per square foot per year. With less maintenance, fewer dollars flow out, and cash flow is introduced into the picture. To understand DCF, we must focus attention on the time value of money as well as on cash flow.

Choosing the correct paint becomes an important business decision that relates to what the architect or the engineer expects from the coatings system. A plant that experiences corrosion due to paint failures faces downtime costs that far exceed the cost of a painting system. A hotel or an office building, on the other hand, will not face shutdown, but it can deteriorate, face loss of business, and experience heavy maintenance costs if incorrect coatings are specified. While the differences between discounted and undiscounted returns can be small, the architect and the engineer should know when to use simple return on investment, which does not take into consideration the value of capital, and when to use DCF.

Money has value with respect to time. An expenditure for a capital improvement such as a painting job calls for the spending of money. If that money had not been expended, it could have been invested at a given rate of return and earned interest. Since it had been expended, it was not available for other investments. For example, a dollar received and invested today is worth more than a dollar to be received 1 year from today (see Tables 20.13 and 20.14). Invested at 8 percent, the dollar will be worth \$1.08 in 1 year. Expressed mathematically, the future amount a of \$1 invested today equals:

$$a = (1 + i)^n$$

where i = interest rate
n = number of years

In light of the above, these are benefits paid back or, as expressed by

TABLE 20.13   Future Amounts at Various Interest Rates

Periods	Percent						
	2	3	4	5	6	7	8
0	1.	1.	1.	1.	1.	1.	1.
1	1.02	1.03	1.04	1.05	1.06	1.07	1.08
2	1.0404	1.0609	1.0816	1.1025	1.1236	1.1449	1.1664
3	1.0612	1.0927	1.1249	1.1576	1.1910	1.2250	1.2597
4	1.0824	1.1255	1.1699	1.2155	1.2625	1.3108	1.3605
5	1.1041	1.1593	1.2167	1.2763	1.3382	1.4026	1.4693
6	1.1262	1.1941	1.2653	1.3401	1.4185	1.5007	1.5869
7	1.1487	1.2299	1.3159	1.4071	1.5036	1.6058	1.7138
8	1.1717	1.2668	1.3686	1.4775	1.5938	1.7182	1.8509
9	1.1951	1.3048	1.4233	1.5513	1.6895	1.8385	1.9990
10	1.2190	1.3439	1.4802	1.6289	1.7908	1.9672	2.1589

accountants, cash-flow annuities. Less maintenance or better plant operation or benefits paid back for any reason are examples. These may be expressed as the present value of these benefits or annuities. They may be expressed mathematically as

$$P = \frac{1 - \dfrac{1}{(1 + i)^n}}{i}$$

TABLE 20.14 Present Value of Annuity of $1

Periods	Percent							
	6	8	10	12	14	16	18	20
1	0.943	0.926	0.909	0.893	0.877	0.862	0.847	0.833
2	1.833	1.783	1.736	1.690	1.647	1.605	1.566	1.528
3	2.673	2.577	2.487	2.402	2.322	2.246	2.174	2.106
4	3.465	3.312	3.170	3.037	2.914	2.798	2.690	2.589
5	4.212	3.993	3.791	3.605	3.433	3.274	3.127	2.991
6	4.917	4.623	4.355	4.111	3.889	3.685	3.498	3.326
7	5.582	5.206	4.868	4.564	4.288	4.039	3.812	3.605
8	6.210	5.747	5.335	4.968	4.639	4.344	4.078	3.837
9	6.802	6.247	5.759	5.328	4.946	4.607	4.303	4.031
10	7.360	6.710	6.145	5.650	5.216	4.833	4.494	4.192
11	7.887	7.139	6.495	5.938	5.453	5.029	4.656	4.327
12	8.384	7.536	6.814	6.194	5.660	5.197	4.793	4.439
13	8.853	7.904	7.103	6.424	5.842	5.342	4.910	4.533
14	9.295	8.244	7.367	6.628	6.002	5.468	5.008	4.611
15	9.712	8.559	7.606	6.811	6.142	5.575	5.092	4.675

where P = present value of annuity of $1
      i = interest rate
      n = number of years

Let us consider a case in which management requires a 10 to 15 percent return on its investment after taxes. If a proposed investment will not yield this rate of return, it is rejected. How is the rate of return calculated? It includes the moneys initially expended and their value recognized as increasing with time:

Present time		Future amount
$1	$n = 2$	$1.17
	$i = 8$ percent	

It includes savings or inflow of cash, evaluated at its present value although anticipated in future payments (savings). These may be expressed as:

Present-value annuity		Future time
	$1	$1
$1.78	$n = 2$	
	$i = 8$ percent	

These tables assume a constancy and a repetitive occurrence of a given action. This is not usually the case in a plant where many factors may cause costs and benefits to vary from year to year. This may be illustrated by the following schematic diagram. For a 10-year period $n = 1, 2, 3 \ldots$ 10, moneys being expended are represented by an upward arrow, and moneys or benefits received by a downward arrow. The length of the arrow may be used to approximate the size of the money flow.

n years: 0 1   2   3   4   5   6   7   8   9   10

The net flow of cash for each year is treated as an expense or as a benefit, being discounted and evaluated in terms of present value. The sum of the present values can then be evaluated in terms of the corporate profit requirements. As stated above, for most companies these requirements usually are 10 to 15 percent per year after taxes.

Example 2 presents cost data on a three-coat epoxy system and a three-coat alkyd system. Both systems were applied in the field, the epoxy to a sandblasted surface at a total cost of $1.099 per square foot and the alkyd to a wire-brushed surface at a total cost of $0.719 per square foot. Both total costs included materials, surface preparation, and application.

Let us compare these alternative systems by using DCF techniques for an industrial-plant coating job. Let us assume that the job calls for 1 million ft² (92,900 m²) to be coated and that the corrosive nature of the jobsite will require that the alkyd system be repainted every 3 years while the epoxy system will need touch-up painting after 7 years. Cost for the alkyd repaint jobs is assumed to be $0.719 per square foot and for the epoxy touch-up $0.50 per square foot. Although a partial plant shutdown may be required for each painting or touch-up job, no costs will be entered for shutdowns.

Let us assume also that the corporate profit requirement is 15 percent for a given investment. The present value of $1, P, invested for n years at an interest rate of i equals

$$P = \frac{1}{(1 + i)^n}$$

Assuming an interest rate of 15 percent,

$$P = \frac{1}{(1.15)^n}$$

n	P
1 year	0.870
2 years	0.756
3 years	0.658
4 years	0.572
5 years	0.497
6 years	0.432
7 years	0.376
8 years	0.327
9 years	0.284
10 years	0.247

These factors are applied to the net cash flow after taxes and are known as present-value DCF factors.

Table 20.15 compares the two coating systems by using DCF procedures. For this example we will not include depreciation, which varies from job to job depending upon the life expectancy of the investment, tax rulings, and, in some cases, tax incentives. Also, depreciation may not be straight-line. If depreciation is to be included, an appropriate column must be set up following column 3, Table 20.7, "Cost of column 2 over column 1." Enter in this column the depreciation values year by year, summed for years in which moneys are being carried for investments being depreciated over a number of years. Carrying the calculations

TABLE 20.15 Discounted-Cash-Flow (DCF) Comparative-Cost Analysis of an Epoxy Coating System versus an Alkyd Coating System

	(1) Cost of epoxy system	(2) Cost of alkyd system	(3) Cost of column 2 over column 1	(4) Income (expense; column 3 plus column 4)	(5) Assumed tax of 50 percent	(6) Net cash flow after tax	(7) Present value of $1 at 15 percent DCF factor	(8) DCF (from columns 6 and 7)
Year								
New plant	$1,099,000	$719,000	−$280,000	...	...	−$280,000	1	−$280,000
1	0	0	0	0	0	0	0.870	0
2	0	0	0	0	0	0	0.756	0
3	0	719,000	+ 719,000	$719,000	$359,500	+ 359,500	0.658	+ 236,551
4	0	0	0	0	0	0	0.572	0
5	0	0	0	0	0	0	0.497	0
6	0	719,000	+ 719,000	719,000	359,500	+ 359,500	0.432	+ 155,304
7	500,000	0	− 500,000	(500,000)	(250,000)	− 250,000	0.376	− 94,000
8	0	0	0	0	0	0	0.327	0
9	0	719,000	+ 719,000	719,000	359,500	+ 359,500	0.284	102,098
10	0	0	0	0	0	0	0.247	0
								+$119,953*

SOURCE: Presented at the NACE Annual Conference, Atlanta, Ga., March 1979.
* Net benefit of epoxy system over alkyd system.

through and summing by years show that the initial investment in an epoxy coating system over a sandblasted surface is an acceptable investment compared with an alkyd coating system, returning $119,953 in excess of a 15 percent return on investment.

If there are other expense items such as intermediate touch-up coats or plant shutdown costs for the two coating systems being compared, they should be added at the appropriate years. Shutdown costs are not expenses normally incurred by office buildings or other jobs, but economic painting criteria are obviously more important for plants that could incur major costs because of corrosion failures.

Table 20.16 (Work Sheet C) presents a list of items to be included in a DCF analysis. This list may be expanded to include other items of expense or income as may be required for a specific job evaluation. Table 20.19 may be used as an illustrative basis of Work Sheet C. Year-by-year costs and benefits may be compared for the investments being considered, and the procedure may be carried beyond 10 years if management so requires.

A high-cost, long-term coating system is usually justified by an analysis of the cost per square foot per year. The high initial cost is more than offset by reducing maintenance that involves periodic touch-up, repair, or replacement costs. It is hard to justify the more expensive system by an analysis of capitalized costs alone.

In choosing the correct corrosion protection you must also consider the cost effectiveness of cathodic systems. It is common to protect buried pipelines, storage tanks, etc., from corrosion galvanically. The equipment to be protected is electrically coupled to sacrificial anodes com-

TABLE 20.16   Work Sheet C: Discounted-Cash-Flow (DCF) Analysis

For year n	
1. Coating system A cost	_____
2. Coating system B cost	_____
3. Cost difference	_____
4. Depreciation	_____
5. Gross income (expense)	_____
6. Taxes	_____
7. Net cash flow after tax	_____
8. Present-value DCF factor	_____
9. DCF (item 7 times item 8)	_____

SOURCE: Presented at the NACE Annual Conference, Atlanta, Ga., March 1979.

INSTRUCTIONS:
1. Complete the above work sheet for each year n for the total period of comparison.
2. Sum the DCF values for each year.

monly made of magnesium, aluminum, or zinc or to a rectifier or some other source of direct current through an inert anode. When applicable, galvanic or cathodic protection does an effective job for a fraction of the cost connected with surface preparation and the application of a protective coating system. For instance, cathodic protection for the immersed areas of an offshore platform costs an estimated 5 cents per square foot per year. However, there are drawbacks. First, the systems require checking and continuous maintenance. Second, they will not protect against atmospheric corrosion or provide complete protection to surfaces which are alternately wet and dry. Third, there is some hazard connected with their application under certain conditions because hydrogen gas is liberated at the cathodes.

20.5 Maintenance Costs   Paint maintenance prevents premature paint failure from easily corrected causes. It lengthens the time between painting cycles and reduces the cost of keeping a surface painted (when this cost is calculated on an annual basis over the lifetime of the coating).

Repainting constitutes the next cycle of applying a paint system. It comes after the failure of one or more of the functions of paint: decoration, protection, or a more specialized function. The failure of only one function may call for repainting while the other functions are still being served.

Until recently it was not unusual for a coating job to require four or five separate coats. With the advent of high-build coatings, coating companies are offering single-coat systems. Each coat, or layer of paint, that is eliminated saves application costs. Former coatings systems required three, four, or even five layers of coating to attain the required thickness of coating film. Today we have numerous high-build vinyls, high-build epoxies, multimil alkyds, high-build urethanes, and others that build to 10 mils (0.254 mm) in one or two coats. These coatings systems, which were exceptional a few years ago, are now commonplace.

Each coat costs approximately 10 to 25 cents per square foot to apply, excluding the cost of materials. The cost savings of eliminating a coat are obvious. Tables 20.17 and 20.18 present data on costs for a few coatings systems, and Table 20.19 compares the economics and performance characteristics of several coatings systems. Not only does the application of fewer coats afford economies in application, but it can enhance the quality of the coating job.

Multicoat systems require curing and drying between coats. As the coating system cures and dries, the outer surface normally picks up impurities such as dust, smog, and smog chemicals from the environment. These impurities between coats result in what the industry terms "intercoat contamination." Such contamination decreases bond strength and adhesion between coats.

TABLE 20.17   1978 Coating Costs for Steel Tank Plate at Jobsite

Material	Cost per gallon	Percent of solids by volume	Applied cost per square foot	
			50,000 ft²	100,000 ft²
Primers				
Inorganic zinc at 2.5 mils	$24.00	50	$ .225	$ .200
Zinc chromate at 1.5 mils	10.00	50	.125	.112
Red lead at 1.5 mils	12.50	50	.137	.125
Finishes				
High-build epoxy at 5.0 mils	15.00	50	$ .237	$ .225
High-build urethane at 5.0 mils	24.00	40	.387	.362
Alkyd enamel at 3.0 mils	10.00	50	.200	.175

Painters, because they have fewer coats to apply, can choose their painting conditions preferentially. This freedom allows for better weather for painting or an extension of the days in which a coating can be applied.

Maintenance coating costs are frequently much higher than initial coating costs. Since the major cost of any coating job is usually associated with surface preparation and priming, maintenance programs should be designed to protect the investment in these operations.

The application of a fresh topcoat is not an involved or necessarily expensive job, but if surface preparation is required, costs can rise astronomically. Preparation costs on the ground by automatic blasting versus manual blasting run from 15 to 38 cents per square foot, depending on the type of surface preparation required and the type of steel being blasted.

TABLE 20.18   1978 Coating Costs for Ladders, Handrails, Guardrails, Cages, Etc., at Jobsite

Material	Cost per gallon	Percent of solids by volume	Applied cost per square foot	
			50,000 ft²	100,000 ft²
Primers				
Inorganic zinc at 2.5 mils	$24.00	50	$ .450	$ .400
Zinc chromate at 1.5 mils	10.00	50	.250	.225
Red lead at 1.5 mils	12.50	50	.275	.250
Finishes				
High-build epoxy at 5.0 mils	15.00	50	$ .475	$ .450
High-build urethane at 5.0 mils	24.00	40	.525	.500
Alkyd enamel at 3.0 mils	10.00	50	.400	.350

TABLE 20.19   Cost per Year per Square Foot for Coatings for Structural Steel on the Ground

East Coast; field painting

System	Surface preparation	Total cost	Mild environment		Moderate environment		Severe environment	
			Life (years)	Cost per year	Life (years)	Cost per year	Life (years)	Cost per year
Two-coat alkyd	SP-2; SP-3	$ .597	2	$ .299	1	$ .597	½	$1.194
Three-coat alkyd	SP-2; SP-3	.719	4	.18	2	.36	1	.719
Three-coat alkyd	SP-6	.819	7	.125	5	.164	3	.273
Two-coat latex	SP-2; SP-3	.618	2	.309	1	.618	½	1.236
Two-coat latex	SP-2; SP-3	.751	4	.188	2	.376	1	.751
Three-coat latex	SP-6	.851	7	.122	5	.17	3	.284
Two-coat epoxy	SP-6	.87	7	.124	6	.145	5	.174
Three-coat epoxy	SP-6	1.099	10	.11	8	.137	6	.183
Three-coat zinc, epoxy	SP-10	1.295	12	.108	10	.13	7	.185
Three-coat zinc, epoxy, urethane	SP-10	1.275	10	.128	9	.142	8	.159
Three-coat vinyl	SP-10	1.178	10	.118	9	.131	8	.147
Three-coat chlorinated rubber	SP-6	.988	12	.082	10	.099	7	.141
Three-coat coal–tar epoxy	SP-6	.914	8	.114	7	.131	6	.152

SOURCE: Presented at the NACE Annual Conference, Atlanta, Ga., March 1979.

NOTE: A lower cost per year is important, but the effective service life, or the time until repainting is required, must also be considered in evaluating and selecting a coatings system.

It should be emphasized that surface-preparation costs for coating work in place are significantly higher. Greasy kitchens, dirt collection on walls, chalking surfaces, and other difficult surfaces present potential problems. Often it is the repaint job that creates the biggest problem for the architect because paint failure can easily occur.

There is even greater potential for repaint failure in a plant environment, and costs can rise astronomically. The surface preparation of tanks or structures in place requires the building of scaffolding or the arrangement of spiders or climbers for the painters. This in itself is expensive. Running sandblast hoses to the areas requiring surface preparation also is expensive. Frequently small areas will fail in different locations of a plant. To repair these areas, equipment and scaffolding must be moved. Moreover, sandblasting on a jobsite creates undesirable dust conditions for workers from other crafts in the same area. When sandblasting takes place, the other crews and crafts frequently must be pulled off the job. In fact, it may be necessary to shut down the plant temporarily while blasting is taking place. These costs are all related to maintenance painting on the jobsite in an operating plant.

If surface preparation is required in maintenance, costs can be $4 to $5 per square foot. For this reason, a good inspection program must be an integral part of the maintenance program.

For all these reasons, an effective maintenance program is an important part of a coatings program. It must include routine inspection schedules with the appropriate inspection reporting forms.

Figures 20.2 and 20.3 show maintenance cost data. The importance of suitable maintenance is emphasized in Fig. 20.3, which shows that the small temporary savings gained by a no-maintenance program were lost when the plant suffered coating failure. An even greater loss, not shown here, is that of plant equipment.

20.6 Environmental Costs   Local, state, and federal regulations have had an economic impact on the coating industry. They have affected coating manufacturing, surface preparation, and coating application. The most significant economic effects are related to:

- Blasting restrictions
- Safety requirements
- Solvent-system restrictions

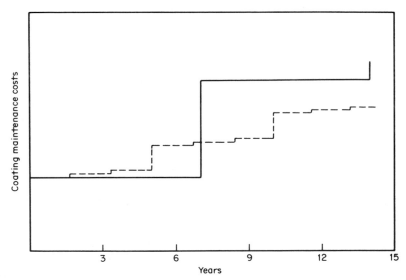

Fig. 20.2 Maintenance costs versus repair schedule (only the trend is shown). The broken line indicates inspected and maintained coating that is retopcoated after 5 years; reblasting to base metal is *not* required. The solid line indicates coating failure requiring reblasting to base metal and recoating after 7 years.

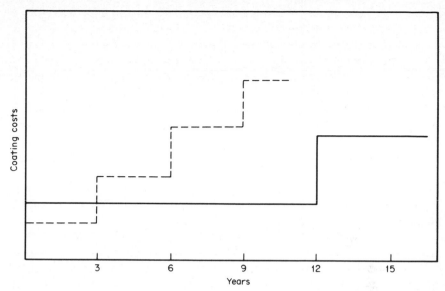

Fig. 20.3 Maintenance costs (only the trend is shown). The broken line indicates an alkyd coating system; the solid line, an inorganic zinc–epoxy coating system.

In some counties and states, local air pollution agencies are requiring that abrasive blasting, particularly sandblasting, be carried out in enclosed areas and that exhaust air be filtered. For this reason, automatic and semiautomatic blasting procedures are expanding rapidly. They offer a means of controlling dust, and their speed and efficiency also give them an economic advantage.

Numerous safety requirements have also been imposed on application practices.

To discuss current regulatory guides in detail is beyond the scope of this chapter. Suffice it to say that the economic effect of the requirements is to increase costs. The exact amount of the increase depends on the type and location of the particular job.

20.7 Galvanizing versus Zinc-Rich Coatings   Galvanizing and zinc-rich coatings with their sacrificial cathodic protection mechanism have revolutionized steel protection. In galvanizing 1½ oz of zinc per square foot (4.6 kg/m²) is equivalent in thickness to 2.5 mils (0.0635 mm) dry in a zinc-rich coating. From a protection standpoint, they are about the same. Galvanizing by "bath" treatment is more easily applied to small parts, gratings, etc. (Table 20.20). It readily forms a continuous film on small members. Galvanizing facilities are not always close to the job, and vat size can be a limiting factor. Heat resistance is somewhat below the melting point of zinc. Zinc-rich coatings are more easily applied to existing

TABLE 20.20 Cost Comparison: Priming versus Galvanizing for Structural-Steel Members (June 1978)

Nominal number size	Weight per foot (pounds)	Hot-dip galvanizing* Gulf Coast — Dollars per foot						West Coast — Dollars per foot			Square feet per foot	Shop priming† All-automatic equipment	Gulf Coast manual		West Coast manual		
Dollars per ton		120.00	130.00	140.00	150.00	160.00	170.00	180.00	190.00	200.00		.40	.50	.55	.60	.70	.80
													Dollars per foot				
Wide-flange shapes																	
12 WF	58	3.48	3.77	4.05	4.35	4.62	4.91	5.21	5.52	5.80	5.4	2.16	2.70	2.97	3.24	3.78	4.32
12 WF	53	3.18	3.44	3.70	3.97	4.23	4.50	4.91	5.04	5.30	5.3	2.16	2.70	2.97	3.24	3.78	4.32
12 WF	50	3.00	3.25	3.50	3.75	4.00	4.25	4.50	4.75	5.00	4.7	1.88	2.35	2.59	2.82	3.29	3.76
12 WF	45	2.70	2.92	3.15	3.38	3.60	3.83	4.05	4.27	4.50	4.7	1.88	2.35	2.59	2.82	3.29	3.76
12 WF	40	2.40	2.60	2.80	3.00	3.20	3.40	3.60	3.80	4.00	4.7	1.88	2.35	2.59	2.82	3.29	3.76
12 WF	36	2.16	2.34	2.51	2.70	2.87	3.06	3.25	3.42	3.60	4.2	1.68	2.10	2.31	2.52	2.94	3.36
12 WF	31	1.86	2.02	2.17	2.32	2.45	2.63	2.78	2.94	3.10	4.2	1.68	2.10	2.31	2.52	2.94	3.36
12 WF	27	1.62	1.75	1.88	2.02	2.16	2.30	2.42	2.56	2.70	4.2	1.68	2.10	2.31	2.52	2.94	3.36
10 WF	77	4.61	5.00	5.40	5.80	6.17	6.55	6.84	7.50	7.70	5.2	2.08	2.60	2.86	3.12	3.64	4.16
10 WF	72	4.33	4.68	5.04	5.40	5.75	6.13	6.49	6.84	7.20	5.1	2.04	2.55	2.81	3.06	3.57	4.08
10 WF	66	3.96	4.28	4.61	4.95	5.28	5.61	5.96	6.26	6.60	5.1	2.04	2.55	2.81	3.06	3.57	4.08
10 WF	60	3.60	3.90	4.20	4.60	4.80	5.10	5.40	5.70	6.00	5.1	2.04	2.55	2.81	3.06	3.57	4.08
10 WF	54	3.24	3.50	3.78	4.04	4.31	4.58	4.85	5.13	5.40	5.0	2.00	2.50	2.75	3.00	3.50	4.00
10 WF	49	2.94	3.18	3.43	3.66	3.92	4.16	4.39	4.65	4.90	5.0	2.00	2.50	2.75	3.00	3.50	4.00
10 WF	39	2.34	2.54	2.74	2.93	3.12	3.32	3.51	3.71	3.90	4.3	1.72	2.15	2.37	2.58	3.01	3.44
10 WF	33	1.98	2.14	2.30	2.48	2.65	2.80	2.97	3.14	3.30	4.3	1.72	2.15	2.37	2.58	3.01	3.44
10 WF	29	1.24	1.89	2.03	2.18	2.32	2.46	2.61	2.76	2.90	3.6	1.44	1.80	1.98	2.16	2.52	2.88
10 WF	25	1.50	1.63	1.75	1.88	2.00	2.12	2.25	2.38	2.50	3.6	1.44	1.80	1.98	2.16	2.52	2.88
10 WF	21	1.26	1.37	1.47	1.58	1.68	1.78	1.89	2.00	2.10	3.6	1.44	1.80	1.98	2.16	2.52	2.88
8 WF	67	4.02	4.35	4.69	5.04	5.35	5.70	6.03	6.37	6.70	4.3	1.72	2.15	2.37	2.58	3.01	3.44
8 WF	58	3.42	3.77	4.05	4.35	4.62	4.91	5.21	5.52	5.80	4.2	1.68	2.10	2.31	2.52	2.94	3.36
8 WF	48	2.88	3.12	3.36	3.60	3.84	4.08	4.32	4.55	4.80	4.1	1.64	2.05	2.26	2.46	2.87	3.28
8 WF	40	2.40	2.60	2.80	3.00	3.20	3.40	3.60	3.80	4.00	4.1	1.64	2.05	2.26	2.46	2.87	3.28
8 WF	35	2.10	2.27	2.45	2.62	2.80	2.97	3.15	3.32	3.50	4.0	1.60	2.00	2.20	2.40	2.80	3.20
8 WF	31	1.86	2.02	2.17	2.32	2.45	2.63	2.78	2.94	3.10	4.0	1.60	2.00	2.20	2.40	2.80	3.20

Wide-flange (WF) beams

Size	Wt	C1	C2	C3	C4	C5	C6	C7	C8	C9	C10	C11	C12	C13	C14	C15	C16
8 WF	28	2.80	2.45	2.10	1.93	1.75	1.40	3.5	2.80	2.65	2.52	2.38	2.24	2.10	1.96	1.82	1.68
8 WF	24	2.80	2.45	2.10	1.93	1.75	1.40	3.5	2.40	2.28	2.16	2.04	1.92	1.80	1.68	1.56	1.44
8 WF	20	2.48	2.17	1.86	1.71	1.55	1.24	3.1	2.00	1.90	1.80	1.70	1.60	1.50	1.40	1.30	1.20
8 WF	17	2.48	2.17	1.86	1.71	1.55	1.24	3.1	1.70	1.62	1.53	1.44	1.36	1.27	1.19	1.11	1.02
6 WF	25	2.48	2.17	1.86	1.71	1.55	1.24	3.1	2.50	2.38	2.25	2.13	2.00	1.88	1.75	1.63	1.50
6 WF	20	2.40	2.10	1.80	1.65	1.50	1.20	3.0	2.00	1.90	1.80	1.70	1.60	1.50	1.40	1.30	1.20
6 WF	15.5	2.40	2.10	1.80	1.65	1.50	1.20	3.0	1.55	1.47	1.40	1.32	1.24	1.16	1.09	1.01	.93
5 WF	18.5	2.00	1.75	1.50	1.38	1.25	1.00	2.5	1.85	1.76	1.67	1.57	1.48	1.39	1.30	1.21	1.11
5 WF	16	2.00	1.75	1.50	1.38	1.25	1.00	2.5	1.60	1.52	1.44	1.36	1.28	1.20	1.12	1.04	.96
4 WF	13	1.60	1.40	1.20	1.10	1.00	.80	2.0	1.30	1.24	1.17	1.11	1.04	.98	.91	.85	.78

I beams

Size	Wt	C1	C2	C3	C4	C5	C6	C7	C8	C9	C10	C11	C12	C13	C14	C15	C16
12 I	50	3.04	2.66	2.28	2.09	1.90	1.52	3.8	5.00	4.75	4.50	4.25	4.00	3.75	3.50	3.25	3.00
12 I	40.8	3.04	2.66	2.88	2.09	1.90	1.52	3.8	4.08	3.88	3.67	3.47	3.26	3.06	2.86	2.65	2.45
12 I	35	2.96	2.59	2.22	2.04	1.85	1.48	3.7	3.50	3.33	3.15	2.98	2.80	2.63	2.45	2.28	2.10
12 I	31.8	2.96	2.59	2.22	2.04	1.85	1.48	3.7	3.18	3.02	2.86	2.70	2.55	2.39	2.23	2.07	1.81
10 I	35	2.64	2.31	1.98	1.82	1.65	1.32	3.3	3.50	3.33	3.15	2.98	2.80	2.63	2.45	2.28	2.10
10 I	25.4	2.56	2.24	1.92	1.76	1.60	1.28	3.2	2.54	2.41	2.29	2.16	2.03	1.91	1.78	1.65	1.52
8 I	23	2.16	1.89	1.62	1.49	1.35	1.08	2.7	2.30	2.19	2.07	1.96	1.84	1.73	1.61	1.50	1.38
8 I	18.4	2.16	1.89	1.50	1.38	1.35	1.08	2.7	1.84	1.75	1.66	1.56	1.47	1.38	1.29	1.20	1.10
7 I	20	2.00	1.75	1.44	1.32	1.25	1.00	2.5	2.00	1.90	1.80	1.70	1.60	1.50	1.40	1.30	1.20
7 I	15.3	1.92	1.68	1.32	1.21	1.20	.96	2.4	1.53	1.45	1.38	1.30	1.22	1.15	1.07	1.00	.92
6 I	17.25	1.76	1.54	1.32	1.16	1.10	.88	2.2	1.73	1.64	1.55	1.47	1.38	1.29	1.21	1.12	1.04
6 I	12.5	1.68	1.47	1.26	1.05	1.05	.84	2.1	1.25	1.19	1.13	1.06	1.00	.94	.88	.81	.75
5 I	14.75	1.52	1.33	1.14	.99	.95	.76	1.9	1.48	1.40	1.33	1.25	1.18	1.11	1.03	.96	.89
5 I	10	1.44	1.26	1.08	.88	.90	.72	1.8	1.00	.95	.90	.85	.80	.75	.70	.65	.60
4 I	9.5	1.28	1.12	.96	.88	.80	.64	1.6	.95	.90	.86	.81	.76	.71	.67	.62	.57
4 I	7.7	1.28	1.12	.96	.72	.80	.64	1.6	.77	.73	.69	.66	.62	.58	.54	.50	.46
3 I	7.5	1.04	.91	.78	.72	.65	.52	1.3	.75	.71	.68	.64	.60	.56	.53	.49	.45
3 I	5.7	1.04	.91	.78	.72	.65	.52	1.3	.57	.54	.51	.49	.46	.43	.40	.37	.34
6 × 4 I	16	1.76	1.54	1.32	1.21	1.10	.88	2.2	1.60	1.52	1.44	1.36	1.28	1.20	1.12	1.04	.96
6 × 4 I	12	1.68	1.47	1.26	1.16	1.05	.84	2.1	1.20	1.14	1.08	1.02	.96	.90	.84	.78	.72
8 × 4 I	15	2.16	1.89	1.62	1.49	1.35	1.08	2.7	1.50	1.43	1.35	1.28	1.20	1.13	1.05	.98	.90
8 × 4 I	13	2.16	1.89	1.62	1.49	1.35	1.09	2.7	1.30	1.24	1.17	1.11	1.04	.98	.91	.85	.78
10 × 4 I	19	2.56	2.24	1.92	1.76	1.60	1.28	3.2	1.90	1.81	1.71	1.62	1.52	1.43	1.33	1.24	1.14
10 × 4 I	17	2.56	2.24	1.92	1.76	1.60	1.28	3.2	1.70	1.62	1.53	1.45	1.36	1.28	1.19	1.11	1.02
10 × 4 I	15	2.56	2.24	1.92	1.76	1.60	1.28	3.2	1.50	1.43	1.35	1.28	1.20	1.13	1.05	.98	.90
12 × 4 I	22	2.96	2.59	2.22	2.04	1.85	1.48	3.7	2.20	2.09	1.98	1.87	1.76	1.65	1.54	1.43	1.32
12 × 4 I	19	2.96	2.59	2.22	2.04	1.85	1.48	3.7	1.90	1.81	1.71	1.62	1.52	1.43	1.33	1.24	1.14
12 × 6 I	36	3.36	2.94	2.52	2.31	2.10	1.68	4.2	3.60	3.42	3.24	3.06	2.88	2.70	2.52	2.34	2.16
14 × 6¾	38	3.68	3.22	2.76	2.53	2.30	1.84	4.6	3.80	3.61	3.42	3.23	3.04	2.85	2.66	2.47	2.28
16 × 8½	58	4.40	3.85	3.30	3.03	2.75	2.20	5.5	5.80	5.51	5.22	4.93	4.64	4.35	4.06	3.77	3.48

TABLE 20.20 Cost Comparison: Priming versus Galvanizing for Structural-Steel Members (June 1978) (*Continued*)

Channels

Nominal number size	Weight per foot (pounds)	Hot-dip galvanizing* Gulf Coast — Dollars per foot				Hot-dip galvanizing* West Coast — Dollars per foot					Shop priming† Square feet per foot	Shop priming† All-automatic equipment	Shop priming† Gulf Coast manual — Dollars per foot			Shop priming† West Coast manual — Dollars per foot	
	Dollars per ton →	120.00	130.00	140.00	150.00	160.00	170.00	180.00	190.00	200.00		.40	.50	.55	.60	.70	.80
18	58	3.48	3.77	4.06	4.35	4.64	4.93	5.22	5.51	5.80	4.4	1.76	2.20	2.42	2.64	3.08	3.52
18	51.9	3.11	3.37	3.63	3.90	4.15	4.41	4.67	4.93	5.19	4.4	1.76	2.20	2.42	2.64	3.08	3.52
18	45.8	2.75	2.98	3.20	3.44	3.66	3.89	4.12	4.35	4.58	4.3	1.72	2.15	2.37	2.58	3.01	3.44
18	42.7	2.56	2.78	2.99	3.20	3.42	3.63	3.84	4.06	4.27	4.3	1.72	2.15	2.37	2.58	3.01	3.44
15	50	3.00	3.25	3.50	3.75	4.00	4.25	4.50	4.75	5.00	3.7	1.48	1.85	2.04	2.22	2.59	2.96
15	40	2.40	2.60	2.80	3.00	3.20	3.40	3.60	3.80	4.00	3.7	1.48	1.85	2.04	2.22	2.59	2.96
15	33.9	2.03	2.20	2.37	2.54	2.71	2.88	3.05	3.22	3.39	3.6	1.44	1.80	1.98	2.16	2.52	2.88
13	50	3.00	3.25	3.50	3.75	4.00	4.25	4.50	4.75	5.00	3.6	1.44	1.80	1.98	2.16	2.52	2.88
13	31.8	1.91	2.07	2.23	2.39	2.54	2.70	2.86	3.02	3.18	3.5	1.40	1.75	1.93	2.10	2.45	2.80
12	30	1.80	1.95	2.10	2.25	2.40	2.55	2.70	2.85	3.00	3.1	1.24	1.55	1.71	1.86	2.17	2.48
12	25	1.50	1.63	1.75	1.88	2.00	2.13	2.25	2.38	2.50	3.0	1.20	1.50	1.65	1.80	2.10	2.40
12	20.7	1.24	1.35	1.45	1.55	1.66	1.76	1.86	1.97	2.07	3.0	1.20	1.50	1.65	1.80	2.10	2.40
10	30	1.80	1.95	2.10	2.25	2.40	2.55	2.70	2.85	3.00	2.7	1.08	1.35	1.49	1.62	1.89	2.16
10	25	1.50	1.63	1.75	1.88	2.00	2.13	2.25	2.38	2.50	2.6	1.04	1.30	1.43	1.56	1.82	2.08
10	20	1.20	1.30	1.40	1.50	1.60	1.70	1.80	1.90	2.00	2.6	1.04	1.30	1.43	1.56	1.82	2.08
10	15.3	.92	1.00	1.07	1.15	1.22	1.30	1.38	1.45	1.53	2.5	1.00	1.25	1.38	1.50	1.75	2.00
9	20	1.20	1.30	1.40	1.50	1.60	1.70	1.80	1.90	2.00	2.4	.96	1.20	1.32	1.44	1.68	1.92
9	15	.90	.98	1.05	1.13	1.20	1.28	1.35	1.43	1.50	2.3	.92	1.15	1.27	1.38	1.61	1.84
9	13.4	.80	.87	.94	1.01	1.07	1.14	1.21	1.27	1.34	2.3	.92	1.15	1.27	1.38	1.61	1.84
8	18.75	1.13	1.22	1.31	1.41	1.50	1.59	1.69	1.78	1.88	2.2	.88	1.10	1.21	1.32	1.54	1.76
8	13.75	.83	.89	.96	1.03	1.10	1.17	1.24	1.31	1.38	2.1	.84	1.05	1.16	1.26	1.47	1.68
8	11.5	.69	.75	.81	.86	.92	.98	1.04	1.09	1.15	2.1	.84	1.05	1.16	1.26	1.47	1.68
7	14.75	.89	.96	1.03	1.11	1.18	1.25	1.33	1.40	1.48	1.9	.76	.95	1.05	1.14	1.33	1.52
7	12.25	.74	.80	.86	.92	.98	1.04	1.10	1.16	1.23	1.9	.76	.95	1.05	1.14	1.33	1.52
7	9.8	.59	.64	.69	.74	.78	.83	.88	.93	.98	1.9	.76	.95	1.05	1.14	1.33	1.52
6	13.0	.78	.85	.91	.98	1.04	1.11	1.17	1.24	1.30	1.7	.68	.85	.94	1.02	1.19	1.36
6	10.5	.63	.68	.74	.79	.84	.89	.95	1.00	1.05	1.7	.68	.85	.94	1.02	1.19	1.36

6	8.2	.49	.53	.57	.62	.66	.70	.74	.78	.82	[illegible]	[illegible]	[illegible]	[illegible]	[illegible]	[illegible]	[illegible]
5	9.0	.54	.59	.63	.68	.72	.77	.81	.86	.80	1.5	.60	.75	.83	.90	1.05	1.20
5	6.7	.40	.44	.47	.50	.54	.57	.60	.64	.67	1.4	.56	.70	.77	.84	.98	1.12
4	7.25	.44	.47	.51	.54	.58	.62	.65	.69	.73	1.2	.48	.60	.66	.72	.84	.96
4	5.4	.32	.35	.38	.41	.43	.46	.49	.51	.54	1.2	.48	.60	.66	.72	.84	.96
3	6.0	.36	.39	.42	.45	.48	.51	.54	.57	.60	1.0	.40	.50	.55	.60	.70	.80
3	5.0	.30	.33	.35	.38	.40	.43	.45	.48	.50	1.0	.40	.50	.55	.60	.70	.80
3	4.1	.25	.27	.29	.31	.33	.35	.37	.39	.41	1.0	.40	.50	.55	.60	.70	.80
Equal-leg angles																	
8 × 8½	26.4	1.58	1.72	1.85	1.98	2.11	2.24	2.38	2.51	2.64	2.7	1.08	1.35	1.49	1.62	1.89	2.16
6 × 6 × 5/16	12.5	.75	.81	.88	.94	1.00	1.06	1.13	1.19	1.25	2.0	.80	1.00	1.10	1.20	1.40	1.60
5 × 5 × 5/16	10.3	.62	.67	.72	.77	.82	.88	.93	.98	1.03	1.7	.68	.85	.94	1.02	1.19	1.36
4 × 4 × ¼	6.6	.40	.43	.46	.50	.53	.56	.59	.63	.66	1.3	.52	.65	.72	.78	.91	1.04
3½ × 3½ × ¼	5.8	.35	.38	.41	.44	.46	.49	.52	.55	.58	1.2	.48	.60	.66	.72	.84	.96
3 × 3 × 3/16	3.71	.22	.24	.26	.28	.30	.32	.33	.35	.37	1.0	.40	.50	.55	.60	.70	.80
2½ × 2¼ × 3/16	3.07	.18	.20	.22	.23	.25	.26	.28	.29	.31	0.8	.32	.40	.44	.48	.56	.64
2 × 2 × ⅛	1.65	.10	.11	.12	.12	.13	.14	.15	.16	.17	0.7	.28	.35	.39	.42	.49	.56
1½ × 1¼ × ⅛	1.23	.07	.08	.19	.09	.10	.11	.11	.12	.12	0.5	.20	.25	.28	.30	.35	.40
1 × 1 × ⅛	0.80	.05	.05	.06	.06	.06	.07	.07	.08	.08	0.3	.12	.15	.17	.18	.21	.24
Unequal-leg angles																	
8 × 6 × 1½	23.0	1.38	1.50	1.61	1.73	1.84	1.96	2.07	2.19	2.30	2.3	.92	1.15	1.27	1.38	1.61	1.84
8 × 4 × ½	19.6	1.18	1.27	1.37	1.47	1.57	1.67	1.76	1.86	1.96	2.0	.80	1.00	1.10	1.20	1.40	1.60
7 × 4 × ⅜	13.6	.82	.88	.95	1.02	1.09	1.16	1.22	1.29	1.36	1.8	.72	.90	.99	1.08	1.26	1.44
6 × 4 × 5/16	10.3	.62	.67	.72	.77	.82	.88	.93	.98	1.03	1.7	.68	.85	.94	1.02	1.19	1.36
6 × 3½ × 5/16	9.8	.59	.64	.69	.74	.78	.83	.88	.93	.98	1.6	.64	.80	.88	.96	1.12	1.28
5 × 3½ × 5/16	8.7	.52	.57	.61	.65	.70	.74	.78	.83	.87	1.4	.56	.70	.77	.84	.98	1.12
5 × 3½ × ¼	6.6	.40	.43	.46	.50	.53	.56	.59	.63	.66	1.3	.42	.65	.72	.78	.91	1.04
4 × 3½ × ¼	6.2	.37	.40	.43	.47	.50	.53	.56	.59	.62	1.25	.50	.63	.69	.75	.88	1.00
4 × 3 × ¼	5.8	.35	.38	.41	.44	.46	.49	.52	.55	.58	1.17	.47	.59	.64	.71	.82	.94
3½ × 3 × ¼	5.4	.32	.35	.38	.41	.43	.46	.49	.51	.54	1.08	.43	.54	.59	.65	.76	.86
3 × 2½ × ¼	4.5	.27	.29	.32	.34	.36	.38	.41	.43	.45	0.92	.37	.46	.50	.55	.64	.74
3 × 2 × 3/16	3.07	.18	.20	.22	.23	.25	.26	.28	.29	.31	0.83	.33	.42	.46	.50	.58	.66
2½ × 2 × 3/16	2.75	.17	.18	.19	.21	.22	.23	.25	.26	.28	0.75	.30	.38	.41	.45	.53	.60
2½ × 1½ × 3/16	2.44	.15	.16	.17	.18	.20	.21	.22	.23	.24	0.67	.27	.34	.37	.40	.47	.54
2 × 1½ × ⅛	1.44	.09	.09	.10	.11	.12	.12	.13	.14	.14	0.58	.23	.29	.32	.35	.41	.46
1½ × 1¼ × 3/16	1.67	.10	.11	.12	.13	.13	.14	.15	.16	.17	0.31	.12	.16	.17	.19	.22	.25
1 × ¾ × ⅛	0.70	.04	.05	.05	.05	.05	.06	.06	.07	.07	0.15	.06	.08	.08	.09	.11	.12
1 × ⅝ × ⅛	0.64	.04	.04	.05	.05	.05	.05	.06	.06	.06	0.14	.06	.07	.08	.08	.10	.11

* Per ASTM standards.
† Surface preparation, SSPC SP-10; prime coat, 3.0 to 5.0 mils dry film thickness of inorganic zinc.

structures in place, weather better in marine and coastal environments, and accept topcoats more readily. They are less expensive on large structural members and have a heat resistance somewhat above the melting point of zinc.

The use of a zinc-rich primer serves several purposes:

- *Simplified construction schedules.* Zinc-rich primers can be applied in the shop as preconstruction primers, eliminating most or all field surface preparation.
- *Minimized field touch-up.* Zinc-rich primers are extremely tough and abrasion-resistant. Damage due to shipping and handling is usually less than 5 percent. Alkyd primer damage due to shipping and handling can exceed 50 percent and frequently is in the 20 to 30 percent range. Costs for field touch-up are high, sometimes exceeding $2 per square foot when rigging and surface preparation are included.
- *Simplified coatings systems.* Zinc-rich primers can be topcoated with a wide variety of industrial topcoats.
- *Corrosion protection.* Zinc-rich primers offer excellent resistance to corrosion both as preconstruction primers and as topcoated primers in coatings systems.

20.8   Economics of Special and General Painting Projects   Roof coatings and marine paints deserve special consideration.

The economics of roof coatings relates to energy savings which go beyond previous economic analysis. Just as new environmental regulations demand white paint on tanks containing organic liquids to avoid vapor emissions, so does the proper choice of color and film thickness avoid energy losses in buildings and homes. Several excellent books and articles (see Bibliography) enable an architect to calculate the optimum thickness of coatings or insulation to achieve energy savings for cooling

TABLE 20.21   Cost of Painting Lines on Pavement (1980 Dollars)

Color	Size (inches)	Cost per linear foot
White	4	$ .09
	8	.15
	12	.20
Yellow	4	.10
	8	.25
	12	.30

SOURCE: *McGraw-Hill's Dodge Manual for Building Construction Pricing and Scheduling*, 1980 ed.

TABLE 20.22   Painting Costs for General and Special Coatings (1980 Dollars)

Description	Crew	Output Per day	Unit	Unit costs Labor	Material	Total
		General coatings				
Exterior work on walls						
Hand sanding and puttying	One laborer	1600	Ft²	$ .07	.05	.12
Galvanized work preparation acid wash gutters		1600	Ft		.04	.04
Oil paint all sheens						
Wood and composition siding, brush one coat	One painter	840	Ft²	.15	.17	.32
Wood and composition siding, brush two coats	One painter	880	Ft²	.15	.18	.33
Wood and composition siding, brush three coats	One painter	1000	Ft²	.13	.20	.33
Wood and composition siding, roll one coat	One painter	1200	Ft²	.11	.24	.35
Wood and composition siding, roll two coats	One painter	1360	Ft²	.09	27.00	27.09
Wood and composition siding, roll three coats	One painter	1600	Ft²	.08	.32	.40
Flat or gloss latex						
Asbestos shingles, brush one coat	One painter	600	Ft²	.22	.15	.37
Asbestos shingles, brush two coats	One painter	760	Ft²	.17	.15	.32
Asbestos shingles, roll one coat	One painter	1520	Ft²	.08	.15	.23
Asbestos shingles, roll two coats	One painter	200	Ft²	.66	.15	.81
Asbestos shingles, spray one coat	One painter	2800	Ft²	.04	.14	.18
Asbestos shingles, spray two coats	One painter	2800	Ft²	.04	.14	.18
Staining brush						
Shingles and rough-sawn siding, one coat	One painter	1200	Ft²	.11	.13	.24
Shingles and rough-sawn siding, two coats	One painter	1240	Ft²	.10	.13	.23
Spraying						
Shingles and rough-sawn siding, one coat	One painter	2800	Ft²	.04	.14	.18
Shingles and rough-sawn siding, two coats	One painter	3200	Ft²	.04	.16	.20
Stucco, medium-texture						
Latex, roll on						
One coat	One painter	2240	Ft²	.06	.22	.28
Two coats	One painter	2680	Ft²	.05	.26	.31
Latex, spray						
One coat	One painter	3200	Ft²	.04	.18	.22
Two coats	One painter	4000	Ft²	.03	.27	.30
Interior walls, brush-applied on						
Plaster or drywall, two coats flat						
Three coats flat	One painter	600	Ft²	.22	.13	.35
Two coats enamel	One painter	480	Ft²	.27	.17	.44
Three coats enamel	One painter	540	Ft²	.24	.13	.37
Wood, two coats enamel	One painter	450	Ft²	.29	.17	.46
Three coats enamel	One painter	490	Ft²	.27	.16	.43
Filler, stain, and varnish, two coats	One painter	380	Ft²	.34	.18	.52
Filler, stain, and varnish, three coats	One painter	280	Ft²	.47	.22	.69
Masonry or concrete, two coats latex	One painter	230	Ft²	.57	.13	.70
Three coats latex	One painter	130	Ft²	1.01	.17	1.18

TABLE 20.22   Painting Costs for General and Special Coatings (1980 Dollars) (*Continued*)

Description	Crew	Output Per day	Unit	Unit costs La-bor	Ma-terial	Total
Interior painting on Sheetrock or plaster, roll						
One coat, including openings, latex	One painter	1600	Ft²	.08	.08	.16
One coat, openings only, latex	One painter	1100	Ft²	.12	.08	.20
Two coats, including openings, latex	One painter	860	Ft²	.15	.11	.26
Two coats, openings only, latex	One painter	520	Ft²	.25	.11	.36
Ceilings, brush-applied on						
Plaster or drywall, two coats flat	One painter	600	Ft²	.22	.12	.34
Three coats flat	One painter	500	Ft²	.26	.16	.42
Two coats enamel	One painter	560	Ft²	.23	.12	.35
Three coats enamel	One painter	460	Ft²	.28	.16	.44
Wood, two coats enamel	One painter	500	Ft²	.26	.15	.41
Three coats enamel	One painter	350	Ft²	.37	.16	.53
Two coats filler, stain, and varnish	One painter	280	Ft²	.47	.25	.72
Three coats filler, stain, and varnish	One painter	220	Ft²	.60	.27	.87
Masonry or concrete, two coats latex	One painter	500	Ft²	.26	.12	.38
Three coats latex	One painter	370	Ft²	.35	.16	.51
Wood floors, two coats enamel	One painter	700	Ft²	.19	.15	.34
Three coats enamel	One painter	540	Ft²	.24	.16	.40
Stain and varnish, two coats	One painter	600	Ft²	.22	.21	.43
Stain and varnish, three coats	One painter	460	Ft²	.28	.22	.50
Concrete, two coats enamel	One painter	560	Ft²	.23	.16	.39
Three coats enamel	One painter	400	Ft²	.33	.19	.52
Piping, two coats, to 3-in diameter	One painter	420	Ft	.31	.11	.42
3 to 6 in	One painter	300	Ft	.44	.16	.60
6 to 9 in	One painter	200	Ft	.66	.26	.92
Over 9 in	One painter	330	Ft²	.40	.16	.56
Ductwork, two coats	One painter	300	Ft²	.44	.19	.63
Silicone surface sealer	One painter	600	Ft²	.22	.06	.28
Lacquer spray on wood panel	One painter	1600	Ft²	.08	.04	.12
Polyurethane ⅛-in coat, brush on	One painter	800	Ft²	.16	.78	.94
Specialty graphics-design painting (labor)	One painter	320	Ft²	.41		.41
Gold-leaf application, labor only	One painter	10	Ft²	13.23		13.23
Aluminum-leaf application, labor only	One painter	60	Ft²	2.20		2.20
Industrial-commercial, large areas						
Wood and fiber-wood deck and siding						
Airless-spray latex, one coat	One painter	7200	Ft²	.01	.04	.05
Airless-spray latex, second coat	One painter	8000	Ft²	.01	.04	.05
Metal decking and siding, corrugated						
Airless-spray latex, one coat	One painter	8000	Ft²	.01	.04	.05
Airless-spray latex, second coat	One painter	9600	Ft²	.01	.04	.05
Special coatings						
Vermiculite, sprayed 1/16 in thick	One painter	338	Ft²	.39	.13	.52
Perlite, sprayed 1/16 in thick	One painter	338	Ft²	.39	.13	.52

Description	Crew	Output Per day	Unit	Unit costs La- bor	Ma- terial	Total
Vitreous-cement enamel	One painter	175	Ft²	.75	.69	1.44
Glazed elastomeric coatings						
Acrylic filler and glaze	One painter	220	Ft²	.60	.43	1.03
Epoxy	One painter	220	Ft²	.60	.35	.95
Exposed aggregate, ¼ in	One painter	90	Ft²	1.47	.35	1.82
Troweled ½ to 1 in	One painter	38	Ft²	3.48	1.92	5.40
Urethane on rough surface, three coats	One painter	288	Ft²	.46	.34	.80
Colored-vinyl glaze	One cement mason	102	Ft²	1.37	1.16	2.53
With chips	One cement mason	90	Ft²	1.55	1.36	2.91
Fireproof intumescent coating, 1-h rating	One painter	330	Ft²	.40	1.28	1.68
2-h rating	One painter	225	Ft²	.58	2.72	3.30
Fire-retardant paints	One painter	450	Ft²	.29	.26	.55

SOURCE: *McGraw-Hill's Dodge Manual for Building Construction Pricing and Scheduling,* 1980 ed.

or heating purposes. These values can be entered into life-cycle costing calculations to recoup investments.

Marine coatings by their nature have esoteric problems. To thwart hull encrustation, antifouling paints based on cuprous oxide or toxic organometallic polymers are used. These require special handling during manufacture and application. Costs therefore are different from those of conventional coatings. Both cost and detailed application data are often covered in internal documents published by the coating manufacturer.

On the other hand, general cost information valuable to the architect or cost engineer is often available through construction service data banks that cover everything from painting lines on pavement (Table 20.21) to general and specialized application costs (Table 20.22).

BIBLIOGRAPHY

"B431 Architects Qualification Statement," C141a Standard Agreement Form," Handbook Supplement Service, American Institute of Architects, January 1980 mailing.

L. C. Choate: "Results of T.G.T. Research on Coatings for the Arctic," *Pipe Line Industry,* March 1975. p. 39.

"Cleanliness and Anchor Patterns Available through Centrifugal Blast of New Steel," Technical Practices Committee Task Group T-6G-13, National Association of Corrosion Engineers, October 1975.

Davidson, Sidney, and J. S. Schindler: *Fundamentals of Accounting,* 5th ed., The Dryden Press, Inc., New York, 1975.

—— and W. L. Weil: *Handbook of Modern Accounting,* 2d ed., McGraw-Hill Book Company, New York, 1977.

Griffin, C. W.: "Plug the Energy Leaks in Your Roof," *Buildings,* November 1977, pp. 90–94.

Holmes, Arthur W., Robert A. Meier, and Donald F. Pabst: *Accounting for Control and Decisions,* Business Publications, Austin, Tex., 1970.

Johnson, G. L., and J. A. Gentry: *Principles of Accounting,* Prentice-Hall, Inc., Englewood Cliffs, N.J., 1974.

Linsley, Jerald: "Return on Investment: Discounted and Undiscounted," *Chemical Engineering,* May 21, 1979, pp. 201–204.

Miller, D.: "Corrosion and the Materials Shortage," *Pipe Line Industry,* March 1975, p. 27.

Munger, C. G.: "Petroleum Industry Use of Zinc-Rich Coatings," Zinc Institute National Zinc-Rich Coatings Conference, Chicago, Dec. 4, 1974, pp. 77–81.

Nelson, W. L.: "Refinery-Construction Costs," *Oil and Gas Journal,* Jan. 6, 1975, p. 146.

Pamer, R. I.: "Corrosion Protection of Chemical Industry Facilities with Zinc Rich," Zinc Institute National Zinc-Rich Coatings Conference, Chicago, Dec. 4, 1974, pp. 34–37.

Rodgers, John: "Understanding Economic Aspects of Coatings as Vital to Engineers," *Materials Protection,* vol. 9, no. 5, May 1970, p. 26.

Roebuck, A. H., and D. L. McCage: "Coating Economics," *Materials Protection,* vol. 15, no. 10, October 1976, p. 30.

Swandby, R. K.: "How to Analyze Costs of Painting a New Plant," *Chemical Engineering,* vol. 62, May 28, 1962, p. 115.

Tator, K. B.: "Engineered Painting Pays Off," *Chemical Engineering,* Dec. 27, 1971, p. 84.

Weaver, P. E.: "How to Specify Coatings," *Hydrocarbon Processing,* February 1970, p. 127.

Index